CONTRIBUTIONS TO NONLINEAR
FUNCTIONAL ANALYSIS

Publication No. 27
of the Mathematics Research Center
The University of Wisconsin

# Contributions to Nonlinear Functional Analysis

Edited by Eduardo H. Zarantonello

Proceedings of a Symposium
Conducted by the Mathematics Research Center,
The University of Wisconsin, Madison
April 12–14, 1971

Academic Press
New York · London   1971

COPYRIGHT © 1971, BY ACADEMIC PRESS, INC.
ALL RIGHTS RESERVED
NO PART OF THIS BOOK MAY BE REPRODUCED IN ANY FORM,
BY PHOTOSTAT, MICROFILM, RETRIEVAL SYSTEM, OR ANY
OTHER MEANS, WITHOUT WRITTEN PERMISSION FROM
THE PUBLISHERS.

ACADEMIC PRESS, INC.
111 Fifth Avenue, New York, New York 10003

*United Kingdom Edition published by*
ACADEMIC PRESS, INC. (LONDON) LTD.
24/28 Oval Road, London NW1 7DD

LIBRARY OF CONGRESS CATALOG CARD NUMBER: 70-182611

PRINTED IN THE UNITED STATES OF AMERICA

# Contents

Numbers in parentheses refer to AMS (MOS) 1970 subject classifications.

FOREWORD . . . . . . . . . . . . . . . . . . . . . ix
PREFACE . . . . . . . . . . . . . . . . . . . . . . xi

Generalized Degree and Nonlinear Problems (47H15, 58B05) . . . 1
   L. Nirenberg
      Courant Institute of Mathematical Sciences
      New York University, New York, New York

A Global Theorem for Nonlinear Eigenvalue Problems
and Applications (47H15, 34B15, 35J25) . . . . . . . . . . . 11
   Paul H. Rabinowitz
      University of Wisconsin, Madison, Wisconsin

Transversality in Nonlinear Eigenvalue Problems . . . . . . . . 37
   R. E. L. Turner
      University of Wisconsin, Madison, Wisconsin

Multiple Eigenvalue Bifurcation for Holomorphic Mappings (47H15) . 69
   Klaus Kirchgässner
      Rurh-Universität Bochum Mathematisches Institut
      Bochum-Querenburg, Germany

Monotonicity Methods in Hilbert Spaces and Some Applications
to Nonlinear Partial Differential Equations (47H05, 35J60) . . . . 101
   Haim Brezis
      Institut de Mathématique, Faculté des Sciences
      Paris, France

Semigroups of Nonlinear Transformations in Banach
Spaces (47D05, 47H05) . . . . . . . . . . . . . . . . . 157
   Michael G. Crandall
      University of California, Los Angeles, California

## CONTENTS

Weak and Strong Solutions of Dual Problems (49A25, 49A50) . . . 181
   J. J. Moreau
      Université de Montpellier, Montpellier, France

Convex Integral Functionals and Duality (28A20, 49A50) . . . . 215
   R. Tyrrell Rockafellar
      University of Washington, Seattle, Washington

Projections on Convex Sets in Hilbert Space and
Spectral Theory (52A50, 47H05, 47B40) . . . . . . . . . . . . 237
      Part I. Projections on Convex Sets . . . . . . . . . . 239
      Part II. Spectral Theory . . . . . . . . . . . . . . . 343
   Eduardo H. Zarantonello
      Mathematics Research Center, University of Wisconsin
      Madison, Wisconsin

Nonlinear Functional Analysis and Nonlinear Integral Equations
of Hammerstein and Urysohn type (45G99, 47H15) . . . . . . . 425
   Felix E. Browder
      University of Chicago, Chicago, Illinois

Asymptotic Behavior of Bounded Solutions
of Some Functional Equations (45M05) . . . . . . . . . . . . 501
   J. J. Levin and D. F. Shea
      University of Wisconsin, Madison, Wisconsin
      Purdue University, Lafayette, Indiana

Singular Perturbations and Singular Layers in
Variational Inequalities (35B25) . . . . . . . . . . . . . . . 523
   J. L. Lions
      Faculté des Sciences de Paris, Paris, France

Gradient Estimates for Solutions of Nonlinear Elliptic
and Parabolic Equations (35B45) . . . . . . . . . . . . . . . 565
   James Serrin
      University of Minnesota, Minneapolis, Minnesota

Shock Waves and Entropy (76L05) . . . . . . . . . . . . . . . 603
   Peter Lax
      Courant Institute of Mathematical Sciences
      New York University, New York, New York

## CONTENTS

The Penalty Method and Some Nonlinear Initial
Value Problems (35A25, 35Q10) . . . . . . . . . . . . . . 635
   *Hiroshi Fujita*
      University of Tokyo, Tokyo, Japan

INDEX . . . . . . . . . . . . . . . . . . . . . 667

# Foreword

This volume contains the complete texts of the fifteen addresses to the Symposium on Nonlinear Functional Analysis, held in Madison on April 12-14, 1971, under the sponsorship of the Mathematics Research Center, University of Wisconsin. There were six sessions covering the following areas: I. Topological degree and bifurcation, II. Monotonicity, III. Convexity, IV. Integral equations, V. Evolution equations, and VI. Partial differential equations. The sessions were chaired by:

    Professor Erich H. Rothe, University of Michigan
    Professor Eduardo H. Zarantonello, Mathematics Research Center,
        University of Wisconsin
    Professor Victor Klee, University of Washington
    Professor John Nohel, University of Wisconsin
    Professor Jürgen K. Moser, New York University
    Professor Charles C. Conley, University of Wisconsin

The program committee consisted of Professors P. H. Rabinowitz, R. E. L. Turner, and L. Rall, with the editor as the chairman. Mrs. Gladys Moran was the symposium secretary, and it is to her experience, intelligent dedication, and inexhaustable enthusiasm that this conference owed its perfect organization. The preparation of the manuscripts for publication was in the able hands of Mrs. Dorothy Bowar. To them both I wish to extend my appreciation for their invaluable assistance.

                                              Eduardo H. Zarantonello

# Preface

Linearity is such a deep-seated notion among mathematicians that any outside venture, especially in functional analysis, is immediately qualified as "nonlinear," as if linearity were the normal way of life in mathematics. Such ventures beyond the linear are seldom strictly nonlinear, for they also apply to linear situations, and more often than not it is there where they are at their best, if only a banal best. Strictly speaking, it is only in a context in which linearity makes sense that one can speak of nonlinearity. As the outlying field of nonlinearity is being explored and developed, it is becoming clear that linearity is just one of many parcels of mathematical territory, the first to be settled and at that a thin and narrow one, and that the tribute paid to it is no longer unquestionable. Hopefully, the term nonlinear will disappear from analysis, to be replaced by a host of new names making for a more precise and representative nomenclature, and one can foresee the time—not so far off—when linearity rather than its absence will have to be qualified.

Two of the main ideas in the contemporary scene of functional analysis—as distinguished from linear functional analysis—are topological degree and monotonicity. Topological degree, in the form of the homotopy invariance of the topological index, has been a prime source of existence proofs since its formulation by Schauder and Leray in 1934. It is also an important instrument in bifurcation theory. However, the requirement in existence questions that the operators involved be compact considerably restricts its domain of application. This gap was partially filled by monotone operators which need be neither compact nor continuous, and which originated in 1960 out of the need to have something in higher dimensions corresponding to increasing functions on the real line. Monotonicity belongs to the lineage of ideas started by Picard's successive approximations and later represented by the Banach contraction principle, but it goes a good deal beyond. The theory of convex functionals and duality, now being vigorously pursued, falls partly within its realm through the fact that subgradients of convex functionals are monotone mappings. Of course, many other techniques are used in functional analysis. Among these one should mention traditional "hard" analysis, whose role in bringing specific problems into the fold of general ideas is permanently assured in this field. Indeed, it is by means of its sophisticated techniques

PREFACE

that such essentials as *a priori* bounds, estimates, coerciveness, contractiveness, and monotonicity are established.

The papers presented in this volume strongly reflect the above-mentioned tendencies in functional analysis. We shall briefly categorize them within such context: L. Nirenberg presents an extension of Leray-Schauder degree and gives an application to a nonlinear elliptic boundary value problem. P. H. Rabinowitz applies degree theory to prove the existence of global continua of solutions of nonlinear eigenvalue problems. Further results about continua of solutions are obtained by R. E. L. Turner using the notion of transversality. K. Kirchgässner shows how variational structure can be used to study some local questions in bifurcation theory.

A large number of papers touch on the notion of monotone operators: H. Brezis presents a brief survey of monotonicity theory, discusses the maximality of the sum of maximal monotone operators, and gives applications to partial differential equations. M. G. Crandall offers a nonlinear version of the Hille-Yosida theorem. Integral equations of the Hammerstein and Urysohn type are the subject of F. E. Browder's article. J. L. Lions gives an extension of boundary layer theory to variational inequalities of elliptic, parabolic, and hyperbolic type. A version of the penalty method for the Navier-Stokes equations is presented by H. Fujita. Three communications deal with convexity: J. J. Moreau discusses various types of weak solutions for minimizing problems in the spirit of duality theory for convex functionals. The duals of convex integral functionals constructed out of one-parameter families of convex functionals are studied in the article by R. T. Rockafellar. E. H. Zarantonello takes up the study of projections on convex sets in Hilbert space, and develops a spectral theory for a class of operators not necessarily linear, extending the classical one for self-adjoint linear operators.

Analysis in its more classical form is represented by three papers: By use of the maximum principle, J. Serrin obtains *a priori* estimates for gradients of solutions of partial differential equations of parabolic and elliptic type. P. D. Lax discusses recent developments in conservation laws, and J. J. Levin and D. F. Shea investigate the asymptotic behavior of the solutions of certain nonlinear integral equations of Volterra type.

Eduardo H. Zarantonello

# Contributions to Nonlinear Functional Analysis

# Generalized Degree and Nonlinear Problems

## L. NIRENBERG

§1.  In this talk, which is based on [6], we will illustrate the use of some topological techniques in solving nonlinear problems. To start with a simple and well known example, let $T$ be a continuous mapping of the closed unit ball $B$ in $R^d$ into $R^{d^*}$ - we wish to solve the equation

$$T(x) = 0 .$$

The topological techniques yield conditions on the boundary values $T_0$ of $T$ which ensure that for every extension $T$ of $T_0$ inside $B$ the equation $T(x) = 0$ is always solvable. Assume that $T_0(x) \neq 0$ on $\partial B$, then one has the following elementary but basic result expressed in terms of the normalized map $\psi(x) = T_0(x)/|T_0(x)|$ mapping $\partial B = S^{d-1}$ into $S^{d^*-1}$:

<u>Proposition.</u>  A necessary and sufficient condition that for every extension $T$ of $T_0$ the equation $T(x) = 0$ is always solvable is that the homotopy class of $\psi$ be nontrivial

This theorem yields useful results only in case $d^* \leq d$. If $d^* = d$ the homotopy class of $\psi$ being nontrivial means that the degree of the map $\psi$, i.e., the number of times the image sphere is covered (counted algebraically), is different from zero. This number $\nu$ is also equal to the degree of the map $T$ at the origin in the image space, i.e. the number of times the origin is covered (counted algebraically).

Consider now an infinite dimensional Banach space X and a continuous map T of B, the closed unit ball in X, into X, with I - T = a compact operator K. The Leray-Schauder theory, which has been one of the most useful techniques in attacking nonlinear problems, is a generalization of the preceding remarks to this situation (B may be the closure of any open set in X). If $T_0(x) \neq 0$, where $T_0 = T|_{\partial B}$, then the mapping T has, again, an integral valued degree $\nu$ at the origin; if $\nu \neq 0$ then $T(x) = 0$ is solvable in B. The degree $\nu$ depends only on $T_0$, in fact only on the homotopy class of $T_0$ within the class of operators such that $I - T_0$ is compact and $T_0(x) \neq 0$ on $\partial B$.

If the range of T is contained within a linear subspace Y of X, $Y \neq X$, then the degree $\nu$ of T at the origin is necessarily zero - since it is the same for all points in a neighbourhood of the origin and, at a point off Y, and thus not in the range of T, it vanishes. In this lecture we shall describe an extension of the Leray-Schauder theorem to such a situation and an application to a nonlinear elliptic boundary value problem. In recent years extensions of the Leray-Schauder theory have been made in various directions. The result presented here (and in [6]) is part of a more general development (see [7], [3], [1], [2]).

Consider a mapping $T : B \to Y \subset X$ as above, with I - T = K compact, $T(x) \neq 0$ on $\partial B$, and Y a closed subspace having finite codimension i. We wish to present a condition on $T_0 = T|_{\partial B}$ to ensure that the equation $T(x) = 0$ is solvable in B for any extension T of $T_0$ inside B - of the form I - compact, and having range in Y. Maps $T_0$ of $\partial B$ into Y with this property are called "essential". Whether $T_0$ is essential or not depends only on its homotopy class (always of the form I-compact) of maps into $Y^* = Y \setminus \{0\}$. For Y = X this is proved in Granas [4], Theorem 22, and the proof is easily extended for any subspace Y. For T = I - K the compact operator K may be approximated by one with finite dimensional range and hence, as one easily sees, the operator $T_0$ may be deformed within its homotopy class to an operator of the form $I - K_1$, mapping

mapping $\partial B$ into $Y$, with $K_1$ mapping into a finite dimensional space. Thus if we write $X$ as a directed sum

$$X = Y \oplus Z, \quad \dim Z = i,$$

so that any vector $x$ in $X$ has the unique decomposition $x = y+z$, with $y \in Y$, $z \in Z$, we may suppose that $T_0$ has the form

$$T_0(x) = T_0(y+z) = y + z - K_1(x) = y - K_2(x),$$

where $K_2$ is a map of $\partial B$ into a finite dimensional subspace $V$ of $Y$. Decomposing $Y$ as a direct sum

$$Y = W_1 \oplus V,$$

with $W_1$ a closed linear subspace of $Y$, so that any $x \in X$ now has the unique decomposition $x = y + z = w_1 + v + z$, $w_1 \in W_1$, $v \in V$, $z \in Z$, we have

$$T_0(x) = w_1 + v - K_2(x) = w_1 - K_3(x)$$

where the range of $K_3$ is in $V$.

Since $T_0(x) \neq 0$ for $x \in \partial B$ we see that $K_3(v+z) \neq 0$ for $v + z \in \partial B$. Hence we may deform $T_0$ via the deformation

$$T_{0t}(x) = T_{0t}(w_1+v+z) = w_1 - K_3(tw_1+v+z), \quad 0 \leq t \leq 1$$

to the map

$$T_{01}(x) = w_1 - K_3(v+z)$$

lying in the same homotopy class. We may therefore suppose that $T_0$ has this very special form, namely, with $V \oplus Z = W$ so that $x = w_1 + w$, we may suppose that

$$T_0(x) = T_0(w_1 + w) = w_1 + \Phi(w)$$

where $\Phi$ is a continuous map of the closed unit ball in $W$ into the linear subspace $V$ of $W$. We shall express the condition for $T_0$ to be "essential" in terms of the map $\Phi$ which does not vanish for $\|w\| = 1$. Suppose dim $W = d$, dim $V = d^*$, $d - d^* = i$; set

(1) $$\Psi(w) = \frac{\Phi(w)}{\|\Phi(w)\|} \quad \text{for} \quad \|w\| = 1.$$

Then we may consider $\Psi$ as a mapping of the sphere $S^{d-1}$ to $S^{d^*-1}$.

<u>Theorem 1.</u>   $T_0$ is "essential" if and only if the map $\Psi$ has nontrivial stable homotopy (defined by suspension).

A proof is given in [6]; in proving sufficiency one first approximates $I - T$ by an operator mapping into a finite dimensional space - reducing the problem to that for finite dimensional $X$. In this case one then applies the Proposition above by showing that the homotopy class of $\dfrac{T_0(x)}{|T_0(x)|}$, mapping the unit sphere in $X$ into that in $Y$, is obtained from the mapping $\Psi$ by repeated suspensions.

§2.   The application that we present grew out of a result of Landesman and Lazer [5] and we shall first describe their result in a slightly restricted form. It concerns a nonlinear elliptic boundary value problem for a real function $u$ in a bounded domain $\mathcal{D} \subset R^n$ with smooth boundary $\Gamma$. (All functions, coefficient of equations, etc., are assumed to be real and smooth in $\bar{\mathcal{D}}$.) Let $L$ be a linear formally self-adjoint elliptic second order operator in $\bar{\mathcal{D}}$ and consider the problem

(2) $$Lu = f(x) - g(u) \quad \text{in } \mathcal{D}, \quad u = 0 \quad \text{on } \Gamma,$$

with $f$ a given (smooth) function; $g(u)$ is continuous and has limits

# GENERALIZED DEGREE

$$\lim_{u \to \pm\infty} g(u) = g(\pm\infty)$$

with

(3) $\quad g(-\infty) < g(u) < g(\infty)$.

Assume that ker L, i.e. the space of solutions of

(4) $\quad Lu = 0$ in $\emptyset$, $u = 0$ on $\Gamma$,

is one dimensional - spanned by the function w. Then from (3) one easily derives a necessary condition for solvability of (2); taking $L_2$ scalar product ( , ) of (2) with w we find

$$(f - g, w) = (Lu, w) = (u, Lw) = 0$$

and using the bounds (3) we obtain the necessary condition

(5) $\quad g(-\infty) \int_{w>0} w\, dx + g(\infty) \int_{w<0} w\, dx < (f, w) < g(\infty) \int_{w>0} w\, dx$

$$+ g(-\infty) \int_{w<0} dx .$$

The surprising result of [5] is that (5) is also sufficient for solvability of (2).

We shall present a generalization of this result, based on Theorem 1, concerning elliptic systems of N equations for N functions $u = (u^1, \ldots, u^N)$ in $\emptyset$. Let L be a linear elliptic system of order m, and consider vector functions u satisfying homogeneous boundary conditions $Bu = 0$ which are "nice" relative to L, i.e., so called, coercive boundary conditions. We will not describe these in any detail except to say that ker L = the space of functions u satisfying $Lu = 0$, and $Bu = 0$, on $\Gamma$ is finite dimensional, spanned, say, by the (vector) functions $w_1, \ldots, w_d$; furthermore, the range of L (acting on smooth functions satisfying $Bu = 0$) consists of the smooth functions which are

$L_2$-orthogonal to a finite number of smooth functions $w'_1, \ldots, w'_{d^*}$. The elliptic operator has an index

$$i = \text{ind } L = d - d^*,$$

and we shall assume that $i = d - d^* \geq 0$.

We shall also make the following hypothesis concerning ker $L$, the space of functions spanned by $w_1, \ldots, w_d$:

(UC)  $w = 0$ <u>is the only function in</u> ker $L$ <u>which vanishes on a set of positive measure in</u> $\bar{\mathcal{D}}$.

The nonlinear system to be solved is of the form

(6) $\qquad L u = g(x, D^\alpha u)$ in $\mathcal{D}$, $Bu = 0$ on $\Gamma$,

where $g$ is a smooth bounded $N$ vector for $x \in \bar{\mathcal{D}}$ and all values of the other arguments; $g$ depends on $u$ and its derivatives $D^\alpha u$ up to order $m-1$. For all arguments $\eta = \{\eta^\alpha\} \neq 0$ with $|\alpha| \leq m-1$ (symmetric in the indices $\alpha_i$ of $(\alpha = \alpha_1 \ldots \alpha_n)$) we suppose that

(7) $\qquad h(x, \eta) = \lim_{r \to \infty} g(x, r\eta)$

and that the convergence is uniform on $\bar{\mathcal{D}} \times \{|\eta| = 1\}$. We shall give sufficient conditions on $h$ to ensure the solvability of (6).

For $a \in S^{d-1}$ define the map $\phi : S^{d-1} \to R^{d^*}$ by

$$\phi_\beta(a) = (h(x, D^\alpha \sum a_j w_j(x)), w'_\beta), \quad \beta = 1, \ldots, d^*.$$

As a consequence of the hypothesis (UC) one may prove (as in [6] with the aid of Lemmas 1 and 2 there) that the mapping $\phi$ is continuous. Assume that $\phi(a) \neq 0$ for $a \in S^{d-1}$ and set

$$\psi(a) = \frac{\phi(a)}{|\phi(a)|}, \quad \psi : S^{d-1} \to S^{d^*-1}.$$

**Theorem 2.** If $\psi$ has nontrivial stable homotopy then (6) is solvable.

By a solution we mean a function in $C^{m-1}$ with derivatives of order m in $L_p$ for large p. If g is smooth then using well known regularity theory, it follows that any such solution is smooth.

The proof of the theorem is the same as that of Theorem 2 in [6].

**Remarks.** (i) If $d = d^*$ then "$\psi$ has nontrivial stable homotopy" means simply that $\psi$ is homotopically nontrivial, i.e. has nonzero degree. In this case one proves the result using the Leray-Schauder degree.

In case $N = 1$, $d = d^* = 1$, and $g = g(x, u)$ depends only on u and not on its derivatives, then $h(x, \eta)$ corresponds to

$$h_\pm(x) = h(x, \pm 1) = \lim_{u \to \pm \infty} g(x, u) .$$

In this case the condition that $\psi$ be homotopically nontrivial means that

$$A_1 = \int_{w>0} h_+ w'\, dx + \int_{w<0} h_- w'\, dx$$

and

$$A_2 = \int_{w<0} h_+ w'\, dx + \int_{w>0} h_- w'\, dx$$

have opposite signs. Theorem 2 then contains the result of Landesman and Lazer described above as a special case.

(ii) In the theorem, $\Omega$ may be a manifold, and the system of vectors u(x) may be replaced by cross sections of a vector bundle in which L acts; g is then also required to take its values there. The vector bundle is supposed to have a Hermitian metric, and the maps $\phi$ and $\psi$ may be defined as before. Their definitions depend on choice of bases $w_i$ and $w'_\alpha$ and so are not canonical. However the condition on the stable homotopy of $\psi$ is independent of these choices.

(iii)   Since it is not known how to determine whether a map $\psi$ has nontrivial stable homotopy, the theorem is not readily applicable.

## REFERENCES

1. K. D. Elworthy. Some problems in algebraic topology. Fredholm maps and $GL_c(E)$ structures. Ph. D. Thesis, Oxford Univ. 1967.

2. K. D. Elworthy, A. Tromba. Differential structures and Fredholm maps on Banach manifolds. Global Analysis. Proc. Symp. Pure Math. 15 Amer. Math. Soc. 1970, p. 45-94.

3. K. Geba. Algebraic topology methods in the theory of compact fields in Banach spaces. Fund. Math. 54 (1964), p. 177-209.

4. A. Granas. The theory of compact vector fields and some of its applications to topology of functional spaces. I. Rozprawy Mat. 30 (1962).

5. E. M. Landesman and A. C. Lazer. Nonlinear perturbations of linear elliptic boundary value problems at resonance. J. Math. Mech. 19 (1970), p. 609-623.

6. L. Nirenberg. An application of generalized degree theory to a class of nonlinear problems. Proc. Symp. Functional Analsis, Liège, Sept. 1970; to appear.

7. A. S. Švarc. The homotopic topology of Banach spaces. spaces. Dokl. Akad. Nauk SSSR 154 (1964), p.61-63; Soviet Math. Dokl. 5 (1964), p. 57-59.

GENERALIZED DEGREE

The work for this paper was supported by the U. S. Air Force Contract AF-49 (638)-1719.
Reproduction in whole or in part is permitted for any purpose of the United States Government.

                Courant Institute of Mathematical Sciences
                    New York University
                    New York, New York

                                    Received April 12, 1971

# A Global Theorem for Nonlinear Eigenvalue Problems and Applications

*PAUL H. RABINOWITZ*

Introduction.

Suppose E is a real Banach space with norm, $\|\cdot\|$, and $\mathbb{R} \times E$ has the product topology. By a <u>nonlinear eigenvalue problem</u> we mean an equation of the form

(0.1) $\qquad u = G(\lambda, u)$

where $\lambda \in \mathbb{R}$, $u \in E$, and $G: \mathbb{R} \times E \to E$ is compact, i.e. it is continuous and maps bounded sets to relatively compact sets. A <u>solution</u> of (0.1) is a pair $(\lambda, u) \in \mathbb{R} \times E$.

Equations of the form (0.1) occur in many parts of mathematical physics, in particular in fluid dynamics and in elasticity theory. Thus the nature of the structure of the set of their solution is an important question. For a collection of some recent work in this direction, see [1]. A great deal of earlier theoretical work is contained in [2], [3]. See also [4] for a survey paper.

We will study (0.1) mainly under the hypotheses that $G(\lambda, u) = \lambda L u + H(\lambda, u)$ where $L: E \to E$ is a compact linear map and $H(\lambda, u) = 0(\|u\|)$ near $u = 0$ uniformly on bounded $\lambda$ intervals. Here the phenomenon of interest is <u>bifurcation.</u> Physical examples are e.g. buckling in elasticity and the onset of thermal convection or the appearance of vortices in fluid dynamics.

To describe this precisely mathematically, note that (0.1) possesses the line of solutions $\{(\lambda, 0) | \lambda \in \mathbb{R}\}$ which will henceforth be called the <u>trivial solutions</u>. We call $(\mu, 0)$ a <u>bifurcation point</u> for (0.1) with respect to the line of trivial solutions if every neighborhood of $(\mu, 0)$ contains nontrivial solutions of (0.1). It is a well known theorem of Krasnoselski [2] that if $\mu$ is a characteristic-value of $L$ of odd multiplicity, then $(\mu, 0)$ is a bifurcation point for (0.1). We will obtain a strengthened version of this result which shows that bifurcation in this situation is a <u>global</u> rather than a <u>local</u> phenomenon. Let $\mathcal{S}$ denote the closure of the set of nontrivial solutions of (0.1). Then we will show: If $\mu$ is a characteristic value of $L$ of odd multiplicity, $\mathcal{S}$ contains a component (i.e. a maximal subcontinuum) containing $(\mu, 0)$ which either is unbounded or contains $(\hat{\mu}, 0)$ where $\hat{\mu} \neq \mu$ is another characteristic value of $L$.

The proof of this result is given in §1, the main tool being the theory of topological degree of Leray and Schauder. Applications of the theorem to differential and to integral equations are mentioned in §2. In particular a nonlinear extension of a linear Strum-Liouville theorem for second order ordinary differential equations is given where nodal properties play an important role. Existence results for positive solutions (i.e. $\lambda > 0$, $u \geq 0$) of quasi-linear elliptic partial differential equations are also obtained.

The homotopy methods which go into the proof of our main result can be used to treat (0.1) under other conditions on $G$ and in particular in situations where bifurcation is not involved. Such a case is provided by requiring that $G(0, u) \equiv 0$. Then the solution set of (0.1) contains a component containing $(0, 0)$ and which is unbounded both in $\mathbb{R}^+ \times E$ and $\mathbb{R}^- \times E$. This result is essentially due to Leray and Schauder. A simple proof is given in §3 and some applications to elliptic and hyperbolic partial differential equations are mentioned.

Extensions of many of the results mentioned here and further details can be found in [5] and [6].

# A GLOBAL EIGENVALUE THEOREM

## 1. The Main Theorem.

Let E and G be as in the introduction. We say $\mu \in \mathbb{R}$ is a <u>characteristic value</u> of L if there exists $v \in E$, $v \neq 0$, such that $v = \mu L v$. The multiplicity of a characteristic value $\mu$ is the dimention of $\bigcup_{j=1}^{\infty} \ker(I - \mu L)^j$ where I denotes the identity map and ker A denotes the kernel of the linear map A. Let r(L) denote the set of real characteristic values of L.

It is well known that a necessary condition for $(\mu, 0)$ to be a bifurcation point for (0.1) with respect to the line of trivial solutions is that $\mu \in r(L)$. This follows since if $\mu \notin r(L)$, then $I - \lambda L$ is invertible for all $\lambda$ near $\mu$. Hence for $\lambda$ near $\mu$, (0.1) is equivalent to

(1.1) $$u = (I - \lambda L)^{-1} H(\lambda, u).$$

Since the right hand side of (1.1) is $0(\|u\|)$ for u near 0 while the left hand side is not, $(\lambda, 0)$ is an isolated solution of (0.1) in $\{\lambda\} \times E$ uniformly in $\lambda$ for $\lambda$ near $\mu$. Consequently $(\mu, 0)$ cannot be a bifurcation point.

Simple examples show that not all $\mu \in r(L)$ correspond to bifurcation points $(\mu, 0)$. For example, taking $E = \mathbb{R}^2$, $u = (x, y)$, $L \equiv I$, and $H(\lambda, u) \equiv (-y^3, x^3)$, (0.1) becomes

(1.2) $$\begin{pmatrix} x \\ y \end{pmatrix} = \lambda \begin{pmatrix} x \\ y \end{pmatrix} + \begin{pmatrix} -y^3 \\ x^3 \end{pmatrix}.$$

Multiplying the first equation in (1.2) by y, the second by x and subtracting shows that $\lambda = 1$, $u = 0$ is not a bifurcation point.

Krasnoselski [2, chapter 4] has shown that if $\mu \in r(L)$ is of odd multiplicity, then $(\mu, 0)$ is a bifurcation point for (0.1) with respect to the line of trivial solutions. Using essentially the same tools as Krasnoselski, namely the theory of degree of mapping of Leray and Schauder, we will show a stronger result obtains here. First some terminology and a technical lemma is needed.

Let $\mathcal{S}$ denote the closure of the set of non-trivial solutions of (0.1). Thus the only trivial solutions in $\mathcal{S}$ are bifurcation points. A subcontinuum of $\mathcal{S}$ is a closed connected subset and a component is a maximal (with respect to inclusion) subcontinuum. The boundary of a set $A$ in $E$ or $\mathbb{R} \times E$ will be denoted by $\partial A$. By a $\delta$ neighborhood of $A$ we mean the set of points within a distance $\delta$ of $A$.

**Lemma 1.3:** Let $\mu \in r(L)$ and let $\mathcal{C}$ denote the component of $\mathcal{S} \cup \{(\mu, 0)\}$ to which $(\mu, 0)$ belongs. Suppose that $\mathcal{C}$ (i) is bounded and (ii) does not contain $(\hat{\mu}, 0)$ for any $\hat{\mu} \in r(L)$, $\hat{\mu} \neq \mu$. Then there exists a bounded open set $\mathcal{O} \subset \mathbb{R} \times E$ such that $\mathcal{C} \subset \mathcal{O}$, $\mathcal{S} \cap \partial \mathcal{O} = \emptyset$, and the only trivial solutions contained in $\mathcal{O}$ consist of the segment $\{(\lambda, 0) \mid |\lambda - \mu| < \varepsilon\}$ for some $\varepsilon < \varepsilon_0$, the distance from $\mu$ to $r(L) - \{\mu\}$.

**Proof:** The compactness of $G$ and (i) imply $\mathcal{C}$ is compact. Let $U_\delta$ be a $\delta$ neighborhood of $\mathcal{C}$ where $\delta < \varepsilon_0$. Therefore $\partial U_\delta \cap \mathcal{C} = \emptyset$.

By remarks made above, if $\lambda \notin r(L)$, $(\lambda, 0)$ will contain a neighborhood disjoint from $\mathcal{S}$ and a fortiori $\mathcal{C}$. With the aid of this observation together with (ii) we can assume that the only trivial solutions $U_\delta$ contains are $\{(\lambda, 0) \mid |\lambda - \mu| < \delta\}$.

Let $K = \overline{U}_\delta \cap \mathcal{S}$. Then $K$ is a compact metric space under the induced topology from $\mathbb{R} \times E$. Since $\partial U_\delta \cap \mathcal{C} = \emptyset$, by a lemma from point set topology [7, chapter 1] there exist disjoint compact sets $K_1, K_2$ of $K$ such that $K_1 \supset \mathcal{C}$, $K_2 \supset (\partial U_\delta) \cap \mathcal{S}$, and $K = K_1 \cup K_2$. Thus if $\mathcal{O}$ is an $\varepsilon$ neighborhood (in $\mathbb{R} \times E$) of $K_1$, where $\varepsilon < \delta$ and $\varepsilon$ is less than the distance from $K_1$ to $K_2$, then $\mathcal{O}$ satisfies the requirements of lemma 1.3.

Next suppose that $\Omega \subset E$ is bounded and open, $b \in E$, and $T: \overline{\Omega} \to E$ is compact. Let $\Psi(u) = u - T(u)$. If $b \in \Psi(\partial\Omega)$, the Leray-Schauder degree of $\Psi$ with respect to $\Omega$ and $b$ is well defined and will be denoted by $d(\Psi, \Omega, b)$. For what follows $b = 0$ and therefore for brevity we write $d(\Psi, \Omega)$. The index of an isolated zero, $v$, or $\Psi$ will be denoted by $i(\Psi, v)$. For the properties of Leray-Schauder degree see e.g.

# A GLOBAL EIGENVALUE THEOREM

[2], [8], [9], or [6, Appendix]. Lastly $B_\rho$ denotes a closed ball in E of radius $\rho$ centered at the origin.

We can now prove our main result:

**Theorem 1.4:** If $\mu \in r(L)$ is of odd multiplicity, $\mathcal{S}$ possesses a component $C$ containing $(\mu, 0)$. Moreover either (i) $C$ is unbounded or (ii) $C$ contains $(\hat{\mu}, 0)$ where $\mu \neq \hat{\mu} \in r(L)$.

**Proof:** If not, there exists a bounded open set $\mathcal{O}$ and $\delta > \varepsilon > 0$ as in Lemma 1.3. For $0 < |\lambda - \mu| \leq \delta$, $(\lambda, 0)$ is an isolated solution of (0.1) in $\{\lambda\} \times E$. Therefore there exists $\rho(\lambda) > 0$ such that $(\lambda, 0)$ is the only solution of (0.1) in $\{\lambda\} \times B_{\rho(\lambda)}$. Define $\rho(\lambda) = \rho(\mu + \delta)$ for $\lambda > \mu + \delta$ and $\rho(\lambda) = \rho(\mu - \delta)$ for $\lambda < \mu - \delta$. Let $\mathcal{O}_\lambda = \{u \in E | (\lambda, u) \in \mathcal{O}\}$ and $(\partial \mathcal{O})_\lambda = \{u \in E | (\lambda, u) \in \partial \mathcal{O}\}$. Then by choosing $\rho(\mu \pm \delta)$ small enough, it follows from the properties of $\mathcal{O}$ that $B_{\rho(\lambda)} \cap (\partial \mathcal{O})_\lambda = \emptyset$ if $|\lambda - \mu| \geq \delta$.

Since for $\lambda \neq \mu$ there are no solutions of (0.1) on $\{\lambda\} \times \partial(\mathcal{O}_\lambda - B_{\rho(\lambda)})$, $d(\Phi(\lambda, \cdot), \mathcal{O}_\lambda - B_{\rho(\lambda)})$ is well defined. We will prove that

(1.5)  $d(\Phi(\lambda, \cdot), \mathcal{O}_\lambda - B_{\rho(\lambda)}) = 0$  for $\lambda \neq \mu$

and then show that (1.5) is incompatible with the odd multiplicity assumption for $\mu$. The theorem is then established.

Let $\lambda > \mu$ and $\lambda^* > \lambda$ where $\lambda^* - \mu$ is greater than the diameter of $\mathcal{O}$. Then $\mathcal{O}_{\lambda^*} = \emptyset$. Defining $\rho = \inf \{\rho(\theta) | \theta \in [\lambda, \lambda^*]\}$, it follows from remarks made earlier that $\rho > 0$. Let $U = \mathcal{O} - [\lambda, \lambda^*] \times B_\rho)$. Then $U$ is a bounded open set in $[\lambda, \lambda^*] \times E$ and by construction. $\Phi(\zeta, u) \neq 0$ for $(\zeta, u) \in \partial U$. (Here $\partial U$ refers to the boundary of $U$ in $[\lambda, \lambda^*] \times E$).

By the homotopy invariance of degree [8], [6]:

(1.6)  $d(\Phi(\zeta, \cdot), \mathcal{O}_\zeta - B_\rho) \equiv$ constant, $\zeta \in [\lambda, \lambda^*]$.

Since $\mathcal{O}_{\lambda*} - B_\rho = \emptyset$,

(1.7) $$d(\Phi(\lambda_*, \cdot), \mathcal{O}_{\lambda*} - B_\rho) = 0$$

and from (1.6) - (1.7),

(1.8) $$d(\Phi(\lambda, \cdot), \mathcal{O}_\lambda - B_\rho) = 0 .$$

$\Phi(\lambda, \cdot)$ has no zeroes in $\{\lambda\} \times (B_{\rho(\lambda)} - \overset{\circ}{B}_\rho)$. Hence:

(1.9) $$d(\Phi(\lambda, \cdot), \overset{\circ}{B}_{\rho(\lambda)} - B_\rho) = 0 .$$

The additivity of degree and (1.8)-(1.9) then imply (1.5) for $\lambda > \mu$. The argument for $\lambda < \mu$ is the same.

For $|\lambda - \mu| < \varepsilon$, $d(\Phi(\lambda, \cdot), \mathcal{O}_\lambda)$ is defined and another application of the homotopy invariance of degree yields:

(1.10) $$d(\Phi(\lambda, \cdot), \mathcal{O}_\lambda) \equiv \text{constant}, \quad |\lambda - \mu| < \varepsilon .$$

Select $\underline{\lambda}, \bar{\lambda}$ so that $\mu - \varepsilon < \underline{\lambda} < \mu < \bar{\lambda} < \mu + \varepsilon$. The additivity of degree and the fact that $(\lambda, 0)$ is an isolated zero of $\Phi(\lambda, \cdot)$ in $\{\lambda\} \times E$ for $\lambda \notin r(L)$ gives:

(1.11)
$$d(\Phi(\underline{\lambda}, \cdot), \mathcal{O}_{\underline{\lambda}}) = i(\Phi(\underline{\lambda}, \cdot), (\underline{\lambda}, 0)) + d(\Phi(\underline{\lambda}, \cdot), \mathcal{O}_{\underline{\lambda}} - B_{\rho(\underline{\lambda})})$$
$$d(\Phi(\bar{\lambda}, \cdot), \mathcal{O}_{\bar{\lambda}}) = i(\Phi(\bar{\lambda}, \cdot), (\bar{\lambda}, 0)) + d(\Phi(\bar{\lambda}, \cdot), \mathcal{O}_{\bar{\lambda}} - B_{\rho(\bar{\lambda})})$$

Combining (1.5), (1.10)-(1.11) gives:

(1.12) $$i(\Phi(\underline{\lambda}, \cdot), (\underline{\lambda}, 0)) = i(\Phi(\bar{\lambda}, \cdot), (\bar{\lambda}, 0)) .$$

However since $\mu \in r(L)$ is of odd multiplicity and $\underline{\lambda} < \mu < \bar{\lambda}$, the indices of the solution $(\underline{\lambda}, 0), (\bar{\lambda}, 0)$ are either $\pm 1$, and differ by a factor of $-1$, i.e.

(1.13) $$i(\Phi(\underline{\lambda}, \cdot), (\underline{\lambda}, 0)) = -i(\Phi(\bar{\lambda}, \cdot), (\bar{\lambda}, 0)) \neq 0 .$$

Thus (1.12) and (1.13) are contradictory and the proof is complete.

Remark 1: Both alternatives of Theorem 1.4 are possible. The simplest example of (i) occurs when $H \equiv 0$. Examples of (ii) are more difficult to construct. We give one following Remark 4. Also in a recent paper of Bauer, Keller and Riess [10], a pair of interlocking nonlinear ordinary differential equation arising in the study of buckling of spherical and hemispherical shells was treated numerically. For these equations, which when given an operator formulation satisfy the hypotheses of Theorem 1.4, it was discovered that (ii) occurs at each bifurcation point.

Remark 2: If $\mu$ is a <u>simple</u> characteristic value of L, i.e. $\mu$ is of multiplicity one, with corresponding eigenvector v, H is smooth near $(\mu, 0)$ (and not necessarily compact), and some technical conditions are satisfied, the structure of $S$ near $(\mu, 0)$ is known precisely [3], [11]. Here in addition to the trivial solutions, $S$ consists of a smooth curve $(\lambda(\alpha), u(\alpha))$ where $\lambda(\alpha) = \mu + 0(1)$, $u(\alpha) = \alpha v + 0(|\alpha|)$ near $\alpha = 0$. Even if H is not smooth but $\mu$ is simple, a better result than Theorem 1.4 is available, namely in any neighborhood $\eta$ of $(\mu, 0)$ of sufficiently small diameter, $C \cap \eta$ can be decomposed into two subcontinua which meet only at $(\mu, 0)$ and which both meet $\partial \eta$. These subcontinua have global extensions but may intersect in the large. (see [5]) If $\mu$ is of odd multiplicity but is not simple, this result is not true as an example following Remark 4 shows. Lastly we note that if $\mu$ is of odd multiplicity and H is smooth but not necessarily compact near $(\mu, 0)$, (0.1) can be converted to an equivalent finite dimensional problem near $(\mu, 0)$ and a finite dimensional application of degree theory give gives a subcontinuum of $S$ meeting $(\mu, 0)$.

Remark 3: If we know a priori that alternative (i) of Theorem 1.4 does not occur, then the proof of the theorem can be modified to obtain the following result: Let $\Gamma = \{(\gamma, 0) | (\gamma, 0) \in C$ and $\gamma \neq \mu\}$. Then $\Gamma$ contains an odd number (in particular $\geq 1$) of points $(\gamma, 0)$ where $\gamma \in r(L)$ is of odd multiplicity.

Remark 4: In several physical problems, in particular some involving buckling or involving rotational, thermal, magnetic, or chemical effects on fluids, it is possible to obtain a priori bounds for solutions of (0.1) in the following sense: there exists a continuous function $M: \mathbb{R}^+ \to \mathbb{R}^+$ such that if $(\lambda, u) \in \mathcal{S}$, $\|u\| \leq M(\lambda)$. If this is the case and $\mu \geq 0$, $\mathcal{C}$ cannot both be unbounded and have a bounded projection on $\mathbb{R}^+$.

Next we give two examples mentioned above, namely: (a) a situation where alternative (ii) of Theorem 1.4 occurs; and (b) an equation of the form (0.1) where $\mu \in r(L)$ is of odd multiplicity but not simple and $\mathcal{C} - \{(\mu, 0)\}$ near $(\mu, 0)$ (or globally for that matter) does not consist of a pair of disjoint continua. These examples are due to M. G. Crandall and the author. There is a common idea behind both examples, and we develop it first.

Let $n \in \mathbb{N}$, $n > 1$. Choose $f$ to be a continuous map from $S^{n-1}$ to $\mathbb{R}^n$, i.e. a vector field on $\mathbb{R}^n$, such that $\|f(u)\| \leq 1$ and $(f(u), u) = 0$ for all $u \in S^{n-1}$ (where $(\cdot, \cdot)$ and $\|\cdot\|$ denote the usual Euclidean inner product and norm on $\mathbb{R}^n$. We define an $n \times n$ matrix valued mapping $B$ on $S^{n-1}$ by $B(u)v = \frac{1}{4}[(u, v)f(u) + (v, f(u))u] + v$ for $u \in S^{n-1}$ and $v \in \mathbb{R}^n$. Then it easily follows that:

(1.14) $\qquad (B(u)v, v) \geq \frac{1}{2} \|v\|^2, \quad u \in S^{n-1}, \quad v \in \mathbb{R}^n$.

We extend $B$ to all of $\mathbb{R}^n$ by homogeneity: $B(tu) = t^2 B(u)$ for all $t \geq 0$, $u \in S^{n-1}$. Then $B(u) = 0(\|u\|)$ near $u = 0$. For any $u \neq 0$, $u \in \mathbb{R}^n$

(1.15) $\qquad B(u)u = \frac{1}{4}\|u\|^3 f(\frac{u}{\|u\|}) + u\|u\|^2$.

Thus $u \neq 0$ and $B(u)u = u$ is equivalent to $\|u\| = 1$ and $f(u) = 0$.

To give an example of (a), we choose $n = 2$ and an $f$ which possesses no zeroes on $S^1$. Such an $f$ clearly exists. Then $B(u)u = u$ possesses only $u = 0$ as a solution. Consider the equation (0.1) where $E = \mathbb{R}^2$, $L = \begin{pmatrix} 1 & 0 \\ 0 & \frac{1}{2} \end{pmatrix}$ and $H(\lambda, u) = -\lambda L B(u)u$. Since the characteristic values of $L$ are 1 and 2 and both are simple, Theorem 1.4 applies here.

If alternative (i) of the theorem were to hold for $(1,(0,0))$ or $(2,(0,0))$, we could find an unbounded sequence of nontrivial solutions $(\lambda_n, u_n)$ of (0.1) with $\lambda_n > 0$. (There are no nontrivial solutions for $\lambda = 0$.) In particular either ($\alpha$) $\|u_n\| \to \infty$ or ($\beta$) $\lambda_n \to \infty$ as $n \to \infty$. Multiplying (0.1) by $L^{-1}$ and taking the inner product of the resulting equation with $u_n$ gives:

(1.16) $$0 \le \|u_n\|^2 - (B(u_n)u_n, u_n)$$

Using the homogeneity of $B$ to combine (1.16) with (1.14) yields:

(1.17) $$\tfrac{1}{2}\|u_n\|^4 \le \|u_n\|^2 \quad \text{or} \quad \|u_n\| \le \sqrt{2}$$

Note that the estimate (1.17) is independent of $\lambda$. Thus ($\alpha$) above cannot occur. Multiplying (0.1) by $\lambda_n^{-1} L^{-1}$, using (1.17), and letting $n \to \infty$, we see that a subsequence of $u_n$ converges to a solution $v$ of the limit equation:

(1.18) $$u = B(u)u$$

Our choice of $f$ implies that only $v = 0$ is a solution of (1.18). Hence $u_n \to 0$ as $n \to \infty$. But then multiplying (0.1) by $(\lambda_n \|u_n\|)^{-1} L^{-1}$ and letting $n \to \infty$ leads to the untenable conclusion that a subsequence of $\left(\frac{u_n}{\|u_n\|}\right)$ converges to $w$ satisfying $w = 0$ and $\|w\| = 1$. Thus (i) of Theorem 1.4 cannot hold and hence (ii) does.

A particular $B(u)$ satisfying the above conditions is (see [5]):

$$B(u) = \begin{pmatrix} 4u_1^2 + 6u_2^2 & -2u_1 u_2 \\ -2u_1 u_2 & 6u_1^2 + 4u_2^2 \end{pmatrix}$$

Next we give an example of (b). Let $n \in \mathbb{N}$ be odd, $n > 1$. Hence we can find a vector field $f$ which has exactly one zero, $v$, on $S^{n-1}$. From (1.15) we see that $v$ is the only solution of (1.18) on $S^{n-1}$. The homogeneity of $B$ then implies that for $\gamma > 0$, the only nonzero solution of

$B(u)u = \gamma u$ is $u = \sqrt{\gamma} \, v$. We take $E = \mathbb{R}^n$, $L = I$, and $H(\lambda, u) = -\lambda B(u)u$ in (0.1). Since $n$ is odd, $\mu = 1$ is a characteristic value of $L$ of odd multiplicity. Therefore the hypotheses of Theorem 1.4 are satisfied. We can assume $(\lambda, u) \in C$ implies $\lambda > 0$. Taking the inner product of (0.1) with $u$ gives (for $\lambda > 0$):

(1.19) $\quad \|u\|^2 = \lambda(\|u\|^2 - (B(u)u, u)) \leq \lambda(\|u\|^2 - \tfrac{1}{2}\|u\|^4)$.

Hence $\tfrac{\lambda}{2}\|u\|^4 \leq (\lambda-1)\|u\|^2$ so if $(\lambda, u) \in C$ and $u \neq 0$, then $\lambda > 1$.

Suppose $(\lambda, u) \in C$, $\|u\| \neq 0$. Then (0.1) can be written:

(1.20) $\quad B(u)u = (\tfrac{\lambda-1}{\lambda})u \equiv \gamma(\lambda)u$.

Thus by our above remarks, $C = \{(\lambda, \sqrt{\tfrac{\lambda-1}{\lambda}} \, v) \mid \lambda \geq 1\}$ and in particular any ball about $(1, 0)$ meets $C$ only once.

## §2. Applications.

Some applications of Theorem 1.4 will be given in this section. In all of these results the characteristic values that we deal with will be simple and alternative (i) of the theorem will occur. If the simple eigenvalue result mentioned in Remark 2 is used, somewhat better results can be obtained.

We begin by considering nonlinear Sturm-Liouville eigenvalue problems for second order ordinary differential equations. It was this question that led us to Theorem 1.4. Let

(2.1) $\quad \mathcal{L}u \equiv -(pu')' + qu = F(x, u, u', \lambda)$, $0 < x < \pi$

together with the separated boundary conditions

(2.2) $\quad a_0 u(0) + b_0 u'(0) = 0$, $a_1 u(\pi) + b_1 u'(\pi) = 0$

where $(a_0^2 + b_0^2)(a_1^2 + b_1^2) \neq 0$. Henceforth we denote boundary conditions by B.C. We take $F$ to be continuous in its arguments on $[0,\pi] \times \mathbb{R}^3$ and $F(x,\xi,\eta,\lambda) = \lambda\,a(x)\xi + K(x,\xi,\eta,\lambda)$ where K is $0((\xi^2 + \eta^2)^{\frac{1}{2}})$ near $(\xi,\eta) = (0,0)$ uniformly on bounded $\lambda$ intervals. The functions p, q, a are continuous on $[0,\pi]$ with p and a positive and p continuously differentiable. Because of the form of F, (2.1)-(2.2) always possesses the trivial solutions: $\{(\lambda,0) | \lambda \in \mathbb{R}\}$.

For $K \equiv 0$, (2.1)-(2.2) becomes a linear Sturm-Liouville eigenvalue problem:

(2.3) $\quad \mathcal{L}\,u = \lambda\,a\,u, \quad 0 < x < \pi, \quad u \in B.C.$

It is well known [12] that (2.3) possesses an increasing sequence of simple eigenvalues $(\mu_n)$ with $\mu_n \to \infty$ as $n \to \infty$. Any eigenfunction $v_n$ corresponding to $\mu_n$ has exactly n-1 simple zeros on $(0,\pi)$. ($\xi$ is a simple zero of v if $v(\xi) = 0$ and $v'(\xi) \neq 0$).

To take advantage of these nodal properties, an appropriate family of sets first used in [13] is introduced. Let E be the Banach space $C^1[0,\pi] \cap B.C.$ under the norm:

$$\|u\|_1 = \max_{x \in [0,\pi]} |u(x)| + \max_{x \in [0,\pi]} |u'(x)|.$$

Let $S_k^+$ denote the set of $u \in E$ leaving exactly k-1 simple zeroes in $(0,\pi)$, u positive in a deleted neighborhood of $x = 0$, and all zeroes of u in $[0,\pi]$ as simple zeroes. (This last condition was inadvertantly omitted in [5] and [6]). Define $S_k^- = -S_k^+$ and $S_k = S_k^+ \cup S_k^-$. Then the sets $S_k^\pm, S_k$ are open subsets of E and any eigenfunction $v_k$ of (2.3) corresponding to $\mu_k$ belongs to $S_k$. We make $v_k$ unique by requiring that $v_k \in S_k^+$ and $\|v_k\|_1 = 1$.

With this terminology the solution set of (2.3) can be characterized as consisting of the line of trivial solutions $\{(\lambda,0) | \lambda \in \mathbb{R}\}$ and in addition for each integer $k > 0$, there exists a line of nontrivial solutions $\{(\mu_k, \alpha v_k) | \alpha \in \mathbb{R}\}$ in $(\mathbb{R} \times S_k) \cup \{(\mu_k, 0)\}$. An analogous statement can be made for (2.1)-(2.2). First note that it is meaningful to study solutions of this equation in $\mathbb{R} \times E$ since it can be converted to an equivalent integral equation in E. Again letting $\mathcal{L}$

21

denote the closure of the set of nontrivial solutions of (2.1)-(2.2) in $\mathbb{R} \times E$, the following theorem obtains:

**Theorem 2.4:** For each integer $k > 0$, there exists a component $C_k$ of $\mathcal{S}$ in $(\mathbb{R} \times S_k) \cup \{(\mu_k, 0)\}$ which contains $(\mu_k, 0)$ and is unbounded.

The proof of Theorem 2.4 is accomplished with the aid of Theorem 1.4 and some preliminaries. If zero is a characteristic value of $\mathcal{L}$, some technical problems are encountered. Hence for the moment it will be assumed that zero is not a characteristic value of $\mathcal{L}$. Then $\mathcal{L}$ possesses a Greens' function $\mathcal{G}(x,y)$ [12] such that (2.1)-(2.2) is equivalent to

$$(2.5) \quad u(x) = \int_0^\pi \mathcal{G}(x, y)[\lambda a(y) u(y) + K(y, u(y), u'(y), \lambda)] dy$$

$$\equiv \lambda L u + H(\lambda, u)$$

where

$$L u = \int_0^\pi \mathcal{G}(x, y) a(y) u(y) dy$$

and $H(\lambda, u) = u - \lambda L u$. It is easily verified that L and H are compact maps on $E$, $\mathbb{R} \times E$ into $E$ respectively with L linear and $H(\lambda, u) = 0 (\|u\|_1)$ near $u = 0$ in $E$ uniformly on bounded $\lambda$ intervals. Moreover the characteristic values of L are the eigenvalues $\mathcal{L}$ and hence are simple. Thus the hypotheses of Theorem 1.4 are satisfied and by that theorem for each integer $k > 0$ there exists a component $C_k$ of $\mathcal{S}$ such that $C_k$ contains $(\mu_k, 0)$ and either is unbounded in $\mathbb{R} \times E$ or meets $(\mu_j, 0)$, $j \neq k$.

**Lemma 2.6:** There exists a neighborhood $\eta_i$ of $(\mu_i, 0)$ such that if $(\lambda, u) \in \eta_i \cap \mathcal{S}$ and $u \neq 0$, then $u \in S_i$.

**Proof:** The lemma asserts that all nontrivial solutions of (2.5) near $(\mu_i, 0)$ be in $\mathbb{R} \times S_i$. If not, there exists a sequence $(\lambda_n, u_n) \in \mathcal{S}$ such that $(\lambda_n, u_n) \to (\mu_i, 0)$ as $n \to \infty$ and $u_n \notin S_i$. From (2.5),

$$(2.7) \qquad \frac{u_n}{\|u_n\|_1} = \lambda_n L \frac{u_n}{\|u_n\|_1} + \frac{H(\lambda_n, u_n)}{\|u_n\|} .$$

Since $\{u_n/\|u_n\|_1\}$ is bounded in $E$ and $\lambda_n \to \mu_i$ as $n \to \infty$, the properties of $L$ and $H$ imply that the right hand side of (2.7) converges for some subsequence $(u_{nj})$ of $(u_n)$. Hence, from (2.7), $u_{nj}/\|u_{nj}\|_1$ converges in $E$ to $v$ satisfying $\|v\| = 1$ and:

$$(2.8) \qquad v = \mu_i L v .$$

Consequently $v = v_i$ or $v = -v_i$. In any event, $v \in S_i$. Since $S_i$ is open and $u_{nj}/\|u_{nj}\|_1$ converges to $v$, it follows that $u_{nj}/\|u_{nj}\|$ and therefore $u_{nj} \in S_i$ for all $nj$ large enough contrary to an above assumption. Thus the lemma is proved.

**Lemma 2.9:** If $(\lambda, u)$ is a solution of (2.1) such that $u$ has a double zero, then $u \equiv 0$.

**Proof:** Suppose for $\tau \in [0, \pi]$, $u(\tau) = 0 = u'(\tau)$. Let $w = pu'$. Then (2.1) can be written as a first order system of the form:

$$(2.10) \qquad \begin{pmatrix} u \\ w \end{pmatrix}' = \begin{pmatrix} w/p \\ qu - F(x, u, \frac{w}{p}, \lambda) \end{pmatrix} \qquad u(\tau) = 0 = w(\tau)$$

Multiplying the equations of (2.10) by $u, w$ respectively, using the Schwarz inequality and the properties of $K$ near $x = \tau$ yields:

$$(2.11) \qquad \frac{d}{dx}(u^2 + w^2) \leq c(u^2 + w^2) \qquad x \text{ near } \tau$$

where $c$ is a constant. Therefore $u^2(x) + p^2(x)(u'(x))^2 \leq (u^2(\tau) + p^2(\tau)(u'(\tau))^2) e^{cx} \equiv 0$ for $x$ near $\tau$ and for all $x \in [0, \pi]$ by continuation.

**Remark:** If $F(x,\xi,\eta,\lambda)$ were Lipschitz continuous in $\xi$, $\eta$ near $(0,0)$, Lemma 2.9 would be an immediate consequence of the basic uniqueness theorem for the initial value problem for ordinary differential equations.

**Proof of Theorem 2.4:** Fix $k > 0$. By what has been shown above, there exists a component $C_k \subset \mathcal{S}$ such that $(\mu_k, 0) \in C_k$ and $C_k$ is either unbounded or meets $(\mu_j, 0)$, $j \neq k$. By lemma 2.6, $C_k \cap \eta_k \subset (\mathbb{R} \times S_k) \cup \{(\mu_k, 0)\}$. Moreover if $C_k \subset (\mathbb{R} \times S_k) \cup \{(\mu_k, 0)\}$, then $C_k$ must be unbounded since it cannot meet $(\mu_j, 0)$, $j \neq k$ for $C_j \cap \eta_j \subset (\mathbb{R} \times S_j) \cup \{(\mu_j, 0)\}$.

To show that $C_k \subset (\mathbb{R} \times S_k) \cup \{(\mu_k, 0)\}$, we argue indirectly. If not, there exists $(\lambda, u) \in C_k \cap (\mathbb{R} \times \partial S_k)$ with $(\lambda, u) \neq (\mu_k, 0)$ and $(\lambda, u) = \lim_{n \to \infty} (\lambda_n, u_n)$ where $u_n \in S_k$. Since $u \in \partial S_k$, $u$ has a double zero and by Lemma 2.9, $u \equiv 0$. Hence $\lambda = \mu_i$ is a characteristic value of $L$. But this is not possible by Lemma 2.6 again.

To complete the proof of Theorem 2.4, an approximation argument is used to eliminate the assumption that zero is not an eigenvalue of $\mathcal{L}$. Observe that even when $\mathcal{L}$ is not invertible, by using e.g. the Greens function for $-\frac{d}{dx} p \frac{d}{dx}$, (2.1)-(2.2) can be converted to an equivalent equation in $E$ so it is still meaningful to consider solutions of (2.1)-(2.2) in $\mathbb{R} \times E$. Unfortunately this equation is not of the form (0.1).

If $\mathcal{L}$ is not invertible, replace $\mathcal{L}$ by $\mathcal{L}_\varepsilon = \mathcal{L} + \varepsilon a$, $\varepsilon > 0$. For $\varepsilon$ sufficiently small, $\mathcal{L}_\varepsilon$ is invertible and the eigenvalues of

(2.12) $\qquad \mathcal{L}_\varepsilon v = \mu a v, \quad 0 < x < \pi, \quad v \in \text{B.C.}$

are $\mu_k(\varepsilon) = \mu_k + \varepsilon$. Consider the equation

(2.13) $\qquad u = \lambda L_\varepsilon u + H_\varepsilon(\lambda, u) \equiv G_\varepsilon(\lambda, u)$

where $L_\varepsilon, H_\varepsilon$ are the integral operators obtained from (2.1)-(2.2) on replacing $\mathcal{L}$ by $\mathcal{L}_\varepsilon$. Let $\mathcal{S}_\varepsilon$ be the analogue of $\mathcal{S}$ for (2.13). By what has been shown above, for each $\varepsilon > 0$ and sufficiently small, there exists an unbounded component

$C_{k,\varepsilon} \subset (\mathbb{R} \times S_k) \cup \{(\mu_k(\varepsilon), 0)\}$ of $S_\varepsilon$ which meets $(\mu_k(\varepsilon), 0)$. Let $\mathcal{O}$ be any bounded open set in $\mathbb{R} \times E$ with $(\mu_k, 0) \in \mathcal{O}$. Then $(\mu_k(\varepsilon), 0) \in \mathcal{O}$ for $\varepsilon$ sufficiently small and therefore there exists $(\lambda_{k,\varepsilon}, u_{k,\varepsilon}) \in C_{k,\varepsilon} \cap \partial \mathcal{O}$. The equation (2.13) or actually its equivalent form as a second order ordinary differential equation provides us with uniform bounds for $|u''_{k,\varepsilon}|$. Therefore a subsequence of $(\lambda_{k,\varepsilon}, u_{k,\varepsilon})$ converges in $\mathbb{R} \times E$ to $(\lambda, u) \in (\mathbb{R} \times \bar{S}_k) \cap \partial \mathcal{O} \cap S$. Since $(\lambda, u) \neq (\mu_k, 0)$, it follows from Lemma 2.9 that $(\lambda, u) \in (\mathbb{R} \times S_k) \cap \partial \mathcal{O} \cap S$. Since $\mathcal{O}$ was arbitrary, the lemma from point set topology mentioned in the proof of Lemma 1.3 implies that there exists $C_k$ as in the statement of Theorem 2.4.

Remark: An improved version of Theorem 2.4 shows $C_k$ possesses unbounded components in each of $(\mathbb{R} \times S_k^+) \cup \{(\mu_k, 0)\}$, $(\mathbb{R} \times S_k^-) \cup \{(\mu_k, 0)\}$ as in the linear case. For this and other generalizations such as permitting $f$ to depend on u in a nonlinear fashion, see [6]. An instructive example giving some idea as to what $C_k$ may look like is given by the equation:

(2.14) $\quad -u'' = (\lambda + f(u^2 + (u')^2, \lambda))u, \quad 0 < x < \pi, \quad u(v)=0 = u(\pi)$

Here $f(0, \lambda) \equiv 0$, $\mu_k = k^2$, and $v_k$ is proportional to $\sin kx$. Trying for a solution of the form $(\lambda, \alpha \sin x)$ shows $C_1$ contains $\{(\lambda, \alpha \sin x) | 1 = \lambda + f(\alpha^2, \lambda)\}$. The function $f(\alpha^2, \lambda) = \lambda \alpha^2 \sin\frac{1}{\alpha^2}$ is an interesting special case.

Nodal properties played an important role in the results obtained above for (2.1)-(2.2). Similarly these properties can be exploited to study a class of nonlinear integral equations. The type of equation that can be treated is:

(2.15) $\quad u(x) = \lambda \int_0^1 K(x, y) F(y, u(y)) u(y) dy \equiv G(\lambda, u)$

where K is a continuous symmetric oscillation kernel on $[0,1]^2$ and $F(y, z)$ is positive, and continuous, on $[0,1] \times \mathbb{R}$. The related linear equation is

(2.16) $\quad v(x) = \mu \int_0^1 K(x, y) F(y, 0) v(y) dy \equiv \mu L v$.

25

Equation (2.16) possesses an increasing sequence of simple positive characteristic values ($\mu_n$) with corresponding eigenfunctions $v_n$ having exactly n-1 nodal zeroes (i.e. $v_n$ changes sign at a zero) in (0,1). [14]. Here the appropriate Banach space is $E = C[0,1]$ under the usual maximum norm. Unbounded components of $\mathfrak{S}$ meet the bifurcation points ($\mu_n, 0$) and lie in sets having nodal properties associated with the eigenfunctions of the linear theory. Equations such as (2.15)-(2.16) can be obtained by converting certain Sturm-Liouville problems for ordinary differential equations (not necessarily of second order) to integral equations [15], [16].

We conclude this section with an application to a class of quasilinear elliptic partial differential equations. In dealing with eigenvalue problems for elliptic partial differential equations, one does not in general have analogues of the nodal properties we have used earlier except for positivity and this we shall exploit in a similar fashion to the ordinary differential equations case.

Let $\Omega$ be a smooth bounded domain in $\mathbb{R}^n$. Consider the boundary value problem:

$$(2.17)\begin{cases} \mathcal{L}u \equiv -\sum_{i,j=1}^{n} a_{ij}(x,u,Du)u_{x_i x_j} + \sum_{i=1}^{n} b_i(x,u,Du)u_{x_i} \\ \qquad + c(x,u,Du)u = \lambda(a(x)u + F(x,u,\lambda)), \quad x \in \Omega \\ u = 0 \quad \text{on } \partial\Omega. \end{cases}$$

Here $Du$ denotes arbitrary first partial derivatives of $u$. All functions appearing in (2.17) are assumed to be continuously differentiable functions of their arguments. It is further assumed that $c \geq 0$, $a \geq a_0 > 0$ ($a_0$ a constant), $F(x, \xi, \lambda) = 0(|\xi|)$ near $\xi = 0$ uniformly on bounded $\lambda$ intervals. Lastly (2.17) is assumed to be uniformly elleptic, i.e. there exists a constant $\beta > 0$ such that

$$(2.18) \qquad \sum_{i,j=1}^{n} a_{ij}(x,\eta,p)\xi_i \xi_j \geq \beta \sum_{i=1}^{n} \xi_i^2$$

for all $x \in \Omega$, $\eta \in \mathbb{R}$, $p, \xi \in \mathbb{R}^n$.

## A GLOBAL EIGENVALUE THEOREM

To formulate (2.17) as an operator equation, let $\alpha \in (0,1)$ and

$$E = \{u \in C^{1+\alpha}(\bar{\Omega}) \mid u = 0 \text{ on } \partial\Omega\}.$$

Here $C^{1+\alpha}(\bar{\Omega})$ denotes the space of functions continuously differentiable in $\bar{\Omega}$ having first derivatives Hölder continuous with exponent $\alpha$. As norm in $E$ we take

$$\|u\|_{1+\alpha} \equiv \max_{x \in \Omega} |u(x)| + \max_{1 \le i \le n} \max_{x \in \Omega} |u_{x_i}(x)| +$$

$$+ \max_{1 \le i \le n} \max_{\substack{x,y \in \Omega \\ x \ne y}} \frac{|u_{x_i}(x) - u_{x_i}(y)|}{|x-y|^{\alpha}}$$

Thus $E$ is a Banach space under $\|\cdot\|_{1+\alpha}$. Spaces $C^{\alpha}(\bar{\Omega})$, $C^{2+\alpha}(\bar{\Omega})$ are defined in a similar fashion with associated norms $\|\cdot\|_{\alpha}$, $\|\cdot\|_{2+\alpha}$. As analogues in $E$ of our earlier sets $S_1^{\pm}$, we have

$$P^+ = \{u \in E \mid u > 0 \text{ in } \Omega, \frac{\partial u}{\partial \nu} < 0 \text{ on } \partial\Omega\}$$

and $P^- = -P^+$ where $\frac{\partial}{\partial \nu}$ denotes the outward pointing normal derivative to $\partial\Omega$. Note that $P^{\pm}$ and $P = P^+ \cup P^-$ are open subsets of $E$.

A mapping of $\mathbb{R} \times E \to E$ can now be defined as follows: For $(\lambda, u) \in \mathbb{R} \times E$, let $v \equiv G(\lambda, u)$ denote the solution of:

(2.19)
$$\begin{cases} -\sum_{i,j=1}^{n} a_{ij}(x, u, Du)v_{x_i x_j} + \sum_{i=1}^{n} b_i(x, u, Du)v_{x_i} \quad c(x, u, Du)v \\ \qquad = \lambda(a(x)u + F(x, u, \lambda)) \qquad x \in \Omega \\ v = 0 \qquad \text{on } \partial\Omega. \end{cases}$$

Since $u \in E$, (2.18) implies that (2.19) is a linear uniformly elliptic equation for $v$ with coefficients in $C^{\alpha}(\bar{\Omega})$. Therefore

27

by the linear existence theory for such equations [17], there exists a unique $v \in C^{2+\alpha}(\bar{\Omega})$ satisfying (2.19). Moreover by the Schauder estimates [17]:

(2.20) $\qquad \|v\|_{2+\alpha} \leq M \|\lambda a(x)u + F(x,u,\lambda)\|_\alpha$

where M is a constant depending on $\alpha, \beta, n$, the diameter of $\Omega$, and on $\|\cdot\|_\alpha$ bounds for $a_{ij}(x,u,Du)$, $b_i(x,u,Du)$, $c(x,u,Du)$. Thus $G(\lambda,u)$ maps bounded subsets of $\mathbb{R} \times E$ into bounded subsets of $C^{2+\alpha}(\bar{\Omega})$ which in turn are relatively compact in E. With the aid of the observations just made, it easily follows that $G$ is continuous. Hence $G: \mathbb{R} \times E \to E$ is compact. It is clear that (2.17) is equivalent to the equation $u = G(\lambda,u)$ in E. As usual the closure of the set of nontrivial solutions of this equation will be denoted by $S$.

Associated with G is a linear mapping L obtained as follows: For $u \in E$, let $w \equiv Lu \in C^{2+\alpha}(\bar{\Omega})$ denote the unique solution of

(2.21) $\qquad \begin{cases} \mathcal{L}w \equiv -\sum_{i,j=1}^{n} a_{ij}(x,0,0) w_{x_i x_j} + \sum_{i=1}^{n} b_i(x,0,0) w_{x_i} + \\ \qquad + c(x,0,0) w = au \qquad x \in \Omega \\ \qquad u = 0 \qquad x \in \partial\Omega. \end{cases}$

As above $L: E \to E$ is a compact linear map. If $K \equiv \overline{P^+}$, K is a closed cone in E with non-empty interior. Let $u \in K$ $u \neq 0$. Then $\mathcal{L}w = au \geq 0$ and by the strong maximum principle $w > 0$ in $\Omega$ and $\frac{\partial w}{\partial \nu} < 0$, i.e. $w \in P^+$, or $w \equiv 0$. Since $u \neq 0$, the second alternative is not possible so $w = Lu \in P^+$. Thus L is a strongly positive operator on K in the sense of Krein-Rutman [2], [18] and by a theorem of Krein-Rutman L possesses a unique eigenvector $v^+$ in $P^+$ with $\|v^+\|_{1+\alpha} = 1$. The corresponding eigenvalue $\mu^+ > 0$, is the smallest characteristic value of L, and is simple.

Next we shall show if $H(\lambda,u) \equiv G(\lambda,u) - \lambda L u$, then H is $0(\|u\|_{1+\alpha})$ near $u = 0$ uniformly on bounded $\lambda$ intervals. For if not, there exists an $\varepsilon > 0$ and a sequence

$(\lambda_n, u_n) \to (\lambda, 0)$ in $\mathbb{R} \times E$ as $n \to \infty$ such that:

(2.22) $\qquad \|H(\lambda_n, u_n)\|_{1+\alpha} > \varepsilon \|u_n\|_{1+\alpha}$

Let $v_n = G(\lambda_n, u_n)$ and $w_n = \lambda_n L u_n$. Then $V_n = v_n / \|u_n\|_{1+\alpha}$, $W_n = w_n / \|u_n\|_{1+\alpha}$ satisfy respectively

$$-\sum_{i,j=1}^{n} a_{ij}(x, u_n, Du_n) V_{n x_i x_j} + \sum_{i=1}^{n} b_i(x, u_n, Du_n) V_{n x_i} +$$

(2.23) $\qquad + c(x, u_n, Du_n) V_n = \lambda_n (a\, u_n + F(x, u_n, Du_n, \lambda_n)) \cdot$

$$\cdot (\|u_n\|_{1+\alpha})^{-1} \qquad x \in \Omega$$

(2.24) $\qquad \mathcal{L} W_n = \lambda_n a\, u_n \|u_n\|_{1+\alpha}^{-1} \qquad x \in \Omega$

(2.25) $\qquad V_n = W_n = 0 \qquad x \in \partial\Omega$.

The right hand sides of (2.23) and (2.24) are uniformly bounded with respect to $n$ in $C^{\alpha}(\bar{\Omega})$ and therefore by the Schauder estimates (2.20) $V_n$ and $W_n$ are uniformly bounded in $C^{2+\alpha}(\bar{\Omega})$. Hence by the Arzela-Ascoli theorem a subsequence of $u_n \|u_n\|_{1+\alpha}^{-1}$ converges in $C^1(\bar{\Omega})$ to $\varphi$ and a corresponding subsequence of $V_n$, $W_n$ converges in $C^2(\bar{\Omega})$ to $W, V$ both of which satisfy

(2.26) $\qquad \mathcal{L}\psi = \lambda a \varphi \qquad x \in \Omega$

$\qquad\qquad \psi = 0 \qquad x \in \partial\Omega$.

Thus $V = W$ and hence

(2.27) $\qquad \|V_n - W_n\|_{1+\alpha} = \dfrac{\|H(\lambda_n, u_n)\|_{1+\alpha}}{\|u_n\|_{1+\alpha}} \to 0$

along this subsequence contrary to (2.22). Therefore $H(\lambda, u) = o(\|u\|_{1+\alpha})$ near $u = 0$ uniformly on bounded $\lambda$ intervals.

We have now verified that $\mu^+$ and $G(\lambda, u)$ satisfy the hypotheses of Theorem 1.4. Hence by that theorem there exists a component $C$ of $S$ which meets $(\mu^+, 0)$ and is either unbounded in $\mathbb{R} \times E$ or meets $(\hat{\mu}, 0)$ where $\mu^+ \neq \hat{\mu} \in r(L)$. A better result actually obtains here, namely:

<u>Theorem 2.28</u>: There exists a component $C$ of $S$ in $(\mathbb{R} \times P) \cup \{(\mu^+, 0)\}$ which contains $(\mu^+, 0)$ and is unbounded.

The proof of Theorem 2.28 uses the following two lemmas analogous to Lemmas 2.6 and 2.9.

<u>Lemma 2.29</u>: There exists a neighborhood $\eta$ of $(\mu^+, 0)$ such that if $(\lambda, u) \in \eta \cap S$ and $u \neq 0$, then $u \in P$. Moreover if $\hat{\mu} \in r(L)$, $\hat{\mu} \neq \mu^+$, there exists a neighborhood $\hat{\eta}$ of $(\hat{\mu}, 0)$ such that if $(\lambda, u) \in \hat{\eta} \cap S$ and $u \neq 0$, then $u \notin P$.

<u>Proof</u>: The proof of the first statement is the same as that of Lemma 2.6 and will be omitted. If the second statement is not true, there exists $\{(\lambda_n, u_n)\} \subset S \cap (\mathbb{R} \times P)$ with $(\lambda_n, u_n) \to (\hat{\mu}, 0)$ as $n \to \infty$. As in (2.8)-(2.9) a subsequence of $(\lambda_n, u_n \|u_n\|_{1+\alpha}^{-1})$ converges to $(\hat{\mu}, \hat{v})$ satisfying

(2.30) $\qquad \hat{v} = \hat{\mu} L \hat{v}, \quad \|\hat{v}\|_{1+\alpha} = 1$.

Since $\hat{v}$ is the limit of points in $P$, $\hat{v} \in K \cup (-K)$. But $\pm v^+$ is the unique normalized eigenfunction of $L$ in $\pm K$. Hence since $\hat{\mu} \neq \mu^+$, there exists a neighborhood $\hat{\eta}$ as above.

<u>Lemma 2.31</u>: Suppose $(\lambda, u) \in C$ (as given by Theorem 1.4) with $(\lambda, u) \neq (\mu^+, 0)$ and $(\lambda, u) = \lim_{n \to \infty} (\lambda_n, u_n)$ where $(\lambda_n, u_n) \in (\mathbb{R} \times P) \cap S$. Then $(\lambda, u) \in \mathbb{R} \times P$.

<u>Proof</u>: Observe first that $(\lambda, u) \in C$ implies that $\lambda > 0$ for otherwise since $C$ is connected, there is a solution $(0, \bar{u})$

of (2.17) with $\bar{u} \neq 0$. But then zero is an eigenvalue of the uniformly elliptic operator

$$\bar{\mathcal{L}} = -\sum_{i,j=1}^{n} a_{ij}(x, \bar{u}, D\bar{u}) \frac{\partial^2}{\partial x_i \partial x_j} + \sum_{i=1}^{n} b_i(x, \bar{u}, D\bar{u}) \frac{\partial}{\partial x_i} +$$

$$+ c(x, \bar{u}, D\bar{u}).$$

However as was the case with $\mathcal{L}$, its smallest eigenvalue is positive.

If $(\lambda, u) \in \mathbb{R} \times \partial P$, then $u \in \partial P^+ \cup \partial P^-$. The argument is the same for either case so we assume the former. Since $u \in \partial P^+$, either (i) there exists $\xi \in \Omega$ such that $u(\xi) = 0$ or (ii) there exists $\eta \in \partial \Omega$ such that $\frac{\partial u}{\partial \nu}(\eta) = 0$. If (i) occurs, there is a neighborhood $\Omega_0 \subset \Omega$ of $\xi$ such that $|F(x, u(x), \lambda)| \leq \frac{a_0 u(x)}{2}$ in $\Omega_0$. The maximum principle and continuity of $u$ implies $u \equiv 0$ in $\Omega_0$. A continuation agreement then implies $u \equiv 0$ in $\bar{\Omega}$. Since $\lambda \neq \mu^+$, (i) is impossible. A similar combined maximum principle - continuation argument shows that (ii) cannot occur. Hence $u \in P$.

<u>Proof of Theorem 2.28:</u> Lemma 2.29 implies $c \cap \eta \subset (\mathbb{R} \times P) \cup \{(\mu^+, 0)\}$. If $c \not\subset (\mathbb{R} \times P) \cup \{(\mu^+, 0)\}$, there exists $(\lambda, u) \in c \cap \mathbb{R} \times \partial P$ with $(\lambda, u) \neq (\mu^+, 0)$ and $(\lambda, u) = \lim_{n \to \infty}(\lambda_n, u_n)$ where $(\lambda_n, u_n) \in (\mathbb{R} \times P) \cap c$. But this is not possible by Lemma 2.31. Hence $c$ is an unbounded subset of $(\mathbb{R} \times P) \cup \{(\mu^+, 0)\}$.

Remarks: A better version of the theorem shows both $\overline{c \cap (\mathbb{R} \times P^+)}$ and $\overline{c \cap (\mathbb{R} \times P^-)}$ is unbounded. Likewise more general F's and B. C. can be permitted. See [5].

## §3. Continua without bifurcation.

Continua of solutions of (0.1) can be obtained in situations which do not involve bifurcation in any direct sense as in §1. In this section we briefly prove such a result and mention some applications.

Consider:

(3.1) $$u = \mathcal{G}(\lambda, u)$$

where now it is assumed that $\mathcal{G}: \mathbb{R} \times E \to E$ is compact and $\mathcal{G}(0, u) \equiv 0$. Then $(0, 0)$ is a solution of (3.1). Since $\mathcal{G}(0, u) \equiv 0$, it readily follows from the Schauder fixed point theorem that (3.1) possesses a solution for all $|\lambda|$ sufficiently small. By using the homotopy ideas of §1, a stronger result can be obtained. Let $S$ denote the set of solutions of (3.1), $\mathbb{R}^+ = \{\lambda \in \mathbb{R} \mid \lambda \geq 0\}$ and $\mathbb{R}^- = -\mathbb{R}^+$.

**Theorem 3.2:** If $\mathcal{G}: \mathbb{R} \times E \to E$ is compact and $\mathcal{G}(0, u) \equiv 0$, then $S$ contains a pair of unbounded components $C^+, C^-$ in $\mathbb{R}^+ \times E$, $\mathbb{R}^- \times E$ respectively and $C^+ \cap C^- = \{(0, 0)\}$.

**Proof:** Let $C^+, C^-$ denote respectively the components of $S$ in $\mathbb{R}^+ \times E$, $\mathbb{R}^- \times E$ which contain $(0, 0)$. The second assertion of the theorem follows since $(0, 0)$ is the unique solution of (3.1) in $\{0\} \times E$. We will prove that $C^+$ is unbounded, the proof for $C^-$ being the same.

Suppose $C^+$ is bounded in $\mathbb{R}^+ \times E$. Since $\mathcal{G}(0, u) \equiv 0$, there exists a $\zeta > 0$ such that $\mathcal{G}: [-\zeta, \zeta] \times B_1 \to \overset{\circ}{B}_1$. Hence $C^+ \cap ([0, \zeta] \times B_1) \subset [0, \zeta] \times \overset{\circ}{B}_1$. Essentially as in Lemma 1.3, there exists a bounded open set $\mathcal{O} \subset \mathbb{R}^+ \times E$ such that $C^+ \subset \mathcal{O}$, $\partial\mathcal{O} \cap S = \emptyset$, and $\mathcal{O} \cap [0, \zeta] \times B_1 = [0, \zeta] \times \overset{\circ}{B}_1$.

Let $\Phi(\lambda, u) = u - \mathcal{G}(\lambda, u)$ and $\mathcal{O}_\lambda = \{u \in E \mid (\lambda, u) \in \mathcal{O}\}$. By our above remarks, $d(\Phi(\lambda, \cdot), \mathcal{O}_\lambda)$ is well defined for $\lambda \in \mathbb{R}^+$. The construction of $\mathcal{O}$ and homotopy invariance of degree implies

(3.3) $$d(\Phi(\lambda, \cdot), \mathcal{O}_\lambda) \equiv \text{constant} = c, \quad \lambda \in \mathbb{R}^+.$$

Since $\Phi(0, \cdot) = I$, $c = 1$. However if $\bar{\lambda}$ is greater than the diameter of $\mathcal{O}$, $\mathcal{O}_{\bar{\lambda}} = \emptyset$ and

(3.4) $$d(\Phi(\bar{\lambda}, \cdot), \mathcal{O}_{\bar{\lambda}}) = 0$$

a contradiction. Hence $C^+$ is unbounded.

Remark: Of course if also $G(\lambda, 0) \equiv 0$ as in the bifurcation case, we have the trivial solutions $\{(\lambda, 0) | \lambda \in \mathbb{R}\}$ so Theorem 3.2 gives us nothing new.
 If we no longer assume $G(0, u) \equiv 0$, then an analogue of Theorem 3.2 can still be obtained provided that more a priori information is known about the solution set $S$.

Theorem 3.5: Suppose $G : \mathbb{R} \times E \to E$ is compact and there exists $M > 0$ such that $u = G(0, u)$ implies $\|u\| < M$. If $d(\Phi(0, \cdot), B_M) \neq 0$, then $S$ possesses unbounded components $C^+$, $C^-$ in $\mathbb{R}^+ \times E$, $\mathbb{R}^- \times E$ respectively which meet $(0, \bar{u})$ for some $\bar{u} \in B_M$.

Proof: The argument is similar to that of Theorem 3.2 and will be omitted.

Remark: Theorems 3.2 and 3.5 are essentially due to Leray and Schauder [8]. For their analogue of Theorem 3.5 however they require that $\{0\} \times B_M$ contain only finitely many solutions.
 As an application of Theorem 3.2, consider the quasilinear elliptic boundary value problem:

(3.6) $\quad \mathcal{L} u = A(x, u, Du, \lambda) \qquad x \in \Omega$

$\qquad\qquad u = 0 \qquad\qquad\qquad x \in \partial\Omega$

where $\Omega, \mathcal{L}$ are as in (2.17). We assume $A$ is a continuously differentiable function of its arguments, $A(x, u, p, 0) \equiv 0$, and $A(x, 0, 0, \lambda) > 0$ for $x \in \Omega$ if $\lambda > 0$. As in §2, (3.6) can be converted to an equivalent operator equation of the form (3.1) in $E = C^{1+\alpha}(\bar{\Omega}) \cap$ B.C. Moreover $G$ is compact and $G(0, u) \equiv 0$.

Theorem 3.7: Under the above conditions on $\Omega$, $\mathcal{L}$, $A$, (3.6) possesses unbounded components $C^+$, $C^-$ of solutions in $\mathbb{R}^+ \times E$, $\mathbb{R}^- \times E$ respectively. Moreover $C^+ \subset (\mathbb{R}^+ \times P^+) \cup \{(0,0)\}$.

Proof: The existence of $C^+$, $C^-$ follow from Theorem 3.2. If $(\lambda, u) \in C^+$, $\lambda > 0$, and $u$ is near $0$, the properties of $A$ and (3.6) imply $\mathcal{L}u > 0$ in $\Omega$ and hence by the maximum principle, $u \in P^+$. Thus $C^+ \cap \eta \subset (\mathbb{R}^+ \times P^+) \cup \{(0,0)\}$ for any open set $\eta \subset \{(0,0)\}$ and of small diameter. The properties of $A$ and the argument of Lemma 2.31 then implies $C^+ \subset (\mathbb{R}^+ \times P^+) \cup \{(0,0)\}$ and the proof is complete.

Theorems 3.2 and 3.5 can also be used to treat some questions involving time periodic solutions of hyperbolic partial differential equations. In particular we mention the nonlinear wave equation:

(3.8) $\qquad u_{tt} - u_{xx} = \lambda F(x,t,u) \quad 0 < x < \pi, \; 0 \leq t \leq 2\pi$

together with the boundary and periodicity conditions:

(3.9) $\qquad u(0,t) = 0 = u(\pi, t) \quad 0 \leq t \leq 2\pi$
(3.9)
$\qquad u(x, t + 2\pi) = u(x,t) \quad 0 \leq x \leq \pi$

For details see [5].

## REFERENCES

1. Keller, J. B. and S. Antman (editors), "Bifurcation Theory and Nonlinear Eigenvalue Problems", Benjamin, New York, 1969.

2. Krasnoselski, M. A., "Topological Methods in the Theory of Nonlinear Integral Equations", Macmillan, New York, 1965.

3. Krasnoselski, M. A., "Positive Solutions of Operator Equations", P. Noordhoff, Ltd., Groningen, 1964.

4. Prodi, G., Problemi di diramazione per equazioni funzionali, Boll U. M. I., XXII, 1967, pp. 413-433.

5. Rabinowitz, P. H., Some global results for nonlinear eigenvalue problems, to appear in Journal of Functional Analysis.

6. Rabinowitz, P. H., Nonlinear Sturm-Liouville eigenvalue problems for second order ordinary differential equations, Comm. Pure Applied Math. $\underline{23}$, 1970, pp. 939-962.

7. Whyburn, G. T., "Topological Analysis, Princeton University Press, Princeton, 1958.

8. Leray, J. and J. Schauder, Topologie et equationes functionelles, Ann. Sci. Ecole Norm., Sup. 3, $\underline{51}$, 1934, pp. 45-78.

9. Schwartz, J. T., "Nonlinear Functional Analysis", Lecture notes, Courant Institute of Mathematical Sciences, New York University, 1965.

10. Bauer, L, E. L. Riess, and H. B. Keller, Antisymmetric buckling of hollow spheres and hemispheres, Comm. Pure Applied Math, $\underline{23}$, 1970, pp. 529-568.

11. Crandall, M. G. and P. H. Rabinowitz, Bifurcation from simple eigenvalues, to appear in Journal of Functional Analysis.

12. Coddington, E. A. and N. Levinson, "Theory of Ordinary Differential Equations", McGraw-Hill, New York, 1955.

13. Crandall, M. G. and P. H. Rabinowitz, Nonlinear Sturm-Liouville eigenvalue problems and topological degree, J. Math. Mech., $\underline{19}$, 1970, pp. 1083-1102.

14. Gantmacher, F. R. and M. G. Krein, "Oszillationsmatrizen, Oszillationkerne, und Kleine Schwingungen mechanischen Systeme", Berlin, Akademie-Verlag, 1960.

15. Karlin, S., "Total Positivity", Vol. I (and II to appear), Stanford University Press, Stanford, 1968.

16. Turner, R. E. L., Nonlinear eigenvalue problems with nonlocal operators, Comm. Pure Applied Math., 23, 1970, pp. 963-972.

17. Ladyzhenskaya, O. A. and N. Uraltseva, "Linear and Quasilinear Elliptic Equations", Academic Press, New York, 1968.

> Department of Mathematics
> University of Wisconsin
> Madison, Wisconsin

> Received March 23, 1971

# Transversality in Nonlinear Eigenvalue Problems

*R. E. L. TURNER*

## 1. Introduction.

In this paper we show that the use of transversality or general position arguments together with traditional techniques for studying nonlinear eigenvalue problems enables one to obtain new results on the geometric structure and multiplicity of solutions of nonlinear eigenvalue problems. The author's interest in the question of multiplicity of solutions grew out of the paper [1] where a class of problems of the form

(1.1) $$u = \lambda G_0 u + G_1(\lambda, u)u$$

was studied. In (1.1) $\lambda \in \mathbb{R}$, the reals, and $u$ is a vector in a real Banach space $\mathcal{X}$ having a norm denoted by $\| \ \|$. The map $G_0$ is compact and linear while $G_1(\lambda, u)u$ is compact and nonlinear (cf. [1] for details). Suppose we let $\mathcal{P}'$ denote the set of pairs $(\lambda, \|u\|)$ corresponding to solutions $(\lambda, u)$ of (1.1) with $u \neq 0$, and set $\mathcal{P} = \overline{\mathcal{P}'}$ (the closure). In [1] it was shown that if $\lambda_0$ is a characteristic value of $G_0$ of odd multiplicity and if $G_1(\lambda, u)u$ is $o(\|u\|)$ near $u = 0$, then $\mathcal{P}$ contains two continua, counting multiplicity, each of which connects $(\lambda_0, 0)$ to "$\infty$" or to a point $(\lambda_i, 0)$ where $\lambda_i$ is a characteristic value of $G_0$ different from $\lambda_0$. If $G_1(\lambda, u)u$ is not $o(\|u\|)$, then one obtains such continua emanating from an interval $[\lambda_0 - \eta_1, \lambda_0 + \eta_2]$ about $\lambda_0$ (with a corresponding $u = 0$).

In this paper we consider a more general equation:

(1.2) $$u = \lambda G_0 u + G_1(\lambda, u)$$

in $\mathbb{R} \times \mathfrak{X}$ (cf. Section 2 for a precise description of $G_1$) and obtain results concerning continua of solutions in $\mathbb{R} \times \mathfrak{X}$ and their multiplicities. The structure of the solution set of an equation of the form (1.2) is of interest as it is a form which arises in many problems of elasticity and fluid mechanics. To obtain our results we use the transversality density theorem ([2], p. 48) and the accompanying structure of transversal intersections. We believe that the use of these tools in nonlinear eigenvalue problems is new.

For the purpose of stating the results and for use in following sections we let $\mathfrak{S}'$ denote the set of solutions $(\lambda, u)$ of (1.2) (or (1.3) below) with $u \neq 0$ and let $\mathfrak{S} = \bar{\mathfrak{S}}'$. The main theorem (2.4) states that if $u = \lambda_0 G_0 u$ where $\lambda_0$ has odd algebraic multiplicity, and if $Q$ is a bounded open set in $\mathbb{R} \times \mathfrak{X}$ with boundary $M$ such that $(\lambda_0, 0)$ is in $Q$ and all other pairs $(\lambda_i, 0)$, $\lambda_i$ a characteristic value of $G_0$, are outside $Q \cup M$, then either there are at least two points in $M \cap \mathfrak{S}$ which can be connected to $(\lambda_0, 0)$ in $\mathfrak{S}$ or there is one such point which, if $M$ is smooth, can be assigned multiplicity two. As a corollary of the theorem we obtain an alternate proof of Theorem 1.3 of [3], showing the existence of one geometric continuum in $\mathfrak{S}$ which meets $(\lambda_0, 0)$ and is either unbounded or meets a point $(\lambda_i, 0)$ where $\lambda_i \neq \lambda_0$. That there may be just one geometric continuum emanating from $(\lambda_0, 0)$ is shown in an example by P. Rabinowitz (these proceedings) and here we include a similar example which seems simpler than the one cited. In concluding Section 2 we state a theorem which is an analogue of Theorem 2.4 (cf. [1] in the situation where $G_1$ is not necessarily small relative to $G_0$ for $u$ near zero.

In Section 3 we show how transversality can be used to obtain a continuum of solutions for an eigenvalue problem involving a cone preserving map. Here we do not need odd multiplicity, but use the cone preservation to obtain an unbounded continuum of solutions of an equation

(1.3) $\qquad u = \lambda F(u)$

where F is a compact map which preserves the cone of non-negative functions in an $L^2$ space and which has a derivative at zero with a single eigenvalue $\lambda_0$ corresponding to eigenvectors in the cone. In this case, of course, the continuum emanates from the point $(\lambda_0, 0)$. We apply the result to a simple example involving a pair of linked differential equations.

2. Global continua.

We begin this section with a definition that will be used in the statement of the main theorem and a pair of lemmas that will be used in its proof. We wish to count the number of zeros, with multiplicity, of an equation

(2.1) $\qquad u - G(\lambda, u) = 0$

where $G(\lambda, u)$ is a compact continuous map of $\mathbb{R} \times \mathfrak{X}$ into $\mathfrak{X}$. To that end we wish to assign a multiplicity to a zero of (2.1) which is isolated on a $C^1$ submanifold of $\mathbb{R} \times \mathfrak{X}$ of codimension 1 (cf. [2], p. 44). If $p' = (\lambda', u')$ is an isolated zero on the submanifold M, then if M is a $C^1$ submanifold, there is a $C^1$ function $\alpha$ mapping a neighborhood $V_1$ of $p'$ in $\mathbb{R} \times \mathfrak{X}$ into $\mathbb{R}$ such that $V_1 \cap M$ is the set $\alpha^{-1}(0)$ and $p'$ is the only zero of (2.1) in $V_1 \cap M$. Moreover, the derivative $D\alpha$ can be assumed to be nonzero at $p'$ and if N is the nullspace of $D\alpha$, we can use N to parametrize M near $p'$. That is, if $\pi$ is a vector complementary to N, satisfying $D\alpha(p')\pi = 1$, then using the implicit function theorem one finds a $C^1$ function $\gamma$ from N to $\mathbb{R}$ such that in a neighborhood $V \subset V_1$ of $p'$, the set M consists of points of the form $p' + n + \gamma(n)\pi$ where n ranges over a neighborhood W of 0 in N. Using this parametrization of M we can define a multiplicity for the zero $p'$ Since both N and $\mathfrak{X}$ have codimension one in $\mathbb{R} \times \mathfrak{X}$, $\mathfrak{X} \cap N = Z$

will have codimension 1 or 2. If it is 2 then there are vectors $x_1$ and $n_1$ such that $\mathcal{X} = Z \oplus \mathbb{R} x_1$ and $N = Z \oplus \mathbb{R} n_1$ where $\mathbb{R} x_1$ denotes the one dimensional subspace spanned by $x_1$. We let $T$ be the linear map taking $Z + \beta x_1$ to $Z + \beta n_1$. If $\mathcal{X} = N$, we let $T$ be the identity. Further, let $Q_1$ be the projection of $\mathbb{R} \times \mathcal{X} \to \mathcal{X}$ taking $(\lambda, u)$ to $u$. We consider the composite map

$$(2.2) \quad x \to Q_1(p' + Tx + \gamma(Tx)\pi) - G(p' + Tx + \gamma(Tx)\pi)$$

from $W_1 = T^{-1}(W) \subset \mathcal{X}$ into $\mathcal{X}$. The only zero of the map (2.2) in $W_1$ is $x = 0$ which corresponds to $p'$, the isolated zero of (2.1) on $M$. The map $QTx$ differs from the identity on $\mathcal{X}$ by a finite rank map and thus the map in (2.2) has the form $I + K$ where $K$ is compact. As such, the isolated zero $x = 0$ has a Leray-Schauder index (cf. [4], p. 187). If we compute the index using a $C^1$ function $\tilde{\alpha}$ for which $\tilde{\alpha}^{-1}(0) = M$ near $p'$, then the nullspace $\tilde{N}$ of $D\tilde{\alpha}$ will coincide with $N$. If $\tilde{\pi}$ satisfies $D\alpha(p')\tilde{\pi} > 0$ and $M = p' + Tx + \tilde{\gamma}(Tx) \tilde{\pi}$ near $p'$ we can form a map similar to (2.2) above and it will have an index at $x = 0$. We can as well represent $M$ near $p'$ as $p' + Tx + \gamma_t(Tx)((1-t)\pi + t \cdot \tilde{\pi})$ where $\gamma_t$ depends continuously on $t$, $\gamma_0 = \gamma$, and $\gamma_1 = \tilde{\gamma}$. This $t$ dependent parametrization of $M$ can be put in (2.2) yielding a homotopy under which $x = 0$ remains an isolated zero. Then the homotopy invariance of the Leray-Schauder degree implies that the index of $x = 0$ will be the same for $t = 0$ and $t = 1$, hence the same using $\alpha$ or $\tilde{\alpha}$. If we use a map $\tilde{T}$ in place of $T$ above, the orientation of $N$ may be changed so we neglect the sign and use the following definition of multiplicity.

<u>Definition 2.1.</u>  If the index of the map (2.2) at $x = 0$ is $\pm m$, we say $p'$ has multiplicity $m$.

We will need the following lemma.

<u>Lemma 2.2.</u>  Let $S$ be a compact metric space and let $S_1$, $S_2$ be closed disjoint set in $S$. Suppose that for each

positive integer $n$, $C_n$ is a continuum (i.e., a closed connected set) in $S$ which has nonempty intersection with $S_1$ and $S_2$. Let $C$ be the set of limit points of the collection $C_n$; i.e., $p \in C$ if and only if there are integers $n_i \to \infty$ and points $p_i \in C_{n_i}$ such that $p_{n_i}$ converges to $p$ as $i \to \infty$. Then $C$ is a continuum having nonempty intersection with $S_1$ and $S_2$.

Proof. First, $C$ is closed for if $p_k \in C$ and $p_k \to p$ then there are sets $C_{n_k}$ with $n_k > k$ and points $p_{n_k} \in C_{n_k}$ satisfying $d(p_{n_k}, p_k) < 1/k$ where $d$ is the distance function. It follows that $p_{n_k}$ converges to $p$ as $k \to \infty$ so $p \in C$. According to [5], p. 12 either the compact set $C$ in $S$ contains a subcontinuum $C'$ meeting $S_1 \cap C$ and $S_2 \cap C$ or $C = K_1 \cup K_2$ where $K_1$ and $K_2$ are disjoint compact subsets of $C$ containing $S_1 \cap C$ and $S_2 \cap C$ respectively. Suppose the latter alternative occurs. Let $\alpha > 0$ be the distance between $K_1$ and $K_2$ and let $Z_1$ be the set of points in $S$ at a distance less than $\alpha/2$ from $K_1$ and let $\partial Z_1$ be the boundary of $Z_1$ in $S$.

If $V_1 \subset Z_1$ is a neighborhood of a point $p_1 \in S_1 \cap C$, there will be a sequence $n_k \to \infty$ such that $C_{n_k} \cap V \neq \phi$. Since each continuum $C_{n_k}$ meets $S_2$, the sets $C_{n_k} \cap S_2$ will have a limit point $p_2 \in C \cap S_2$ and, taking a subsequence $C_{m_k}$ of $C_{n_k}$ and a neighborhood $V_2 \subset S - \bar{Z}_1$ of the part $p_2$ we may assume that $C_{m_k} \cap \partial Z_1 \neq 0$ since $C_{m_k}$ is connected, and $C \cap \partial Z_1 \neq \phi$ results.

Lemma 2.3. Let $\sigma$ be a $C^1$ curve in $\mathbb{R}^{n+1}$ parametrized by $\sigma(t)$, $-1 \leq t \leq 1$. Let $T_t$ be the unit tangent vector to $\sigma(t)$ at $t$. Then the perpendicular subspace $T_t^\perp$ can be represented as $V_t X_n$ where $V_t$ is a continuous map of $[-1,1]$ into $SO_{n+1}$ (orthogonal maps with determinant 1) and $X_n$ is the subspace of $\mathbb{R}^{n+1}$ consisting of points of the form $(0, x_1, \ldots, x_n)$.

**Proof.** We can choose $V_0: (1, 0, 0, \ldots, 0) \to T_0$ and hence $V_0 X_n = T_0^\perp$. Since $T_t$ is continuous, the perpendicular projection $F_t$ on $T_t$ is continuous as is the projection $E_t = I - F_t$ on $T_t^\perp$. Since $E_t$ will be uniformly continuous on $[-1, 1]$ there are points $t_{-r}, t_{-r+1}, \ldots, t_0 = 0, t_1, \ldots, t$ so that $\|E_{t_{k+1}} - E_{t_k}\| < 1$, the norm being the usual operator norm in $\mathbb{R}^{n+1}$. Suppose we have defined $V_t$ on $[0, t_k]$ to satisfy the requirements of the lemma. Then (cf. [6], p. 268). The map

$$(2.3) \qquad W_t = E_t(I + E_t(E_t - E_{t_k})E_{t_k})^{-1/2} E_{t_k}$$

is an isometric map of $T_{t_k}^\perp$ to $T_t^\perp$ for $t_k \leq t \leq t_{k+1}$. We let

$$W_t' = \begin{cases} v \to W_t v, & \text{if } v \in U_{t_k} X_n \\ T_{t_k} \to T_t \end{cases}$$

and define $U_t$ for $t_k \leq t \leq t_{k+1}$ by

$$U_t = W_t' U_{t_k}$$

By this process one extends $U_t$ to the domain $[0, 1]$ and similarly to $[-1, 0]$.

A number $\lambda$ is called a characteristic value of a compact map $G_0$ if there exists $u \neq 0$ such that $u - \lambda G_0 u = 0$. The characteristic values will be countable in number and we denote them by $\lambda_i$, where $i$ runs over some set of integers. We now state the main theorem.

**Theorem 2.4.** Let $G(\lambda, u) = \lambda G_0 u + G_1(\lambda, u)$ where $\lambda$ is a real parameter, $G_0$ is a compact linear map of a real Banach space $\mathcal{X}$ into itself, and $G_1(\lambda, u)$ is a continuous map of $\mathbb{R} \times \mathcal{X}$ into $\mathcal{X}$ which maps bounded sets into precompact sets. Suppose $G(\lambda, 0) = 0$ and

$$\lim_{\|u\| \to 0} \frac{\|G_1(\lambda, u)\|}{\|u\|} = 0$$

uniformly on bounded $\lambda$ sets. Let $\lambda_0$ be a characteristic value of $G_0$ of odd algebraic multiplicity. Let Q be a bounded open set in $\mathbb{R} \times \mathcal{X}$ having a boundary M which is a $C^1$ submanifold of $\mathbb{R} \times \mathcal{X}$ of codimension one. Suppose $(\lambda_0, 0) \in Q$ while $(\lambda_i, 0) \in \bar{Q}^c$ (the complement) for $i \neq 0$. Then for equation (2.1) either there are at least two points $p_1$ and $p_2$ in $M \cap \mathcal{S}$ which are connected to $(\lambda_0, 0)$ in $\mathcal{S}$ or there is one such point having multiplicity two. If no smoothness is imposed upon M, it still contains at least one point which is connected to $(\lambda_0, 0)$ in $\mathcal{S}$.

Proof. Let Q be contained in a set

$$B_R = \{(\lambda, u) \mid (\lambda - \lambda_0)^2 + \|u\|^2 \leq R^2\}$$

where $R > 4$ and let $\lambda_i$, $m_1 \leq i \leq m_2$, denote the characteristic values of $G_0$ in the interval $[\lambda_0 - 2R, \lambda_0 + 2R]$. Let $n_0$ be a positive integer for which $n_0^{-1} < 1/2 \min|\lambda_i - \lambda_{i+1}|$, $m_1 \leq i \leq m_2$. It is well known that a point $(\lambda, 0)$ cannot be a bifurcation point of (2.1) if $\lambda$ is not a characteristic value of $G_0$ (cf [4], p. 192); that is, the only solutions near $(\lambda, 0)$ have $u = 0$. Using compactness one sees that for each $n > 2n_0$ there must be a $\delta_n > 0$, $\delta_n < 1/n$, such that the set

(2.4) $\quad L_n = \{(\lambda, u) \in B_{2R} \mid |\lambda - \lambda_i| \geq 1/n, m_i \leq i \leq m_2, 0 < \|u\| \leq \delta_n\}$

contains no zeros of $u - G(\lambda, u)$. Let $\Delta_n = L_n \cap \{(\lambda, u) \mid \|u\| = \delta_n\}$ and $H_n = \Delta_n \cup (L_{n_0} - L_n)$. The set $H_n$ is closed and contains no seros of $u - G(\lambda, u)$. Since G is compact and continuous, there exists an $\epsilon_n > 0$, $\epsilon_n < \frac{1}{n}$, such that $\|u - G(\lambda, u)\| \geq \epsilon_n$ on $H_n$. Let $G(B_{2R})$ be the image of $B_{2R}$ under G and let $z_1, \ldots, z_{k_n}$ be an $\frac{\epsilon_n}{3}$ net for $G(B_{2R})$ consisting of elements of $G(B_{2R})$.

Let $P_n$ be the Leray-Schauder projection taking $G(B_{2R})$ into the span $S_n$ of $\{z_1, z_2, \ldots, z_{k_n}\}$. The map $P_n$ is defined as

$$(2.5) \qquad P_n z = \frac{\sum_{i=1}^{k_n} \mu_i(z) z_i}{\sum_{i=1}^{k_n} \mu_i(z)}$$

where

$$\mu_i(z) = \begin{cases} \frac{\epsilon_n}{2} - \|z - z_i\|; & \text{if } \|z - z_i\| \leq \frac{\epsilon_n}{2} \\ 0 & ; \text{if } \|z - z_i\| > \frac{\epsilon_n}{2} \end{cases}$$

The map $P_n$ is clearly continuous and since $P_n z$ is a convex combination of points within distance $\epsilon_n/3$ of $z$, we have $\|P_n z - z\| \leq \epsilon_n/3$ for $z \in G(B_{2R})$. The map $G_n \equiv P_n G$ takes $\beta_{2R}^n \equiv (\mathbb{R} \times S_n) \cap B_{2R}$ into $S_n$ and is continuous. On $H_n \cap \beta_{2R}^n$ we have

$$\|u - G_n(\lambda, u)\| = \|u - G(\lambda, u) + G(\lambda, u) - P_n G(\lambda, u)\|$$

$$\geq \|u - G(\lambda, u)\| - \|G(\lambda, u) - P_n G(\lambda, u)\|$$

$$\geq \epsilon_n - \epsilon_n/3 = 2/3 \epsilon_n .$$

By using coordinates $x_1, x_2, \ldots, x_m$ with respect to a basis $u_1, \ldots, u_m$ for $S = S_n$ (we suppress the subscript n) we can analyze a problem in Euclidean space. If $J$ takes $u = \sum_{i=1}^{m} x_i u_i$ to $x = (x_1, x_2, \ldots, x_m)$ and $J_1$ takes $(\lambda, u)$ to $(\lambda, x)$ then the map $P_n G$ will correspond to a map $f = J P_n G J_1^{-1}$ represented by $m$ functions $f_k(\lambda, x_1, x_2, \ldots, x_m)$; $k = 1, 2, \ldots, m$. Let $B \subset \mathbb{R}^m$ be the image of the closed unit ball in $S$ under $J$ and let $\Sigma$ be its boundary. Further, let

$U_{2R} = J_1 \beta_{2R}^n$ and $H = J_1 H_n$. We know that $\|x - f(\lambda, x)\| \geq \epsilon = \|J^{-1}\|^{-1} \epsilon_n$ for $(\lambda, x)$ in $H$ where the norm of $x - f$ is the standard one in $\mathbb{R}^m$ and the norm of $J^{-1}$ is the operator norm for a map from $\mathbb{R}^m$ to $S$. On a compact set inside $U_{2R}$, say on $U_{2R-1}$, we can use mollification, i.e, convolution with smooth kernel having small support, to obtain a twice continuously differentiable ($C^2$) map $\tilde{f} = \tilde{f}(\lambda, x)$ which satisfies

$$\|\tilde{f}(\lambda, x) - f(\lambda, x)\| \leq \min(\epsilon/4, \epsilon \|J_1\|^{-1})$$

for $(\lambda, x)$ in $U_{2R-1}$. Then $\|x - \tilde{f}(\lambda, x)\|$ will be bounded below by $3\epsilon/4$ on $H$.

We now use transversality to obtain a map which has a tractable zero set. While one could use Sard's theorem in this application it is not immediately applicable in later parts of the paper, whereas the transversality density theorem is well suited to a variety of applications. We refer the reader to [2], Chapter 4 for the necessary material on transversality. Let $a = (a_1, a_2, \ldots, a_m)$ be a point in $\mathbb{R}^m$ and define $\tilde{f}_a$ to be $\tilde{f}(\lambda, x) + a$. Let $X = \mathbb{R}^{m+1}$, $Y = \mathbb{R}^{m+1} \times \mathbb{R}^m$, and $W = \mathbb{R}^{m+1} \times \{0\} \subset Y$. Let

(2.6) $\qquad \rho_a : X \to Y$

be given by

$$\rho_a : (\lambda, x) \to (\lambda, x, x - \tilde{f}_a(\lambda, x))$$

for $(\lambda, x)$ in the interior of $U_{2R-1}$. Note that the evaluation map

(2.7) $\qquad ev_\rho = \mathbb{R}^m \times X \to Y$

given by $\qquad ev_\rho : (a, \lambda, x) \to (\lambda, x, x - \tilde{f}_a(\lambda, x))$

is a $C^2$ map. Further, the map $ev_\rho$ is transversal to $W$; that is, the image of the tangent map of $ev_\rho$ at any point $(a, \lambda, x)$ provides a complement to $W$ in $Y$. In fact the image under the tangent map of vectors of the form $(\dot{a}, 0, 0)$ form a complementary subspace. Then by the transversality density theorem ([2], p. 48), $\rho_a$ is transversal to $W$ for a dense set of values of $a$. We choose an $a$ with $\|a\| < \min(\frac{\epsilon}{4}, \epsilon \|J_1\|^{-1})$ for which $\rho_a$ is transversal to $W$. Then (cf. [2], p. 45) $\rho_a^{-1}(W)$ is a $C^2$ submanifold of $\mathbb{R}^{m+1}$ having dimension 1 and its intersection with the set $U_{2R-2}$ has a finite number of components. As such the zero set $\tilde{Z}$ of $x - \tilde{f}_a$ in $U_{2R-2}$ consists of a finite collection of $C^2$ curves which start and finish in the boundary $\partial U_{2R-2}$ or which form closed loops in $U_{2R-2}$. Each component can be represented as a $C^2$ map $\sigma(t)$ from a real interval into $\mathbb{R}^{m+1}$.

We define the open discs

(2.8)
$$\begin{cases} D_0^- = \{(\lambda, x) \mid \lambda = \lambda_0 - \frac{1}{n}, \; x \in \delta(B - \Sigma)\} \\ D_0^+ = \{(\lambda, x) \mid \lambda = \lambda_0 + \frac{1}{n}, \; x \in \delta(B - \Sigma)\} \end{cases}$$

where $\delta = \delta_n$ was the number occuring in (2.4). If we denote the vector $(1, 0, 0, \ldots, 0)$ in $\mathbb{R}^{m+1}$ by $e_0$ then using the map

$$\rho_\alpha = D_0^- \cup D_0^+ \to (D_0^- - \alpha e_0) \cup (D_0^+ + \alpha e_0) \quad (\alpha \in \mathbb{R})$$

which displaces the discs one finds, using transversality, that for some $\alpha$, $0 < \alpha < 1/n$, the image of $\rho_a$; i.e., the union of the discs $D_0^- - \alpha e_0$ and $D_0^+ + \alpha e_0$ are transversed to the zero set of $x - \tilde{f}_a$. Then there are only a finite number of zeros in each displaced disc and at such a zero the tangent vector to the curve of zeros in $\mathbb{R}^{m+1}$ does not lie in the disc. In this situation the map

(2.9)
$$x \to x - \tilde{f}_a(\lambda, x)$$

with $\lambda$ fixed, say at $\lambda_0 - \frac{1}{n} - \alpha$, will have a nonsingular derivative at a zero $z$. Were this not true, the derivative of the map with respect to the $x$ variables would have a null vector. However, letting $\lambda$ vary we already have a null-vector of the full map

(2.10) $\quad g:(\lambda,x) \to x - \tilde{f}_a(\lambda,x)$

in the direction of the tangent to the zero curve passing through through z. The derivative of the map (2.10) would then have a nullspace of dimension at least two and a range of dimension at most m-1. Transversality of the map $\rho_a$ (cf. (2.6)), however, requires the range to have dimension m.

The nonsingularity of the derivative of (2.9) at $\lambda = \lambda_0 - \frac{1}{n} - \alpha$ and likewise at $\lambda = \lambda_0 + \frac{1}{n} + \alpha$ implies that all zeros in the displaced discs have Leray-Schauder index equal to $\pm 1$. We let $D^-(D^+)$ denote $D_0^- - \alpha\, e_0 (D_0^+ + \alpha\, e_0)$ together with its boundary. Since the boundary lies in H, it contains no zeros of (2.10). We now proceed to analyze the finite number of zeros in each of the closed discs and the curves in Z leaving these points.

For fixed $\lambda$, the solution $u = 0$ of $u - G(\lambda, u) = 0$ is isolated if $\lambda$ is not a characteristic value of $G_0$. The Leray-Schauder index of $u = 0$ is then $(-1)^\beta$ where $\beta$ is the sum of the algebraic multiplicities of the characteristic value of $G_0$ between 0 and $\lambda$. If, for $\lambda_{-1} < \lambda < \lambda_0$, the index of zero is $+1$, then for $\lambda_0 < \lambda < \lambda_1$, the index is $-1$. We suppose that this situation prevails - the case with signs reversed can be treated in the same manner. For $\lambda = \lambda_0 - \frac{1}{n} - \alpha$, $u = 0$ is the only solution of $u - G(\lambda, u) = 0$ in $\|u\| \le \delta_n$ and hence by additivity, the degree of $\Phi \equiv \Phi(\lambda) \equiv u - g(\lambda, u)$ on the set $\{u \mid \|u\| < \delta_n\}$ is $+1$. If we let $\Phi_n \equiv u - G_n(\lambda, u)$, considering $\Phi_n$ to act in $\mathcal{X}$, then

$$\|t\Phi + (1-t)\Phi_n\| = \|u - tG - (1-t)P_n G\|$$

$$= \|u - G - (1-t)(P_n - I)G\|$$

$$\ge \|u - G\| - \|(1-t)(P_n - I)G\|$$

$$\ge 2/3\, \epsilon_n$$

for $\|u\| = \delta_n$ and hence, by the homotopy invariance of degree, the degree of $\Phi_n$ on the set where $\|u\| < \delta_n$ is also +1. The degree of $\Phi_n$ on the set of u in $S_n$ satisfying $\|u\| < \delta_n$ must then also be +1 (cf [4], p. 106). By similar arguments we have

$$\deg(x-f) = \deg(x-\tilde{f}) = \deg(x-\tilde{f}_a) = +1$$

where the degree is with respect to zero, in the set $\delta(B - E)$.
If we let $p_1, p_2, \ldots, p_\ell$ be the zeros of $x - \tilde{f}_a$ in $D^-$ and $i(p_k)$ the index of $p_k$ for the map (2.9), then aditivity of the degree yields

(2.11) $$\sum_{k=1}^{\ell} i(p_k) = +1 \ .$$

In a similar manner, letting the zeros of (2.9) in $D^+$ be $q_1, q_2, \ldots, q_r$ and their indices $i(q_k)$, we have

(2.12) $$\sum_{k=1}^{r} i(q_k) = -1 \ .$$

Given a curve of zeros of $x - \tilde{f}_a$ passing through $D^-$ and meeting $D^-$ at $p_k$ with $i(p_k) = +1$ we orient the curve so that near $p_k$, $\lambda$ increases when the curve is traversed in the positive sense. If $i(p_k) = -1$, we reverse the orientation. At $D^+$ we adopt the same convention.
Suppose we let

(2.13) $$T = \{(\lambda, x) \mid |\lambda - \lambda_0| \geq \frac{1}{n} + \alpha, \ x \in \delta B\} \ .$$

Then a path in Z entering $U_{2R-2} - T$ at a point $p \in D^-$ with $i(p) = +1$ may remain in $U_{R-2} - T$ until reaching $\partial U_{2R-2}$. Such a point p is said to be of type 1. A point $p \in D^-$ with $i(p) = -1$ which, followed in the backward direction, connects to $\partial U_{2R-2}$ in $U_{2R-2} - T$, is designated type 2. If a zero path starting from $p \in D^-$ with $i(p) = 1$ contains a point other than p in T, then since the path through p is transversal to $D^-$, there must be a first such point distinct from p, if one follows the zero curve from p in the positive sense. If the first point is in $D^-$ we say p is of type 3. If $i(p) = -1$ and the first such point in the backward direction is in $D^-$, p is of type 4. If $i(p) = +1$ and the first point in T is in $D^+$,

TRANSVERSALITY IN EIGENVALUE PROBLEMS

p is of type 5; if $i(p) = -1$ and the first point is in $D^+$, going backwards, the type is 6. The remaining possibility is that with $i(p) = +1$ (or $-1$) the path first meets T in the set

$$K_n = \{(\lambda,x) \mid |\lambda-\lambda_i| \leq \frac{1}{n}, \text{ some } i \neq 0, x \in \delta \Sigma\},$$

all other points on the boundary of T being prescribed as zeros. In the last event we assign p type 1 (or 2), respectively. We make a similar classification of zeros in $D^+$ calling a point q with $i(q) = -1$ $(+1)$ type 1 (type 2) if it reaches $\partial U_{2R-2}$ in $U_{2R-2} - T$ in the forward (backward) direction; type 3 or 4 when returning to $D^+$; type 5 or 6 when first meeting $D^-$; and type 1 or 2 when first meeting $K_n$ in the forward or backward direction respectively. We let $n_i$ ($m_i$) be the number of zeros of type i ($1 \leq i \leq 6$) in $D^-$ ($D^+$).

Suppose $p \in D^-$ is a zero of type 3 and let the curve through p be parametrized by $\sigma(t)$ with t in a bounded interval I. Let $\sigma(t_0) = p$ and let $t_1 > t_0$ be the smallest parameter value for which $\sigma$ returns to $D^-$. If $T_t$ is the tangent to $\sigma(t)$ at t then by Lemma 2.3, $T_t^\perp$ can be represented by $U_t X_m$ where $U_t$ is a continuous map from I into $SO_{m+1}$ and $X_m$ is the subspace of $\mathbb{R}^{m+1}$ consisting of vectors $(0,x)$. Consider the family of maps $\phi_t : X_m \to X_m$ defined by

(2.14) $\qquad \phi_t = x \to g(\sigma(t) + U_t x)$

where g denotes the map (2.10). The map $\phi_t$ has $x = 0$ as an as an isolated zero for all t and the index is constant in t by the homotopy invariance of degree. Since $\sigma(t)$ remains in $U_{2R-2} - T$ for $t_0 < t < t_1$, the $\lambda$ component of $\sigma(t)$ must be decreasing for increasing t as t approaches $t_1$. Thus the $\lambda$ component of $T_{t_0}$ is positive and that of $T_{t_1}$ is negative.

Let $\phi_{t_0} \equiv \phi_0$, and let $T^s$ be the unit vector in the direction of $(1-s) T_{t_0} + s e_0$ where $e_0 = (1,0,\ldots,0) \in \mathbb{R}^{m+1}$.

49

Arguments like those in the proof of lemma 2.3 show that there is a continuous map $V_s$ of $0 \le s \le 1$ into $SO_{m+1}$ such that $V_0 = U_{t_0}$ and $V_1$ is an orthogonal map satisfying $V_1 e_0 = e_0$. Moreover $T_{t_0} \notin V_s X_m$ for $0 \le s \le 1$ since $T^s$ is never perpendicular to $T_{t_0}$. Thus $V_s X_m$ remains transversal to $\sigma(t)$ at $\sigma(t_0)$ and hence $x = 0$ remains an isolated zero of the maps

$$(2.15) \qquad \psi_s : x \to g(\sigma(t_0) + V_s x) \qquad 0 \le s \le 1 .$$

The index of $x = 0$ for $\psi_s$, $0 \le s \le 1$ is then constant. Since $V_1 e_0 = e_0$, and $V_1 \in SO_{m+1}$, the degree of $V_1$ on $\mathbb{R}^m$ is $+1$. One sees that the index of $\psi_1$ at $x = 0$ is the same as the index of $g(\sigma(t_0) + x)$, which we are assuming is $+1$. Following the homotopies $\psi_s$ and $\phi_t$ one sees that $\phi_1(x) = g(\sigma(t_1) + U_{t_1} x)$ has index $+1$

Since $U_t$ is continuous, if we assume $U_{t_0} e_0 = T_{t_0}$ then $U_{t_1} e_0 = T_{t_1}$ where $T_{t_1}$ has a negative $\lambda$ component. One then constructs a map $W_s = [0,1] \to SO_{m+1}$ satisfying: $W_0 = U_{t_1}$ and $W_1 e_0 = -e_0$ so that $\sigma(t) + W_s X_n$ remains transversal to $\sigma(t)$ at $\sigma(t_1)$. The map

$$(2.16) \qquad \theta_s : x \to g(\sigma(t_1) + W_s x), \qquad 0 \le s \le 1$$

has $x = 0$ as an isolated zero for $0 \le s \le 1$ implying that $\theta_1$ along with $\theta_0 = \phi_1$ has index $+1$. However, since $W_1 e_0 = -e_0$ and $W_1 \in SO_{m+1}$, the restriction of $W_s$ to $\mathbb{R}^m$ must have determinant $-1$. Since the index of $g$ at $0$ is the product of the index of $W_s$ at $0 \in X_n$ and the index of $g$ at $\sigma(t_1)$, considering the $\lambda$ coordinate as fixed, the latter index must be $-1$. Thus a point of type 3 in $D^-$ leads to another point in $D^-$ of type 4. Likewise, one of type 4 leads back to one of type 3 and we have a pairing of such points. That is, $n_3 = n_4$. A similar argument shows that points in $D^-$ of type 5 pair with points in $D^+$ of type 6, giving $n_5 = m_6$. Similarly $m_3 = m_4$ and $m_5 = n_6$.

Now the degree of $x - \tilde{f}_a$ on $D^-$ is $+1$ and on $D^+$, is $-1$. Using the additivity of degree we have

$$(2.17) \quad \begin{array}{l} n_1 - n_2 + n_3 - n_4 + n_5 - n_6 = 1 \\ -m_1 + m_2 - m_3 + m_4 - m_5 + m_6 = -1 \end{array}$$

Subtracting the two equations and using the equations derived from the pairings we obtain

$$(2.18) \quad n_1 + m_1 - n_2 - m_2 = 2 \ .$$

Thus there are at least two zero curves leaving a neighborhood of $(\lambda_0, 0)$ and going either to the boundary $\partial U_{2R-2}$ or to a neighborhood of a point $(\lambda_i, 0)$ $i \neq 0$.

Returning to the space $\mathbb{R} \times S_n \subset \mathbb{R} \times \mathfrak{X}$ and letting

$$B((\mu, w); r) = \{(\lambda, u) \mid |\lambda - u|^2 + \|w - u\|^2 \leq r^2\}$$

we see that outside the set $\{(\lambda, u) \mid \|u\| \leq \delta_n\} \cup L_{n_0}$ there are at least two curves $\dot{C}_n^1$ and $\dot{C}_n^2$ which meet the ball $B((\lambda_0, 0); 3/n)$ and go either to the boundary of $B_{2R-2}$ or to a ball $B((\lambda_i, 0), 2/n)$ for some $i \neq 0$. Each point $(\lambda, u)$ on $\dot{C}_n^1$ or $\dot{C}_n^2$ is a solution of the equation

$$(2.19) \quad u + e_n(\lambda, u) - P_n G(\lambda, u) = 0$$

where $e_n(\lambda, u)$ is the continuous map $J^{-1}(\tilde{f}_a - f) J_1$ which had its dependence on $n$ suppressed. Our choice of the approximation $\tilde{f}_a$ was such that $\|e_n(\lambda, u)\| \leq \epsilon_n < 1/n$ for $(\lambda, u)$ in $\beta_{2R-2}^n$. If $(\lambda_n^i(t), u_n^i(t))$, $t \in I$, is a parametrization of $\dot{C}_n^i$ from the boundary $\partial B(\lambda_0, 0), \frac{3}{n})$ to $\partial B_{2R-2}$ or to $\partial B((\lambda_i, 0), \frac{2}{n})$ then the curve

$$(2.20) \quad \tilde{C}^i = \{\lambda_n^i(t), u_n^i(t) + e_n(\lambda_n^i(t), u_n^i(t))\} \quad t \in I$$

meets $B((\lambda_0, 0); \frac{4}{n})$ and either $\partial \beta_{2R-3}$ or $\partial B((\lambda_i, 0); \frac{3}{n})$, since $\|e_n\| < \frac{1}{n}$. Moreover, using (2.19) we see that the curves $\tilde{C}_n^i$ lie in $E_{2R} = [\lambda_0 - 2R, \lambda_0 + 2R] \times \widehat{G(B_{2R})}$ where the roof indicates the closed convex hull. Since, in a

Banach space, the closed convex hull preserves compactness, $E_{2R}$ is a compact metric space. We obtain continuous curves (continua) $C_n^i$ ($i = 1, 2$) which connect $(\lambda_0, 0)$ with $\partial B_{2R-3}$ or to $(\lambda_j, 0)$, $j \neq 0$, by adjoining line segments to $\tilde{C}_n^i$, one lying within $B((\lambda_0, 0); \frac{4}{n})$ and going from $(\lambda_0, 0)$ to a point on $\tilde{C}_n^i$ and, if $\tilde{C}_n^i$ does not meet $\partial B_{2R-3}$, another segment in $B((\lambda_j, 0); \frac{3}{n})$ from $(\lambda_j, 0)$ to $\tilde{C}_n^i$ for a suitable j. If Q (cf. p. ) is an open set with boundary M (not necessarily smooth) then since each curve $C_n^i$ joins a point in Q to a point in $\bar{Q}^c$ it must meet $M = \partial Q$. Let $\{p_n'\}$ be a sequence of points with $p_n' \in C_n^1 \cap M \subset E_{2R} \cap M$. Using compactness we can extract a subsequence $p_{n_k}'$ converging to a point $p \in M$. If $p = (\lambda, u) \in E_{2R}$ is a limit point of points $p_n \in C_n^1$ then it is also a limit point for $\tilde{C}_n^1$. Using compactness one can extract a subsequence $p_{n_k} \in \tilde{C}_n^1$ converging to p. Since $P_n G$ converges to G and $e_n$ converges to zero, p is a zero of $u - G(\lambda, u) = 0$. Using Lemma 2.2 with $S_1 = (\lambda_0, 0)$ and $S_2 = M \cap E_{2R}$ we see that there is a continuum of zeros of $u - G(\lambda, u) = 0$ connecting $(\lambda_0, 0)$ with M. While the continuum might contain some solutions not in $S$ it cannot connect through trivial solutions, for no curve $C_n^i$ meets the closure of $L_{n_0}$. One can thus remove solutions not in $S$ and still have a continuum in $S$ connecting $(\lambda_0, 0)$ to a point $p_1$ in M. If $M \cap E_{2R}$ contains more than one limit point of the curves $C_n^i$ ($i = 1$ or 2), say it contains $p_2 \neq p_1$, then arguments similar to those given show that $p_2$ is connected to $(\lambda_0, 0)$ in $S$.

If $M \cap S$ contains a single point p' and M is smooth, then we must show that p' has multiplicity two. Suppose we have parametrized M in a neighborhood of p' as described at the beginning of section 2; that is, we have a neighborhood $W_1$ of 0 in $X$ and a map T so that for $x \in W_1$, $p = p' + Tx + \gamma(Tx)\pi$ describes a neighborhood V of p' in M. Let r be chosen so that the set $M \cap B(p', r)$ is contained in V. For sufficiently large n, M will not intersect any ball $B((\lambda_j, 0), \frac{4}{n})$; $m_1 \leq j \leq m_2$. Choosing such an n we let

(2.21) $$\overline{M_1 = (M - B(p', r) - L_n)},$$

where the bar denotes the closure and $L_n$ is defined by (2.4). On $M_1$, $u - G(\lambda, u)$ has no zeros and hence for $(\lambda, u) \in M_1$, $\|u - G(\lambda, u)\| \geq \epsilon$ for some $\epsilon > 0$. Note that for $x \in \partial W_1$ its image $y$ under the map (2.2) must also satisfy $\|y\| \geq \epsilon$. We choose an $\epsilon/3$ net for $G(B_{2R})$ as before and let $P$ (cf. 2.5) project onto its span $S$. We can assume that $\mathbb{R} \times S$ contains $p'$ and $\pi$ as well as the vectors $x_1$ and $n_1$ used to define $T$. Then a map like (2.2) with $G$ replaced by $PG$ and $x$ restricted to $S$ will have the same degree on $W_1 \cap S$ as the original map (2.2) had on $W_1$ (cf. [4], p. 106). Since $p'$, $\pi$, $x_1$ and $n_1$ are in $\mathbb{R} \times S$, the intersection of $V \subset M$ with $\mathbb{R} \times S$ will again be parametrized as $p' + Tx + \gamma(Tx)\pi$ for $x \in W_1 \cap S$.

As we saw earlier, the problem in $\mathbb{R} \times S$, with the dimension of $S$ equal to $m$, is equivalent to a problem in $\mathbb{R} \times \mathbb{R}^m$. By choosing a new complementary vector $\pi$ if necessary we may also assume that in the Euclidean space the correspondents of $\pi$ and $N \cap S$ are perpendicular. Let us suppose $S = \mathbb{R}^m$ with $\pi \perp N \cap S$ to avoid introducing more notation. We can obtain the map $x - \tilde{f}_a$ as before, making sure that it does not vanish on $M_1 \cap (\mathbb{R} \times S)$. We then have the following situation: There are at least two curves of zeros of $x - \tilde{f}_a$ leaving the ball $B((\lambda_0, 0), \frac{4}{n})$ in $\mathbb{R} \times S$ and reaching either the boundary of the ball of radius $2R-2$ centered at $(\lambda_0, 0)$ or a ball $(B(\lambda_i, 0), \frac{3}{n})$ for some $i \neq 0$. In either case the curves must cross $M \cap (\mathbb{R} \times S)$ and can do so only near $p'$.

The index at 0 for the map (2.2) is the degree of the map

(2.22) $$x \to Q(p' + Tx + \gamma(Tx)\pi) - \tilde{f}_a(p' + Tx + \gamma(Tx)\pi)$$

with respect to zero on the set $W_2 = W_1 \cap S$. Since all maps involved are continuous the degree will not be changed if we replace $p'$ in (2.22) by $p''$ sufficiently close to $p'$ (using a homotopy). Again, using transversality we can choose such a $p''$ so that the manifold described by $p'' + Tx + \gamma(Tx)\pi$

as $x$ varies in $W_2$, is transversal to the zero set of $x - \tilde{f}_a$. At this point, without loss of generality, we assume that $M \cap (\mathbb{R} \times S)$ is transversal to the zero set and continue with (2.22). Again, the transversality of $M \cap (\mathbb{R} \times S)$ to the zero set means that the zeros of the map (2.22) must be ones at which the derivative is nonsingular. The degree of the map is then merely the sum of the number of zeros with Jacobian determinant positive minus the number having it negative (cf. [7], Chapter III).

Suppose $\sigma(t), t \in I$, represents a zero curve which leaves a point in $D^-$ of type 1. As before, the map (2.14) has constant index at 0 as $t$ varies. If $\sigma(t_1)$ is the point at which the path first reaches $M \cap (\mathbb{R} \times S)$, then as we saw with the analysis of a zero in $D^-$, the map (2.14) will have index +1 at $t=t_1$ as well as the map

$$(2.23) \qquad x \to g(\sigma(t_1) + V_1 x)$$

where $V_1 \in SO_{m+1}$ takes $e_0$ to a unit normal $n$ to $M$ when translated to $\sigma(t_1)$, where the direction of $n$ is chosen so that the angle between the curve tangent $T_{t_1}$ and $n$ is less than $\pi/2$. In this case $n$ will point into $\overline{Q}^c \cap (\mathbb{R} \times S)$. We can make $V_1$ one end of a path $V_s$ from $[0,1]$ into $SO_{m+1}$ such that $V_s$ maps $e_0$ to the normal to $M$ at the point $p(s) = p' + T(sx') + \gamma(T(sx'))$ where $p(1) = \sigma(t_1) = p' + Tx' + \gamma(Tx')\pi$. Using $g$, introduced above, the map (2.22) can be written

$$(2.24) \qquad x \to g(p' + Tx + \gamma(Tx)\pi).$$

Since the zero of (2.23) at 0, which corresponds to a zero of (2.24) at the point $x'$, is nondegenerate, the index of each map will be that of its derivative at the appropriate point. That is, the index of (2.23) will be that of

$$(2.25) \qquad D\,g(p(1)) \cdot V_1$$

and the index of (2.24) will be that of

(2.26) $\qquad D\, g(p(1)) \cdot (T + (D\, \gamma(Tx'), \circ)\pi)$ .

The maps $V_1$ and $T + (D\gamma(Tx'), \circ)\pi$ each map onto the tangent space of M at $\sigma(t_1)$ and to compare the indices it suffices to compare the orientations that the two maps give to the tangent space to $M \cap (\mathbb{R} \times S)$ at $p(1)$. Letting

$$T_s = T + (D\, \gamma\,(Ts\,x'), \circ)\,\pi$$

we see that $V_s^{-1}T_s$ is nonsingular as $s$ varies and hence $V_s^{-1}T_s$ and $V_0^{-1}T_0$ have determinants of the same sign. Suppose that the determinant is positive; in the negative case all signs will be reversed and the index of the original problem at $p'$ will change sign. If we carry out the scheme above for any zero curve $\tilde{\sigma}(t)$ arriving at $M \cap (\mathbb{R} \times S)$ from $D^-$, with index $+1$ at $D^-$, then we will arrive at a map $\tilde{V}_0$ such that $\tilde{V}_0^{-1}T_0$ has positive determinant since $\tilde{V}_0 e_0$ is the normal to $N$ which, from $p'$, points into $\bar{Q}^c \cap (\mathbb{R} \times S)$. Each such zero on $M \cap (\mathbb{R} \times S)$ thus contributes $+1$ to the degree of (2.24) on $W_2$. By a similar computation a curve leaving $D^-$ with index $-1$ and reaching $M$ will contribute $-1$ to the degree of (2.24). A curve from $D^+$ contributes $+1$ if it has index $-1$ at $D^+$ and $-1$ if it has index $+1$ at $D^+$. These zeros then give a total degree of $+2$ according to (2.18). There may also be zero curves which enter $Q \cap (\mathbb{R} \times S)$ near $p'$ and then leave again without meeting $D^-$ or $D^+$. Arguments similar to those already given show that the indices contributed by the two ends of such a curve are of opposite signs so that the total degree for (2.24) is $+2$. This completes the proof of the Theorem.

As a corollary of Theorem 2.4 we obtain the following result due to P. H. Rabinowitz [3]

<u>Corollary 2.5.</u> Let the conditions of Theorem 2.4 be satisfied. Then if $B_R$ is the set $\{(\lambda, u) \mid (\lambda - \lambda_0)^2 + \|u\|^2 = R^2\}$, there is a continuum in $S$ joining $(\lambda_0, 0)$ either to $\partial B_R$ or to a point $(\lambda_i, 0)$ for $i \neq 0$.

__Proof.__ Let $Q_n$ be the set $B_R$ minus a ball of radius $1/n$ about each point $(\lambda_i, 0)$, $i \neq 0$. Then for $n$ large the final assertion of Theorem 2.4 tells us there will be a continuum $C_n$ in $S$ joining $(\lambda_0, 0)$ to $\partial B_R$ or to the boundary of one of the balls about the points $(\lambda_i, 0)$. If, in the latter case, we join the intersection of $C_n$ with such a ball to the corresponding center $(\lambda_i, 0)$ we have a continuum from $(\lambda_0, 0)$ to $(\lambda_i, 0)$. Now Lemma 2.2 in conjunction with arguments already used in the proof of theorem (2.4) show that the limit set of the $C_n$ is a continuum in $S$ joining $(\lambda_0, 0)$ to $\partial B_R$ or to a point $(\lambda_i, 0)$, $i \neq 0$.

We give an example to show that the boundary $M$ of a region $Q$ as described in the theorem may contain a single zero of multiplicity two. The example is a simplified version of one due to P. H. Rabinowitz. (These proceedings). One needs a map on the two sphere $S^2$ having just one fixed point. One can obtain such a map by stereographic projection from the planar vector field $v(z) = i z^2$ where $z = x + i y$. The corresponding differential equation is $\frac{dz}{dt} = i z^2$ and the integral curves have the form $z(t) = (a - it)^{-1}$. One obtains a planar map by letting each point be an initial point at time $t = 0$ for the corresponding flow and taking as its image the position at time $t = 1$. The corresponding map on $S^2$ leaves just one point fixed. Identifying each point in $S^2$ with a unit vector in $\mathbb{R}^3$ we obtain a map $C$ which takes a unit vector $u$ to another unit vector $C(u)$ and which leaves just one vector $u_0$ fixed. It is also the case that no vector is mapped to $-u$. Let $C$ be extended to all of $\mathbb{R}^3$ to be a map homogeneous of degree 3 and consider the problem

(2.27) $$w = \lambda(w + C(w))$$

where $w \in \mathbb{R}^3$. The linearized problem at $w = 0$ is $w = \lambda w$ and $\lambda = 1$ is an eigenvalue of multiplicity 3. Since $C(w) \neq 0$ for $w \neq 0$ we cannot have $(1, w)$ as a nontrivial solution. Hence (2.27) is equivalent to

(2.2 ) $$w = \frac{\lambda}{1-\lambda} C(w) .$$

Since $C(w)$ is not colinear with $w$ unless $w = \alpha u_0$ the only nontrivial solution of (2.27) has the form $(\lambda, \alpha u_0)$ where, using $C(\alpha u_0) = \alpha^3 C(u_0)$, we must have $1 = \lambda(1-\lambda)^{-1}\alpha^2$ or $\lambda = (1+\alpha^2)^{-1}$.

We see that one can produce various types of bifurcating solutions by choosing flows on $S^2$ with differing fixed points. We also see that any flow on $S^2$ (or on $S^{2n}$ for $n = 1, 2, \ldots$) must have at least one fixed point for otherwise we could use the flow to construct an example of an eigenvalue of odd multiplicity from which no bifurcation occurred. Taking a flow $C$ on $S^1$ which merely rotates through an angle $\pi/2$, one can extend $C$ to be homogeneous of degree 3 in $\mathbb{R}^2$. This gives rise to a standard example $w = \lambda(w + Cw)$ for $w = (w_1, w_2)$ in $\mathbb{R}^2$ for which no bifurcation occurs from the point $\lambda = 1$, $w = 0$. Using the linear map $A$ which takes $w_1 \to w_1$ and $w_2 \to 2w_2$ one easily verifies that $w = \lambda(Aw + Cw)$ has nontrivial solutions connecting the points $(1/2, 0)$ and $(1, 0)$. This phenomenon was discussed in [3].

Theorem 2.4 has an analogue for operators not satisfying the condition $\|G_1(\lambda, u)\| = o(\|u\|)$. If, as in the paper [1], we define

$$(2.29) \qquad n(\lambda^*) = \lim_{\substack{r \to 0 \\ \|u\| < r \\ |\lambda - \lambda^*| \leq r}} \sup \frac{\|(I - \lambda G_0)^{-1} G_1(\lambda, u)\|}{\|u\|}$$

with $n(\lambda^*) = +\infty$ if $I - \lambda^* G_0$ has no inverse, then the set of trivial solutions from which bifurcation may occur are those of the form $(\lambda, 0)$ where

$$(2.30) \qquad \lambda \in \tilde{\sigma} = \{\lambda \in \mathbb{R} \mid n(\lambda) \geq 1\} .$$

Letting $\tilde{\rho} = \mathbb{R} - \tilde{\sigma}$ we have the following theorem the proof of which is very similar to that of Theorem 2.4.

Theorem 2.6. Let the hypotheses of Theorem 2.4 be satisfied with $o(\|u\|)$ replaced by $O(\|u\|)$ for $\|G_1(\lambda, u)\|$ as $\|u\| \to 0$. Let $[a, b]$ be a real interval with a and b in $\tilde{\rho}$, and suppose that the total algebraic multiplicity associated with characteristic values of $G_0$ in the interval $[a, b]$ is odd. Then either there are at least two points $p_1$ and $p_2$ in $M \cap S$ which are connected in $S$ to the set $[a, b] \cap \tilde{\sigma} \times \{0\}$ in $\mathbb{R} \times \mathcal{X}$ or one such point having multiplicity two. If M is not smooth, there is still one such point $p_1$.

## 3. Cone Maps.

Here we give an application of the use of transversality in finding continua of solution pairs $(\lambda, u)$ in eigenvalue problems involving cone preserving maps. We do not aim for the greatest generality, but merely illustrate the way in which transversality enters.

Theorem 3.1. Let $\mathcal{X} = L^2(\Omega)$ where $\Omega$ is a bounded open subset of $\mathbb{R}^n$ and let $K_0$ be the cone of functions in $L^2$ which are nonnegative a.e. let F be a compact continuous map of $\mathcal{X}$ into $\mathcal{X}$ which takes $K_0$ into itself. Suppose F has a derivative $A_0$ at $u = 0$. Suppose $\lambda A_0 v = v$, $v \in K_0 - \{0\}$ implies $\lambda = \lambda_0$, where $\lambda_0^{-1}$ is the spectral radius of $A_0$. If $B_R = \{(\lambda, u) \mid |\lambda|^2 + \|u\|^2 = R^2\}$ then $S$ for the equation

(3.1) $$u - \lambda F(u) = 0$$

contains a continuum in $\mathbb{R} \times K_0$ which meets $(\lambda_0, 0)$ and $\partial B_R$.

Proof. Since F is compact $DF(0) = A_0$ will also be compact and we know from standard results on linear cone maps (cf [8], p. 1016) that $\lambda_0 A_0 v = v$ for some $v \in K_0 - \{0\}$. Let $\beta_R$ be the ball $\{u \mid \|u\| \le R\}$ in $\mathcal{X}$. Given $\epsilon > 0$ and assuming $R > \lambda_0 + 1$ we let $z_1, \ldots, z_\ell$ in $K_0$ be an $\epsilon$-net for $F(\beta_R)$ and for $A_0(\beta_R)$. Suppose $\xi_1, \ldots, \xi_n$ are the coordinates in $\Omega \subset \mathbb{R}^n$ and that we cover $\Omega$ with a mesh of closed cubes of side length $2^{-m}$, obtained by subdividing each axis $\xi_i$ into intervals of length $2^{-m}$. The characteristic functions of these cubes (intersected with $\Omega$), as m, ranges over positive integers, have a linear span which is dense in $L^2(\Omega)$. Thus for a sufficiently large fixed m, the vectors $z_i$ will be within $L^2$ distance $\epsilon$ of the span S of the characteristic functions of the $2^{-m}$ mesh. If $x_1 \ldots, x_N$ are the normalized characteristic functions (i.e. with $\|x_i\| = 1$) then the perpendicular projection onto S will be

$$(3.2) \qquad P = \sum_{i=1}^{N} (\circ, x_i) x_i$$

and since Pz is the point in S closest to z, we have

$$\|Pz - z\| \leq 2\epsilon$$

for each z in $A_0(\beta_R)$ or $F(\beta_R)$. Since for $u \in B_1$, $\|PA_0 u - Au\| \leq 2\epsilon$, we see that for the operator norm we have $\|PA_0 - A_0\| \leq 2\epsilon$. Given $\eta > 0$ there is a number $C_\eta > 0$ such that $|\lambda - \lambda_0| \geq \eta$, $0 \leq \lambda \leq R$ and $u \in K_0$ imply

$$(3.3) \qquad \|\lambda A_0 u - u\| \geq C_\eta \|u\|.$$

Otherwise, we could find a sequence of unit vectors $u_j \in K_0$ and $\mu_j$ with $|\mu_j - \lambda_0| \geq \eta$, $0 \leq \mu_j \leq R$ such that $\mu_j A_0 u_j - u_j$ converged to zero which, with compactness of $A_0$, would give an eigenvector of $A_0$ in $K_0$ for a characteristic value other than $\lambda_0$. We may assume that the constant $C_\eta$ satisfies $C_{\eta_1} \leq C_{\eta_2} \leq \eta_2$ if $\eta_1 \leq \eta_2$ since if that is not the case we can always use the constants

$$C'_\eta = \min \{ \min_{\mu \geq \eta} C_\mu, \eta \}.$$

If we choose $\epsilon < C_\eta/2R$ above then (3.3) yields

$$\|u - \lambda PA_0 u\| = \|u - \lambda A_0 u + \lambda A_0 u - \lambda PA_0 u\|$$
$$\geq C_\eta \|u\| - \lambda \frac{C_\eta}{2R} \|u\|$$
$$\geq 1/2 \, C_\eta \|u\|.$$

The projection P maps $K_0$ into the cone K consisting of the span of $\mathfrak{X}_1, \ldots, \mathfrak{X}_n$, with nonnegative coefficients. Thus the map PF will map K into K and will approximate F to within $2\epsilon$ on $\beta_R$. Moreover, PF has a derivative $PA_0$ at $u = 0$ and we see from the inequalities above that any characteristic value of $PA_0$ on the interval $(0, R)$ which corresponds to a characteristic vector in K, must be within distance $\eta$ of $\lambda_0$. For $\epsilon$ sufficiently small there must, in fact, be such a characteristic value. To see this we appeal first to standard perturbation theory which tells us that there will be at least one characteristic value within (complex) distance $\eta$ of $\lambda_0$, for $\epsilon$ small. Then, since $PA_0$ is a cone map, its spectral radius must be an eigenvalue $\mu$ corresponding to an eigenvector in the cone; i.e. $\mu^{-1}$, which lies on $(0, \lambda_0 + \eta]$, must be a characteristic value for $PA_0$ with a corresponding vector in K.

We let f denote PF restricted to S and let A be its derivative at $u = 0$ in S. As we did earlier we may consider S to be the space $\mathbb{R}^N$ by identifying $u = \sum_{i=1}^{n} x_i \mathfrak{X}_i$ with the vector $X = (x_1, \ldots, x_N)$. We henceforth consider $x \in \mathbb{R}^N$ as the vector variable and have $K = \{x \mid x_i \geq 0, \text{ all } i\}$. We write $x \geq y$ if and only if $x - y \in K$. Since $F(u)$ has the form $A_0 u + e_0(u)$ where $e_0(u)$ is $o(\|u\|)$ as $\|u\| \to 0$ we see that

$(\mu.4)$ $\qquad f(x) = Ax + e(x)$

where $e(x) = o(\|x\|)$ as $\|x\| \to 0$.

Suppose we find a map $\tilde{e}(x)$ defined on $\beta_R$ which satisfies:

    i)   $\tilde{e}(x)$ is $C^2$.

    ii)  there exists an $r > 0$ such that $\tilde{e}(x) = 0$ for $\|x\| \leq r$.

    iii) for $x \in K \cap \beta_R$ either $\tilde{e}(x) \geq e(x)$ or $\tilde{e}(x) \geq 0$.

    iv) $\|\tilde{e}(x) - e(x)\| \leq C_\eta / 4R \|x\|$ for $x \in K \cap \beta_R$.

Then the map $\tilde{f}(x) = Ax + \tilde{e}(x)$ (which depends upon $\eta$) will satisfy:

1) $\tilde{f}$ is $C^2$.
2) $\tilde{f}(x) \equiv Ax$ for $\|x\| \leq r$.
3) for $x \in K \cap \beta_R$, $\tilde{f}(x) \in K$.
4) $\|\tilde{f}(x) - f(x)\| \leq C_\eta / 4$ for $x \in K \cap \beta_R$.
5) There is a $\delta_\eta > 0$ such that $\|x\| \leq \delta_\eta$, $x \in K$ $|\lambda - \lambda_0| > \eta$, $|\lambda| \leq R$ implies

(3.5) $$\|\lambda \tilde{f}(x) - x\| \geq \frac{C_\eta}{2} \|x\|.$$

That is, knowing that $e(x)$ is $o(\|x\|)$ near $x = 0$, we can find $\delta_\eta > 0$ such that $\|x\| < \delta_\eta$ implies $\|e(x)\| \leq \frac{C_\eta}{4R} \|x\|$ and thus

$$\|\lambda \tilde{f}(x) - x\| \geq \|\lambda Ax - x\| - \|\lambda \tilde{e}(x) - \lambda e(x)\| - \|\lambda e(x)\|$$
$$\geq \frac{C_\eta}{2} \|x\|.$$

Further, since $C_\eta$ is monotone, for $|\lambda - \lambda_0| > 1$, $|\lambda| \leq R$, $x \in K$ and $\|x\| \leq \delta_1$.

(3.6) $$\|\lambda \tilde{f}(x) - x\| \geq \frac{C_1}{2} \|x\|.$$

To construct $\tilde{e}$ we first choose $r_1$ so that $\|x\| \leq r_1$ implies $\|e(x)\| \leq C_\eta [16 R(1 + \sqrt{N})]^{-1} \|x\|$. Then on $\beta_{R}$ approximate $e(x)$ to within $r_1 C_\eta [16 R(1 + \sqrt{N})]^{-1}$ by a $C^2$ map $\tilde{e}^1(x)$. Let $\tilde{e}_i^1(x)$ be the coordinate functions corresponding to the map $\tilde{e}^1$ and let $\tilde{e}_i^2(x) = \tilde{e}_i^1(x) + r_1 C_\eta [16 R(1 + \sqrt{N})]^{-1}$. Then the corresponding map $\tilde{e}^2$ on K will satisfy $\tilde{e}^2(x) \geq e(x)$ and where $\|\tilde{e}^1(x)\| \leq r_1 C_\eta [16 R (1 + \sqrt{N})]^{-1}$, i.e. on $\|x\| \leq r_1$, we have $\tilde{e}^2(x) \geq 0$.

Let $\eta(t)$ be an increasing $C^\infty$ function on $[0,\infty)$ which is zero for $0 \leq t \leq 1/2\, r_1$ and identically one for $t \geq r_1$. Letting $\tilde{e}(x) = \tilde{e}^2(x) \cdot \eta(\|x\|)$ and setting $r = 1/2\, r_1$ we have

    i) $\tilde{e}$ is $C^2$

    ii) $\tilde{e}(x) = 0$ for $\|x\| \leq r$

Suppose $x \in K$. Then

    iii) $\tilde{e}(x) \geq e(x)$ for $\|x\| \geq 2r$ and $\tilde{e}(x) \geq 0$ for $\|x\| \leq 2r$

    iv) for $\|x\| \leq r$, $\dfrac{\|\tilde{e}(x) - e(x)\|}{\|x\|} = \dfrac{\|e(x)\|}{\|x\|} \leq \dfrac{C_\eta}{4R}$;

for $r \leq \|x\| \leq 2r$,

$$\frac{\|\tilde{e}(x) - e(x)\|}{\|x\|} \leq \frac{\|\tilde{e}(x)\| + \|e(x)\|}{\|x\|}$$

$$\leq \frac{\|e(x)\|}{\|x\|} + \frac{r_1}{\|x\|} \cdot \frac{C_\eta}{16\, R(1+\sqrt{N})}(1 + \sqrt{N})$$

$$+ \frac{\|e(x)\|}{\|x\|}$$

$$\leq 2\frac{C_\eta}{16R(+\sqrt{N})} + \frac{C_\eta}{8R} \leq \frac{C_\eta}{4R};$$

for $\|x\| \geq 2r$,

$$\frac{\|\tilde{e}(x) - e(x)\|}{\|x\|} \leq \frac{r_1}{r_1}\, \frac{C_\eta}{16R(1+\sqrt{N})} \cdot (1+\sqrt{N}) = \frac{C_\eta}{16R}.$$

The map $\tilde{e}$ thus satisfies i) - iv) from p.   .

    Having the map $\tilde{f}$ we choose a set of linearly independent vectors $w_1, w_2, \ldots, w_N$ in $K^0$, the interior of $K$, and a similar set in the dual cone; i.e., the cone of linear functionals which are nonnegative on $K$. In this case we can use $w_1, \ldots, w_N$ for the second set as well. Let $a = (a_1, a_2, \ldots, a_N)$ be a point in $G \equiv K^0 \subset \mathbb{R}^N$ and define a map

$$(3.7) \qquad h_a : x \to \sum_{k=1}^{N} a_k (x, w_k) w_k .$$

Let $\tilde{f}_a = \tilde{f} + h_a$ and with $\mathbb{R}^0 = (0, R)$, $K_R^0 = K^0 \cap \beta_R$ let $\rho_a$ be the map of $\mathbb{R}^0 \times K_R^0$ into $\mathbb{R}^0 \times K_R^0 \times \mathbb{R}^N$ defined by

$$(3.8) \qquad \rho_a : (\lambda, x) \to (\lambda, x, x - \lambda \tilde{f}_a(x)).$$

Since for each $x \in K_R^0$, $(x, w_k) \neq 0$, the evaluation map (cf. p.     ) $ev_\rho$ is transversal to $W = \mathbb{R}^0 \times K_R^0 \times \{0\}$, there are, by the transversality density theorem, arbitrarily small values of $a \in K^0$ for which the zero set of $x - \lambda \tilde{f}_a(x)$ is a $C^2$ submanifold of $\mathbb{R}^0 \times K_R^0$. We choose such an $a$, making sure that $\|\lambda \tilde{f}_a(x) - x\| \geq \frac{C_\eta}{4} \|x\|$ for $|\lambda - \lambda_0| \geq \eta$, $|\lambda| \leq R$, and $\|x\| \leq \delta_\eta$; and that $\|\lambda \tilde{f}_a(x) - x\| \geq \frac{C_1}{4} \|x\|$ for $|\lambda - \lambda_0| \geq 1$, $|\lambda| \leq R$, and $\|x\| \leq \delta_1$.

In the part of $K_R^0$ where $\|x\| < r$, $\tilde{f}_a$ coincides with the linear map $A + h_a$ which maps $K$ to $K^0$. Standard results from the theory of linear cone preserving maps tell us that the zero set of $x - \lambda \tilde{f}_a$ for $\|x\| < r$ and $0 < \lambda < R$ consists of a segment: $\lambda = \lambda_1$, $x = t x_1$ where $x_1 \in K^0$ with $\|x_1\| = r$ and $0 < t < 1$. Since the segment is part of a one dimensional $C^2$ submanifold of $\mathbb{R}^0 \times K_R^0$, it must, starting in the direction of increasing $t$, reach the boundary of the region $\mathbb{R}^0 \times K_R^0$. Since the segment described exhausts the zero set for $\|x\| < r$, the zero curve cannot reenter the part of $\mathbb{R}^0 \times K_R^0$ where $\|x\| < r$. Further, for $\|x\| \geq r$, we cannot have zeros of $x - \lambda \tilde{f}_a(x) = 0$ with arbitrarily small $\lambda$ so it cannot reach the part of the boundary where $\lambda = 0$. Since $\tilde{f}_a$ maps to the interior of $K$ we cannot have zeros $(\lambda, x)$ with $x$ arbitrarily close to $\partial K$, for continuity of the maps would yield a zero $(\lambda, x)$ with $x \in \partial K$, $\|x\| \geq r$, and $\lambda > 0$, an impossibility. The zero curve must then reach the part of the boundary where $\lambda^2 + \|x\|^2 = R^2$. We note that the zero curve must also avoid the region where $|\lambda - \lambda_0| \geq 1$ and $\|x\| \leq \delta_1$.

We now let $\eta$ take the values $1/n$, $n = 1, 2, 3, \ldots$, obtaining curves $C_n'$ which run from $B((\lambda_0, 0), \frac{1}{n})$ to a point $(\lambda, x)$ satisfying $\lambda^2 + \|x\|^2 = R^2$ in $\mathbb{R} \times \mathcal{X}$ and which consists of solutions of an equation

(3.9) $\qquad x - \lambda(f(x) + r_n(x)) = 0$

where $\|r_n(x)\| \to 0$ as $n \to \infty$, uniformly for $\|x\| \leq R$. The remainder of the argument is essentially the same as the corresponding one in the proof of Theorem 2.4. We obtain a continuum of solutions connecting $(\lambda_0, 0)$ with $\partial B_R$ and not intersecting the set $|\lambda - \lambda_0| > 1$, $\|x\| < \delta_1$. As before, there must then be a connecting continuum in $\mathcal{S} \cap (\mathbb{R} \times K_0)$.

Let $L^2(\Omega, \mathbb{R}^N)$ consist of elements $f = (f_1, \ldots, f_N)$ where each $f_i$ is an $L^2(\Omega)$ function. We let $\|f\| = (\sum_{i=1}^{N} \|f_i\|^2)^{1/2}$, the norm in the sum being the $L^2(\Omega)$ norm. We say $f \in K_0$ if and only if each $f_i$ is a.e. nonnegative.

<u>Corollary 3.2.</u> In the statement of Theorem 3.1 let $L^2(\Omega)$ be replaced by $L^2(\Omega, \mathbb{R}^N)$ with its corresponding cone of nonnegative functions. Then the conclusions of that theorem hold.

<u>Proof.</u> We do the case $N = 2$, the general one being much the same. Given $\Omega$ we let $\Omega_1 = \Omega$ and let $\Omega_2$ be a translation $\Omega + v$ where $v \in \mathbb{R}^n$ is chosen so that $\Omega_1 \cap \Omega_2 = \phi$. Let $\Omega' = \Omega_1 \cup \Omega_2$. Then given $x_1(\xi)$, $x_2(\xi)$ in $L^2(\Omega)$ we let

$$x(\xi) = \begin{cases} x_1(\xi) & \text{if } \xi \in \Omega_1 \\ x_2(\xi - v) & \text{if } \xi \in \Omega_2 \end{cases}.$$

Then $x \in L^2(\Omega')$ and is a.e. nonnegative if and only if $x_1$ and $x_2$ are. If $f_1(x_1, x_2)$ and $f_2(x_1, x_2)$ are the components of a map in $L^2(\Omega, \mathbb{R}^2)$, we let

$$(f(x))(\xi) = \begin{cases} f_1(x(\xi), x(\xi+v)), & \xi \in \Omega_1 \\ f_2(x(\xi-v), x(\xi)), & \xi \in \Omega_2 \end{cases}$$

We now have a problem in $L^2(\Omega')$ to which Theorem 3.1 applies and the recovery of a continuum of solutions in $\mathbb{R} \times L^2(\Omega, \mathbb{R}^2)$ is immediate.

As an application of the corollary we consider a pair of linked differential equations. Let

$$\tau y = -\frac{d}{dx}(p(x))\frac{dy}{dx} + q(x) y$$

where $p$ is a positive function in $C^1[0,1]$ and $q(x)$ is a nonnegative function in $C[0,1]$. Letting $H^2[0,1]$ denote the collection of functions in $L^2[0,1]$ with two distribution derivatives in $L^2$, we let

$$\mathcal{D}(L) = \{y \in H^2[0,1] \mid y(0) = y(1) = 0\}$$

and for $y \in \mathcal{D}(L)$ let

$$Ly = \tau y.$$

It is well-known that $L$ has an integral operator inverse with a nonnegative piecewise continuously differentiable kernel. We denote the inverse operator by $G$.

Let $\alpha = (\alpha_1, \alpha_2)$ be an element of $\mathbb{R}^2$ with $\|\alpha\| = (\alpha_1^2 + \alpha_2^2)^{1/2}$. Let $h_1$ and $h_2$ be continuous maps from $\mathbb{R}^2$ to $\mathbb{R}^1$ which are nonnegative on the cone $\{\alpha \mid \alpha_1 \geq 0, \alpha_2 \geq 0\}$ in $\mathbb{R}^2$. Suppose $h_1$ and $h_2$ are $o(\|\alpha\|)$ as $\|\alpha\| \to 0$. We look for solutions of the pair of equations

(3.10)
$$L y_1 = \lambda (y_1 + h_1(y_1, y_2))$$
$$L y_2 = \lambda (y_2 + h_2(y_1, y_2))$$

where $y_1$ and $y_2$ are in $\mathcal{D}(L)$ and the pair $(y_1, y_2)$, which we will also denote by y, is such that $h_1(y_1,y_2)$ and $h_2(y_1,y_2)$ are in $L^2(\Omega)$. We call a pair $(\lambda,y)$ positive if $\lambda > 0$, $y_1 \geq 0$, and $y_2 \geq 0$. Again, $\mathcal{S}$ denotes the $R \times L^2$ closure of solutions $(\lambda, y)$ with $y \neq 0$.

<u>Theorem 3.3.</u>  Let $\lambda_0$ be the smallest eigenvalue of L. Then for any $R > \lambda_0$ the equation (3.10) has a continuum of positive solutions in $\mathcal{S}$ joining $(\lambda_0, 0)$ to the set where $\lambda^2 + \|y\|^2 = R^2$.

<u>Proof.</u>  We let $Ly_1 = \phi_1$ and $Ly_2 = \phi_2$ or $y_1 = G\phi_1$ and $y_2 = G\phi_2$. The equations (3.10) become

(3.11)
$$\phi_1 = \lambda[G\phi_1 + h_1(G\phi_1, G\phi_2)]$$
$$\phi_2 = \lambda[G\phi_2 + h_2(G\phi_1, G\phi_2)]$$

Using $\|\phi_1\|$ for the $L^2$ norm and $\|\phi\| = (\|\phi_1\|^2 + \|\phi_2\|^2)^{1/2}$ for the norm of $\phi = (\phi_1, \phi_2)$ we show that the terms $h_i(G\phi_1, G\phi_2)$ in (3.11) are $o(\|\phi\|)$ as $\|\phi\| \to 0$. Since G is a bounded map of $L^2[0,1]$ into $C = C[0,1]$, we have $\|G\phi_1\|_C \leq b\|\phi_1\|$ for some constant $b > 0$. Given $\epsilon > 0$ we choose r so that $\alpha_1^2 + \alpha_2^2 \leq r$ implies (for $i = 1$ or 2) $|h_i(\alpha_1, \alpha_2)| \leq \epsilon/b \cdot \|\alpha\|$. Then if $\|\phi\| \leq r/2b$,

$$\|h_i(G(\phi_1), G(\phi_2))\| \leq \|h_i(G(\phi_1), G(\phi_2))\|_C$$
$$\leq \sup_{|\alpha_i| \leq b\|\phi_i\|} |h(\alpha_1, \alpha_2)|$$
$$\leq \frac{\epsilon}{b} \cdot b\|\phi\|.$$

Similarly one shows that the maps $h_i(G\phi_1, G\phi_2)$ are continuous.

Since the terms $h_i$ are $o(\|\phi\|)$ we see that the derivative of the right hand side of (3.11) is the map

$$\lambda A_0 : \begin{pmatrix} \phi_1 \\ \phi_2 \end{pmatrix} \to \lambda \begin{pmatrix} G\phi_1 \\ G\phi_2 \end{pmatrix}.$$

It is known that the smallest eigenvalue $\lambda_0$ of L corresponds to a nonnegative eigenvector $\psi$ and that no other eigenvector of L is of one sign. Thus the map $A_0$ has $\lambda_0$ as a characteristic value of multiplicity two with eigenvectors $(\psi, 0)$ and $(0, \psi)$, and no other eigenvectors in the cone of nonnegative pairs. The assertion of the theorem now follows from Corollary 3.2.

## REFERENCES

1. R. E. L. Turner, Nonlinear eigenvalue problems with applications to elliptic equations, to appear in Arch. Rat. Mech. Anal.

2. R. Abraham and J. Robbin, Transversal mappings and flows, W. A. Benjamin, Inc., New York, 1967.

3. P. H. Rabinowitz, Some global results for nonlinear eigenvalue problems, to appear in J. Functional Anal.

4. M. A. Krasnoselski, Topological methods in the theory of nonlinear integral equations, MacMillan, New York, 1965.

5. G. T. Whyburn, Analytic topology, Amer. Math. Soc. Colloq. Publ., Vol. 28, Providence, 1942.

6. F. Riesz and B. Sz.-Nagy, Functional analysis, F. Ungar Publ. Co., New York, 1955.

7. J. T. Schwartz, Nonlinear Functional Analysis, lecture notes, New York University, New York, 1965.

8. H. Schaefer, Some spectral properties of positive linear operators, Pac. J. Math. 10 (1960), 1009-1019.


This work was supported by NSF Grant GP-21078.


Department of Mathematics
University of Wisconsin
Madison, Wisconsin


Received May 18, 1971.

# Multiple Eigenvalue Bifurcation for Holomorphic Mappings

*KLAUS KIRCHGÄSSNER*

## §1. Introduction

The existence of nontrivial solutions of the equation

(1')  $$\lambda u - Lu - V(u) = \theta$$

is studied in $H \times \mathbb{R}$, where $H$ is a real Hilbert space, $L$ a bounded, compact, symmetric linear operator acting in $H$, and $V$ is a nonlinear operator which is holomorphic in a neighborhood of $\theta$. Thus, $V$ has locally a representation as a series of continuous homogeneous operators. It is assumed that in this representation the homogeneous operator of the lowest order $V_k$, $k > 1$, is the strong gradient (F-derivative) of a $C^1$-functional $\ell$. $\ell$ is not assumed to be weakly continuous (cf. [18], [19]).

Obviously $(\theta, \lambda)$-$\theta$ denotes the zero element in $H$ - is a solution for all $\lambda \in \mathbb{R}$. The existence of nontrivial solutions of (1) can be guaranteed, if $V$ is completely continuous, for every eigenvalue $\lambda_1 \neq 0$ of odd multiplicity, even if $L$ is not symmetric (see [6], [14] for the local theory and [15] for the global result). There exist continua of nontrivial solutions emanating from $(\theta, \lambda_1)$ which either are unbounded in $H \times \mathbb{R}$ or meet another point $(\theta, \tilde{\lambda}_1)$, $\tilde{\lambda}_1 \neq \lambda_1$, $\tilde{\lambda}_1$ eigenvalue of L. Simple examples show that if $\lambda_1$ is of even multiplicity then $(\theta, \lambda_1)$ may not be a point of bifurcation, i.e. a limit point of nontrivial solution of (1') in $H \times \mathbb{R}$.

If V is the gradient of a sequentially weakly continuous functional, then for every eigenvalue $\lambda_1 \neq 0$, $(\theta, \lambda_1)$ is a point of bifurcation (cf. [13], [14]). If in addition, V is invariant under some finite group, then the set of nontrivial solutions is richer. The number of nontrivial solutions emanating from $(\lambda_1, \theta)$ can be bounded below using the variational theory of Ljusternik-Schnirelmann (cf. [3], [4], [12]).

The holomorphy suggests satisfactory results for the case of a complex space. Indeed, bifurcation has been proved in [23] with rather weak nondegeneracy assumptions for complex solutions of (1').

In [7], Cronin obtained results, for the case when V is a polynomial operator, which are analogous to the fundamental theorem of algebra. However, in order to obtain real solutions, V has to be odd.

Since in applications the main interest is in real solutions the different results presently known suggest the consideration of (1') under the assumptions stated. By its nature (1') is a perturbation problem. Some of the phenomena of linear analytic perturbation theory have analogues in the nonlinear case (cf. the Remarks to Theorem 3). A closely related problem has been studied in [20], [22] and [25]. Apart from the special problem treated in [22] the results obtained do not require a non-degeneracy assumption only for $k = 2$, a case which is not discussed in detail in this paper. During the symposium, where this contribution was presented, the author learned of a still unpublished result of D. Sather [21] which contains Theorem 4 of this paper.

Using Schmidt-Lyapunov's method (1') is transformed into

(1) $\qquad u - T(u) - \mu S(u, \mu) = \theta$

in a finite dimensional real Hilbert space H. T is homogeneous of degree k and continuous and is the gradient of a $C^1$-functional $\ell$ in H. S is bounded and continuous in some neighborhood of $(\theta, 0)$. The set of fixed points $u_0$ of T and the set of critical points $U_0$ of $\ell$ on the unit sphere $S_1$ with $\ell(U_0) = \sigma_0/(k+1) > 0$ have the same cardinality via

$u_0 = U_0 \sigma_0^{-\frac{1}{(k-1)}}$. In general, a fixed point of T cannot be expected to be continuable to a solution of (1) if the corresponding critical point is unstable in the sense of [1]. However, local extrema of $\ell$ on $S_1$ are stable and should be continuable. For $k = 3, 4$ this is proved for local maxima and for general $k \geq 3$, $k \in \mathbb{N}$ for local minima in Theorem 2. The only unnatural assumption is, that the fixed point should be simply degenerate, i.e. the dimension of the kernel of $1 - DT(u_0)$ is at most 1. The results for (1') are stated in Theorem 3. In view of the assumption of simple degeneracy these are satisfactory especially in the case of an eigenvalue $\lambda_1 \neq 0$ of multiplicity two of L in (1') (Theorem 4).

The case $k = 2$ can be treated for arbitrary multiplicity of $\lambda_1$ with the method applied here. The technical details are given in a forthcoming paper.

## 2. Notation

Let H be a real Hilbert space, $T: H \to H$ a continuous homogeneous mapping of degree k, $T(\alpha u) = \alpha^k T(u)$, $k \in \mathbb{N}$, $k > 1$, and $S: H \times \mathbb{R} \to H$ a continuous and bounded mapping with $S(\theta, \mu) = \theta$. $\theta$ denotes the zero element of H. Moreover, we set $B_\delta(u_0) = \{u / u \in H, \|u - u_0\| \leq \delta\}$, $B_\delta = B_\delta(\theta)$.

We are interested in solutions of the equation

(1) $\qquad u - T(u) - \mu S(u, \mu) = 0$.

The trivial solution $(\theta, \mu)$, $\mu \in \mathbb{R}$ satisfy (1). Nontrivial solutions are sought for in a neighborhood of $(u_0, 0)$, where $u_0$ is any nonzero fixed point of T. A further assumption is made for T: T is assumed to be the strong gradient (F-derivative) of the functional $\ell$ defined by:

$$\ell(u) = \frac{1}{k+1}(T(u), u).$$

This assumption is equivalent to the requirement that for all $u \in H$ the F-derivative of T, $DT(u)$, is a symmetric linear operator (cf. [24], p. 56).

As a continuous power, $T$ is bounded uniformly in bounded sets. Moreover, $T$ defines a continuous symmetric k-linear form, its polar-form $T^{(k)} \equiv T^{(k)}(u,\ldots,u) = T(u)$ (cf. [10], pp. 760 ff.). For later use we introduce some additional notations:

$$T^{(j)}(u; v_1, \ldots, v_j) = \binom{k}{j} T^{(k)}(u, \ldots, u, v_1, \ldots, v_j)$$

$$T^{(j)}(u;v) = T^{(j)}(u; v, \ldots, v).$$

Then the following identities are valid:

$$T(u+v) = \sum_{j=0}^{k} T^{(j)}(u;v), \quad T^{(0)} = T$$

(2)
$$T^{(j)}(u; u, v_1, \ldots, v_{j-1}) = \frac{k-j+1}{j} T^{(j-1)}(u; v_1, \ldots, v_{j-1}).$$

$T$ possesses for every $u \in H$ a F-derivative $DT(u)$. More generally we define:

(3) $\qquad DT^{(j)}(u;v) = jT^{(j)}(u; v, \ldots, v, \cdot)\ j = 1, \ldots, k$

with $DT(u) = DT^{(1)}(u;v)$.

Setting

$$\Delta T'(u;v) = \sum_{\nu=2}^{k} \frac{1}{\nu} DT^{(\nu)}(u;v)$$

(4)
$$L(u;v) = 1 - DT(u) - \Delta T'(u;v)$$

one obtains the identity:

(5) $\qquad v - T(u+v) + T(u) = L(u;v)v \quad \text{for all } u, v \in H.$

Since $T$ is a gradient operator, the $k+1$-linear form $\ell^{(k+1)}$ defined by:

$$\ell^{(k+1)}(u_1, \ldots, u_{k+1}) = \frac{1}{k+1} (T^{(k)}(u_1, \ldots, u_k), u_{k+1})$$

is symmetric and continuous. It represents the polar-form of the functional $\ell$. Thus, for every $v \in H$, $L(u;v)$ is a bounded symmetric linear operator.

For later use we list some identities which can be obtained by elementary calculations from (2)-(4). Setting $v = \eta\phi + \xi u_0$ ; $\phi, u_0 \in H$ ; $\eta, \xi \in \mathbb{R}$, we obtain

$$\Delta T'(u_0; \eta\phi + \xi u_0)(\eta\phi + \xi u_0) = \sum_{\nu=1}^{k} F_\nu(\xi)\eta^\nu T^{(\nu)}(u_0;\phi)$$

$$F_\nu(\xi) = \sum_{j=\nu_0}^{k-\nu} \binom{k-\nu}{j} \xi^j ,$$

(6)
$$\Delta T'(u_0; \eta\phi + \xi u_0) = \sum_{\nu=1}^{k} f_\nu(\xi)\eta^{\nu-1} DT^{(\nu)}(u_0;\phi)$$

$$f_\nu(\xi) = \sum_{j=\nu_0}^{k-\nu} \binom{k-\nu}{j} \frac{\xi^j}{\nu+j}, \quad \nu_0 = \max(2-\nu, 0).$$

Moreover $g(x) = 0(h(x))$, $x \in \mathbb{R}^n$, is used to indicate that $|g(x)/h(x)|$ is bounded in some neighborhood of $\theta$.

## 3. Existence of Solutions of Equation (1).

In this section $H$ is assumed to be finite dimensional.

**Lemma 1:** Let $u \in H$ be fixed, $\theta \in A$, $A$ compact and $0$ not in the spectrum $\Sigma L(u;v)$ for all $v \in A \setminus \{\theta\}$. Then there exists in a suitable interval $(0, \delta]$ a positive continuous function $f$ such that for $\lambda_0(v)$, the smallest eigenvalue in modulus,

$$|\lambda_0(v)| \geq f(\|v\|), \quad v \in B_\delta \cap A \setminus \{\theta\}$$

holds.

**Proof:** Denote by $S_q$ the sphere $S_q = \{v \mid \|v\| = q\}$ and consider the form $Q$

$$Q(v,x) = |(L(u;v)x, x)|.$$

Since for every $v$, $L(u;v)$ is a symmetric linear operator

$$\inf_{(S_q \cap A) \times S_1} Q(v,x) = f(q)$$

is the smallest eigenvalue in absolute value for $v \in S_q \cap A$ which in view of the compactness of $(S_q \cap A) \times S_1$ does not vanish for $q \neq 0$ and depends continuously on $q$; q.e.d.

We define the total eigenprojections corresponding to the group of eigenvalues which emanate from 0 when $v$ is chosen in a neighborhood of the origin. To this end the complexification $\mathcal{H}$ of $H$ is introduced which is isometrically isomorphic to $H \times H$. The symmetric linear operator $L(u;v)$ is extended onto $\mathcal{H}$ as a selfadjoint operator in a natural way and will be denoted again by $L$.

The two oriented semicircles $\Gamma^{\pm} \subset \mathbb{C}$ with the parametrization

$$\zeta^{\pm}(t) = \pm i\gamma \begin{cases} -e^{it}, & t \in [0, \pi] \\ (1 - \frac{2}{\pi}(t-\pi)), & t \in [\pi, 2\pi] \end{cases}$$

are chosen to define the projections $P^{\pm}(v)$ by (cf. [11], p. 178):

(7) $$P^{\pm}(v) = -\frac{1}{2\pi i} \int_{\Gamma^{\pm}} (L(u;v) - \zeta\, 1)^{-1} d\zeta$$

for fixed $u \in H$ and all $v \in B_{q_0} = \{v / \|v\| \leq q_0\}$ for which the inverse exists. The circle $\sigma_\gamma = \{z/z \in \mathbb{C}, |z| = \gamma\}$ should separate $\Sigma L(u;0)$ so that no nonzero eigenvalue lies in its interior; $q_0$ is required to be so small that $\text{dist}(\Sigma L(u;v), \sigma_\gamma) \geq \alpha > 0$ holds for all $v \in B_{q_0}$.

The situation of interest in the following analysis is described by Lemma 1. Therefore, $P^{\pm}(v)$ are defined for all $v \in A \setminus \{0\} \cap B_{q_0}$ and are continuous. Since 0 may be an eigenvalue of $L(u;\theta)$, $P^{\pm}(v)$ may not be defined for $v = \theta$.

Since $L(u;v)$ is selfadjoint $P^{\pm}(v)$, if defined, are orthogonal projections which commute with $L(u;v)$ and reduce it; it follows $\|P^{\pm}(v)\| = 1$. The projection $P(v) \equiv P^{+}(v) + P^{-}(v)$ is defined for all $v$ with $\Sigma L(u;v) \cap \sigma_Y = \phi$, even if $0 \in \Sigma L(u;v)$; therefore $Q(v) = 1 - P(v)$ is the complementary projection for all such $v$. Let $\tilde{Q}^{\pm}(v)$ denote the total eigen-projections of $L|_{QH}$ corresponding to all positive (negative) eigenvalues, then $\tilde{Q}(v) = \tilde{Q}^{+}(v) + \tilde{Q}^{-}(v)$.

If the assumptions of Lemma 1 are satisfied we can, for fixed $u$, define the following mapping $M: A \cap B_{q_0} \to H$:

(8')   $\qquad M(v) = 1 - \rho N(v), \qquad \rho > 0,$

where, if $v \neq \theta$

(8")
$$P^{\pm}(v)N(v) = \pm \|v\| P^{\pm}(v) \{L(u;v)v - \mu S(u+v, \mu)\}$$
$$\tilde{Q}^{\pm}(v)N(v) = \pm \quad \tilde{Q}^{\pm}(v)\{L(u;v)v - \mu S(u+v, \mu)\}$$

and if $v = \theta$:

(8'")   $\qquad N(\theta) = -\mu \tilde{Q}^{+}(\theta)S(u,\mu) + \mu \tilde{Q}^{-}(\theta)S(u,\mu).$

Since $P^{\pm}(v)$ is uniformly bounded, $\|v\| P^{\pm}(v)$ is continuous at $v = \theta$ if appropriately defined for $v = \theta$; thus, $M(v)$ is continuous. Let $u_0$ be a fixed point of $T$. Then (5) implies

(9)   $\qquad u_0 + v - T(u_0 + v) = L(u_0;v)v.$

The Euler identity which is a special case of (2) shows that $u_0$ is an eigenelement of $DT(u_0)$ for the eigenvalue $k$, and therefore $\tilde{Q}^{-}(\theta)u_0 = u_0$. Now, if $(S(u_0,\mu), u_0) \neq 0$, then for $u = u_0$ and given $\mu$ every fixed point $\tilde{v}$ of $M$ given via $\tilde{u} = u_0 + \tilde{v}$ is a solution of equation (1). Indeed, $\tilde{v} = \theta$ implies $\mu \tilde{Q}^{-}(\theta)S(u_0,\mu) = 0$, but since $u_0 \in \tilde{Q}^{-}(\theta)H$ this yields $\mu = 0$. Thus, if $\mu \neq 0$ we have $\tilde{v} \neq \theta$ and in view of (8") and (5) $u_0 + \tilde{v}$ is a solution of (1).

**Lemma 2:** Let the assumption of Lemma 1 hold. If $u_0$ is a fixed point of $T$, $\mu \in \mathbb{R}$ such that $(S(u_0, \mu), u_0) \neq 0$ and $\tilde{v} \in A$ is a fixed point of $M$; then $u_0 + \tilde{v}$ solves equation (1).

We shall use Brouwer's theorem to establish the existence of a fixed point of $M$. For this reason $M$ was constructed so that for $\mu = 0$ and $v \neq \theta$ the spectrum of $P^{\pm}(v)L(u;v)$, $Q^{\pm}(v)L(u;v)$ is positive, which for sufficiently small $\rho$ will imply the applicability of this theorem.

**Theorem 1:** Let $T$ and $S$ be mappings satisfying the assumptions of Section 2; $u_0$ be a fixed point of $T$ with $(S(u_0, 0), u_0) \neq 0$, and $0 \notin \Sigma L(u_0; v)$ for $v \neq \theta$ and $\|v\|$ sufficiently small. Then there exists a $\mu_0 > 0$ such that for all $|\mu| < \mu_0$, equation (1) has at least one solution.

**Proof:** We will show for appropriately chosen $\delta_p$ and $\delta_q > 0$ that $M$ maps the closed convex set

$$B_\alpha = \{v / \|Pv\| \leq \alpha \delta_p, \ \|Qv\| \leq \alpha \delta_q, \ \alpha \in (0,1]\}$$

with $P = P^+(v) + P^-(v)$, $Q = Q^+(v) + Q^-(v)$ into itself. We choose $\delta_p, \delta_q$ so small that the projections $P^{\pm}(v)$ are defined in $B_1 \backslash \{\theta\}$. Since $Q^+L - Q^-L$ is positive definite, there exists a constant $\beta > 0$ such that

(10) $$((Q^+(v) - Q^-(v))L(u_0;v)v, v) \geq \beta \|Qv\|^2.$$

Let $C_{1/2} = B_1 \backslash B_{1/2}$ and

$$f_\delta = \min_{v \in C_{1/2}} f(\|v\|)$$

where $f$ is given by Lemma 1, which may be assumed to be strictly increasing in $\|v\|$. Thus, $\delta_p$ and $\delta_q$ can be determined to satisfy

(11) $$\delta_p f_\delta = \beta \delta_q.$$

## MULTIPLE EIGENVALUE BIFURCATION

In view of the assumptions on $S$ in section 2, $N$ and $S$ satisfy

(12)
$$\|PN(v)\| \leq c_1(\|Pv\| + |\mu|)$$
$$\|QN(v)\| \leq c_1(\|Qv\| + |\mu|)$$
$$\|S(u_0+v,\mu)\| \leq c_2$$

for bounded sets in $H \times \mathbb{R}$. Setting

(13)
$$\mu_0 = \frac{f_\delta \delta_p}{8c_2} = \frac{\beta \delta_q}{8c_2}$$
$$\rho \leq \min\left(\frac{\tilde{\delta}}{c_1(\tilde{\delta}+\mu_0)}, \frac{\beta \tilde{\delta}^2}{4c_1^2(\tilde{\delta}+\mu_0)^2}\right)$$

where $\tilde{\delta}$ is either $\delta_p$ or $\delta_q$, it is easily verified that $M$ maps $B_{1/2}$ into $B_1$. Consider $v \in C_{1/2}$, then

$$(PN(v),v) \geq \|v\|(f_\delta \|Pv\|^2 - \mu_0 c_2 \|Pv\|) \geq \|v\| \frac{\delta_p^2}{8} f_\delta$$

$$(QN(v),v) \geq (\beta \|Qv\|^2 - \mu_0 c_2 \|Qv\|) \geq \frac{\beta \delta_q^2}{8} .$$

The inequality (13) for $\rho$ implies that $M$ maps $C_{1/2}$ into $B_1$. Thus $M(B_1) \subseteq B_1$, $B_1$ compact and convex; there exists a fixed point $\tilde{v} \in B_1$ of $M$. Since $S$ is continuous, the assumptions of Lemma 2 are valid for small $|\mu|$ and $A = B_1$. The theorem is proved.

The condition $0 \in \Sigma L(u_0;\theta)$ for $v \neq \theta$ is too strong to obtain significant results. For later applications we formulate a new version of Theorem 1. For this purpose let $0 \in \Sigma L(u_0;\theta)$ and $P_0$ be the corresponding projection, $R_0$ the projection onto the span of $u_0$ and $Q_0 = 1 - P_0 - Q_0$.

Define

(14a)
$$M_0(v) = v - \rho N_0(v)$$

and

(14b) $$\hat{M}_0(v) = v - \rho \hat{N}_0(v)$$

where

(14') 
$$N_0(v) = (-P_0-Q_0-R_0)(L(u_0;v)v - \mu S(u_0+v;\mu))$$
$$\hat{N}_0(v) = (P_0+Q_0-R_0)(L(u_0;v)v - \mu S(u_0+v;\mu))$$

$f_\delta$ is as in Lemma 1, $B_\alpha$ and $C_{1/2}$ as in the proof of Theorem 1, $\delta = (\delta_p^2 + \delta_q^2)^{1/2}$.

<u>Corollary:</u> Let $u_0, T$ and $S$ be as in Theorem 1. Let $C \subset H$ be such that $B_1 \cap C$ is homoeomorphic to the closed unit ball. If for $v \in C_{1/2} \cap C$ every eigenvalue of $L(u_0;v)$ is negative (respectively positive, except a simple eigenvalue near 1-k); if furthermore $M_0(C \cap B_1) \subseteq C$ (resp. $\hat{M}_0(C \cap B_1) \subseteq C$) and $0 < \mu_0 \leq f_\delta \delta_p/8c_2$, then for all $\mu: |\mu| \leq \mu_0$, equation (1) has at least one solution in $B_\delta(u_0) \cap C$.

The proof is analogous to that of Theorem 1.

## 4. Solutions in the Neighborhood of Critical Points of the Functional $\ell$.

In this section the existence of a solution of (1) is proved in the neighborhood of a fixed point $u_0$ which corresponds to a positive relative minimum of $\ell$ on the unit sphere $S_1$. The crucial point is the investigation of the spectrum of $L(u_0;v)$ near the origin. The space $H$ is again assumed to be finite dimensional.

<u>Lemma 3:</u> Let $U_0$ be a positive relative minimum of $\ell$ on $S_1$; $\sigma_0 = \ell(U_0)/(k+1)$, $u_0 = \sigma_0^{-1/(k-1)} U_0$. Introducing the abbreviations

$a_1 = 1$

$a_{2j-1} = \dfrac{(k-1)(k-3)\cdots(k-2j+3)}{(j-1)!\, 2^{j-1}\, \|u_0\|^{2j-2}}$, $j = 2, 3, \ldots, [\dfrac{k+1}{2}]$, $k > 2$,

we have

(1) $\Sigma L(u_0, \theta) \subset \mathbb{R}_0^- = \{x/x \in \mathbb{R},\ x \leq 0\}$ ;

(2) if for some $\phi \in S_1$ and $j_0 \leq [\dfrac{k-1}{2}]$

$(T^{(2j-1)}(u_0;\phi), \phi) = a_{2j-1}, j = 1, \ldots, j_0$ ,

it follows that

(i) $(T^{(2j)}(u_0;\phi), \phi) = 0$ ,  $j = 1, 2, \ldots, j_0$ ,

(ii) $Q_0 T^{(j)}(u_0;\phi) = 0$ ,  $j = 1, 2, \ldots, 2j_0$ ,

(iii) $(T^{(2j_0+1)}(u_0;\phi)\phi) > a_{2j_0+1}$

**Proof:** Since $T$ is homogeneous of degree $k$, $u_0$ is a fixed point of $T$ with $\|u_0\| = \sigma_0^{-1/(k-1)}$. Setting $u = \alpha U_0 + \beta v$, $\alpha^2 + \beta^2 = 1$, $v$ orthogonal to $U_0$, one obtains by means of the minimum property for $U_0$ and (2):

$\ell(U_0) + \sum_{j=2}^{k} \dfrac{x^j}{j} T^{(j-1)}(U_0;v) + x^{k+1} \ell(v) \geq \ell(U_0)(1+x^2)^{\frac{k+1}{2}}$

with $x = \beta/\alpha$. Assertions (1) and (2) (i) and (iii) follow immediately by comparing terms of equal order in $x$, the fact that $L(u_0;\theta)u_0 = (1-k)u_0$ and that $T^{(j)}(u_0;v)$ is homogeneous of degree $k-j$ in $u_0$. Setting $v = \gamma w + \delta\phi$, $\gamma^2 + \delta^2 = 1$, $w \in Q_0 H$ and $\phi$ as assumed in (2), then the terms which are linear in $\gamma$ must vanish for $j = 2, \ldots, 2j_0$, which yields (2), (ii).

The critical point $U_0$ is isolated on $S_1$, and likewise the fixed point $u_0$ in $H$ if and only if strict inequality holds in (2) (iii) for some $j_0$.

**Remark 1:** If in Lemma 3, $U_0$ is a positive maximum of $\ell$ on $S_1$, then

$$\sum L(u_0;\theta) \subset \mathbb{R}_0^+ \cup \{1-k\}; \mathbb{R}_0^+ = \{x/x \in \mathbb{R}, x \geq 0\}.$$

Moreover assertion (2) of Lemma 3 holds if in (iii) the inequality is reversed.

**Remark 2:** If $k$ is even and $U_0$ as in Lemma 3, then $U_0$ is isolated on $S_1$.

**Remark 3:** If $k$ is even and $U_0$ a negative maximum (resp. minimum) of $\ell$ on $S_1$, then $u_0 = U_0 \sigma_0^{-1(k-1)}$ is a fixed point of $T$ and Lemma 3 (resp. Remark 1) holds. However the fixed points corresponding to a positive maximum (minimum) and a negative minimum (maximum) may not be different.

**Remark 4:** If $k$ is odd and $U_0$ is a negative maximum (minimum) of $\ell$ on $S_1$, then $u_0 = U_0(-\sigma_0)^{-1/(k-1)}$ satisfies $u + T(u) = \theta$ and

$$\sum L(u_0;\theta) \subset \mathbb{R}_0^+ \; (\sum L(u_0;\theta) \subset \mathbb{R}_0^- \cup \{1+k\})$$

and the relations of Lemma 3(2) (Remark 1) hold if the $a_k$'s are multiplied by $-1$.

In the following lemma the behaviour of $\sum L(u_0;v)$ is investigated for $v$ in some set $A$ which will later shown to be invariant under $M_0$.

**Lemma 4:** Let be $k > 2$, $U_0$ be a positive relative minimum of $\ell$ on $S_1$ which is isolated if $k = 3$, and $u_0$ be the corresponding fixed point of $T$. Let $j_0$ denote the largest index for which the relations in Lemma 3, (2) hold. Moreover, 0 is assumed to be a simple eigenvalue of $L(u_0;\theta)$.

Set

$$\Lambda_\alpha(\eta) = \frac{\alpha_1}{2}\eta^2 + \frac{\alpha_2}{\|u_0\|^2}\eta^4$$

where

$$1 < \alpha_1 < \frac{k-1}{4\|u_0\|^2}(T^{(3)}(u_0;\phi),\phi)-1, \quad \alpha_2 = 0, \text{ if } j_0 = 1,$$

$$\alpha_1 = 1, \quad 0 < \alpha_2 < \frac{1}{24}\min(2k+2, 3k-9), \text{ if } j_0 > 1.$$

$$C_\alpha = \{v/(v, u_0) \geq -\Lambda_\alpha(\|P_0 v\|)\}$$

$$C_\beta = \{v/\|Q_0 v\| \leq \beta(\|P_0 v\|^{2j_0+1} + |\mu|)\}, \beta > 0$$

$$C = C_\alpha \cap C_\beta .$$

Then there exists a ball $B_\delta$ and a constant $\gamma > 0$, depending only on $\alpha_1, \alpha_2$ and $\beta$ such that for all $\lambda \in \Sigma L(u_0;v)$ and for all $v \in C \cap B_\delta$

$$\lambda \leq -\gamma \|P_0 v\|^{\min(2j_0, 4)} + \gamma |\mu|^2 .$$

**Proof:** Let $\phi$ be the eigenfunction belonging to $0 \in \Sigma L(u_0;\theta)$, $\|\phi\| = 1$, and set $v = \eta\phi + \xi u_0 + Q_0 v$. In view of Lemma 3, (1) all eigenvalues of $L(u_0;\theta)$ are strictly negative except 0. Thus for $v \in B_\delta$, $\delta$ small, only the behaviour of the largest eigenvalue has to be investigated; denote this eigenvalue by $\lambda$ and by $\varphi$ the corresponding eigenfunction. Since $L(u_0;v)|_{(Q_0+R_0)H}$ is continuously invertible, $P_0\varphi = \theta$ would imply $Q_0\varphi = R_0\varphi = \theta$. Thus, $P_0\varphi \neq \theta$ and $\varphi$ can be normalized so that $P_0\varphi = \phi$. We set $\varphi = \phi + Q_0\varphi + \zeta u_0$. According to (4), $\varphi$ must satisfy

(15) $$\lambda\varphi = \varphi - DT(u_0)\varphi - \sum_{\nu=2}^{k} T^{(\nu)}(u_0;v,\ldots,v,\varphi) .$$

Lemma 3, (2) (ii) shows that all terms of the form $Q_0 T^{(\nu)}(u_0;\phi)$ vanish for $\nu = 1, 2, \ldots, 2j_0$. Since by (2)

$$T^{(\nu)}(u_0; v, \ldots, v, R_0 \varphi) = \frac{k-\nu+1}{\nu \|u_0\|} (\varphi, \frac{u_0}{\|u_0\|}) T^{(\nu-1)}(u_0; v)$$

holds, one obtains by applying $Q_0$ to (15), that all terms either contain $Q_0 \varphi$ or $Q_0 v$ or are of order $0(|\eta|^{2j_0+1})$ at least. This yields:

$$\|Q_0 \varphi\| \le c_1 (\|Q_0 v\| + |\eta|^{2j_0+1}).$$

Let us estimate the error which is made if $Q_0 v$ and $Q_0 \varphi$ are neglected in (15). Again, Lemma 3, (2) (ii) gives:

$$P_0 T^{(\nu)}(u_0; \phi, \ldots, \phi, Q_0 v) = P_0 T^{(\nu)}(u_0; \phi, \ldots, \phi, Q_0 \varphi) = \theta,$$

$$\nu = 1, 2, \ldots, 2j_0.$$

Therefore, if $P_0$ or $R_0$ are applied to (15), neglecting $Q_0 v$ and $Q_0 \varphi$ causes an error of the order $0(\|Q_0 v\|^2)$. With the abbreviations

$$A_0(\eta, \xi) = \sum_{\nu=2}^{k} \nu f_\nu(\xi) \eta^{\nu-1} (T^{(\nu)}(u_0; \phi), \phi)$$

$$A_1(\eta, \xi) = \sum_{\nu=2}^{k} (k-\nu+1) f_\nu(\xi) \eta^{\nu-1} (T^{(\nu-1)}(u_0; \phi), \phi)$$

$$A_2(\eta, \xi) = \sum_{\nu=3}^{k} \frac{(k-\nu+1)(k-\nu+2)}{\nu-1} f_\nu(\xi) \eta^{\nu-1} (T^{(\nu-2)}(u_0; \phi), \phi)$$

where the functions $f_\nu$ are given by (6), one obtains by applying $P_0$ and $R_0$ to (15) and eliminating $\zeta$:

(16) $\quad -\lambda = f_1(\xi) + A_0(\eta, \xi) - \dfrac{A_1^2(\eta, \xi)}{\|u_0\|^2 ((k-1) + \lambda + f_1(\xi)) + A_2(\eta, \xi))}$

$$+ 0(\|Q_0 v\|^2).$$

Equation (16) determines $\lambda$ implicitly as a function of $\eta, \xi$ and $Q_0 v$. Using (6), elementary calculations yield:

$$-\lambda = \frac{k-1}{2}\xi + \eta^2\{(T^{(3)}(u_0;\phi),\phi) - \frac{k-1}{4\|u_0\|^2}\}$$

$$+\eta^4\{(T^{(5)}(u_0;\phi),\phi) - \frac{k-3}{4\|u_0\|^2}(T^{(3)}(u_0;\phi),\phi)$$

$$+\frac{(k-1)(k-2)}{24\|u_0\|^4} - \frac{1}{4\|u_0\|^2}[(T^{(3)}(u_0;\phi),\phi) - \frac{k-1}{4\|u_0\|^2}]\}$$

$$+ \eta^2\xi\{\frac{3(k-3)}{4}(T^{(3)}(u_0;\phi),\phi) - \frac{(k-1)(k-2)}{3\|u_0\|^2}\}$$

$$+ \xi^2 \frac{(k-1)(k-2)}{6} + 0((|\xi|+|\eta|^2)^3 + \|Q_0 v\|^2) .$$

The coefficient of $\eta^2$ is positive. Thus, if $\xi \geq 0$, a $\delta_1 > 0$ can be found such that for $v \in C_\beta \cap B_{\delta_1}$ the inequality holds. Therefore, only the case $\xi \leq 0$ has to be considered.

Case $j_0 = 1$. Let $k = 3$; since $u_0$ is isolated, we have strict inequality in Lemma 3, (2) (iii) which implies that the coefficient of $\eta^2$ is strictly greater than $(k-1)/4\|u_0\|^2$. If $k = 4$ this follows from the second remark to Lemma 3, and if $k \geq 5$ it is implied by the maximum property of $j_0$. Hence we have:

$$-\lambda \geq \frac{k-1}{2}(\xi + \frac{\alpha_1\eta^2}{2\|u_0\|^2}) + \eta^2\{(T^{(3)}(u_0;\phi),\phi) - \frac{k-1}{4\|u_0\|^2} - \frac{(k-1)\alpha_1}{4\|u_0\|^2}\}$$

$$+ 0((|\xi|+\eta^2)^2 + \|Q_0 v\|^2) .$$

Now, a positive $\delta_2(\alpha_1)$ can be found such that the inequality asserted holds for $v \in B_\delta \cap C_\beta, \delta = \min(\delta_1, \delta_2)$.

Case $j_0 > 1$. The coefficient of $\eta^2$ equals $(k-1)/4 \|u_0\|^2$ and $\alpha_1 = 1$. Hence, higher order terms have to be considered. We can restrict ourselves to the set $\xi \leq 0$. If $\xi \geq -\eta^2/(2\sqrt{2}\|u_0\|^2)$ we obtain:

$$-\lambda \geq \frac{k-1}{4\|u_0\|^2}\eta^2(1-\frac{1}{\sqrt{2}})\eta^2 + 0(\eta^4 + \|Q_0 v\|^2).$$

Therefore, a $\delta_3 > 0$ exists such that for $v \in B_\delta \cap C$ the inequality holds.

Thus, only $\xi \leq -\eta^2/(2\sqrt{2}\|u_0\|^2)$ is left to be considered. But in this case one derives easily from (17):

$$-\lambda \geq \frac{k-1}{2}\xi + \frac{k-1}{4\|u_0\|^2}\eta^2 + \frac{a(k-1)}{48\|u_0\|^4}\eta^4 + 0(\eta^6 + \|Q_0 v\|^2),$$

where

$$a = \begin{cases} 3k-9, & \text{if } k \leq 11 \\ 2k+2, & \text{if } k \geq 11. \end{cases}$$

Thus, $\xi \geq -\Lambda_\alpha(\eta)/\|u_0\|^2$ implies:

$$-\lambda \geq \frac{k-1}{2}(-\alpha_2 + \frac{a}{24})\frac{\eta^4}{\|u_0\|^4} + 0(\eta^6 + \|Q_0 v\|^2).$$

Since $\alpha_2 < a/24$ we have for a suitably chosen $\delta_4(\alpha_2)$ for all $v \in B_{\delta_4(\alpha)} \cap C_\beta$ the inequality asserted. The lemma is proved.

<u>Corollary:</u> Let $k > 2$ and $U_0$ be a positive relative maximum of $\ell$ on $S_1$ which is isolated if $k=3$, and $u_0$ be the corresponding fixed point of $T$. Let $j_0$ denote the largest index for which the relations in Lemma 3, (2) hold and let 0 be a simple eigenvalue of $L(u_0;\theta)$.

Set

$$\Lambda_\alpha(\eta) = \frac{\alpha_1}{2}\eta^2 + \frac{\alpha_2}{\|u_0\|^2}\eta^2$$

where

$$1 > \alpha_1 > \frac{4\|u_0\|^2}{k-1}(T^{(3)}(u_0;\phi),\phi) - 1, \quad \alpha_2 = 0, \quad \text{if } j_0 = 1,$$

$$\alpha_1 = 1, \quad \alpha_2 > \frac{1}{24}\max\{(3k-9) - \frac{1}{\sqrt{2}}(k-11), 3k-9\}, \quad \text{if } j_0 > 1,$$

$$\hat{C}_\alpha = \{v/(v,u_0) \leq -\Lambda_\alpha(\|P_0 v\|)\}$$

$$\hat{C} = \hat{C}_\alpha \cap C_\beta,$$

with $C_\beta$ and $B_\delta$ as in Lemma 4. Then there exists a $\hat{\gamma} > 0$ such that for all $\lambda \in \Sigma L(u_0;v)$, except one simple eigenvalue near $1-k$, we have:

$$\lambda \geq \hat{\gamma} \|P_0 v\|^{\min(2j_0, 4)} - \hat{\gamma}|\mu|^2.$$

The proof is analogous to the proof of Lemma 4 with obvious changes. ($1-k$ is a simple eigenvalue of $L(u_0;\theta)$ with eigenfunction $u_0$.)

The next Lemma shows that the set $C$ is mapped into itself by $M_0$ (see (14)) if $\|v\|$ is sufficiently small and the projection of $S$ onto $u_0$ does not vanish.

<u>Lemma 5:</u> Let the assumptions of Lemma 4 be satisfied and the mapping $M_0$ be given by (14). Moreover, let $\gamma_0 = (S(u_0, 0), u_0) \neq 0$. Choose $\alpha_2$ so that $3/16 < \alpha_2 < 1/4$. Then, there exist positive constants $\mu_0, \delta$ such that for the set $A = C \cap B_\delta$, $M_0(A) \subseteq C$ holds for all $\mu \in [-\mu_0, 0]$ if $\gamma_0 > 0$ and for all $\mu \in [0, \mu_0]$ if $\gamma_0 < 0$.

<u>Proof:</u> First we prove the assertion for $C_\beta$. Setting $v' = M_0(v)$ and $v = \eta\phi + \zeta u_0 + Q_0 v$ we have:

$$Q_0v' = Q_0v + \rho Q_0\{L(u_0;v)v - S(u_0+v,\mu)\}.$$

Observe that in view of Lemma 3, (2) (ii) and equation (2), all terms $Q_0 T^{(\nu)}(u_0;v)$ vanish if $v \in (P_0 + R_0)H$ and $\nu \leq 2j_0$. Since the eigenvalues of $Q_0 L(u_0;v)$ are negative for small $\|v\|$, the mapping $Q_0 M_0$ is locally contractive in $Q_0 v$ for fixed $P_0 v + R_0 v$, $\rho$ sufficiently small, and we obtain:

$$\|Q_0 v'\| \leq q\|Q_0 v\| + \rho c \{|\eta|^{2j_0+1} + |\mu|\}$$

for some constants $q \in (0,1)$ and $c > 0$. But this immediately implies $M_0(C_\beta) \subseteq C_\beta$ if $\rho c \leq (1-q)\beta$.

Consider now $C_\alpha$. Setting $v' = \eta'\phi + \xi'u_0 + Q_0 v'$ one obtains from (6) and (14):

$$\eta' = \eta - \rho\{F_1(\xi)\eta + \sum_{\nu=2}^{k} F_\nu(\xi)\eta^\nu (T^{(\nu)}(u_0;\phi),\phi)$$

(18)
$$+ \mu(S(u_0+v,\mu),\phi) + 0(\|Q_0 v\|^2)\}$$

$$\xi' = \xi - \rho\{(k-1)\xi + \frac{1}{\|u_0\|^2} \sum_{\nu=2}^{k} \frac{k-\nu+1}{\nu} F_\nu(\xi)\eta^\nu (T^{(\nu-1)}(u_0;\phi),\phi)$$

$$+ \sum_{\nu=2}^{k} \binom{k}{\nu}\xi^\nu + \frac{\mu}{\|u_0\|^2}(S(u_0+v,\mu),u_0) + 0(\|Q_0 v\|^2)\}.$$

Case $j_0 = 1$. Considering the lowest order terms in (18) yields:

$$\xi' + \frac{\alpha_1 \eta'^2}{2\|u_0\|^2} = (1-\rho(k-1))(\xi + \frac{\alpha_1 \eta^2}{2\|u_0\|^2}) + \rho(k-1)(\alpha_1-1)\frac{\eta^2}{2\|u_0\|^2}$$

$$-\rho\frac{\mu}{\|u_0\|^2}(S(u_0+v),u_0) + \rho 0((\eta^2+|\mu|)((\eta^2+|\xi|)+\|Q_0 v\|^2+\mu^2)).$$

Choose $\rho > 0$ so small that $\rho(k-1) < 1$. Let us suppose that $\gamma_0 = (S(u_0,0), u_0) < 0$; the case $\gamma_0 > 0$ is handled analogously. Then $\delta_1$ and $\mu_0$ (independent of $\delta_1$) can be selected such that for arbitrary $\mu \in [0, \mu_0]$ the term $-\gamma_0 \mu / \|u_0\|^2$ dominates all other terms containing $\mu$ and

$$\xi' + \frac{\alpha_1 \eta'^2}{2\|u_0\|^2} \geq (1-\rho(k-1))(\xi + \frac{\alpha_1 \eta^2}{2\|u_0\|^2}) + \rho(k-1)(\alpha_1-1)\frac{\eta^2}{2\|u_0\|^2}$$

$$+ \rho 0(\eta^4 + \xi^2) \geq 0$$

holds. Thus, if $v \in C_\alpha, v' \in C_\alpha$.

Case $j_0 > 1$. Here all terms up to the order $(\eta^2 + |\xi|)^2$ have to be considered. Since $(T^{(3)}(u_0; \phi), \phi) = (k-1)/2 \|u_0\|^2$, equations (18) and (6) yield:

$$G(\xi', \eta') = \xi' + \frac{\eta'^2}{2\|u_0\|^2} + \frac{\alpha_2 \eta'^4}{\|u_0\|^4} = (\xi + \frac{\eta^2}{2\|u_0\|^2} + \frac{\alpha_2 \eta^4}{\|u_0\|^4})(1-\rho(k-1))$$

(19)

$$+ \rho(k-1)(\alpha_2 - \frac{k+1}{8})\frac{\eta^4}{\|u_0\|^4} - \rho(\frac{k}{2})\xi\frac{\eta^2}{\|u_0\|^2} - \rho(\frac{k}{2})\xi^2 - \rho\frac{\mu}{\|u_0\|^2}(S(u_0+v,\mu), u_0)$$

$$+ 0((|\xi|+\eta^2)^3 + (|\xi|+\eta^2)|\mu| + |\mu|^2 + \|Q_0 v\|^2).$$

Again assuming $\gamma_0 < 0$, $\delta_2$ and $\mu_0$ (independent of of each other) can be chosen such that all terms containing $\mu$ are dominated by $-\rho\mu\gamma_0 / \|u_0\|^2$ for $v \in B_{\delta_2}$.

If $\xi \geq 0$ choose $0 < q < 1-\rho(k-1)$ and select $\delta_3(q)$ so that for $v \in B_{\delta_3(q)} \cap C_\alpha$

$$G(\xi', \eta') \geq q\, G(\xi, \eta) \geq 0$$

holds.

If $0 \geq \xi \geq -(1-\frac{1}{4k})\eta^2/2\|u_0\|^2$ we can find a $\delta_4(q)$ for which

$$G(\xi',\eta') \geq \frac{q}{8k\|u_0\|^2}\eta^2 \geq 0$$

holds for $v \in B_{\delta_4}(q)$.

If $\xi \leq -(1-\frac{1}{4k})\eta^2/2\|u_0\|^2$ one obtains from (19):

$$G(\xi',\eta') \geq (1-\rho(k-1))G(\xi,\eta) + \frac{\rho(k-1)\eta^4}{\|u_0\|^4}(\alpha_2 - \frac{3}{16}) + \rho0(\eta^6).$$

According to Lemma 4, $\alpha_2$ is an arbitrary positive number less than $\min(2k+2, 3k-9)/24$. But, since $j_0 > 1$ we have $k \geq 5$; and thus, if $3/16 < \alpha_2 < 1/4$, the coefficient of $\eta^4$ is strictly positive. Hence, there is a $\delta_5$ such that for $v \in B_{\delta_5}$, $G(\xi',\eta') \geq 0$. Choose $\delta = \min(\delta_2, \delta_3(q), \delta_4(q), \delta_5)$ then $v \in B_\delta \cap C_\alpha$ implies $v' = M_0(v) \in C_\alpha$. q.e.d.

<u>Corollary:</u> Under the hypotheses of the Corollary of Lemma 4 with $k = 3$ or $4$, there exists a ball $B_\delta$ and a constant $\mu_0 > 0$ and independent of $\delta$ such that $\hat{M}_0$ defined in (14) maps $\hat{C} \cap B_\delta$ into $\hat{C}$ for all $\mu \in [0, \mu_0]$ if $(S(u_0,0), u_0) > 0$, and for all $\mu \in [-\mu_0, 0]$ if $(S(u_0,0), u_0) < 0$.

The proof is literally the same as the first part of the proof of Lemma 5, since only the case $j_0 = 1$ has to be considered.

For $k \geq 5$ we cannot exclude $j_0 > 1$. But then, in view of the lower bound for $\alpha_2$ in the Corollary of Lemma 4, $\hat{C}_\alpha$ is not invariant under $\hat{M}_0$.

<u>Definition:</u> Let $U_0$ be a critical point of $\ell$ on $S_1$ with $T(U_0) = \sigma_0 U_0$. $U_0$ is called simply degenerate if $\dim N(\sigma_0 1 - DT(U_0)) \leq 1$ holds. A fixed point $u_0$ of $T$ is called simply degenerate if $\dim N(1 - DT(u_0)) \leq 1$.

Theorem 2: Assume that T and S satisfy the assumptions of section (2).

(i) If $k = 3$ or $4$, $U_0$ is an isolated, simply degenerate positive maximum, $u_0$ denotes the corresponding fixed point of T, and $\gamma_0 = (S(u_0,0), u_0) > 0$ (resp. $< 0$), then there exists a $\mu_0 > 0$ such that for all $\mu \in [0, \mu_0]$ (resp. $\mu \in [-\mu_0, 0]$) equation (1) has a solution $\tilde{u}(\mu) \in B_{\delta(\mu)}(u_0)$ where $\delta(\mu) \to 0$ as $\mu \to 0$.

(ii) If $k \geq 3$, $U_0$ is a simply degenerate positive minimum of $\ell$ on $S_1$ which is isolated for $k = 3$, $u_0$ denotes the corresponding fixed point, and $\gamma_0 = (S(u_0, 0), u_0) > 0$ (resp. $< 0$), then there exists a $\mu_0 > 0$ such that for all $\mu \in [-\mu_0, 0]$ (resp. $\mu \in [0, \mu_0]$) equation (1) has a solution $\tilde{u}(\mu) \in B_{\delta(\mu)}(u_0)$, where $\delta(\mu) \to 0$ as $\mu \to 0$.

Proof: If $0 \not\in \Sigma L(u_0; \theta)$ then the assertions follow from the implicit function theorem. If 0 is a simple eigenvalue of $L(u_0; \theta)$ the Lemmas 4 and 5 and their Corollaries can be applied.

(i) According to the Corollary of Lemma 4, $f_\delta$ (see Corollary of Theorem 1) can be estimated below by:

$$f_\delta \geq \hat{\gamma} \left( \frac{\delta_p^2}{4} - \mu^2 \right).$$

Moreover, choose $\mu_0$ and $\delta$ as in the Corollary of Lemma 5. Then, if $\hat{M}_0$ is defined by (14) we have $M_0(\hat{C} \cap B_\delta) \subseteq \hat{C}$ for $\delta$ sufficiently small. But if in addition

$$\mu_0 \leq \frac{\hat{\gamma} \delta_p}{8c_2} \left( \frac{\delta_p^2}{4} - \mu_0^2 \right)$$

where $\delta^2 = \delta_p^2 + \delta_q^2$, the Corollary of Theorem 1 can be applied since $\hat{C} \cap B_\delta$ is homoeomorphic to the closed unit ball, yielding a solution $\tilde{u}(\mu)$ of (1), where the sign of $\mu$ is restricted due to the Corollary of Lemma 5. It follows from the above inequality for $\mu_0$:

$$\lim_{\mu \to 0} \tilde{u}(\mu) = u_0$$

(ii) For $k = 3$ the proof is analogous to (i). If $k = 4$, $U_0$ is isolated by Corollary 2 of Lemma 3 and (i) can be applied again. If however, $k \geq 5$, we have by Lemma 4:

$$f_\delta \geq \gamma \left\{ \left(\frac{\delta_p}{2}\right)^{\min(2j_0, 4)} - \mu^2 \right\}.$$

if $\mu_0$ satisfies

$$\mu_0 \leq \frac{\gamma \delta_p}{8 c_2} \left\{ \left(\frac{\delta_p}{2}\right)^{\min(2j_0, 4)} - \mu_0^2 \right\},$$

the Corollary of Theorem 1 can be applied yielding the existence of a solution $\tilde{u}(\mu)$ of (1) with the required properties.
q.e.d.

The case $k = 2$ can be studied with the same method even if the fixed point has a higher order degeneracy. The technical details will be given in a forthcoming paper.

The asymmetry of the results for maxima and minima of $\ell$ on $S_1$ is somewhat surprising and is very likely due to the method applied here. However, if $H$ is 2-dimensional more can be said about the case of a positive maximum.

Let $H$ be 2-dimensional. In our previous terminology only the case $P_0 + R_0 = 1$ is of interest. Since $H$ is isomorphic to $\mathbb{R}^2$ we may assume $U_0 = u_0 = (1, 0)$, $\phi = (0, 1)$. If $U_0$ is a nonisolated positive maximum or minimum of $\ell$, which implies that $k$ is odd, then $\ell$ is constant on $S_1$ having the form:

$$\ell(x, y) = \frac{1}{k+1}(x^2+y^2)^{\frac{k+1}{2}}, \quad (x, y) \in \mathbb{R}^2.$$

It is easily checked that $u_0$ can be continued to a solution of equation (1) if $(S(u_0, 0), u_0) \neq 0$. If $U_0$ is isolated, then $\ell$ is of the form given above up to an error term $0(y^{2j_0+2})$; $j_0$ is given by Lemma 3, (2) (iii), where strict inequality holds and equation (1) can be written as follows:

$$x = x(x^2 + y^2)^{\frac{k-1}{2}} + g_1(x,y) + \mu S_1$$

$$y = y(x^2 + y^2)^{\frac{k-1}{2}} + g_2(x,y) + \mu S_2$$

where

$$|g_1(x,y)| = 0(y^{2j_0+2})$$

$$g_2(x,y) = cx^{k-2j_0-1} y^{2j_0+1} + 0(y^{2j_0+2}), \quad c \neq 0,$$

$$S(u_0 + v, \mu) = (S_1, S_2).$$

From the first equation one obtains $x$ as a function of $y$ and $\mu$, $x = F(y, \mu) = G(y) + \mu \tilde{G}(y, \mu)$ where $G$ is holomorphic in $y$, $\tilde{G}$ continuous. The above system is solvable if and only if

$$xg_2(x,y) + \mu x S_2 - yg_1(x,y) - \mu y S_1 = cy^{2j_0+1} + \mu S_2$$
$$+ 0(y^{2j_0+2} + |\mu||y|) = 0$$

is solvable in $y$ near $(1, 0)$ with $x = F(y, \mu)$. But, since $c \neq 0$, solvability follows from the implicit function theorem for all $\mu$ with $|\mu|$ sufficiently small.

<u>Corollary:</u> Let $H$ be 2-dimensional, $k \geq 3$, $U_0$ a positive maximum or minimum of $\ell$ on $S_1$, and $u_0$ be the corresponding fixed point of $T$. If $Y_0 = (S(u_0, 0), u_0) \neq 0$ then there exists a constant $\mu_0 > 0$ such that (1) has a solution $\tilde{u}(\mu) \in B_{\delta(\mu)}(u_0)$ for $|\mu| \leq \mu_0$.

## 5. Multiple Eigenvalues and Bifurcation.

In this section the results of the preceding paragraph are applied to prove the existence of bifurcating solutions. The space $H$ is now an arbitrary real Hilbert space.

Let $L: H \to H$ be a linear, compact and symmetric operator and $V: H \to H$ holomorphic in a neighborhood of $\theta$. Then $V$ locally has a power series representation (cf. [10], pp. 760 ff.) which in a neighborhood of the origin is assumed to have the form:

$$(20) \qquad V(w) = \sum_{\nu=k}^{\infty} V_\nu(w), \quad k \in \mathbb{N}, \; k > 1,$$

where $V_\nu$ is homogeneous of degree $\nu$ and continuous. The following problem is considered in $H \times \mathbb{R}$:

$$(21) \qquad (\lambda 1 - L)w - V(w) = 0.$$

For $\lambda \neq 0$ the trivial solution $(0, \lambda)$ can be a limit point of nontrivial solutions only if $\lambda$ is an eigenvalue of $L$ (cf. [14], pp. 191). In this case $(\theta, \lambda)$ is said to be a bifurcation point. If $V$ is the strong gradient of a weakly continuous functional then every nonzero eigenvalue is a bifurcation point. Lower bounds for the number of solutions emanating from such a point have been derived in ([13], [24], [3]).

We consider the case where only $V_k$ is supposed to be the strong gradient of a continuous functional $\ell$; the higher order perturbation terms are arbitrary within the regularity assumptions.

In the following Lemma the problem is reduced to a finite dimensional problem of the form (1). The number of solutions of (1) gives for some special cases the exact number of bifurcation solutions of (21) in a neighborhood of $(0, \lambda_1)$ where $\lambda_1$ is an eigenvalue of $L$ (cf. [9] and [20]).

**Lemma 7:** Let $\lambda_1 \neq 0$ be an eigenvalue of $L$ and $\mu = (\lambda - \lambda_1)^{1/(k-1)} \in \mathbb{R}$ - which implies $\lambda > \lambda_1$ if $k$ is odd.

Denote by $\Pi$ the orthogonal eigenprojection of $\lambda_1$, then there are operators $T:\Pi H \to \Pi H$ and $S: \Pi H \times \mathbb{R} \to \Pi H$ with the properties described in Section 2 with $T(u) = \Pi T(\Pi u)$ such that for every solution $\tilde{u}$ of $u - T(u) - \mu S(u,\mu) = 0$: $\mu \tilde{u}$ solves equation (21).

Proof:    Set $Y := 1 - \Pi$, then the equation

(22)        $Y(\lambda 1 - L)w - YV(\Pi w + Yw) = 0$

can be solved by the standard implicit-function theorem (cf. [8], pp. 265 ff.) and one obtains $Yw$ as a holomorphic function in $\Pi w$ and $\lambda$ and it is easily seen that $\|Yw\| \leq c\|\Pi w\|^k$. The definition of $\mu$ then yields:

$$Yw = \mu^k G(\Pi w, \mu),$$

where, for later use, the equation for $G(\Pi w, 0) \equiv G_0(\Pi w)$ will be given explicitly:

$$(\lambda_1 1 - L) Y G_0(\Pi w) = Y V_k(\Pi w).$$

Setting $\mu u = \Pi w$ and applying $\Pi$ to (21) yields:

(23)    $u - \Pi V_k(u) - \mu \Pi V_{k+1}(u) - \mu^2 \Pi V_{k+2}(u)$

$\qquad - \mu^{2k-1} \Pi DV_k(u) G_0(u) + O(|\mu|^3 \|u\|^{k+3}) = \theta.$

Identifying $\Pi V_k$ with $T$ and the remainder term with $S$ shows that for $\mu \neq 0$ (23) is an equation of the form (1) with the required properties of $T$ and $S$. Since a solution of (23) yields via (22) a solution of (21), the Lemma is proven.

Corollary:    Let $k$ be odd. Letting $\mu = (\lambda_1 - \lambda)^{\frac{1}{(k-1)}}$, $\lambda < \lambda_1$, and $T$ and $S$ as in the proof of Lemma 7, then for every solution $\tilde{u}$ of $u + T(u) + \mu S(u,\mu) = \theta$: $\mu \tilde{u}$ solves (21).

Observe that, if $u_0 \in \Pi H$ is a fixed point of $\Pi V_k$, we have:

$$(\Pi DV_k(u_0)G_0(u_0), u_0) = k(u_0, G_0(u_0)) = \theta .$$

Thus, the condition $(S(u_0;0), u_0) \neq 0$ which is vital for the application of Theorem 2, can be verified from $\Pi V_{k+1}$ or, if $\Pi V_{k+1}$ vanishes, from $\Pi V_{k+2}$, without the need to invert (22).

The implications of Theorem 2 and its Corollary for the bifurcation of the solution set of equation (21) are obvious now. It should be pointed out that generalizations are easy to obtain: V may depend on $\lambda$ continuously and the holomorphy of V could be replaced by continuity of $V_k$ and a homogeneousness assumption for the remainder. However, the conditions of Theorem 3 would be less clearcut and harder to verify for an actual problem.

Denote by $\tilde{\ell}$ the functional

$$\tilde{\ell}(u) = \frac{1}{k+1}(V_k(u), u), \quad u \in \Pi H ,$$

and suppose for convenience that for every critical point $U_0$ of $\tilde{\ell}$ on $\tilde{S}_1 \subset \Pi H$, $\gamma_0 = (\Pi V_{k+1}(U_0), U_0) \geq 0$ [†].

**Theorem 3:** Let L be a symmetric, compact linear operator on H; let V be an operator defined and holomorphic in a neighborhood of $\theta$ with the representation (20), $V_k$ be the strong gradient of a functional $\ell$ and $\lambda_1 \neq 0$ be an eigenvalue of L: then there exist constants $\sigma$, $\delta > 0$ such that:

**(1) k = 3:**

the number of nontrivial solutions of (21) for $\sigma < \lambda < \lambda_1$ with $\|w\| \leq \delta(\lambda_1 - \lambda)^{1/(k-1)}$ equals at least the number of isolated, simply degenerate, local negative maxima and positive minima; for $\lambda_1 < \lambda < \sigma$, $\|w\| \leq \delta(\lambda - \lambda_1)^{1/(k-1)}$ the number of nontrivial solutions of (21) is bounded below by the number

---

[†] If $\gamma_0 \leq 0$, all inequalities in Theorem 3 are reversed.

of isolated, simply degenerate, local negative minima and positive maxima;

(2) $k = 4$:

assertion (1) holds, without the restriction to isolated positive minima or negative maxima. The solutions corresponding to positive maxima (minima) and negative minima (maxima) may coincide.

(3) $k \geq 5$:

the number of nontrivial solutions of (21) for $\sigma < \lambda < \lambda_1$ and $\|w\| \leq \delta(\lambda_1 - \lambda)^{1/(k-1)}$ equals at least the number of isolated, simply degenerate, local negative maxima and positive minima. If $k$ is even, the critical points need not be isolated. The solutions corresponding to positive maxima (minima) and negative minima (maxima) may coincide.

The proof is an immediate consequence of Theorem 2, Lemma 7 and its Remark and the Remarks 3 and 4 of Lemma 3.

The restriction to maxima and minima of $\tilde{\ell}$ on $\tilde{S}_1$ is natural, since other critical points may be unstable (cf. [1]), and a continuation need not to exist; if $\gamma_0 = 0$ one easily constructs examples even for $k = 3$ such that the assertion of Theorem 3 is false. This corresponds to a well known phenomena for linear analytic perturbation problems for a selfadjoint operator family of an eigenvalue of multiplicity greater than one (cf. [16] and [17], pp. 35 ff.).

In view of the Corollary of Theorem 2 more can be said if $\lambda_1 \neq 0$ is an eigenvalue of multiplicity two of $L$. Instead of stating the most general result, Theorem 4 is formulated for a special case, which is of importance in applications. (see also [21]).

Theorem 4: Let L and V be as in Theorem 3; $\lambda_1 \neq 0$ be an eigenvalue of L of multiplicity 2.

(1) If k is odd, $(V_k(u), u) > 0$ if $u \neq 0$, $V_{k+1}(u) = \theta$ for all $u \in H$, $(V_{k+2}(u), u) > 0$ for $u \neq \theta$; then there exist constants $\delta, \sigma > 0$ such that for every $\lambda \in [\lambda_1 - \sigma, \lambda_1 + \sigma]$ there exist at least two nontrivial solutions $w_1, w_2$ of (21) with $\|w_j\| \leq \delta |\lambda - \lambda_1|^{1/(k-1)}$, $j = 1, 2$.

(2) If k is even, $\tilde{\ell} \neq 0$ on $\Pi H$ and $(V_{k+1}(u), u) > 0$ for $u \neq 0$; then there exist constants $\delta, \sigma > 0$ such that for every $\lambda \in [\lambda_1 - \sigma, \lambda_1 + \sigma]$ there exists at least one nontrivial solution w of (21) with $\|w\| \leq \delta |\lambda - \lambda_1|^{1/(k-1)}$.

The proof is a consequence of the Corollary of Theorem 2, Lemma 7 and Remark 3 of Lemma 3. The distinction between k even and odd is due to the fact, that $\tilde{\ell}$ changes sign on $\tilde{S}_1$ if k is even.

For the case k=2 more general results hold which are obtainable by the method applied here. No restriction to a double eigenvalue is necessary. The technical details will be given in a forthcoming paper.

## REFERENCES

1. V. I. Arnol'd, Singularities of Smooth Mappings, Russ. Math. Surv. **23** (1968), 1-43. McMillan LTD. London 1968.

2. R. Bartle, Singular Points of Functional Equations, Trans. Am. Math. Soc. **75** (1953), 366-384.

3. M. Berger, On Nonlinear Perturbations of the Eigenvalues of a Compact Selfadjoint Operator, Bull. Am. Math. Soc. **73** (1967), 704-708.

4. F. E. Browder, Nonlinear Eigenvalue Problems and Group Invariance, in "Functional Analysis and Related Fields", Ed. F. E. Browder, Springer-Verlag, Berlin 1970.

5. J. Cronin, Some Mappings with Topological Degree Zero, Proc. Am. Math. Soc. 7 (1956), 1139-1145.

6. J. Cronin, Fixed Points and Topological Degree in Nonlinear Analysis, Am. Math. Soc., Providence, R. I., 1964.

7. J. Cronin, Upper and Lower Bounds for the Number of Solutions of Nonlinear Equations, in "Nonlinear Functional Analysis", Proc. Symp. Pure Math. Vol. XVIII, 1, Ed. F. E. Browder, Am. Math. Soc., Providence, R. I., 1970.

8. J. Dieudonné, Foundations of Modern Analysis, Academic Press, New York, 1960.

9. L. M. Graves, Remarks on Singular Points of Functional Equations, Trans. Am. Math. Soc. 79 (1955), 150-157.

10. E. Hille, R. S. Phillips, Functional Analysis and Semi-Groups, Am. Math. Soc., Providence, R. I., 1927.

11. T. Kato, Perturbation Theory for Linear Operators, Springer-Verlag, Berlin 1966.

12. J. B. Keller, S. Antman (eds.), Bifurcation Theory and Nonlinear Eigenvalue Problems, Benjamin Inc., New York 1969.

13. M. A. Krasnosel'skii, Application of Variational Methods to the Problem of Bifurcation Points, Math. Sb. 33 (1953), 199-214.

14. M. A. Krasnosel'skii, Topological Methods in the Theory of Nonlinear Integral Equations, Pergamon Press, Oxford 1964.

15. P. H. Rabinowitz, A Global Theorem for Nonlinear Eigenvalue Problems and Applications, Symp. Nonlinear Functional Analysis, Madison, Wis. 1971, Ed. E. H. Zarantonello.

16. F. Rellich, Störungstheorie der Spektralzerlegung I, Math. Ann. 113 (1936), 600-619.

17. F. Rellich, Perturbation Theory of Eigenvalue Problems, Gordon and Breach, New York 1969.

18. E. H. Rothe, Completely Continuous Scalars and Variational Methods, Ann. of Math. 49 (1948), 265-278.

19. E. H. Rothe, Critical Points and Gradient Fields in Hilbert Space, Acta Math. 85 (1951), 73-98.

20. D. Sather, Branching of Solutions of an Equation in Hilbert Space, Arch. Rat. Mech. Anal. 36 (1970), 47-64.

21. D. Sather, Nonlinear Gradient Operators and the Method of Lyapunov-Schmidt, to be published.

22. D. Sather, G. H. Knightly, On Nonuniqueness of Solutions of the von Kármán Equations, Arch. Rat. Mech. Anal. 36 (1970), 65-78.

23. J. Schwartz, Compact Analytical Mappings of B-Spaces and a Theorem of Jane Cronin, Comm. Pure Appl. Math. 16 (1963), 253-260.

24. M. M. Vainberg, Variational Methods for the Study of Nonlinear Operators, Holden-Day, Inc., San Francisco 1964.

25. M. M. Vainberg, P. G. Aizengendler, The Theory and Methods of Investigation of Branch Points of Solutions, in "Progress in Mathematics", Vol. 2, Plenum Press, New York 1968.

26. M. M. Vainberg, V. A. Trenogin, The Methods of Lyapunov and Schmidt in the Theory of Nonlinear Equations and Their Further Development, Russ. Math. Surv. $\underline{17}$ (1962), 1-60.

Ruhr-Universität Bochum
Mathematisches Institut
4630 Bochum
Buscheystrasse
NA 2/33

Received May 3, 1971

# Monotonicity Methods in Hilbert Spaces and Some Applications to Nonlinear Partial Differential Equations

*HAIM BREZIS*

We recall first some classical properties of maximal monotone operators in Hilbert spaces. In doing so we concentrate on a particular class of monotone operators, namely those which are gradients of convex functions. We emphasize their specific properties which do not hold for general monotone operators.

Next we consider evolution equations associated with gradients of convex functions: smoothing effect on the initial data, behavior at infinity, etc.. We mention some applications to non linear partial differential equations and also several open problems.

## I. Maximal monotone operators in Hilbert spaces.

Let $H$ be a real Hilbert space. Let $A$ be a mapping from $H$ into $H$ which could eventually be multivalued, i.e., to every $u \in H$ we associate a subset $Au \subset H$ (which may be empty). We set $D(A) = \{u \in H; Au \neq \phi\}$, $R(A) = \bigcup_{u \in H} Au$, $(Au)^{-1} = \{f \in H; u \in Af\}$, $(\lambda A)(u) = \{\lambda f; f \in Au\}$, $(A_1 + A_2)(u) = \{f_1 + f_2; f_1 \in A_1 u, f_2 \in A_2 u\}$. One says that $A$ is a <u>monotone operator</u> (or a monotone graph) if it satisfies

(1) $\quad (f_1 - f_2, u_1 - u_2) \geq 0 \quad \forall u_1, u_2 \in D(A), \forall f_1 \in Au_1,$

$\forall f_2 \in Au_2.$

The following property is clearly equivalent to (1)

(2) $|(u_1 + \lambda f_1) - (u_2 + \lambda f_2)| \geq |u_1 - u_2|$

$$\forall \lambda \geq 0, \forall u_1, u_2 \in D(A), \forall f_1 \in Au_1, \forall f_2 \in Au_2.$$

Inequality (2) just says that $(I + \lambda A)^{-1}$, wherever defined, is a <u>contraction</u> in H.

Definition (1) involves only the notion of scalar product and can be extended to mappings which map a Banach space into its dual space. While property (2) makes sense for mappings which map a Banach space into itself; these mappings are usually called accretive. For simplicity we restrict ourselves to the case of a Hilbert space, but some of the results we are going to discuss could be extended in one or two ways.

Using a well known fixed point theorem for contractions, E. H. Zarantonello, in his pioneering paper [32] proved the following

<u>Theorem 1.</u> <u>Assume A is monotone, single-valued, Lipschitz continuous with D(A) = H. Then</u>

(3) $R(I + A) = H$.

Slightly later, Theorem 1 was extended by F. Browder [10] and G. Minty [24] who showed that (3) holds true if one replaces the Lipschitz by a continuity assumption. Introducing the concept of maximal monotone operators, Minty [24] was able to give a complete characterization of monotone operators satisfying (3).

One says that a monotone operator A is <u>maximal monotone</u> if it is maximal in the sense of inclusion of graphs, i.e., it admits no proper monotone extension.

<u>Theorem 2.</u> <u>Let A be monotone. Then A is maximal monotone if and only if R(I + A) = H (resp. R(I + $\lambda$A) = H for every $\lambda$ > 0).</u>

It is easy to check that a monotone, singlevalued, continuous, everywhere defined operator, is maximal monotone, and so Theorem 2 implies Theorem 1.

An important class of monotone operators consists of gradients of convex functions. More precisely, let $\varphi$ be a convex lower semicontinuous (l. s. c.) function from $H$ into $(-\infty, +\infty]$. We assume $\varphi \not\equiv +\infty$, and let

$$D(\varphi) = \{u \in H \,;\, \varphi(u) < +\infty\} \,.$$

For $u \in D(\varphi)$, the set

$$\partial \varphi(u) = \{f \in H;\, \varphi(v) - \varphi(u) \geq (f, v-u) \quad \forall\, v \in D(\varphi)\}$$

is called the <u>subdifferential</u> of $\varphi$ at $u$. Note that $\partial\varphi(u)$ is closed and convex, and may be empty; however if $\varphi$ is Gateaux differentiable at $u$, then $\partial\varphi(u)$ is reduced to a single point and coincides with the Gateaux differential.

<u>Theorem 3</u> (Minty [25]). <u>The operator</u> $u \mapsto \partial\varphi(u)$ <u>is maximal monotone.</u>

Since the monotonicity of $\partial\varphi$ is immediate, it is sufficient to show that for every $f \in H$, equation

(4) $$u + \partial\varphi(u) \ni f$$

has a solution. One can check easily that $u$ satisfies (4) if and only if the convex function $\psi(v) = \frac{1}{2}|v - f|^2 + \varphi(v)$ achieves its minimum at $u$. But the function $\psi$ is convex l. s. c. and tends to $+\infty$ as $|v| \to +\infty$, thus its minimum is attained.

Maximal monotone operators have simple convexity and topological properties (see [6], [13]). Let A be maximal monotone, then

(5) $$\overline{D(A)} \text{ is convex}$$

(6) for every $u \in D(A)$, $Au$ is closed and convex; so it has a unique element of least norm which we denote by $A^0 u$.

(7) let $f_n \in Au_n$ such that $u_n \to u$ weakly, $f_n \to f$ weakly and $\lim \sup (f_n, u_n) \leq (f, u)$; then $f \in Au$ and $(f_n, u_n) \to (f, u)$.

In the case where $A = \partial \varphi$, one has

$$D(A) \subset D(\varphi) \subset \overline{D(\varphi)} = \overline{D(A)}$$

and in general those 3 sets are distinct (for a simple proof of $\overline{D(\varphi)} = \overline{D(A)}$ see [4] Remark 4).

The Yosida approximation provides a convenient way of approximating maximal monotone operators by monotone operators which are Lipschitz continuous. Let $A$ be maximal monotone; for $\lambda > 0$, $J_\lambda = (I + \lambda A)^{-1}$ is a contraction defined on all of $H$; it is called the <u>resolvent</u> of $A$. The <u>Yosida approximation</u> of $A$ is defined by

$$A_\lambda u = \frac{u - J_\lambda u}{\lambda}$$

(note that $A_\lambda u \in AJ_\lambda u$). $A_\lambda$ is monotone (everywhere defined) Lipschitz continuous (with Lipschitz constant $1/\lambda$). Also for every $u \in H$, $J_\lambda u \to \text{Proj}_{\overline{D(A)}} u$ as $\lambda \to 0$, and for $u \in D(A)$, $|A_\lambda u| \leq |A^0 u|$ with $A_\lambda u \to A^0 u$ as $\lambda \to 0$ (see [13]).

It is of interest to notice that if $A$ is the subdifferential of a convex function, then its Yosida approximation remains in the same class. More precisely

<u>Theorem 4.</u> Let $A = \partial \varphi$; the function $\varphi_\lambda$ defined by

$$\varphi_\lambda(u) = \operatorname*{Min}_{v \in H} \{\tfrac{1}{2\lambda} |u-v|^2 + \varphi(v)\} = \tfrac{\lambda}{2} |A_\lambda u|^2 + \varphi(J_\lambda u) \text{ is con-}$$

<u>vex, Frechet differentiable and</u> $\partial \varphi_\lambda = A_\lambda$. <u>In addition</u> $\varphi_\lambda(u) \uparrow \varphi(u)$ <u>as</u> $\lambda \downarrow 0$.

The convexity and differentiability of $\varphi_\lambda$ were proved by Moreau [26] who made extensive use of the notion of inf-convolution. For $u \in \overline{D(A)}$ we have

$$\varphi(J_\lambda u) \leq \varphi_\lambda(u) \leq \varphi(u)$$

and since $J_\lambda u \to u$, we get $\varphi_\lambda(u) \to \varphi(u)$. If $u \notin \overline{D(A)}$, $\varphi_\lambda(u) \geq \frac{1}{2\lambda} |(u - J_\lambda u)|^2 - c_1 |J_\lambda u| - c_2$ (by Hahn-Banach $\varphi$ is bounded from below by an affine function), so that $\varphi_\lambda(u) \to +\infty$ as $\lambda \to 0$.

One of the main purposes of the theory of monotone operators is to obtain <u>surjectivity</u> results. The following theorem provides a very simple and useful sufficient condition.

<u>Theorem 5.</u> <u>Let A be maximal monotone.</u> <u>Assume</u>

(8) $\quad \lim_{\substack{|u| \to +\infty \\ u \in D(A)}} |A^0 u| = +\infty \quad$ (i.e. $A^{-1}$ is bounded[†])

<u>Then</u> $\quad R(A) = H$.

Note that in the case where $A = \partial\varphi$, then $R(A) = H$ if and only if $\lim_{|u| \to +\infty} \{\varphi(u) - (f, u)\} = +\infty$ for every $f \in H$.
Actually, if one replaces in Theorem 5 the assumption (8) by "$A^{-1}$ is locally bounded" then we get a condition which is both necessary and sufficient for surjectivity; this result, proved independently by F. Browder [11] and Rockafellar [28] is based on the following:

<u>Theorem 6.</u> <u>Let B be a monotone operator; then B is locally bounded at every point of</u> Int $D(B)$.

Using the terminology of partial differential equations,

---

[†] One says that B is <u>bounded</u> if for every bounded set $N \subset H$, $\bigcup_{u \in N} Bu$ is bounded.

Theorem 5 asserts that if $A$ is maximal monotone and if one knows <u>a priori estimates</u> for possible solutions of the equation $Au \ni f$ with a bound of the form $|u| \leq \omega(|f|)$ ($\omega$ continuous) then $R(A) = H$. In many applications such a bound is provided by a <u>coerciveness</u> assumption:

(9)   there exists $u_0 \in H$ such that
$$\lim_{\substack{|u| \to +\infty \\ u \in D(A)}} \frac{(A^0 u, u - u_0)}{|u|} = +\infty .$$

Surprisingly it turns out that in case $A = \partial \varphi$, property (9) is also a necessary condition.

<u>Theorem 7.</u>   Assume $A = \partial \varphi$. <u>The following are equivalent</u>

(10)   <u>for every</u> $u_0 \in D(\varphi)$, $\lim_{\substack{|u| \to +\infty \\ u \in D(A)}} \dfrac{(A^0 u, u - u_0)}{|u|} = +\infty$ .

(11)   <u>there exists</u> $u_0 \in H$ <u>such that</u> $\lim_{\substack{|u| \to +\infty \\ u \in D(A)}} \dfrac{(A^0 u, u - u_0)}{|u|} = +\infty$ .

(12)   $\lim_{\substack{|u| \to +\infty \\ u \in D(A)}} |A^0 u| = +\infty$ .

(13)   $R(A) = H$ <u>and</u> $A^{-1}$ <u>is bounded.</u>

(14)   $\lim_{|u| \to +\infty} \dfrac{\varphi(u)}{|u|} = +\infty$ .

**Remark.** It is clear that Theorem 7 does not hold for general maximal monotone operators; for example a rotation by $\pi/2$ in $H = \mathbb{R}^2$ satisfies (12) and not (11).

**Proof of Theorem 7.** $(10) \Longrightarrow (11) \Longrightarrow (12) \Longrightarrow (13)$ are immediate. To show that $(13) \Longrightarrow (14)$, we can always reduce to the case where $\varphi \geq 0$ since $\varphi$ is bounded from below by an affine function (this amounts to shift $A$ by a constant). Let $r > 0$ be fixed. For every $z \in H$ with $|z| \leq r$ there exists (by (13)) $v \in D(A)$ such that $Av \ni z$ and $|v| \leq M$. Thus

$$\varphi(u) - \varphi(v) \geq (z, u-v) \quad \forall u \in D(\varphi)$$

and thus $(z, u) \leq \varphi(u) + Mr, \forall u \in D(\varphi), \forall z \in H ; |z| \leq r$. Consequently, $r|u| \leq \varphi(u) + Mr$ and $\dfrac{\varphi(u)}{|u|} \geq r - \dfrac{Mr}{|u|}$.

Hence $\lim\inf\limits_{|u| \to +\infty} \dfrac{\varphi(u)}{|u|} \geq r$.

Finally $(14) \Longrightarrow (10)$ since we have

$$\varphi(u_0) - \varphi(u) \geq (A^0 u, u_0 - u)$$

and

$$\dfrac{(A^0 u, u - u_0)}{|u|} \geq \dfrac{\varphi(u) - \varphi(u_0)}{|u|} \to +\infty \quad \text{as} \quad |u| \to +\infty .$$

Another important topic in the theory of monotone operators concerns the <u>sum of maximal monotone operators.</u> If $A$ and $B$ are maximal monotone, $A + B$ need not be maximal monotone. So it is natural to raise the question: when is $A + B$ maximal monotone ? There is no general and convenient answer to this question. However the following criteria may be used to prove that $A + B$ is maximal.

**Theorem 8.** (see [6]) <u>Let</u> $f \in H$; <u>for every</u> $\lambda > 0$ <u>the equation</u> $u_\lambda + A_\lambda u_\lambda + B u_\lambda \ni f$ <u>has a unique solution and</u> $f \in R(I + A + B)$ <u>if and only if</u> $A_\lambda u_\lambda$ <u>is bounded as</u> $\lambda \to 0$. <u>In this case</u> $u_\lambda \to u$ <u>as</u> $\lambda \to 0$ <u>and</u> $f \in u + Au + Bu$.

Sufficient conditions have been given by Rockafellar [29] and Crandall-Pazy [13]:

if $(\text{Int}(D(A))) \cap D(B) \neq \phi$, then $A + B$ is maximal monotone

if B is <u>dominated</u> by A, i.e., $D(A) \subset D(B)$ and $|B^0 u| \leq k |A^0 u| + \omega(|u|)$ for all $u \in D(A)$, where $k < 1$ and $\omega$ is continuous, then $A + B$ is maximal monotone.

One may also ask a more restricted question: when is $A + \partial\varphi$ maximal monotone ? Again, this is an open problem, however the following sufficient condition turns out to be quite useful in applications.

<u>Theorem 9.</u> <u>Let A be a maximal monotone and let $\varphi$ be a convex l.s.c. function from H into $(-\infty, +\infty]$, $\varphi \not\equiv +\infty$. Assume there exists C such that</u>

(15) $\qquad \varphi((I + \lambda A)^{-1} u) \leq \varphi(u) + C\lambda$, $\forall \lambda > 0$, $\forall u \in D(\varphi)$.

<u>Then $A + \partial\varphi$ is maximal monotone.</u> <u>In addition</u>

(16) $\qquad |A^0 u| \leq |(A + \partial\varphi)^0 u| + \sqrt{C} \quad \forall u \in D(A) \cap D(\partial\varphi)$.

<u>Proof of Theorem 9.</u> We apply Theorem 8 with $B = \partial\varphi$. Let $u_\lambda$ be the solution of

$$u_\lambda + A_\lambda u_\lambda + \partial\varphi(u_\lambda) \ni f$$

i.e.,

$$\varphi(v) - \varphi(u_\lambda) \geq (f - u_\lambda - A_\lambda u_\lambda, v - u_\lambda) \quad \forall v \in D(\varphi).$$

Taking $v = (I + \lambda A)^{-1} u_\lambda$ we get

$$C\lambda \geq (f - u_\lambda - A_\lambda u_\lambda, -\lambda A_\lambda u_\lambda)$$

so that $|A_\lambda u_\lambda|^2 \leq |f - u_\lambda||A_\lambda u_\lambda| + C$ and $|A_\lambda u_\lambda| \leq |f - u_\lambda| + \sqrt{C}$. Let $v_0 \in D(A) \cap D(\varphi)$ (such $v_0$ exists by (15)); by the monotonicity of $A_\lambda$ we have

$$\varphi(v_0) - \varphi(u_\lambda) \geq (f - u_\lambda - A_\lambda v_0, v_0 - u_\lambda).$$

Since $\varphi(u_\lambda) \geq -C_1|u_\lambda| - C_2$ and $|A_\lambda v_0|$ is bounded, we conclude that $|u_\lambda|$ is bounded as $\lambda \to 0$. Hence $A + \partial\varphi$ is maximal monotone. For $u \in D(\partial\varphi)$ and $z \in \partial\varphi(u)$ we have

$$\varphi(J_\lambda u) - \varphi(u) \geq (z, J_\lambda u - u)$$

and thus $\lambda C \geq (z, -\lambda A_\lambda u)$. Consequently, if $u \in D(A) \cap D(\partial\varphi)$ we have $(A^0 u, z) \geq -C$, for all $z \in \partial\varphi(u)$.
Let $f = (A + \partial\varphi)^0 u$, so that $f = y + z$ with $y \in Au$ and $z \in \partial\varphi(u)$. We have $(A^0 u, f) = (A^0 u, y) + (A^0 u, z) \geq |A^0 u|^2 - C$. Hence

$$|A^0 u| \leq |f| + \sqrt{C}.$$

Assumption (15) is convenient because it is preserved under sum of maximal monotone operators. More precisely

<u>Theorem 10.</u> <u>Let $A^1$ and $A^2$ be maximal monotone with $A^1 + A^2$ maximal monotone. Assume</u>

$$\varphi(J^1_\lambda u) \leq \varphi(u) + C_1 \lambda \quad \forall \lambda > 0, \; \forall u \in D(\varphi)$$

$$\varphi(J^2_\lambda u) \leq \varphi(u) + C_2 \lambda \quad \forall \lambda > 0, \; \forall u \in D(\varphi).$$

<u>Then</u>

(17) $\qquad \varphi(J_\lambda u) \leq \varphi(u) + (C_1 + C_2)\lambda \quad \forall \lambda > 0, \; \forall u \in D(\varphi)$

<u>where</u> $J^1_\lambda = (I + \lambda A^1)^{-1}$, $J^2_\lambda = (I + \lambda A^2)^{-1}$, $J_\lambda = (I + \lambda(A^1 + A^2))^{-1}$. <u>In particular,</u> $A^1 + A^2 + \partial\varphi$ <u>is maximal monotone.</u>

**Proof of Theorem 10.** Let $\alpha > 0$ and let $u_\alpha$ be the solution of $u \in u_\alpha + \lambda A^1 u_\alpha + \lambda A_\alpha^2 u_\alpha$. By Theorem 8 we know that $u_\alpha \to J_\lambda u$ as $\alpha \to 0$. We have $\alpha u \in \alpha u_\alpha + \alpha \lambda A^1 u_\alpha + \lambda(u_\alpha - J_\alpha^2 u_\alpha)$, or $\dfrac{\alpha u + \lambda J_\alpha^2 u_\alpha}{\alpha + \lambda} \in u_\alpha + \dfrac{\alpha \lambda}{\alpha + \lambda} A^1 u_\alpha$. Thus

$$u_\alpha = J^1_{\alpha\lambda/(\alpha+\lambda)} \left( \frac{\alpha u + \lambda J_\alpha^2 u_\alpha}{\alpha + \lambda} \right)$$

For a fixed $u$, the mapping $z \mapsto J^1_{\alpha\lambda/(\alpha+\lambda)} \left( \dfrac{\alpha u + \lambda J_\alpha^2 z}{\alpha + \lambda} \right)$ maps the closed convex set

$$K = \{x \in H \,;\, \varphi(x) \leq \varphi(u) + \lambda(C_1 + C_2)\}$$

into itself, and it is a strict contraction. Thus, its fixed point $u_\alpha$ belongs to $K$. Consequently, $\varphi(u_\alpha) \leq \varphi(u) + \lambda(C_1 + C_2)$; passing to the limit as $\alpha \to 0$, we get (17).

**Remark.** Let $K$ be a closed convex subset of $H$ and let $\varphi = I_K$ be the indicator function of $K$, i.e.,

$$I_K(u) = \begin{cases} 0 & \text{if } u \in K \\ +\infty & \text{if } u \notin K \end{cases}$$

Assumption (15) just asserts that $(I + \lambda A)^{-1} K \subset K$, for all $\lambda > 0$ (for this case see also Brezis-Pazy [7] Theorems 2.2 and 4.2).

**Theorem 11.** Under the assumptions of Theorem 9 we have

$$\overline{D(A + \partial\varphi)} = \overline{D(A) \cap D(\partial\varphi)} = \overline{D(A) \cap D(\varphi)}$$

Proof. Clearly $\overline{D(A) \cap D(\partial\varphi)} \subset \overline{D(A) \cap D(\varphi)}$. Conversely, we first show that $\overline{D(A)} \cap \overline{D(\varphi)} \subset \overline{D(A) \cap D(\varphi)}$. Indeed let $u \in \overline{D(A)} \cap \overline{D(\varphi)}$ and let $v_\varepsilon \in D(\varphi)$ be such that $v_\varepsilon \to u$ in $H$. Let $u_\varepsilon = J_\varepsilon v_\varepsilon = (I + \varepsilon A)^{-1} v_\varepsilon$. By (15), $u_\varepsilon \in D(A) \cap D(\varphi)$ and

$$|u_\varepsilon - u| \le |u_\varepsilon - (I + \varepsilon A)^{-1} u| + |(I + \varepsilon A)^{-1} u - u|$$

$$\le |v_\varepsilon - u| \le |(I + \varepsilon A)^{-1} u - u|.$$

Thus $u_\varepsilon \to u$ as $\varepsilon \to 0$.

We prove now that $D(A) \cap D(\varphi) \subset \overline{D(A) \cap D(\partial\varphi)}$. Let $u \in D(A) \cap D(\varphi)$ and let $u_\varepsilon \in D(A) \cap D(\partial\varphi)$ be the solution of

$$u_\varepsilon + \varepsilon(A u_\varepsilon + \partial\varphi(u_\varepsilon)) \ni u$$

(which exists by Theorem 9). We have

$$\varphi(u) - \varphi(u_\varepsilon) \ge (\frac{u - u_\varepsilon}{\varepsilon} - A u_\varepsilon, u - u_\varepsilon) \ge \frac{1}{\varepsilon} |u - u_\varepsilon|^2 - (A^0 u, u - u_\varepsilon).$$

Hence $u_\varepsilon \to u$ as $\varepsilon \to 0$.

## II. Some examples of maximal monotone operators.

Let $\Omega$ be a bounded domain in $\mathbb{R}^N$ with smooth boundary $\Gamma$.

Example 1. Let $H = L^2(\Omega)$ and let $j$ be a convex l.s.c. function from $\mathbb{R}$ into $(-\infty, +\infty]$; assume $j \not\equiv +\infty$ and let $\beta = \partial j$. We define for $u \in H$

$$\varphi(u) = \begin{cases} \frac{1}{2} \int_\Omega |\text{grad } u|^2 dx + \int_\Gamma j(u) d\Gamma & \text{if } u \in H^1(\Omega)[†] \text{ and } j(u) \in L^1(\Gamma) \\ +\infty & \text{otherwise}. \end{cases}$$

---

[†] $H^k$ and $H_0^k$ denote the usual Sobolev spaces

It is easy to check that $\varphi$ is convex l. s. c. on H (note that if $u_n \to u$ in $L^2(\Omega)$ and $\varphi(u_n) \leq \lambda$, then $u_n$ is bounded in $H^1(\Omega)$, thus $u_n \to u$ weakly in $H^1(\Omega)$ and by Fatou's lemma, $\liminf \int_\Gamma j(u_n) d\Gamma \geq \int_\Gamma j(u) d\Gamma$).

**Theorem 12.** We have $\partial\varphi(u) = -\Delta u$ with $D(\partial\varphi) = \{u \in H^2(\Omega);\ -\frac{\partial u}{\partial n} \in \beta(u)\ \text{a.e. on}\ \Gamma\}$ where $\partial/\partial n$ <u>is the outward normal derivative.</u> <u>In addition there exist constants</u> $c_1, c_2$ <u>such that</u> $\|u\|_{H^2} \leq c_1 |-\Delta u + u|_{L^2} + c_2$ <u>for every</u> $u \in D(\partial\varphi)$.

**Remark.** The precise description of the domain of a subdifferential often amounts to proving a regularity theorem for some nonlinear elliptic equation, and usually it is not an easy matter. Note that $\partial\varphi$ is "apparently" a linear operator but it has to be considered as a nonlinear operator because its domain is not a linear subspace.

**Proof of Theorem 12.** Let $Au = -\Delta u$ with $D(A) = \{u \in H^2(\Omega);\ -\frac{\partial u}{\partial n} \in \beta(u)\ \text{a.e. on}\ \Gamma\}$. We are going to prove that A is maximal monotone and $A \subset \partial\varphi$. It is clear that A is monotone and also for $u \in D(A)$ and $v \in D(\varphi)$

$$\int_\Omega -\Delta u \cdot (v-u) dx = \int_\Omega \operatorname{grad} u (\operatorname{grad} v - \operatorname{grad} u) dx - \int_\Gamma \frac{\partial u}{\partial n}(v-u) d\Gamma$$

$$\leq \varphi(v) - \varphi(u).$$

Thus $A \subset \partial\varphi$, and it remains to show that $R(I + A) = H$. This is proved in [5] under slightly more general assumptions, but for completeness we sketch here the proof. Let $f \in H$ be given and let $u_\lambda \in H^2(\Omega)$, $\lambda > 0$, be the solution of

(18) $\begin{cases} -\Delta u_\lambda + u_\lambda = f & \text{on } \Omega \\ -\dfrac{\partial u_\lambda}{\partial n} = \beta_\lambda(u_\lambda) & \text{on } \Gamma. \end{cases}$

The existence of $u_\lambda$ can be established for example by a fixed point argument for the map $u \in L^2(\Gamma) \mapsto Tu \in L^2(\Gamma)$ where $Tu = v|_\Gamma$ and $v$ is the solution of the linear equation

$$-\Delta v + v = f \text{ on } \Omega, \quad v + \lambda \frac{\partial v}{\partial n} = (I + \lambda\beta)^{-1} u \text{ on } \Gamma,$$

(note that $T$ is a strict contraction in $L^2(\Gamma)$).

We show that $\|u_\lambda\|_{H^2} \leq c_1 |f|_{L^2} + c_2$ where $c_1$ and $c_2$ are independent of $\lambda$.

Multiplying equation (18) by $u_\lambda - v_0$ where $v_0 \in D(\beta)$ we get easily $\|u_\lambda\|_{H^1} \leq c_1 |f|_{L^2} + c_2$. Interior estimates in $H^2$ are immediate and we have $\|u_\lambda\|_{H^2(\Omega')} \leq c(|f|_{L^2(\Omega)} + \|u_\lambda\|_{H^1(\Omega)})$ for $\overline{\Omega'} \subset \Omega$.

Next we obtain estimates near the boundary by using local charts and "tangential shifts" as in the linear theory. Assume that the equation †

(19) $$-\sum_{k,\ell=1}^{N} \frac{\partial}{\partial x_\ell} \left( a_{k\ell}(x) \frac{\partial u}{\partial x_k} \right) + u = f$$

holds in $\Omega_R = \{x = (x', x_N); |x'| < R \text{ and } 0 < x_N < R\}$ where $x' = (x_1, x_2, \ldots, x_{N-1})$ and the equation

(20) $$-\frac{\partial u}{\partial \nu} = \beta(u)$$

holds in $\Gamma_R = \{x = (x', x_N); |x'| < R \text{ and } x_N = 0\}$ where $\frac{\partial u}{\partial \nu} = -\sum_{k=1}^{N} a_{kN} \frac{\partial u}{\partial x_k}$ and $a_{k\ell}(x)$ are smooth with $\sum_{k,\ell=1}^{N} a_{k\ell} \xi_k \xi_\ell \geq a|\xi|^2$ for all $\xi \in \mathbb{R}^N$, $a > 0$.

Let $\eta(x)$ be a smooth function with $\eta = 1$ on a neighborhood of $\Omega_{R/2}$ and $\eta = 0$ for $|x'| \geq R$ or $x_N \geq R$.

---

†for simplicity we have dropped $\lambda$.

Multiplying equation (19) by $-\sum_{i=1}^{N-1} \frac{\partial}{\partial x_i}(\eta^2 \frac{\partial u}{\partial x_i})$ and integrating by parts we get

$$\sum_{i=1}^{N-1} \int_{\Gamma_R} \beta'(u)\eta^2 d\Gamma + \sum_{i=1}^{N-1}\sum_{k,\ell=1}^{N} \int_{\Omega_R} \frac{\partial}{\partial x_i}(a_{k\ell} \frac{\partial u}{\partial x_k}) \frac{\partial}{\partial x_\ell}(\eta^2 \frac{\partial u}{\partial x_i}) dx$$

$$= -\sum_{i=1}^{N-1} \int_{\Omega_R} f \cdot \frac{\partial}{\partial x_i}(\eta^2 \frac{\partial u}{\partial x_i}) dx.$$

From the monotonicity of $\beta$ we deduce after some rearrangements that

$$\sum_{i=1}^{N-1}\sum_{k=1}^{N} \|\eta \frac{\partial^2 u}{\partial x_i \partial x_k}\|^2_{L^2(\Omega_R)} \leq C(\|f\|_{L^2(\Omega_R)} + \|u\|_{H^1(\Omega_R)})$$

$$(\|u\|_{H^1(\Omega_R)} + \sum_{i=1}^{N-1}\sum_{k=1}^{N} \|\eta \frac{\partial^2 u}{\partial x_i \partial x_k}\|_{L^2(\Omega_R)}).$$

Consequently

$$\sum_{i=1}^{N-1}\sum_{k=1}^{N} \|\eta \frac{\partial^2 u}{\partial x_i \partial x_k}\|_{L^2(\Omega_R)} \leq C(\|f\|_{L^2(\Omega_R)} + \|u\|_{H^1(\Omega_R)}).$$

Using the equation (19) to estimate $\partial^2 u/\partial x_N^2$ we obtain finally

$$\|u\|_{H^2(\Omega_{R/2})} \leq C(\|f\|_{L^2(\Omega_R)} + \|u\|_{H^1(\Omega_R)}).$$

Going back to (18) we conclude that $\|u_\lambda\|_{H^2(\Omega)}$ is bounded as $\lambda \to 0$. Hence $u_{\lambda_n} \to u$ weakly in $H^2(\Omega)$, $u_{\lambda_n} \to u$ in $L^2(\Gamma)$, $(\partial u_{\lambda_n})/\partial n \to \partial u/\partial n$ in $L^2(\Gamma)$.

Also $\|u_\lambda - (I + \lambda\beta)^{-1}u_\lambda\|_{L^2(\Gamma)} \leq \lambda \|\frac{\partial u_\lambda}{\partial n}\|_{L^2(\Gamma)} \to 0$

and thus $(I + \lambda_n\beta)^{-1}u_{\lambda_n} \to u$ in $L^2(\Gamma)$.

Finally, we have $-\Delta u + u = f$ and since $-\frac{\partial u_\lambda}{\partial n} \in \beta((I + \beta)^{-1}u)$ a.e. on $\Gamma$, we have $-\frac{\partial u}{\partial n} \in \beta(u)$ a.e. on $\Gamma$.

Let now $k$ be another convex l.s.c. function from $\mathbb{R}$ into $(-\infty, +\infty]$ and let $\gamma = \partial k$. The function $\psi$ defined on $L^2(\Omega)$ by

$$\psi(u) = \begin{cases} \int_\Omega k(u)\,dx & \text{if } k(u) \in L^1(\Omega) \\ +\infty & \text{otherwise} \end{cases}$$

is convex l.s.c. and it is easy to check that $f \in \partial\psi$ if and only if $f(x) \in \gamma(u(x))$ a.e. on $\Omega$. Also $\overline{D(\partial\psi)} = \overline{D(\psi)} = \{u \in L^2(\Omega); u(x) \in \overline{D(\gamma)}$ a.e. on $\Omega\}$.

<u>Corollary 13.</u> <u>If $D(\beta) \cap D(\gamma) \neq \phi$, then $\partial\varphi + \partial\psi$ is maximal monotone and</u>

$$\overline{D(\partial\varphi) \cap D(\partial\psi)} = \{u \in L^2(\Omega); u(x) \in \overline{D(\gamma)} \text{ a.e. on } \Omega\}.$$

<u>In particular, for every $f \in L^2(\Omega)$ there exists a unique function $u \in H^2(\Omega)$ satisfying</u>

$$\begin{cases} -\Delta u + \gamma(u) + u \ni f & \text{a.e. on } \Omega \\ -\frac{\partial u}{\partial n} \in \beta(u) & \text{a.e. on } \Gamma. \end{cases}$$

Proof. After some changes of variable and shifts we can assume that $0 \in \beta(0)$ and $0 \in \gamma(0)$. By Theorem 9 it is sufficient to show that $\varphi((I + \lambda\partial\psi)^{-1}u) \leq \varphi(u)$ for all $u \in D(\varphi)$, and all $\lambda > 0$; but this clearly holds since $(I + \lambda\gamma)^{-1}$ is a monotone contraction in $\mathbb{R}$. From Theorem 11 we deduce that $\overline{D(\partial\varphi + \partial\psi)} = \overline{D(\partial\varphi)} \cap \overline{D(\partial\psi)}$ and since $\overline{D(\partial\varphi)} = L^2(\Omega)$, (note that $\mathcal{D}(\Omega) \subset D(\partial\varphi)$), we conclude that

$$\overline{D(\partial\varphi)} \cap \overline{D(\partial\psi)} = \overline{D(\partial\varphi)} = \{u \in L^2(\Omega); u(x) \in \overline{D(\gamma)} \text{ a.e. on } \Omega\}.$$

Let now $K$ be the closed convex set $K = \{u \in L^2(\Omega); \psi_1 \le u \le \psi_2 \text{ a.e. on } \Omega\}$ where $\psi_1, \psi_2 \in H^2(\Omega)$, $\psi_1 \le \psi_2$ a.e. on $\Omega$. We use the notation

$$\beta^+(r) = \text{Max}\{z; z \in \beta(r)\}, \quad \beta^-(r) = \text{Min}\{z; z \in \beta(r)\} \text{ if } r \in D(\beta)$$

$$\beta^+(r) = \beta^-(r) = +\infty \quad \text{if } r \notin D(\beta), \; r \ge \text{Sup } D(\beta)$$

$$\beta^+(r) = \beta^-(r) = -\infty \quad \text{if } r \notin D(\beta), \; r \le \text{Inf } D(\beta).$$

**Corollary 14.** Assume

$$(21) \quad \frac{\partial \psi_1}{\partial n} + \beta^-(\psi_1) \le 0 \quad \text{and} \quad \frac{\partial \psi_2}{\partial n} + \beta^+(\psi_2) \ge 0 \quad \text{a.e. on } \Gamma.$$

Then $\partial\varphi + \partial I_K$ is maximal monotone and $\overline{D(\partial\varphi) \cap K} = K$. In particular for every $f \in L^2(\Omega)$, there exists a unique function $u \in H^2(\Omega) \cap K$ satisfying

$$\int_\Omega (-\Delta u + u)(v-u)\,dx \ge \int_\Omega f \cdot (v-u)\,dx \quad \underline{\text{for all}} \quad v \in K$$

and

$$-\frac{\partial u}{\partial n} \in \beta(u) \quad \text{a.e.} \quad \underline{\text{on}} \; \Gamma.$$

**Proof.** Write $K$ as $K = K_1 \cap K_2$ where $K_1 = \{u \in L^2(\Omega); u \ge \psi_1 \text{ a.e. on } \Omega\}$ and $K_2 = \{u \in L^2(\Omega); u \le \psi_2 \text{ a.e. on } \Omega\}$. We first show that $(I + \lambda(\partial\varphi + \Delta\psi_1))^{-1} K_1 \subset K_1$. Indeed let $u \in K_1$ and let $u_\lambda = (I + \lambda(\partial\varphi + \Delta\psi_1))^{-1} u$, i.e., $u_\lambda \in H^2(\Omega)$ satisfies

$$(22) \quad \begin{cases} u_\lambda - \lambda\Delta u_\lambda + \lambda\Delta\psi_1 = u & \text{a.e. on } \Omega \\ -\dfrac{\partial u_\lambda}{\partial n} \in \beta(u_\lambda) & \text{a.e. on } \Gamma. \end{cases}$$

Multiplying (22) by $-(u_\lambda - \psi_1)^-$ and integrating by parts we get

$$\int_\Omega |(u_\lambda - \psi_1)^-|^2 dx + \int_\Omega |\text{grad}(u_\lambda - \psi_1)^-|^2 dx + \int_\Gamma (\frac{\partial u_\lambda}{\partial n} - \frac{\partial \psi_1}{\partial n})(u_\lambda - \psi_1)^- d\Gamma$$

$$= -\int_\Omega (u - \psi_1)(u_\lambda - \psi_1)^- dx \leq 0.$$

Also $(\frac{\partial u_\lambda}{\partial n} - \frac{\partial \psi_1}{\partial n})(u_\lambda - \psi_1)^- \geq 0$ a.e. on $\Gamma$ (note that if $u_\lambda < \psi_1$ at some point, let $\xi$ be such that $u_\lambda < \xi < \psi_1$, and then by (21) we have $-\frac{\partial u_\lambda}{\partial n} \leq \beta^+(u_\lambda) \leq \beta^-(\xi) \leq \beta^-(\psi_1) \leq \frac{\partial \psi_1}{\partial n}$). Consequently $(u_\lambda - \psi_1)^- = 0$ and $u_\lambda \in K_1$; similarly $(I + \lambda(\partial\varphi + \Delta\psi_2))^{-1} K_2 \subset K_2$.

We deduce from Theorem 9 that $\partial\varphi + \Delta\psi_1 + \partial I_{K_1}$ is maximal monotone and so is $\partial\varphi + \partial I_{K_1}$. Also, by Theorem 11, we have $\overline{D(\partial\varphi) \cap D(\partial I_{K_1})} = \overline{D(\partial\varphi) \cap K_1} = \overline{D(\partial\varphi)} \cap K_1 = K_1$. Applying Theorem 10 with $A^1 = \partial\varphi + \Delta\psi_2$ and $A^2 = \partial I_{K_1}$ and $\varphi = I_{K_2}$ (note that $(I + \lambda A^2)^{-1} K_2 \subset K_2$ is obviously satisfied since $\text{Proj}_{K_1}(K_2) \subset K_2$) we obtain that $\partial\varphi + \Delta\psi_2 + \partial I_{K_1} + \partial I_{K_2}$ is maximal monotone and so is $\partial\varphi + \partial I_{K_1} + \partial I_{K_2} = \partial\varphi + \partial I_K$. And finally we have $\overline{D(A^1 + A^2) \cap K_2} = \overline{D(A^1 + A^2)} \cap K_2 = \overline{D(\partial\varphi)} \cap K_1 \cap K_2 = K$.

Remark. Corollary 13 and 14 are related to regularity results (proved in [5] and [8]) for unilateral problems.

Example 2. Let $H = L^2(\Omega)$; we define for $u \in H$

$$\varphi(u) = \begin{cases} \frac{1}{2} \int_\Omega |\text{grad } u|^2 dx + \int_\Omega |\text{grad } u| dx & \text{if } u \in H_0^1(\Omega) \\ +\infty & \text{otherwise} \end{cases}$$

It is easy to check that $\varphi$ is convex and l.s.c. on $H$.

**Theorem 15.** We have $f \in \partial\varphi(u)$ <u>if and only if</u> $u \in H^2(\Omega) \cap H_0^1(\Omega)$ <u>satisfies the inequality</u>

$$(23) \quad \int_\Omega -\Delta u(v-u)dx + \int_\Omega |\text{grad } v|dx - \int_\Omega |\text{grad } u|dx \geq \int_\Omega f(v-u)dx$$

$$\forall\, v \in H_0^1(\Omega).$$

<u>In particular, for every</u> $f \in H$, <u>there is a unique solution</u> $u \in H^2(\Omega) \cap H_0^1(\Omega)$ of (23).

The proof of Theorem 15 is based on the following

**Lemma 1.** Let $u \in H_0^1(\Omega)$ and let $u_\varepsilon$ be the solution of

$$\begin{cases} u_\varepsilon - \varepsilon \Delta u_\varepsilon = u & \text{on } \Omega \\ u_\varepsilon = 0 & \text{on } \Gamma. \end{cases}$$

There is a constant $C$, depending only on $\Omega$, such that

$$(24) \quad \int_\Omega |\text{grad } u_\varepsilon|dx \leq \int_\Omega |\text{grad } u|dx + \varepsilon\, C \|u_\varepsilon\|_{H^2(\Omega)}$$

In particular if $\Omega$ is convex, one can take $C = 0$.

**Proof of Lemma 1.** Let $\zeta$ be a smooth function such that $\zeta > 0$ on $\Omega$, $\zeta = 0$ on $\Gamma$ and $\frac{\partial \zeta}{\partial n} \neq 0$ on $\Gamma$.

We first prove that, if $v$ is a smooth function on $\bar{\Omega}$ which vanishes on $\Gamma$, then

$$(25) \quad \Delta v - \frac{\partial^2 v}{\partial n^2} = \frac{\partial v}{\partial n}\left(\frac{\partial \zeta}{\partial n}\right)^{-1}\left(\Delta \zeta - \frac{\partial^2 \zeta}{\partial n^2}\right) \text{ on } \Gamma.$$

We can assume that $0 \in \Gamma$ and we choose a coordinate system $(\vec{e}_1, \vec{e}_2, \ldots, \vec{e}_N)$ such that $\vec{e}_N = \vec{n}$. Locally, the equation of $\Gamma$ is $x_N = f(x')$. In a neighborhood of $0$ in in $\mathbb{R}^{N-1}$ we have $\zeta(x', f(x')) \equiv 0$ and $v(x', f(x')) \equiv 0$. Thus, for $1 \leq i \leq N-1$

$$\frac{\partial \zeta}{\partial x_i}(x', f(x')) + \frac{\partial \zeta}{\partial x_N}(x', f(x'))\frac{\partial f}{\partial x_i}(x') \equiv 0$$

$$\frac{\partial^2 \zeta}{\partial x_i^2}(x', f(x')) + 2\frac{\partial^2 \zeta}{\partial x_N \partial x_i}(x', f(x'))\frac{\partial f}{\partial x_i}(x') +$$

$$+ \frac{\partial \zeta}{\partial x_N}(x', f(x'))\frac{\partial^2 f}{\partial x_i^2}(x') \equiv 0 \ .$$

Since $\frac{\partial f}{\partial x_i}(0) = 0$, we have $\frac{\partial^2 \zeta}{\partial x_i^2}(0,0) + \frac{\partial \zeta}{\partial n}(0,0)\frac{\partial^2 f}{\partial x_i^2}(0) = 0$,

and hence

$$\Delta\zeta(0,0) - \frac{\partial^2 \zeta}{\partial n^2}(0,0) + \frac{\partial \zeta}{\partial n}(0,0)\sum_{i=1}^{N-1}\frac{\partial^2 f}{\partial x_i^2}(0) = 0 \ .$$

Similarly

$$\Delta v(0,0) - \frac{\partial^2 v}{\partial n^2}(0,0) + \frac{\partial v}{\partial n}(0,0)\sum_{i=1}^{N-1}\frac{\partial^2 f}{\partial x_i^2}(0) = 0 \ .$$

Consequently

$$\frac{\partial v}{\partial n}(\Delta\zeta - \frac{\partial^2 \zeta}{\partial n^2}) = \frac{\partial \zeta}{\partial n}(\Delta v - \frac{\partial^2 v}{\partial n^2}) \quad \text{on } \Gamma \ .$$

Note that if $\Omega$ is convex, $\zeta$ can be chosen to be concave so that $\Delta\zeta - \frac{\partial^2 \zeta}{\partial n^2} \leq 0$ and $\frac{\partial \zeta}{\partial n} < 0$. Define for $\xi \in \mathbb{R}^N$ and $\lambda > 0$, the convex function

$$j_\lambda(\xi) = \begin{cases} \frac{1}{2\lambda}|\xi|^2 & \text{if } |\xi| \leq \lambda \\ |\xi| - \frac{\lambda}{2} & \text{if } |\xi| \geq \lambda \end{cases} \ .$$

By continuity it is sufficient to establish (24) for smooth $u$. We have

$$j_\lambda(\text{grad } u) - j_\lambda(\text{grad } u_\varepsilon) \geq \sum_{k=1}^{N} \frac{\partial j_\lambda}{\partial \xi_k}(\text{grad } u_\varepsilon) \cdot \left(\frac{\partial u}{\partial x_k} - \frac{\partial u_\varepsilon}{\partial x_k}\right)$$

$$= \sum_{k=1}^{N} \frac{\partial j_\lambda}{\partial \xi_k}(\text{grad } u_\varepsilon) \cdot -\varepsilon \Delta \frac{\partial u_\varepsilon}{\partial x_k}.$$

Hence

$$\int_\Omega j_\lambda(\text{grad } u_\varepsilon)\,dx \leq \int_\Omega j_\lambda(\text{grad } u)\,dx$$

$$- \varepsilon \sum_{k,\ell,m} \int_\Omega \frac{\partial^2 j_\lambda}{\partial \xi_k \partial \xi_\ell}(\text{grad } u_\varepsilon) \frac{\partial^2 u_\varepsilon}{\partial x_k \partial x_m} \cdot \frac{\partial^2 u_\varepsilon}{\partial x_\ell \partial x_m}\,dx$$

$$+ \varepsilon \int_\Gamma \frac{\partial j_\lambda}{\partial \xi_k}(\text{grad } u_\varepsilon) \frac{\partial^2 u_\varepsilon}{\partial x_k \partial n}\,d\Gamma$$

$$\leq \varepsilon \int_{[x \in \Gamma;\, |\frac{\partial u_\varepsilon}{\partial n}| < \lambda]} \frac{1}{\lambda} \frac{\partial u_\varepsilon}{\partial n} \frac{\partial^2 u_\varepsilon}{\partial n^2}\,d\Gamma$$

$$+ \varepsilon \int_{[x \in \Gamma;\, |\frac{\partial u_\varepsilon}{\partial n}| \geq \lambda]} \text{sign}\left(\frac{\partial u_\varepsilon}{\partial n}\right) \frac{\partial^2 u_\varepsilon}{\partial n^2}\,d\Gamma.$$

As $\lambda \to 0$ we get

$$\int_\Omega |\text{grad } u_\varepsilon|\,dx \leq \int_\Omega |\text{grad } u|\,dx +$$

$$+ \varepsilon \int_\Gamma \text{sign}\left(\frac{\partial u_\varepsilon}{\partial n}\right) \frac{\partial^2 u_\varepsilon}{\partial n^2}\,d\Gamma.$$

(Note that, by (25), $\frac{\partial^2 u_\varepsilon}{\partial n^2} = 0$ on the set where $\frac{\partial u_\varepsilon}{\partial n} = 0$).

Applying (25) we see that $\left|\dfrac{\partial^2 u_\varepsilon}{\partial n^2}\right| \leq C \left|\dfrac{\partial u_\varepsilon}{\partial n}\right|$ on $\Gamma$ (where C depends only on $\Omega$). In the case where $\Omega$ is convex we have $\dfrac{\partial u_\varepsilon}{\partial n} \cdot \dfrac{\partial^2 u_\varepsilon}{\partial n^2} \leq 0$ on $\Gamma$ (by (25)) so that

$$\int_\Omega |\operatorname{grad} u_\varepsilon| dx \leq \int_\Omega |\operatorname{grad} u| dx .$$

<u>Proof of Theorem 15.</u> It is immediate that if $u \in H^2(\Omega) \cap H^1_0(\Omega)$ satisfies (23) then $f \in \partial\varphi(u)$.

Conversely assume $u \in H^1_0(\Omega)$ and $f \in \partial\varphi(u)$. By definition, we have

(26) $\dfrac{1}{2}\int_\Omega |\operatorname{grad} v|^2 dx + \int_\Omega |\operatorname{grad} v| dx - \dfrac{1}{2}\int_\Omega |\operatorname{grad} u|^2 dx$

$-\int_\Omega |\operatorname{grad} u| dx \geq \int_\Omega f \cdot (v-u) dx \qquad \forall\, v \in H^1_0(\Omega)$.

Taking $v = (1-t)u + tw$ in (26) with $t \in (0,1)$ and $w \in H^2(\Omega) \cap H^1_0(\Omega)$ we get, after letting $t \to 0$,

(27) $\int_\Omega (\operatorname{grad} u, \operatorname{grad} w - \operatorname{grad} u) dx + \int_\Omega |\operatorname{grad} w| dx$

$-\int_\Omega |\operatorname{grad} u| dx \geq \int_\Omega f \cdot (w-u) dx$

$\qquad\qquad\qquad \forall\, w \in H^2(\Omega) \cap H^1_0(\Omega)$.

Thus

$\int_\Omega (\operatorname{grad} w, \operatorname{grad} w - \operatorname{grad} u) dx + \int_\Omega |\operatorname{grad} w| dx$

$-\int_\Omega |\operatorname{grad} u| dx \geq \int_\Omega f \cdot (w-u) dx .$

Taking $w = u_\varepsilon$, as in Lemma 1, we obtain

$$\int_\Omega -\Delta u_\varepsilon \cdot \varepsilon \Delta u_\varepsilon \, dx + \int_\Omega |\text{grad } u_\varepsilon| \, dx - \int_\Omega |\text{grad } u| \, dx$$

$$\geq \int_\Omega f \cdot \varepsilon \Delta u_\varepsilon \, dx,$$

and

$$\int_\Omega |\Delta u_\varepsilon|^2 \, dx \leq C \|u_\varepsilon\|_{H^2(\Omega)} + \int_\Omega |f| |\Delta u_\varepsilon| \, dx.$$

Consequently $|\Delta u_\varepsilon|_{L^2(\Omega)}$ is bounded as $\varepsilon \to 0$ and $u \in H^2(\Omega)$.

Let now $k$ be a convex l.s.c. function from $\mathbb{R}$ into $(-\infty, +\infty]$ and let $\partial k = \gamma$. The function $\psi$ is defined on $L^2(\Omega)$ by

$$\psi(u) = \begin{cases} \int_\Omega k(u) \, dx & \text{if } k(u) \in L^1(\Omega) \\ +\infty & \text{otherwise.} \end{cases}$$

<u>Corollary 16.</u> <u>If</u> $0 \in D(\gamma)$, <u>then</u> $\partial \psi + \partial \psi$ <u>is maximal monotone and</u> $\overline{D(\partial \psi + \partial \psi)} = \{u \in L^2(\Omega); u(x) \in \overline{D(\gamma)} \text{ a.e on } \Omega\}$. <u>In particular, for every</u> $f \in L^2(\Omega)$, <u>there exists a unique function</u> $u \in H^2(\Omega) \cap H^1_0(\Omega)$ <u>satisfying</u>

$$\int_\Omega (-\Delta u + \gamma(u)) \cdot (v-u) dx + \int_\Omega |\text{grad } v| \, dx - \int_\Omega |\text{grad } u| \, dx$$

$$\geq \int_\Omega f \cdot (v-u) dx \qquad \forall v \in H^1_0(\Omega);$$

more precisely there exists $g \in L^2(\Omega)$ <u>such that</u> $g(x) \in \gamma(u(x))$ a.e. on $\Omega$ <u>and</u>

$$\int_\Omega (-\Delta u + g)(v-u) dx + \int_\Omega |\text{grad } v| \, dx - \int_\Omega |\text{grad } u| \, dx$$

$$\geq \int_\Omega f(v-u) dx \qquad \forall v \in H^1_0(\Omega).$$

Proof of Corollary 16. We reduce to the case where $0 \in \gamma(0)$. By Theorem 9 it is sufficient to show that

$$\varphi((I + \lambda\partial\psi)^{-1}u) \leq \varphi(u) \quad \forall u \in D(\varphi), \quad \forall \lambda > 0$$

which clearly holds since $(I + \lambda\gamma)^{-1}$ is a contraction on $\mathbb{R}$.

Remark. Problems of similar nature appear in [27].

Example 3. Let $\Lambda = -\Delta$ be the canonical isomorphism from $H_0^1(\Omega)$ onto $H^{-1}(\Omega)$. Let $H = H^{-1}(\Omega)$ with its usual scalar product $(\bar{\Lambda}^1 f, g)$ for $f, g \in H^{-1}(\Omega)$. Let $j$ be a convex l.s.c. function from $\mathbb{R}$ into $(-\infty, +\infty]$ with $j \not\equiv +\infty$, and let $\beta = \partial j$. We assume that $\lim_{|r| \mapsto +\infty} \frac{j(r)}{|r|} = +\infty$ (or equivalently $R(\beta) = \mathbb{R}$). For $u \in H^{-1}(\Omega)$ we define

$$\varphi(u) = \begin{cases} \int_\Omega j(u)dx & \text{if } u \in L^1(\Omega) \text{ and} \\ & j(u) \in L^1(\Omega) \\ +\infty & \text{otherwise} \end{cases}$$

Theorem 17. The function $\varphi$ is convex l.s.c. on $H^{-1}(\Omega)$. Also $f \in H^{-1}(\Omega)$, $f \in \partial\varphi(u)$ if and only if $(\bar{\Lambda}^1 f)(x) \in \beta(u(x))$ a.e. on $\Omega$.

Proof of Theorem 17. Let $u_n$ be a sequence such that $u_n \in H^{-1}(\Omega) \cap L^1(\Omega)$, $u_n \to u$ in $H^{-1}(\Omega)$ and $\int_\Omega j(u_n(x))dx \leq \lambda$. We first prove that $u_n \to u$ weakly in $L^1(\Omega)$ by using Dunford-Pettis theorem (see e.g. [15] p. 294). The integrals $\int |u_n|$ are uniformly absolutely continuous i.e. $\forall \varepsilon > 0 \ \exists \delta > 0$ such that meas $E < \delta$ implies $\int_E |u_n|dx < \varepsilon$. Indeed, let $A > \frac{2\lambda}{\varepsilon}$ and let $R$ be such that $\frac{j(r)}{|r|} \geq A$ for $|r| > R$. For $\delta < \frac{\varepsilon}{2R}$ we have

$$\int_E |u_n|\,dx \le \int_{[x \in E;\, |u_n(x)| \ge R]} |u_n|\,dx + \int_{[x \in E;\, |u_n(x)| < R]} |u_n|\,dx$$

$$\le \int_\Omega \frac{j(u_n)}{A}\,dx + R\delta \le \frac{\lambda}{A} + R\delta < \varepsilon \quad .$$

By Dunford-Pettis theorem there is a subsequence such that $u_{n_k} \to \tilde{u}$ weakly in $L^1(\Omega)$. Since we already know that $u_n \to u$ in $H^{-1}(\Omega)$, we conclude that $u_n \to u$ weakly in $L^1(\Omega)$. Finally the function $u \to \int_\Omega j(u)\,dx$ is convex l.s.c. on $L^1(\Omega)$ (by Fatou's lemma) and thus it is weakly l.s.c. on $L^1(\Omega)$.

We define the operator $A$ on $H^{-1}(\Omega)$ to be
$Au = \{\Lambda w;\ w \in H_0^1(\Omega)$ and $w(x) \in \beta(u(x))$ a.e. on $\Omega\}$
with $u \in D(A)$ if and only if there is some $w \in H_0^1(\Omega)$ such that $w(x) \in \beta(u(x))$ a.e. on $\Omega$. We prove now that $A \subset \partial\varphi$ and next that $A$ is a maximal monotone. We need the following

<u>Lemma 2.</u>  Let $F \in H^{-1}(\Omega) \cap L^1(\Omega)$ and let $w \in H_0^1(\Omega)$. Let $g \in L^1(\Omega)$ and let $h$ be measurable with

(28) $\qquad F \cdot w \ge h \ge g \qquad$ a.e. on $\Omega$ .

Then $h \in L^1(\Omega)$ and $(F,w) \ge \int_\Omega h\,dx$ (where $(,)$ denotes the scalar product in the duality between $H_0^1$ and $H^{-1}$).

<u>Proof of Lemma 2.</u>  Let

$$w_n = \begin{cases} n & \text{if } w \ge n \\ w & \text{if } |w| \le n \\ -n & \text{if } w \le -n \end{cases}$$

Let $h_n = h \dfrac{w_n}{w}$ and let $g_n = g \dfrac{w_n}{w}$. Multiplying (28) by

$\frac{w_n}{w}$ we get

$$F \cdot w_n \geq h_n \geq g_n \quad \text{a.e. on } \Omega$$

and hence

$$0 \leq h_n - g_n \leq F \cdot w_n - g_n \quad \text{a.e. on } \Omega.$$

The sequence $h_n - g_n \to h - g$ a.e. on $\Omega$ as $n \to +\infty$ and also

$$\int_\Omega (h_n - g_n) dx \leq \int_\Omega F \cdot w_n dx - \int_\Omega g_n dx = (F, w_n) - \int_\Omega g_n \, dx.$$

Since $w_n \to w$ in $H_0^1(\Omega)$ and $g_n \to g$ in $L^1(\Omega)$, we conclude by Fatou's lemma that $h - g \in L^1(\Omega)$ and thus $h \in L^1(\Omega)$ with

$$\int_\Omega (h-g) \, dx \leq (F, w) - \int_\Omega g(x) dx.$$

Proof of Theorem 17. continued.

Let $f \in Au$, i.e. $u \in H^{-1}(\Omega) \cap L^1(\Omega)$, $f = \Lambda w$ with $w \in H_0^1(\Omega)$, $w(x) \in \beta(u(x))$ a.e. on $\Omega$. Let $v \in H^{-1}(\Omega) \cap L^1(\Omega)$ be such that $j(v) \in L^1(\Omega)$. We have $j(v) - j(u) \geq w \cdot (v - u)$ a.e. on $\Omega$. Applying Lemma 2 with $F = u - v$, $h = j(u) - j(v)$ and $g = -C_1|u| - C_2 - j(v)$ ($j(r) \geq -C_1|r| - C_2$), we conclude that $j(u) \in L^1(\Omega)$ and

$$\int_\Omega j(v) dx - \int_\Omega j(u) dx \geq (w, v-u) = (\Lambda^{-1} f, v-u).$$

Hence $f \in \partial\varphi(u)$.

We prove now that $A$ is maximal monotone. For a given $f \in H^{-1}(\Omega)$ we have to find $u \in H^{-1}(\Omega) \cap L^1(\Omega)$ and $w \in H_0^1(\Omega)$ such that $u + \Lambda w = f$ and $w(x) \in \beta(u(x))$ a.e. on $\Omega$. Let $\gamma = \bar{\beta}^1$ so that $D(\gamma) = \mathbb{R}$. Without loss of generality we can assume that $0 \in \gamma(0)$. Let $w_\lambda \in H_0^1(\Omega)$ be the solution of the equation

(29) $$\gamma_\lambda(w_\lambda) + \Lambda w_\lambda = f$$

(which exists by standard results).

Multiplying (29) by $w_\lambda$ and integrating over $\Omega$ we see that $w_\lambda$ is bounded in $H_0^1(\Omega)$ as $\lambda \to 0$. Thus we can find a sequence $\lambda_n \to 0$ such that $w_{\lambda_n} \to w$ weakly in $H_0^1(\Omega)$, $w_{\lambda_n} \to w$ a.e. on $\Omega$, $(I + \lambda_n \gamma)^{-1} w_{\lambda_n} \to w$ a.e. on $\Omega$. The proof of Theorem 17 is completed by using the following general result.

**Theorem 18.** <u>Let $\gamma$ be a maximal monotone graph in $\mathbb{R} \times \mathbb{R}$ such that $D(\gamma) = \mathbb{R}$ and $0 \in \gamma(0)$. Let $f_n$ and $v_n$ be measurable functions on $\Omega$ such that $v_n \to v$ a.e. on $\Omega$, $f_n(x) \in \gamma(v_n(x))$ a.e. on $\Omega$ and $f_n \cdot v_n \in L^1(\Omega)$ with $\int_\Omega f_n \cdot v_n \, dx \leq C$. Then, there is a subsequence $n_k \to +\infty$ such that $f_{n_k} \to f$ weakly in $L^1(\Omega)$ and $f(x) \in \gamma(v(x))$ a.e. on $\Omega$.</u>

**Proof of Theorem 18.** Let $\beta = \gamma^1$ and let $j$ be such that $j(0) = 0$, $\partial j = \beta$. Since $R(\beta) = \mathbb{R}$ we get $\lim\limits_{|r| \to +\infty} \frac{j(r)}{|r|} = +\infty$.
We have a.e. on $\Omega$

$$j(0) - j(f_n(x)) \geq v \cdot -f_n(x) \qquad \text{for every } v \in \beta(f_n(x)).$$

Consequently

$$j(f_n(x)) \leq v_n(x) \, f_n(x) \qquad \text{a.e. on } \Omega,$$

and so

$$\int_\Omega j(f_n(x)) dx \leq \int_\Omega f_n \, v_n \, dx \leq C.$$

Using again Dunford-Pettis theorem (as in the beginning of the proof of Theorem 17) we get a sequence $n_k \to +\infty$ such that $f_{n_k} \to f$ weakly in $L^1(\Omega)$. We conclude the proof of Theorem 19 with the help of the following

**Lemma 3.** <u>Let $M$ be maximal monotone in a Hilbert space $\mathcal{H}$. Let $f_n$ and $v_n$ be measurable functions from $\Omega$ (a finite measure space) into $\mathcal{H}$. Assume $v_n \to v$ a.e. on $\Omega$ and $f_n \to f$ weakly in $L^1(\Omega; \mathcal{H})$. If $f_n(x) \in M v_n(x)$ a.e. on $\Omega$, then $f(x) \in M v(x)$ a.e. on $\Omega$.</u>

Proof of Lemma 3. It is sufficient to show that, for every $N$, $f(x) \in Mv(x)$ a.e. on $\Omega_N = \{x \in \Omega; |v(x)| \leq N\}$. By Egorov's lemma, for every $\delta > 0$ there is a set $E \subset \Omega_N$ such that meas $E < \delta$ and $v_n \to v$ uniformly in $\Omega_N \setminus E$. Thus we are reduced to the case where v is bounded on $\Omega$ and $v_n \to v$ uniformly on $\Omega$. Without loss of generality we may assume now that $D(M)$ is bounded (if $D(M)$ is not bounded, consider $\tilde{M} = M + \partial I_B$, where $I_B$ is the indicator function of a ball centered at $0$ of large radius). Let $\tilde{v} \in L^\infty(\Omega; \mathcal{H})$ and let $\tilde{f} \in L^1(\Omega; \mathcal{H})$ be such that $\tilde{f}(x) \in M\tilde{v}(x)$ a.e. on $\Omega$. By the monotonicity of $M$, we have $(\tilde{f} - f_n, \tilde{v} - v_n) \geq 0$ a.e. on $\Omega$, and thus $\int_\Omega (\tilde{f} - f_n, \tilde{v} - v_n) dx \geq 0$. Consequently $\int_\Omega (\tilde{f} - f, \tilde{v} - v) dx \geq 0$. Let now $\tilde{v} = (I + M)^{-1}(v + f)(\tilde{v} \in L^\infty(\Omega; \mathcal{H})$ since $D(M)$ is bounded). We have $\tilde{v} + M\tilde{v} \ni v + f$ a.e. on $\Omega$. Choosing $\tilde{f} = v + f - \tilde{v}$ we get $\int_\Omega |v - \tilde{v}|^2 dx \leq 0$ so that $\tilde{v} = v$ and $f \in Mv$ a.e. on $\Omega$.

Example 4. Let $\mathcal{H}$ be a Hilbert space and let $H = L^2(0, T; \mathcal{H})$ with its usual Hilbert structure. Given $u_0 \in \mathcal{H}$ we define on $H$ the maximal monotone operator $A$ by $Au = \dfrac{du}{dt}$ with

$$D(A) = \{u \in H; \frac{du}{dt} \in H \text{ in the sense of distributions and}$$

$$u(0) = u_0\} .$$

Let $\varphi$ be a convex l.s.c. function from $\mathcal{H}$ into $(-\infty, +\infty]$ such that $\varphi \not\equiv +\infty$. We introduce on $H$ the function $\Phi$ by

$$\Phi(u) = \begin{cases} \int_0^T \varphi(u(t)) dt & \text{if } \varphi(u) \in L^1(0, T) \\ +\infty & \text{otherwise} \end{cases} .$$

It is easy to check (using Fatou's lemma) that $\Phi$ is l.s.c. and that $f \in \partial \Phi(u)$ if and only if $f(t) \in \partial \varphi(u(t))$ a.e. on $(0, T)$.

**Theorem 19.** If $u_0 \in D(\varphi)$ then $A + \partial \Phi$ is maximal monotone.

**Proof of Theorem 19.** First we reduce to the case where $\varphi \geq 0$ (this amounts to shift $\partial \Phi$ by a constant). We are going to prove that

(30) $\quad \Phi((I + \lambda A)^{-1} u) \leq \Phi(u) + \varphi(u_0) \lambda \qquad \forall u \in H, \quad \forall \lambda > 0.$

Indeed, let $u_\lambda = (I + \lambda A)^{-1} u$ i.e.

$$\lambda \frac{du_\lambda}{dt} + u_\lambda = u \quad \text{and} \quad u_\lambda(0) = u_0.$$

Let $\varphi_\alpha$ be the approximation of $\varphi$ introduced in Theorem 4. We have

$$\varphi_\alpha(u(t)) - \varphi_\alpha(u_\lambda(t)) \geq (\partial \varphi_\alpha(u_\lambda(t)), u(t) - u_\lambda(t)) =$$

$$= (\partial \varphi_\alpha(u_\lambda(t)), \lambda \frac{du_\lambda}{dt}(t)) = \lambda \frac{d}{dt} \varphi_\alpha(u_\lambda(t)).$$

Hence

$$\int_0^T \varphi_\alpha(u_\lambda) dt \leq \int_0^T \varphi_\alpha(u) dt - \varphi_\alpha(u_\lambda(T)) + \varphi_\alpha(u_0) \leq \Phi(u) + \varphi(u_0).$$

Passing to the limit as $\alpha \to 0$, we get (30), which implies by Theorem 9 that $A + \partial \Phi$ is maximal monotone.

**Corollary 20.** For every $f \in L^2(0, T; H)$ and $u_0 \in D(\varphi)$ there exists a unique $u \in C([0, T]; H)$ such that $u(t) \in D(\partial \varphi)$ a.e. on $(0, T)$, $\frac{du}{dt} \in L^2(0, T; H)$,

(31) $\quad \begin{cases} \frac{du}{dt} + \partial \varphi(u) \ni f & \text{a.e. on } (0, T) \\ u(0) = u_0. \end{cases}$

In addition the following estimates holds

(32) $\quad \left( \int_0^T \left| \frac{du}{dt} \right|^2 dt \right)^{1/2} \leq \left( \int_0^T |f|^2 dt \right)^{1/2} + |\varphi(u_0)|^{1/2} + C_1 |u_0|^{1/2} + C_2$

(where $C_1$ and $C_2$ depend only on $\varphi$ ; in particular $C_1 = C_2 = 0$ if $\varphi \geq 0$).

Proof of Corollary 20. By Theorems 5 and 19 it is sufficient to show that $(A + \partial \Phi)^{-1}$ is bounded. Let $f \in (A + \partial \Phi)(u)$ and and let $v_0 \in D(\varphi)$ be fixed. We have

$$\varphi(v_0) - \varphi(u(t)) \geq (f(t) - \frac{du}{dt}(t), v_0 - u(t)) \quad \text{a.e. on } (0,T).$$

Let $\tilde{\varphi}(u) = \varphi(u) + (a,u) + b$, be such that $\tilde{\varphi} \geq 0$ on $H$. Thus $\frac{1}{2} \frac{d}{dt} |u - v_0|^2 \leq \varphi(v_0) + |f||u - v_0| + |a||u| + |b|$
and we get a bound on $|u|_H$ provided $|f|_H$ is bounded. Estimate (32) follows easily from (30) and Theorem 9 applied with $\tilde{\varphi}$ instead of $\varphi$.

This example leads us to evolution equations associated with maximal monotone operators.

III. Evolution equations associated with maximal monotone operators.

Let $A$ be a maximal monotone in a Hilbert space $H$. One of the main results is the following

Theorem 21. Let f be absolutely continuous from $[0,T]$ into $H$ and let $u_0 \in D(A)$. There exists a unique function $u \in C([0,T];H)$ satisfying

(33) $\quad u(t) \in D(A) \qquad \forall t \in [0,T]$

(34) $\quad u(t)$ is Lipschitz continuous on $[0,T]$ (and thus $u(t)$ is differentiable a.e. on $(0,T)$).

(35) $\quad \frac{du}{dt} + Au \ni f \qquad$ a.e. on $(0,T)$

(36) $\quad u(0) = u_0$.

The proofs of Theorem 21 and of the following remarks can be found in Kato [19]. But many authors, including

F. Browder [12], Crandall-Pazy [13], Dorroh [14], Komura [20], have considered related problems.

**Remark.** The solution of (35)-(36) has some additional properties:

(37)  $u$ is differentiable from the right at every $t \in [0, T)$

and $\dfrac{d^+u}{dt}(t) + (Au(t) - f(t))^0 = 0 \qquad \forall\, t \in [0, T)$

(38) $\left|\dfrac{d^+u}{dt}(t)\right| \leq \left|\dfrac{d^+u}{dt}(0)\right| + \int_0^t \left|\dfrac{df}{dt}(s)\right| ds =$

$\left|(Au_0 - f(0))^0\right| + \int_0^t \left|\dfrac{df}{dt}(s)\right| ds \qquad \forall\, t \in [0, T]$.

(39) Given $f, \hat{f}$ and $u_0, \hat{u}_0$, the corresponding solutions $u, \hat{u}$ satisfy

$|u(t) - \hat{u}(t)| \leq |u_0 - \hat{u}_0| + \int_0^t |f(s) - \hat{f}(s)| ds.$

From (39) it is clear that the mapping $\{u_0, f\} \mapsto u$ can be extended by continuity from $\overline{D(A)} \times L^1(0, T; H)$ into $C([0, T]; \overline{D(A)})$. In the case where $f = 0$, the mapping $u_0 \mapsto u(t)$ is called the semigroup of nonlinear contractions generated by $-A$ on $\overline{D(A)}$. It may well happen, even in the linear case, that for $u_0 \in \overline{D(A)}$ and $f \in L^1(0, T; H)$, the corresponding $u(t)$ is nowhere differentiable and $u(t) \notin D(A)$ $\forall\, t \in [0, T]$. Thus, we have to consider $u(t)$ as a generalized solution of (35)-(36). We are going to show that for some special classes of maximal monotone operators $A$, in particular $A = \partial \varphi$ or Int $D(A) \neq \phi$ (which correspond in the linear case to self adjoint or bounded operators), the generalized solution is "almost" a strong solution.

We consider first the case where $A = \partial \varphi$, $\varphi$ being a convex l.s.c. function from $H$ into $(-\infty, +\infty]$ with $\min_H \varphi = 0$. Let $K = \{v \in H;\ \varphi(v) = 0\}$.

**Theorem 22.** *For every* $u_0 \in \overline{D(\varphi)}$ *and* $f \in L^2(0,T;H)$ *there is a unique solution* $u \in C([0,T];H)$ *of* (35)-(36) *satisfying*

(40) $\quad u(t) \in D(A) \quad$ a.e. on $(0,T)$

(41) $\quad \sqrt{t}\, \dfrac{du}{dt} \in L^2(0,T;H)$ .

In addition we have the estimates

(42) $\quad \left(\displaystyle\int_0^T \left|\dfrac{du}{dt}(t)\right|^2 t\, dt\right)^{1/2}$

$$\leq \left(\int_0^T |f(t)|^2 t\, dt\right)^{1/2} + \frac{1}{\sqrt{2}} \int_0^T |f(t)|\, dt + \frac{1}{\sqrt{2}}\, \text{dist}(u_0, K)$$

(43) $\quad \left(\displaystyle\int_\delta^T \left|\dfrac{du}{dt}(t)\right|^2 dt\right)^{1/2}$

$$\leq \left(\int_0^T |f(t)|^2 dt\right)^{1/2} + \frac{1}{\sqrt{2\delta}} \int_0^\delta |f(t)|\, dt + \frac{1}{\sqrt{2\delta}}\, \text{dist}(u_0, K)$$

$$\forall\, \delta \in (0,T) .$$

**Remark.** Note that in Theorem 22 we assume only $u_0 \in \overline{D(\varphi)}$ (instead of $u_0 \in D(\varphi)$ in Corollary 20). Theorem 22 is closely related to Proposition 5 in [4] but we give here a new proof which exploits a suggestion of P. Lax.

**Proof of Theorem 22.** As in [4], the crucial point is to establish the estimate (42) (or (43)) for the Yosida approximation $\dfrac{du_\lambda}{dt} + A_\lambda u_\lambda = f,\ u_\lambda(0) = u_0$ (where $A_\lambda = \partial \varphi_\lambda$, see Theorem 4). Next one can pass to the limit as $\lambda \to 0$ using standard devices. In order to simplify the notations we drop $\lambda$.

**Estimate I.** (energy estimate) Let $v_0 \in K$; we have

$$\varphi(v_0) - \varphi(u(t)) \geq (\partial\varphi(u(t)), v_0 - u(t)) = (f(t) - \dfrac{du}{dt}(t), v_0 - u(t)) .$$

131

Hence
$$\int_0^T \varphi(u(t))dt \le \int_0^T |f(t)| |u(t)-v_0| dt + \frac{1}{2}|u_0 - v_0|^2.$$

But $|u(t) - v_0| \le |u_0 - v_0| + \int_0^t |f(s)|ds$ (apply (39) with $\hat{u} \equiv v_0$ and $\hat{f} \equiv 0$). Consequently

(44) $\quad \int_0^T \varphi(u(t))dt \le \frac{1}{2}(|u_0 - v_0| + \int_0^T |f(t)|dt)^2.$

**Estimate II.** Multiplying (35) by $t\frac{du}{dt}$ we get
$$t|\frac{du}{dt}(t)|^2 + t\frac{d}{dt}\varphi(u(t)) = (f(t), t\frac{du}{dt}(t))$$

Consequently

(45) $\quad \int_0^T |\frac{du}{dt}(t)|^2 t\, dt + T\varphi(u(T))$

$$\le \int_0^T |f(t)| |\frac{du}{dt}(t)| t\, dt + \int_0^T \varphi(u(t))dt.$$

Thus

(46) $\quad (\int_0^T |\frac{du}{dt}(t)|^2 t\, dt)^{1/2}$

$$\le (\int_0^T |f(t)|^2 t\, dt)^{1/2} + (\int_0^T \varphi(u(t))dt)^{1/2}.$$

Combining (44) and (46) we obtain (42). From (45) we deduce also that

(47) $\quad T\varphi(u(T)) \le \frac{1}{4}\int_0^T |f(t)|^2 t\, dt + \int_0^T \varphi(u(t))dt.$

Multiplying (35) by $\frac{du}{dt}$ and integrating on $(\delta, T)$ we get

$$\int_\delta^T |\frac{du}{dt}(t)|^2 dt \le \varphi(u(\delta)) + \int_\delta^T |f(t)| |\frac{du}{dt}(t)| dt.$$

Hence using (47) with $T = \delta$ we have

$$\int_\delta^T (|\frac{du}{dt}(t)| - \frac{1}{2}|f(t)|)^2 dt \le \frac{1}{4}\int_\delta^T |f(t)|^2 dt + \frac{1}{4\delta}\int_0^\delta |f(t)|^2 t\, dt$$

$$+ \frac{1}{2\delta}(|u_0 - v_0| + \int_0^\delta |f(t)|dt)^2.$$

The estimate (43) follows easily.

In the case where f is smooth, u(t) has some further properties.

<u>Theorem 23.</u> <u>Let</u> $u_0 \in \overline{D(\varphi)}$ <u>and let</u> f <u>be absolutely continuous from</u> $[0,T]$ <u>into</u> H. <u>Then the corresponding solution of (35)-(36) satisfies</u>

(48)     $u(t) \in D(A)$            $\forall\, t \in (0,T]$

(49)     $t\,|\frac{du}{dt}(t)| \in L^\infty(0,T)$

(50)     $\frac{d^+u}{dt}(t)$ exists for all $t \in (0,T)$ and

$$\frac{d^+u}{dt}(t) + (Au(t) - f(t))^0 = 0 \qquad \forall\, t \in (0,T).$$

In addition we have the estimate

(51)     $|\frac{d^+u}{dt}(t)| \le \int_0^t |\frac{df}{dt}(s)|\frac{s^2}{t^2}dx + \frac{\sqrt{2}}{t}(\int_0^t |f(s)|^2 s\,ds)^{1/2}$

$$+ \frac{1}{t}\int_0^t |f(s)|ds + \frac{1}{t}\,\text{dist}(u_0, K) \qquad \forall\, t \in (0,T).$$

<u>Proof of Theorem 23.</u> Since $u(t) \in D(A)$ a.e. on $(0,T)$ and f is absolutely continuous, we conclude by Theorem 21 that (48) and (50) hold. In order to establish (51) we use (42) and the fact that the function $t \mapsto |\frac{d^+u}{dt}(t)| - \int_0^t |\frac{df}{dt}(s)|ds$ is nonincreasing in t (see (38)). For $0 < t < T$ we have

$$\left|\frac{d^+u}{dt}(T)\right| \le \left|\frac{d^+u}{dt}(t)\right| + \int_t^T \left|\frac{df}{dt}(s)\right| ds .$$

Consequently

$$\int_0^T \left[\left|\frac{d^+u}{dt}(T)\right| - \int_t^T \left|\frac{df}{dt}(s)\right| ds\right] t\, dt \le \int_0^T \left|\frac{du}{dt}(t)\right| t\, dt$$

$$\le \frac{T}{\sqrt{2}} \left(\int_0^T \left|\frac{du}{dt}(t)\right|^2 t\, dt\right)^{1/2} .$$

Thus

$$\frac{T^2}{2}\left|\frac{d^+u}{dt}(T)\right| \le \int_0^T t\left(\int_t^T \left|\frac{df}{dt}(s)\right| ds\right) dt + \frac{T}{\sqrt{2}}\left(\int_0^T \left|\frac{du}{dt}(t)\right|^2 t\, dt\right)^{1/2}$$

$$= \int_0^T \left|\frac{df}{dt}(t)\right| \frac{t^2}{2} dt + \frac{T}{\sqrt{2}}\left(\int_0^T \left|\frac{du}{dt}(t)\right|^2 t\, dt\right)^{1/2} .$$

We conclude by applying (42).

The estimate (51) provides also informations about the behavior of $u(t)$ as $t \to +\infty$. Let $\varphi$ be a convex l. s. c. function from $H$ into $(-\infty, +\infty]$, $\varphi \not\equiv +\infty$. Let $f$ be a function from $[0, +\infty)$ into $H$ such that $\left|\frac{df}{dt}\right| \in L^1(0, +\infty)$, so that $\lim f(t) = f_\infty$ exists. We assume that $f_\infty \in R(\partial\varphi)$.

**Theorem 24.** Let $u_0 \in \overline{D(\varphi)}$ and let $u$ be the solution of $\frac{du}{dt} + \partial\varphi(u) \ni f$ a.e. on $(0, +\infty)$, $u(0) = u_0$. Then $\lim_{t \to +\infty} \left|\frac{du}{dt}(t)\right| = 0$. If in addition $\left|\frac{df}{dt}(t)\right| = 0(t^{-\alpha})$ with $\alpha > 2$ as $t \to +\infty$, then $\left|\frac{du}{dt}(t)\right| = 0(t^{-1})$ as $t \to +\infty$.

**Proof of Theorem 24.** Let $\tilde\varphi(u) = \varphi(u) - (f_\infty, u) - \underset{H}{\text{Inf}}\{\varphi(u) - (f_\infty, u)\}$, since $f_\infty \in R(\partial\varphi)$, Min $\tilde\varphi = 0$. We have $\frac{du}{dt} + \partial\tilde\varphi(u) \ni f - f_\infty$ so that by (51) we obtain

(52) $\quad |\dfrac{d^+u}{dt}(t)| \le \int_0^t |\dfrac{df}{dt}(s)| \, \dfrac{s^2}{t^2} \, ds + \dfrac{\sqrt{2}}{t} (\int_0^t |f(s)-f_\infty|^2 s\, ds)^{1/2}$

$\qquad + \dfrac{1}{t}\int_0^t |f(s)-f_\infty|\, ds + \dfrac{1}{t}\, \text{dist}\,(u_0, K)\,,$

where now $K = \{v \in H;\ \partial\varphi(v) \ni f_\infty\}$. Let $\varepsilon$ be fixed and let $t_0$ be such that $\int_{t_0}^t |\dfrac{df}{dt}(s)|\, ds < \varepsilon$ for $t \ge t_0$. We have, for $t$ large enough,

$$\int_0^t |\dfrac{df}{dt}(s)|\, \dfrac{s}{t}\, ds \le \dfrac{1}{t}\int_0^{t_0} |\dfrac{df}{dt}(s)|\, s\, ds + \int_{t_0}^t |\dfrac{df}{dt}(s)|\, ds < 2\varepsilon\,.$$

Since $\int_0^t \left|\dfrac{df}{dt}(s)\right| \dfrac{s^2}{t^2}\, ds \le \int_0^t \left|\dfrac{df}{dt}(s)\right| \dfrac{s}{t}\, ds$, we get

$$\lim_{t\to+\infty} \int_0^t \left|\dfrac{df}{dt}(s)\right| \dfrac{s^2}{t^2}\, ds = \lim_{t\to+\infty} \int_0^t \left|\dfrac{df}{dt}(s)\right| \dfrac{s}{t}\, ds = 0\,.$$

On the other hand

$$\dfrac{1}{t}(\int_0^t |f(s)-f_\infty|^2\, s\, ds)^{1/2} \le M\,(\dfrac{1}{t}\int_0^t |f(s)-f_\infty|\, ds)^{1/2}$$

where $M = \sup\limits_{s \ge 0} |f(s)-f_\infty|$.

Thus it remains only to prove that $\lim\limits_{t\to+\infty} \dfrac{1}{t}\int_0^t |f(s)-f_\infty|\, ds = 0$.

But $\dfrac{1}{t}\int_0^t |f(s)-f_\infty|\, ds \le \dfrac{1}{t}\int_0^t ds \int_s^{+\infty}|\dfrac{df}{dt}(\tau)|\, d\tau =$

$\int_t^{+\infty} |\dfrac{df}{dt}(\tau)|\, d\tau + \int_0^t |\dfrac{df}{dt}(s)|\, \dfrac{s}{t}\, ds\,.$

In the case where $|\dfrac{df}{dt}(t)| \le C t^{-\alpha}$, $\alpha > 2$, it is easy to check that $|f(t) - f_\infty| \le \int_t^\infty |\dfrac{df}{dt}(s)|\, ds \le C t^{1-\alpha}$ and that, by (52)

$$\left|\frac{d^+u}{dt}(t)\right| \leq C t^{1-\alpha} + \frac{1}{t} \text{ dist }(u_0, K) \quad \text{as} \quad t \to +\infty.$$

**Remark.** In the case where $(Ax - Ay, x-y) \geq \gamma |x - y|^2$ for some $\gamma > 0$, it is standard that $\left|\frac{du}{dt}(t)\right|$ decays exponentially as $t \to 0$ provided $\left|\frac{df}{dt}\right| \to 0$ fast enough.

**Problem.** Suppose, for simplicity, that $f \equiv 0$. It would be of great interest to determine whether $\lim_{t \to +\infty} u(t)$ exists (one can prove it under some additional assumptions, for example if $\{u \in H \; ; \; |u| \leq C_1 \text{ and } \varphi(u) \leq C_2 \}$ is compact for every $C_1, C_2$; see Remark 6 in [4]). Assuming $\lim_{t \to +\infty} u(t) = u_\infty$ exists, one would like to have further information about the mapping $u_0 \mapsto u_\infty$. For example, is it true that $u_\infty = \lim_{n \to +\infty} (I + \lambda \partial \varphi)^{-n} u_0$?

Evolution equations associated with maximal monotone operators $A$ such that Int $D(A) \neq \phi$ have also remarkable properties. We restrict ourselves to the case where $f$ is absolutely continuous, but further results for the case where $f \in L^1(0, T; H)$ can be found in [2].

**Theorem 25.** <u>Let $A$ be maximal monotone with Int $D(A) \neq \phi$. Let $u_0 \in \overline{D(A)}$ and let $f$ be absolutely continuous on $[0, T]$. There exists a unique function $u \in C([0, T]; H)$ satisfying</u> $\left|\frac{du}{dt}\right| \in L^1(0, T)$, (48), (49), (50) and $u(0) = u_0$.

**Proof of Theorem 25.** Let $u_\lambda$ be the solution of the equation $\frac{du_\lambda}{dt} + A_\lambda u_\lambda = f$, $u_\lambda(0) = u_0$. Let $v_0 \in$ Int $D(A)$; by Theorem 6, there is a constant $C$ such that $|v - v_0| < \rho$ implies $v \in D(A)$ and $|A^0 v| \leq C$. We have $(A_\lambda u_\lambda - A_\lambda v, u_\lambda - v) \geq 0$ which implies, for $v = v_0 + \rho w$, $|w| \leq 1$, $\rho |A_\lambda u_\lambda| \leq (A_\lambda u_\lambda, u_\lambda - v_0) + C|u_\lambda - v_0| + C\rho$ or

$$\rho \left|\frac{du_\lambda}{dt}\right| \leq \rho(|f| + C) + (f - \frac{du_\lambda}{dt}, u_\lambda - v_0) + C|u_\lambda - v_0|.$$

Hence

$$\rho \int_0^T |\frac{du_\lambda}{dt}| dt \leq \rho \int_0^T (|f| + C) dt + \int_0^T (|f| + C) |u_\lambda - v_0| dt$$

$$+ \frac{1}{2} |u_0 - v_0|^2.$$

But $|u_\lambda(t) - v_0| \leq |u_0 - v_0| + \int_0^t (|f(s)| + C) ds$ (apply (39) with $\hat{u} \equiv v_0$). Consequently

(53) $\quad \rho \int_0^T |\frac{du_\lambda}{dt}| dt \leq \rho \int_0^T (|f| + C) dt + |u_0 - v_0| \int_0^T (|f| + C) dt$

$$+ \frac{1}{2} (\int_0^T (|f| + C) dt)^2 + \frac{1}{2} |u_0 - v_0|^2 =$$

$$= \rho \int_0^T (|f| + C) dt + \frac{1}{2} (|u_0 - v_0| + \int_0^T (|f| + C) dt)^2.$$

On the other hand, we have for $t \in [0,T]$

$$|\frac{du_\lambda}{dt}(T)| \leq |\frac{du_\lambda}{dt}(t)| + \int_t^T |\frac{df}{dt}(s)| ds.$$

Finally

(54) $\quad T|\frac{du_\lambda}{dt}(T)| \leq \int_0^T |\frac{du_\lambda}{dt}| dt + \int_0^T (\int_t^T |\frac{df}{dt}(s)| ds) dt$

$$\leq \int_0^T (|f| + C) dt + \frac{1}{2\rho} (|u_0 - v_0| + \int_0^T (|f| + C) dt)^2 +$$

$$+ \int_0^T |\frac{df}{dt}(t)| t \, dt.$$

The estimates (53) and (54) are independent of $\lambda$; (54) actually shows that $t \frac{du_\lambda}{dt}(t)$ is bounded in $L^\infty(0,T;H)$. We conclude the proof by passing to the limit as $\lambda \to 0$.

**Behavior as** $t \to +\infty$. Assume now that $|\frac{df}{dt}| \in L^1(0,+\infty)$ so that $\lim_{t \to +\infty} f(t) = f_\infty$ exists and assume $|f - f_\infty| \in L^1(0,+\infty)$.

**Theorem 26.** *Suppose* Int $(\bar{A}^1 f_\infty) \neq \phi$ *and let* u *be the solution of* (35), (36). *Then* $|\frac{du}{dt}| \in L^1(0,+\infty)$, $\lim_{t \to +\infty} u(t) = u_\infty$ *exists,* $f_\infty \in A u_\infty$ *and* $\lim_{t \to +\infty} |\frac{du}{dt}(t)| = 0$. *If in addition* $t|\frac{df}{dt}(t)| \in L^1(0,+\infty)$, *then* $|\frac{du}{dt}| = 0(t^{-1})$ *as* $t \to +\infty$.

**Proof of Theorem 26.** Let $\tilde{A}u = Au - f_\infty$; we have $\frac{du}{dt} + \tilde{A}u \ni f - f_\infty$ a.e. on $(0,+\infty)$. If we take in the proof of Theorem 25, $v_0 \in \text{Int}(\bar{A}^1 f_\infty)$, we have $0 \in \tilde{A}v$ for $|v - v_0| < \rho$ and thus $C = 0$ is permissible. Hence, by (53) and (54) we have

$$\rho \int_0^T |\frac{du}{dt}| dt \leq \rho \int_0^T |f - f_\infty| dt + \frac{1}{2}(|u_0 - v_0| + \int_0^T |f - f_\infty| dt)^2$$

and

$$|\frac{d^+u}{dt}(t)| \leq \frac{1}{t}\int_0^t |f(s) - f_\infty| ds + \frac{1}{2\rho t}(|u_0 - v_0| + \int_0^t |f - f_\infty| ds)^2$$
$$+ \int_0^t |\frac{df}{dt}(s)| \frac{s}{t} ds.$$

One can also consider sums of operators in the previous classes. Let $\varphi$ be a convex l.s.c. function from H into $(-\infty,+\infty]$, $\varphi \not\equiv +\infty$ and let A be maximal monotone with (Int $D(A)) \cap D(\partial\varphi) \neq \phi$.

**Theorem 27.** *Let* f *be absolutely continuous from* [0,T] *in into* H *and let* $u_0 \in \overline{D(A)} \cap \overline{D(\varphi)}$. *There exists a unique function* $u \in C([0,T];H)$ *satisfying*

$u(t) \in D(A) \cap D(\partial\varphi)$ $\quad \forall t \in (0,T]$

$t\frac{du}{dt} \in L^\infty(0,T)$

$\frac{du}{dt} + Au + \partial\varphi(u) \ni f$ a.e. on $(0,T)$, $u(0) = u_0$.

Remark. The case $f = 0$ has been treated previously by F. Browder using a different and ingenious argument.

Proof of Theorem 27. We know already that $A + \partial\varphi$ is maximal monotone. Also for any maximal monotone $B$ such that $(\text{Int } D(A)) \cap D(B) \ne \phi$ we have

$$\overline{D(A + B)} = \overline{D(A) \cap D(B)} = \overline{D(A)} \cap \overline{D(B)}.$$

Indeed, let $u \in \overline{D(A)} \cap \overline{D(B)}$ and let $v \in \text{Int } D(A)$ such that $|v - u| < \varepsilon$. Let $u_\lambda = (I + \lambda B)^{-1}u$ and $v_\lambda = (I + \lambda B)^{-1}v$. We have $|v_\lambda - u_\lambda| \le |v - u| < \varepsilon$ and $|v_\lambda - u| \le \varepsilon + |u_\lambda - u|$. Choose $\lambda$ small enough so that $|u_\lambda - u| < \varepsilon$ and $v_\lambda \in \text{Int } D(A)$. Let $v_0 \in \text{Int } D(A) \cap D(\partial\varphi)$; we can always assume that $\text{Min } \varphi = \varphi(v_0) = 0$. We consider the approximate equation

$$\frac{du_\lambda}{dt} + A_\lambda u_\lambda + \partial\varphi_\lambda(u_\lambda) = f, \qquad u_\lambda(0) = u_0$$

and then we pass easily to the limit as $\lambda \to 0$, after having established the crucial estimates. In order to simplify the notations we drop $\lambda$.

Estimate I. As in the proof of Theorem 25 we have

$$\rho |Au| \le (Au, u - v_0) + C|u - v_0| + C\rho$$

Hence

$$\rho |Au| \le (f - \frac{du}{dt} - \partial\varphi(u), u - v_0) + C|u - v_0| + C\rho.$$

Therefore

$$\rho |Au| + \varphi(u) \le (|f| + C) |u - v_0| + C\rho - \frac{1}{2}\frac{d}{dt}|u - v_0|^2$$

Consequently

$$\rho \int_0^T |Au| dt + \int_0^T \varphi(u) dt \le \int_0^T (|f| + C)|u - v_0| dt + \frac{1}{2}|u_0 - v_0|^2 + C\rho T.$$

But $|u(t) - v_0| \leq |u_0 - v_0| + \int_0^t (|f| + C) ds$ and thus

$$\rho \int_0^T |Au| dt + \int_0^T \varphi(u) dt \leq \frac{1}{2}(|u_0 - v_0| + \int_0^T (|f| + C) dt)^2 + C \rho T.$$

**Estimate II.** Multiplying the equation $\frac{du}{dt} + Au + \partial\varphi(u) = f$ by $t\frac{du}{dt}$ we get

$$\int_0^T |\frac{du}{dt}(t)|^2 t\, dt \leq \int_0^T (|Au| + |f|) |\frac{du}{dt}| t\, dt + \int_0^T \varphi(u) dt.$$

Using now the estimate

$$|\frac{du}{dt}(T)| \leq |\frac{du}{dt}(t)| + \int_0^T |\frac{df}{dt}| ds \quad \text{for} \quad t \in [0, T]$$

we get

$$|\frac{du}{dt}(T)|^2 \leq 2|\frac{du}{dt}(t)|^2 + 2(\int_0^T |\frac{df}{dt}| ds)^2$$

and

$$\frac{T^2}{2}|\frac{du}{dt}(T)|^2 \leq 2\int_0^T |\frac{du}{dt}(t)|^2 t\, dt + T^2(\int_0^T |\frac{df}{dt}| ds)^2.$$

Therefore

$$T^2 |\frac{du}{dt}(T)|^2 \leq 4 \int_0^T (|Au| + |f|) |\frac{du}{dt}| t\, dt + 4 \int_0^T \varphi(u) dt +$$

$$+ 2 T^2 (\int_0^T |\frac{df}{dt}| ds)^2.$$

In other words we have on $(0, T)$

(55) $$F(t)^2 \leq \int_0^t F(s) G(s) ds + H(t)$$

where $F(t) = t|\frac{du}{dt}|$, $G(t) = 4(|Au| + |f|)$ and $H(t) =$
$= 4 \int_0^t \varphi(t) ds + 2 t^2 (\int_0^t |\frac{df}{dt}| ds)^2$. If $G \geq 0$, (55) implies

that

$$F(t) \leq \frac{1}{2}\int_0^t G(s)\,ds + (\sup_{[0,t]}|H|)^{1/2}.$$

Combined with Estimate I, we get a bound independent of $\lambda$ for $t|\frac{du}{dt}(t)|$ in $L^\infty(0,T)$.

Problem. We have seen that the semigroups $S(t)$ generated by several classes of maximal monotone operators $A$ ($A = \partial\varphi$, Int $D(A) \neq \phi$ or general $A$ if dim $H < +\infty$ by reduction to the previous case) have a <u>smoothing effect</u> on the initial data.

More precisely:

$$(56)\begin{cases} S(t) \text{ maps } \overline{D(A)} \text{ into } D(A) \text{ and for every } x \in \overline{D(A)}, \\ \text{there is a constant } C \text{ such that } t|\frac{d^+}{dt}S(t)x| \leq C \text{ for} \\ t \in (0,1]. \end{cases}$$

Is it possible to find a simple <u>characterization</u> of maximal monotone operators $A$ such that the semigroup $S(t)$ generated by $-A$ satisfies (56)? Note that in the <u>linear</u> case, we get exactly the generators of <u>analytic</u> semigroups of contractions.

## IV Some applications to nonlinear partial differential equations.

Example 5. Let $\gamma$ be a maximal monotone graph in $\mathbb{R} \times \mathbb{R}$ such that $0 \in D(\gamma)$.

Corollary 28. Let $u_0(x) \in L^2(\Omega)$ be such that $u_0(x) \in \overline{D(\gamma)}$ a.e. <u>on $\Omega$ and let</u> $f$ <u>be absolutely continuous from</u> $[0,T]$ <u>into</u> $L^2(\Omega)$. There exists a unique solution $u$ of the equation

$$\frac{\partial u}{\partial t} - \Delta u + \gamma(u) \ni f \qquad \text{a.e. on } \Omega \times (0,T)$$

$$u(x,t) = 0 \qquad \text{a.e. on } \Gamma \times (0,T)$$

$$u(x,0) = u_0(x) \qquad \text{a.e. on } \Omega$$

satisfying $u \in C([0,T]; L^2(\Omega))$, $tu \in L^\infty(0,T;H^2(\Omega) \cap H_0^1(\Omega))$, $t\frac{\partial u}{\partial t} \in L^\infty(0,T;L^2(\Omega))$, $u(x,t) \in H^2(\Omega)$ $\forall t \in (0,T]$.

In addition we have for every $t \in (0,T)$

$$(57) \begin{cases} \dfrac{\partial^+ u}{\partial t} - \Delta u + \gamma(u) = f \quad \text{a.e. on } \{x \in \Omega; \gamma \text{ is singlevalued at } u(x,t)\} \\ \dfrac{\partial^+ u}{\partial t} + \text{Proj}_{\gamma(u)} f = f \quad \text{a.e. on } \{x \in \Omega; \gamma \text{ is multivalued at } u(x,t)\} \end{cases}$$

(where $\text{Proj}_I r$ = projection of $r \in \mathbb{R}$ on the closed interval $I \subset \mathbb{R}$).

Proof of Corollary 28. The first part is immediate by combining Theorem 23 and Corollary 13 (applied with $\beta(0) = \mathbb{R}$, $\beta(r) = \phi$ for $r \neq 0$). Thus we have only to prove (57). In view of (50) we consider $(-\Delta u + \gamma(u) - f)^0$ for a fixed $t$. Let $D = \{r \in \mathbb{R}; \gamma(r) \text{ contains more than 1 point}\}$; obviously $D$ is denumerable. On the set $\{x \in \Omega; u(x,t) \notin D\}$ we have clearly $(-\Delta u + \gamma(u) - f)^0 = -\Delta u + \gamma(u) - f$. We use now the fact that if $u(x) \in H^2(\Omega)$, then

(58) $\Delta u = 0$ a.e. on $\{x \in \Omega; u(x) \in D\}$

Indeed, it is standard that for $u \in H^1(\Omega)$ and $c \in \mathbb{R}$, grad $u = 0$ a.e. on $\{x \in \Omega; u(x) = c\}$ (this is based on the following properties: $u$ is differentiable a.e. in the classical sense and, in each direction, a set of positive measure has one as linear density a.e.) Applying again this result to grad $u$ we see that $\Delta u = 0$ a.e. on $\{x \in \Omega; u(x) = c\}$. Considering a denumerable union of such sets we get (58). Therefore a.e. on $\{x \in \Omega; u(x,t) \in D\}$ we have

$$(-\Delta u + \gamma(u) - f)^0 = (\gamma(u) - f)^0 = \text{Proj}_{\gamma(u)} f - f.$$

Remark. One can prove that $A = -\Delta + \gamma$ with domain $D(A) = \{u \in W^{2,p}(\Omega) \cap W_0^{1,p}(\Omega);$ there is a $g \in L^p(\Omega)$ such that $g(x) \in \gamma(u(x))$ a.e. on $\Omega\}$ is m-accretive in the space $L^p(\Omega)$ (in the sense of Kato [19]; see [9]). But it is not known whether the semigroup generated by $-A$ has a smoothing effect in $L^p(\Omega)$ i.e. is it true that $u(x,t) \in W^{2,p}(\Omega)$ for $t > 0$, assuming $u_0(x) \in L^p(\Omega)$? (The answer is likely to be positive!). More generally which m-accretive operators in Banach spaces play the same role (with regard to smoothing effects) as subdifferentials of convex functions in Hilbert spaces?

Example 6. Let $\beta$ be a maximal monotone graph in $\mathbb{R} \times \mathbb{R}$.

Corollary 29. <u>Let $u_0(x) \in L^2(\Omega)$ and let $f$ be absolutely continuous from $[0,T]$ into $L^2(\Omega)$. There exists a unique solution $u$ of the equation</u>

(59) $\begin{cases} \dfrac{\partial u}{\partial t} - \Delta u = f & \text{a.e.} \quad \text{on } \Omega \times (0,T) \\ -\dfrac{\partial u}{\partial n} \in \beta(u) & \text{a.e.} \quad \text{on } \Gamma \times (0,T) \\ u(x,0) = u_0(x) & \text{a.e.} \quad \text{on } \Omega \end{cases}$

<u>satisfying</u> $u \in C([0,T]; L^2(\Omega))$, $t u \in L^\infty(0,T; H^2(\Omega))$, $t \dfrac{\partial u}{\partial t} \in L^\infty(0,T; L^2(\Omega))$, $u(x,t) \in H^2(\Omega) \quad \forall t \in (0,T]$.

Proof of Corollary 29. It is a direct application of Theorem 23, taking in $H = L^2(\Omega)$ the function $\varphi$ of Example 1 (Theorem 12).

<u>Behavior as</u> $t \to +\infty$. Assume $\dfrac{df}{dt} \in L^1(0,+\infty; L^2(\Omega))$ so that $\lim_{t \to +\infty} f(t) = f_\infty$ exists in $L^2(\Omega)$, and suppose

(60) $\qquad f_\infty \in R(\partial\varphi)$

(a necessary and sufficient condition for (60) to hold is that

$$\int_\Omega f_\infty(x)dx \in (\text{meas } \Gamma) R(\beta) ; \text{ see [30]).}$$

<u>Theorem 30.</u> <u>If</u> $f - f_\infty \in L^1(0, +\infty; L^2(\Omega))$, <u>then</u> $\lim_{t \to +\infty} u(x,t)$
$= u_\infty(x)$ <u>exists in</u> $L^2(\Omega)$, <u>and</u> $u_\infty \in H^2(\Omega)$ <u>satisfies</u>

(61) $\quad -\Delta u_\infty = f_\infty$ a.e. on $\Omega$, $-\dfrac{\partial u_\infty}{\partial n} \in \beta(u_\infty)$ a.e. on $\Gamma$.

<u>Proof of Theorem 30.</u> It is based on estimates which appear in the proof of Theorem 22, and also on a <u>compactness</u> argument. Let $v_\infty$ be such that $f_\infty \in \partial\varphi(v_\infty)$. We have

$$|u(t) - v_\infty|_{L^2(\Omega)} \leq |u_0 - v_\infty|_{L^2(\Omega)} + \int_0^t |f(s) - f_\infty|_{L^2(\Omega)} ds .$$

On the other hand, let

$$\tilde{\varphi}(u) = \varphi(u) - \varphi(v_\infty) - \int_\Omega f_\infty (u - v_\infty) dx .$$

Thus

$$\tilde{\varphi}(u) = \frac{1}{2} \int_\Omega |\text{grad } u|^2 dx + \int_\Gamma j(u) d\Gamma - \frac{1}{2} \int_\Omega |\text{grad } v_\infty|^2 dx$$

$$- \int_\Gamma j(v_\infty) d\Gamma - \int_\Omega f_\infty (u - v_\infty) dx$$

$$= \frac{1}{2} \int_\Omega |\text{grad } u - \text{grad } v_\infty|^2 dx + \int_\Omega \text{grad } u \cdot \text{grad } v_\infty dx +$$

$$+ \int_\Omega \Delta v_\infty (u - v_\infty) dx + \int_\Gamma j(u) - j(v_\infty) d\Gamma$$

$$= \frac{1}{2} \int_\Omega |\text{grad } u - \text{grad } v_\infty|^2 dx + \int_\Omega |\text{grad } v_\infty|^2 dx +$$

$$\int_\Gamma \frac{\partial v_\infty}{\partial n} (u - v_\infty) + j(u) - j(v_\infty) d\Gamma \geq \frac{1}{2} \int_\Omega |\text{grad } u - \text{grad } v_\infty|^2 dx.$$

Clearly we have $\frac{du}{dt} + \partial\tilde{\varphi}(u) \ni f - f_\infty$. Therefore we deduce from (47) that

$$t\,\tilde{\varphi}(u(t)) \leq \frac{1}{4}\int_0^t |f(s) - f_\infty|_{L^2}^2\,ds + \frac{1}{2}(|u_0 - v_\infty|_{L^2} + \int_0^t |f(s) - f_\infty|\,ds)^2.$$

Consequently $\|u(x,t)\|_{H^1(\Omega)}$ is bounded as $t \to +\infty$, and there is a sequence $t_n \to +\infty$ such that $u(x,t_n) \to u_\infty(x)$ in $L^2(\Omega)$. Since $f - \frac{du}{dt} \in \partial\varphi(u)$ and $f - \frac{du}{dt} \to f_\infty$ as $t \to +\infty$ in $L^2(\Omega)$ we have $f_\infty \in \partial\varphi(u_\infty)$. In addition, for $t_1 \geq t_2$, we have

$$|u(t_1) - u_\infty|_{L^2} \leq |u(t_2) - u_\infty|_{L^2} + \int_{t_2}^{t_1} |f(s) - f_\infty|_{L^2}\,ds$$

which shows that $\lim_{t \to +\infty} u(x,t) = u_\infty(x)$ in $L^2(\Omega)$.

Remark. In general, the solution of (61) need not be unique. It would be of interest to "recognize" $u_\infty$ among all solutions of (61).

Comments. Equations of the form (59) are of physical interest and were studied by many authors. The case where $\beta$ is a continuous (or even Lipschitz continuous) function was considered by A. Friedman in [17] (results concerning the behavior as $t \to +\infty$ are given only under the restrictive assumption $\beta(r_1) - \beta(r_2) \cdot (r_1 - r_2) \geq \alpha |r_1 - r_2|^2 \quad \alpha > 0)$.

The case where $\beta(r) = \begin{cases} 0 & \text{for } r > 0 \\ (-\infty, 0] & \text{for } r = 0 \\ \phi & \text{for } r < 0 \end{cases}$

corresponds to variational inequalities introduced in [23].

The case where $\beta(r) = \begin{cases} +1 & \text{for } r > 0 \\ [-1, +1] & \text{for } r = 0 \\ -1 & \text{for } r < 0 \end{cases}$

appears in [16]. Further regularity results in $L^p$ spaces for the solution of (59) have been obtained in [5] (but again, the question of smoothing effect in $L^p$ has not been settled).

Example 7. Let $\beta$ be a monotone and continuous (for simplicity!) function from $\mathbb{R}$ onto $\mathbb{R}$ i.e. $R(\beta) = \mathbb{R}$.

<u>Corollary 31.</u> <u>Let $u_0(x) \in H^{-1}(\Omega)$ and let $f$ be absolutely continuous from $[0, T]$ into $H^{-1}(\Omega)$. There exists a unique solution $u$ of the equation</u>

$$(62) \begin{cases} \dfrac{\partial u}{\partial t} - \Delta\beta(u) = f & \text{on } \Omega \times (0, T) \\ \beta(u(x,t)) = 0 & \text{on } \Gamma \times (0, T) \\ u(x, 0) = u_0(x) & \text{on } \Omega \end{cases}$$

<u>satisfying</u> $u \in C([0,T]; H^{-1}(\Omega))$, $t\beta(u) \in L^\infty(0, T: H_0^1(\Omega))$, $t\dfrac{\partial u}{\partial t} \in L^\infty(0, T; H^{-1}(\Omega))$, $tu \in L^\infty(0, T; L^1(\Omega))$, $u(x,t) \in L^1(\Omega)$ <u>and</u> $\beta(u(x,t)) \in H_0^1(\Omega)$ $\forall t \in (0, T]$.

Proof of Corollary 31. It is a direct application of Theorem 23 taking in $H = H^{-1}(\Omega)$ the function $\varphi$ of Example 3 (Theorem 17). Note that $L^\infty(\Omega) \subset D(\varphi)$ and thus $D(\varphi)$ is dense in $H^{-1}(\Omega)$.

<u>Behavior as $t \to +\infty$</u>. Assume $\dfrac{df}{dt} \in L^1(0, +\infty; H^{-1}(\Omega))$ so that $\lim\limits_{t \to +\infty} f(t) = f_\infty$ exists in $H^{-1}(\Omega)$.

<u>Corollary 32.</u> We have $\lim\limits_{t \to +\infty} \|\beta(u(x,t)) - g_\infty(x)\|_{H_0^1} = 0$ <u>where</u>

$g_\infty \in H_0^1(\Omega)$ is the solution of $-\Delta g_\infty = f_\infty$. In addition if
$$\left\| \frac{df}{dt}(t) \right\|_{H^{-1}(\Omega)} = O(t^{-\alpha}) \text{ for some } \alpha > 2 \text{ then}$$
$$\| \beta(u(x,t)) - g_\infty(x) \|_{H_0^1} = O(t^{-1}) \text{ as } t \to +\infty.$$

Corollary 32 is a direct application of Theorem 24.

Comments. Equations of the form (62) have been extensively studied in the literature, but the properties we present here (smoothing effect and behavior at infinity) seem to be new. Sharp regularity results were obtained for the case where $\beta(r) = |r|^{m-1} r$ and $\Omega = \mathbb{R}$ by Aronson [1]; cf also [3] and [21] for additional references.

After some transformations (see [3]), the Stefan free boundary value problem can be written in the form (62) with

$$\beta(r) = \begin{cases} a_1 r & \text{for } r \leq 0 \\ 0 & \text{for } 0 < r < k \\ a_2(r-k) & \text{for } r \geq k \end{cases}$$

$(a_1, a_2 > 0)$; cf also [18] for another approach.

Example 8. Let $\beta$ be a maximal monotone graph in $\mathbb{R} \times \mathbb{R}$ with $D(\beta) = \mathbb{R}$ and $0 \in \beta(0)$. Let $j: \mathbb{R} \to \mathbb{R}$ be such that $\partial j = \beta$.

Theorem 33. Let $u_0 \in H_0^1(\Omega)$ with $j(u_0) \in L^1(\Omega)$ and let $v_0 \in L^2(\Omega)$. Let $f \in L^1(0,T; L^2(\Omega))$. There exist two functions $u$ and $g$ satisfying

$$\begin{cases} \dfrac{\partial^2 u}{\partial t^2} - \Delta u + g = f & \text{in the sense of distribution on } \Omega \times (0,T) \\ u(x,t) = 0 & \text{on } \Gamma \times (0,T) \\ u(x,0) = u_0(x), \ \dfrac{\partial u}{\partial t}(x,0) = v_0(x) & \text{on } \Omega \\ g(x,t) \in \beta(u(x,t)) & \text{a.e. on } \Omega \times (0,T) \end{cases}$$

$u \in L^\infty(0, T; H_0^1(\Omega))$, $\frac{\partial u}{\partial t} \in L^\infty(0, T; L^2(\Omega))$, $g \in L^1(\Omega \times (0, T))$.

<u>Proof of Theorem 33.</u> Let $u_{0\lambda} \in H^2(\Omega) \cap H_0^1(\Omega)$ be such that $u_{0\lambda} \to u_0$ in $H_0^1(\Omega)$ and $j(u_{0\lambda}) \to j(u_0)$ in $L^1(\Omega)$ as $\lambda \to 0$. Let $v_{0\lambda} \in H_0^1(\Omega)$ be such that $v_{0\lambda} \to v_0$ in $L^2(\Omega)$ as $\lambda \to 0$. Let $f_\lambda$ with $\frac{\partial f_\lambda}{\partial t} \in L^1(0, T; L^2(\Omega))$ be such that $f_\lambda \to f$ in $L^1(0, T; L^2(\Omega))$ as $\lambda \to 0$. Let $u_\lambda$ be the solution of the equation

(63) $\begin{cases} \dfrac{\partial^2 u_\lambda}{\partial t^2} - \Delta u_\lambda + \beta_\lambda(u_\lambda) = f_\lambda & \text{on } \Omega \times (0, T) \\ u_\lambda(x, t) = 0 & \text{on } \Gamma \times (0, T) \\ u_\lambda(x, 0) = u_{0\lambda}(x), \quad \dfrac{\partial u_\lambda}{\partial t}(x, 0) = v_{0\lambda}(x) & \text{on } \Omega. \end{cases}$

Note that (63) can be written in the form of a system in $H = H_0^1(\Omega) \times L^2(\Omega)$ as

$$\frac{dU_\lambda}{dt} + A U_\lambda + B U_\lambda = F_\lambda, \quad U_\lambda(0) = U_{0\lambda}$$

where

$$U = \begin{pmatrix} u \\ \frac{\partial u}{\partial t} \end{pmatrix}, \quad A = \begin{pmatrix} 0, & -I \\ -\Delta, & 0 \end{pmatrix}, \quad B = \begin{pmatrix} 0 \\ \beta_\lambda \end{pmatrix}, \quad F = \begin{pmatrix} 0 \\ f_\lambda \end{pmatrix}, \quad U_{0\lambda} = \begin{pmatrix} u_{0\lambda} \\ v_{0\lambda} \end{pmatrix}$$

Since $B$ is Lipschitz on $H$ and $U_{0\lambda} \in D(A)$, $\frac{dF}{dt} \in L^1(0, T; H)$ there is a strong solution $u_\lambda$ of (63)

Multiplying (63) by $\frac{\partial u_\lambda}{\partial t}$ and integrating over $\Omega$ we get

$$\frac{1}{2} \frac{d}{dt} \left|\frac{\partial u_\lambda}{\partial t}\right|_{L^2}^2 + \frac{d}{dt} |\text{grad } u_\lambda|_{L^2}^2 + \frac{d}{dt} \int_\Omega j_\lambda(u_\lambda) \, dx \leq |f_\lambda|_{L^2} \left|\frac{\partial u_\lambda}{\partial t}\right|_{L^2}$$

which implies that $\left|\frac{\partial u_\lambda}{\partial t}\right|_{L^2}$ and $|\text{grad } u_\lambda|_{L^2}$ are bounded in $L^\infty(0,T)$ as $\lambda \to 0$.

Next we multiply (63) by $v_\lambda = (I + \lambda \beta)^{-1} u_\lambda$ and we integrate over $\Omega \times (0,T)$. We obtain

$$\int_0^T\!\!\int_\Omega \beta_\lambda(u_\lambda) v_\lambda \, dxdt \le \int_0^T\!\!\int_\Omega |f_\lambda||u_\lambda| dxdt + \int_0^T\!\!\int_\Omega \left|\frac{\partial u_\lambda}{\partial t}\right|^2 dxdt$$

$$+ \int_\Omega \left|\frac{\partial u_\lambda}{\partial t}(x,T)\right| |u_\lambda(x,T)| dx + \int_\Omega |v_{0\lambda}(x)| |u_{0\lambda}(x)| dx$$

which is bounded as $\lambda \to 0$.

Let $\lambda_n \to 0$ be such that

$$u_{\lambda_n} \to u \qquad \text{weakly in } L^\infty(0,T; H_0^1(\Omega))$$

$$\frac{\partial u_{\lambda_n}}{\partial t} \to \frac{\partial u}{\partial t} \qquad \text{weakly in } L^\infty(0,T; L^2(\Omega))$$

$$u_{\lambda_n} \to u \qquad \text{a. e. on } \Omega \times (0,T)$$

$$\beta_{\lambda_n}(u_{\lambda_n}) \to g \qquad \text{weakly in } L^1(\Omega \times (0,T)) \; ;$$

such a $\lambda_n$ exists by Theorem 18 and also $g(x,t) \in \beta(u(x,t))$ a. e. on $\Omega \times (0,T)$.

Example 9. Let $\beta$ be a maximal monotone graph in $\mathbb{R} \times \mathbb{R}$ with $D(\beta) = \mathbb{R}$ and $0 \in \beta(0)$. Let $j: \mathbb{R} \to \mathbb{R}$ be such that $\partial j = \beta$.

Theorem 34. _Let_ $u_0 \in H^1(\Omega)$ _with_ $j(u_0) \in L^1(\Gamma)$ _and let_ $v_0 \in L^2(\Omega)$. _Let_ $f \in L^1(0,T;L^2(\Omega))$. _There exist two functions_ u _and_ g _satisfying_

$$-\int_0^T \int_\Omega \frac{\partial u}{\partial t} \cdot \frac{\partial \zeta}{\partial t} dx\, dt + \int_0^T \int_\Omega \operatorname{grad} u \cdot \operatorname{grad} \zeta \, dx\, dt + \int_0^T \int_\Gamma g \zeta \, d\Gamma \, dt$$

$$= \int_0^T \int_\Omega f \cdot \zeta \, dx\, dt + \int_\Omega v_0(x) \zeta(x,0) dx$$

$$\forall \zeta; \frac{\partial \zeta}{\partial t} \in L^2(0,T;H^1(\Omega)), \quad \zeta \in C(\bar{\Omega} \times [0,T]), \quad \zeta(x,T) = 0$$

$$g(x,t) \in \beta(u(x,t)) \quad \text{a.e.} \quad \text{on } \Gamma \times (0,T)$$

$$u(x,0) = u_0(x) \quad \text{a.e.} \quad \text{on } \Omega$$

$$u \in L^\infty(0,T;H^1(\Omega)), \quad \frac{\partial u}{\partial t} \in L^\infty(0,T;L^2(\Omega)), \quad g \in L^1(\Gamma \times (0,T)).$$

**Proof of Theorem 34.** We consider the approximate equation

$$(64) \begin{cases} \dfrac{\partial^2 u_\lambda}{\partial t^2} - \Delta u_\lambda = f & \text{on } \Omega \times (0,T) \\[6pt] -\dfrac{\partial u_\lambda}{\partial n} = \beta_\lambda(u_\lambda) + \lambda \dfrac{\partial u_\lambda}{\partial t} & \text{on } \Gamma \times (0,T) \\[6pt] u_\lambda(x,0) = u_0(x), \quad \dfrac{\partial u_\lambda}{\partial t}(x,0) = v_0(x) & \text{on } \Omega \end{cases}$$

which we write in the form of a system in $H^1(\Omega) \times L^2(\Omega)$ as

$$\frac{dU_\lambda}{dt} + A U_\lambda = F, \quad U_\lambda(0) = U_0$$

where $A\begin{pmatrix} u \\ v \end{pmatrix} = \begin{pmatrix} -v \\ -\Delta u \end{pmatrix}$ and

$D(A) = \{\begin{pmatrix} u \\ v \end{pmatrix}; u \in H^2(\Omega), v \in H^1(\Omega) \text{ and } -\frac{\partial u}{\partial n} = \beta_\lambda(u) + \lambda v \text{ on } \Gamma\}$.

There is a constant $\gamma(\lambda)$ such that $A + \gamma I$ is maximal monotone.

Indeed

$$\langle AU_1 - AU_2, U_1 - U_2 \rangle = -\int_\Omega \text{grad}(v_1 - v_2) \cdot \text{grad}(u_1 - u_2)\, dx$$

$$- \int_\Omega (v_1 - v_2)(u_1 - u_2)\, dx - \int_\Omega \Delta(u_1 - u_2)(v_1 - v_2)\, dx$$

$$= \int_\Gamma (\beta_\lambda(u_1) - \beta_\lambda(u_2))(v_1 - v_2) + \lambda|v_1 - v_2|^2\, d\Gamma$$

$$- \int_\Omega (v_1 - v_2)(u_1 - u_2)\, dx$$

$$\geq \int_\Gamma \lambda|v_1 - v_2|^2 - \frac{1}{\lambda}|u_1 - u_2||v_1 - v_2|\, d\Gamma$$

$$- \int_\Omega |v_1 - v_2||u_1 - u_2|\, dx$$

$$\geq -\gamma(\|u_1 - u_2\|_{H^1}^2 + |v_1 - v_2|_{L^2}^2).$$

Thus (64) has a generalized solution for $\begin{pmatrix} u_0 \\ v_0 \end{pmatrix} \in \overline{D(A)} = H^1(\Omega) \times L^2(\Omega)$ and $f \in L^1(0, T; L^2(\Omega))$. We multiply first (64) by $\dfrac{\partial u_\lambda}{\partial t}$ and integrate over $\Omega$:

$$\frac{1}{2} \frac{d}{dt} \left|\frac{\partial u_\lambda}{\partial t}\right|_{L^2}^2 + \frac{1}{2} \frac{d}{dt} |\text{grad } u_\lambda|_{L^2}^2 + \int_\Gamma (\beta_\lambda(u_\lambda) + \lambda \frac{\partial u_\lambda}{\partial t}) \frac{\partial u_\lambda}{\partial t}\, d\Gamma$$

$$= \int_\Omega f \cdot \frac{\partial u_\lambda}{\partial t}\, dx,$$

which leads to a bound independent of $\lambda$ for $\dfrac{\partial u_\lambda}{\partial t}$ in $L^\infty(0, T; L^2(\Omega))$ and for $u_\lambda$ in $L^\infty(0, T; H^1(\Omega))$. Next we multiply (64) by $v_\lambda = (I + \lambda\beta)^{-1} u_\lambda$ and integrate over $\Omega \times (0, T)$:

$$\int_0^T \int_\Gamma (\beta_\lambda(u_\lambda) + \lambda \frac{\partial u_\lambda}{\partial t}) v_\lambda \, d\Gamma \, dt \leq \int_\Omega |\frac{\partial u_\lambda}{\partial t}(x, T)| |u_\lambda(x, T)| dx$$

$$+ \int_\Omega |v_0(x)| |u_0(x)| dx + \int_0^T \int_\Omega |\frac{\partial u_\lambda}{\partial t}|^2 dx \, dt + \int_0^T \int_\Omega |f| |u_\lambda| dx \, dt,$$

which leads to a bound on $\int_0^T \int_\Gamma \beta_\lambda(u_\lambda) v_\lambda \, d\Gamma \, dt$ independent of $\lambda$.

At last we multiply (64) by a function $\zeta$ satisfying the assumptions of Theorem 34 and we integrate on $\Omega \times (0, T)$:

$$-\int_0^T \int_\Omega \frac{\partial u_\lambda}{\partial t} \cdot \frac{\partial \zeta}{\partial t} dx \, dt + \int_0^T \int_\Omega \operatorname{grad} u_\lambda \cdot \operatorname{grad} \zeta \, dx \, dt$$

$$-\int_0^T \int_\Gamma \beta_\lambda(u_\lambda) \zeta \, d\Gamma \, dt - \int_0^T \int_\Gamma \lambda \frac{\partial u_\lambda}{\partial t} \zeta \, d\Gamma \, dt$$

$$= \int_0^T \int_\Omega f \cdot \zeta \, dx \, dt + \int_\Omega v_0(x) \zeta(x, 0) dx$$

But

$$\int_0^T \int_\Gamma \lambda \frac{\partial u_\lambda}{\partial t} \zeta \, d\Gamma \, dt = -\lambda \int_\Gamma u_0(x) \zeta(x, 0) d\Gamma - \lambda \int_0^T \int_\Gamma u_\lambda \frac{\partial \zeta}{\partial t} dx \, dt.$$

Finally we let $\lambda \to 0$ as in the proof of Theorem 33. Notice that all the integrations by parts we have done can be justified by using an approximation of $\binom{u_0}{v_0}$ by $\binom{u_{0\alpha}}{v_{0\alpha}} \in D(A)$ and of $f$ by $f_\alpha$ with $\frac{\partial f_\alpha}{\partial t} \in L^1(0, T; L^2(\Omega))$ so that the corresponding solution $u_{\lambda\alpha}$ of (64) is a strong solution.

<u>Comments.</u>  Results related to Theorem 33 were proved by W. Strauss [31] for the case where $\beta$ is singlevalued and continuous (but not necessarily monotone) and by J. L. Lions [22] for the case where

$$\beta(r) = \begin{cases} +1 & \text{if } r > 0 \\ [-1, +1] & \text{if } r = 0 \\ -1 & \text{if } r < 0 \end{cases}$$

Theorem 34 answers a question raised in [22].

<u>Problem</u>.   More generally it would be of interest to solve the equation $\dfrac{d^2 u}{dt^2} + \partial \varphi(u) \ni f$, $u(0) = u_0$, $\dfrac{du}{dt}(0) = v_0$.
In the particular case where $\varphi = I_K$ is the indicator function of a closed convex set K, the solution u represents, roughly speaking, the trajectory of an optical ray caught in K and reflecting at the boundary of K.

## REFERENCES

1. D. G. Aronson, Regularity properties of flows through porous media. SIAM J. Applied Math. <u>17</u> (1969) p. 461-467, Archive Rat. Mech. Anal. <u>37</u> (1970) p. 1-10.

2. P. Benilan - H. Brezis, to appear. Ann. Inst. Fourier.

3. H. Brezis, On some degenerate nonlinear parabolic equations. Nonlinear Functional Analysis. p. 28-38 Proc. Symp. Pure Math. Vol. 18 AMS (1970).

4. H. Brezis, Propriétés régularisantes de certains semi-groupes nonlinéaires. Israel J. Math. <u>9</u> (1971) p. 513-534.

5. H. Brezis, Problèmes unilatéraux. J. Math. Pures et Appliquées (to appear).

6. H. Brezis - M. Crandall - A. Pazy, Perturbations of nonlinear maximal monotone sets in Banach spaces. Comm. Pure Appl. Math. <u>23</u> (1970) p. 123-144.

7. H. Brezis - A. Pazy, Semigroups of nonlinear contractions on convex sets. J. Funct. Anal. **6** (1970) p. 237 - 281.

8. H. Brezis - G. Stampacchia, Sur la régularité de la solution d'inéquations elliptiques. Bull. Soc. Math. France **96** (1968) p. 153-180.

9. H. Brezis - W. Strauss, (to appear).

10. F. Browder, Variational boundary value problems for quasilinear elliptic equations of arbitrary order. Proc. Nat. Acad. Sci. U.S.A. **50** (1963) p. 31-37.

11. F. Browder, Nonlinear monotone and accretive operators in Banach spaces. Proc. Nat. Acad. Sci. U.S.A. **61** (1968) p. 388-393.

12. F. Browder, Nonlinear operators and nonlinear equations of evolution in Banach spaces. Nonlinear Functional Analysis. Proc. Symp. Pure Math (to appear).

13. M. Crandall - A. Pazy, Semigroups of nonlinear contractions and dissipative sets. J. Funct. Anal. **3** (1969) p. 376-418.

14. J. R. Dorroh, A nonlinear Hille-Yosida-Phillips theorem. J. Funct. Anal. **3** (1969) p. 345 - 353.

15. N. Dunford - J. Schwartz, Linear operators Interscience, New York, 1958.

16. G. Duvaut - J. L. Lions, Sur de nouveaux problèmes d'inéquations variationnelles posées par la Mecanique. C. R. Acad. Sci. Paris **269** (1969) p. 510 -513.

17. A. Friedman, Generalized heat transfer between solids and gases under nonlinear boundary conditions. J. Math. Mech. 8 (1959) p. 161-184.

18. A. Friedman, The Stefan problem in several space variables. Trans. Amer. Math. Soc. 132 (1968) p. 51-87.

19. T. Kato, Accretive operators and nonlinear evolution equations in Banach spaces. Nonlinear Functional Analysis p. 138-161. Proc. Symp. Pure Math. Vol. 18, AMS (1970).

20. Y. Komura, Nonlinear semigroups in Hilbert spaces. J. Math. Soc. Japan 19 (1967) p. 493-507.

21. J. L. Lions, Quelques méthodes de résolution des problèmes aux limites nonlinéaires. Dunod, Gauthier Villars.

22. J. L. Lions, Sur les inéquations variationnelles d'évolution pour les opérateurs du $2^{eme}$ ordre en t. Symposium sur les problèmes d'évolution Istituto Naz. di Alta Mat. Rome (1970).

23. J. L. Lions - G. Stampacchia, Variational inequalities. Comm. Pure Appl. Math. 20 (1967) p. 493 - 519.

24. G. Minty, Monotone (nonlinear) operators in Hilbert space. Duke Math. J. 29 (1962) p. 341-346.

25. G. Minty, On the monotonicity of the gradient of a convex function. Pac. J. Math. 14 (1964) p. 243-247.

26. J. J. Moreau, Proximité et dualité dans un espace hilbertien. Bull. Soc. Math. France 93 (1965) p. 273 - 299.

27. P. P. Mosolov - V. P. Miasnikov, Variational methods in the theory of the fluidity of a viscous-plastic medium. P. M. M. Vol. 29 (1965) p. 545-577.

28. R. T. Rockafellar, Local boundedness of nonlinear monotone operators. Michigan Math. J. 16 (1969) p. 397 - 407.

29. R. T. Rockafellar, On the maximality of the sums of nonlinear monotone operators. Trans. Amer. Math. Soc. (to appear).

30. M. Schatzman, (to appear).

31. W. Strauss, On weak solutions of semilinear hyperbolic equations (to appear).

32. E. H. Zarantonello, Solving functional equations by contractive averaging. Tech. Rep. no. 160, Mathematics Research Center, Univ. of Wisconsin, Madison, 1960.

Part of this paper was written while the author was visiting at the University of Chicago under NSF Grant GP-23564.

> Institut de Mathématique
> Faculté des Sciences
> 9 Quai Saint Bernard
> Paris 5 France

> Received May 19, 1971

# Semigroups of Nonlinear Transformations in Banach Spaces

*MICHAEL G. CRANDALL*

Let $X$ be a real Banach space and $C$ be a subset of $X$. By a <u>quasi-contraction semigroup</u> $S$ on $C$ we mean a function $S$ on $[0,\infty)$ such that $S(t): C \to C$ for $t \geq 0$, which satisfies the following conditions:

(i) $\quad S(t)S(\tau) = S(t+\tau)$ for $t, \tau \geq 0$.

(ii) $\quad$ There is a real number $\omega$ such that
$$\|S(t)x - S(t)y\| \leq e^{\omega t} \|x-y\|$$
for $t \geq 0$ and $x, y \in C$,

and

(iii) $\quad \lim_{t \downarrow 0} S(t)x = S(0)x = x$ for $x \in C$.

If $S$ satisfies the above conditions we write $S \in Q_\omega(C)$. If $S \in Q_0(C)$, then $S$ is called a semigroup of contractions on $C$. A substantial theory concerning quasi-contraction semigroups has developed in the last few years. Here we provide a summary of the main results obtained by the author and T. M. Liggett in [14] and [15], some comments on other developments, and some new examples.

Section 1 discusses the notion of a generator of a nonlinear semigroup and the principle result giving sufficient conditions that an operator generate a semigroup. Section 2 reviews the results concerning existence of generators and Section 3 discusses some results in special spaces. Section 4 contains examples of nonlinear operators satisfying the conditions of Section 1. The reader interested only in possible applications of the semigroup theory to concrete differential equations can restrict his attention to Sections 1 and 4. With some small exceptions, proofs are given only in Section 4 and then rather briefly. This paper is self-contained in other respects

1. <u>Sufficient conditions for a generator.</u>  Two classical notions concerning "generators" must be abandoned in the general nonlinear theory. The first is that of "infinitesimal generator" and the second is that generators are functions. What survives intact from the linear theory is Hille's exponential formula:

(1.1) $$S(t) = \lim_{n \to \infty} (I + \frac{t}{n} A)^{-n}.$$

It is well-known that if $S \in Q_\omega(X)$ is linear (i.e., each $S(t)$ is linear), then (1.1) holds if $-A$ is the infinitesimal generator of $S$. We will adopt (1.1) as a <u>definition</u> of "$-A$ generates $S$" in the nonlinear case. The class of $A$'s to be considered is defined below.

<u>Definition 1.2.</u>  Let $A \subseteq X \times X$ and $\omega$ be a real number. Then $A \in G(\omega)$ if

$$\|(x_1 + \lambda y_1) - (x_2 + \lambda y_2)\| \geq (1 - \lambda\omega)\|x_1 - x_2\|$$

for $[x_i, y_i] \in A$, $i = 1, 2$ and $\lambda \geq 0$. $G(\omega)$ is the class of subsets $A$ of $X \times X$ such that $A + \omega I$ is <u>accretive</u>. $G(0)$ is the class of accretive sets.

Functions are to be identified with their graphs, so Definition 1.2 applies to functions as a special case. According to Definition 1.2, if $A \in G(\omega)$ then setting

$$(I + \lambda A)^{-1}(x + \lambda y) = x$$

for $[x, y] \in A$ defines a function $(I + \lambda A)^{-1}$ on

$$D((I + \lambda A)^{-1}) = R(I + \lambda A) = \{x + \lambda y : [x, y] \in A\}$$

provided $\lambda \geq 0$ and $\lambda \omega < 1$. Moreover, writing $J_\lambda = (I + \lambda A)^{-1}$,

$$\|J_\lambda u - J_\lambda v\| \leq (1 - \lambda \omega)^{-1} \|u - v\|$$

for $u, v \in D(J_\lambda)$, $\lambda \geq 0$ and $\lambda \omega < 1$.

<u>Definition 1.3.</u>  Let $C \subseteq X$, $S \in Q_\omega(C)$ and $A \in G(\omega)$. Then $S$ is <u>generated by</u> $-A$ if

$$S(t)x = \lim_{n \to \infty} (I + \frac{t}{n}A)^{-n} x$$

for $t > 0$ and $x \in C$.

The reasons for this definition will be explained later. The main result concerning sufficient conditions that $-A$ generate a semigroup demands little more than $A \in G(\omega)$.

<u>Theorem 1.4.</u>  Let $A \in G(\omega)$ and $\lambda_0 > 0$. <u>If</u> $R(I + \lambda A)$ contains $\overline{D(A)}$ <u>for</u> $0 < \lambda < \lambda_0$, <u>then</u>

(1.5) $$S(t)x = \lim_{n \to \infty} (I + \frac{t}{n}A)^{-n} x$$

<u>exists for</u> $t > 0$, $x \in \overline{D(A)}$ <u>and, if</u> $S$ <u>is defined by</u> (1.5), $S \in Q_\omega(\overline{D(A)})$.

Theorem 1.4 is Theorem I of [14]. Above, $D(A) = \{x : x \in X \text{ and } Ax \neq \emptyset\}$ where $Ax = \{y : [x, y] \in A\}$. Actually, more is established in the course of proving Theorem 1.4. It is shown that

159

(1.6) $$\|J_\lambda^m x - J_\mu^n x\| \leq \{[(n\mu - \lambda m)^2 + n\mu(\lambda-\mu)]^{1/2}$$
$$\cdot \exp[2\omega(n\mu + m\lambda)] + [m\lambda(\lambda-\mu) + (m\lambda-n\mu)^2]^{1/2}$$
$$\cdot \exp[4\omega n\mu]\} |Ax|$$

for $x \in D(A)$, $|Ax| = \inf\{|y|: y \in Ax\}$, $\lambda \geq \mu \geq 0$, $\omega > 0$ and $\lambda\omega \leq (1/2)$. If $\omega \leq 0$, then one may use $\omega = 0$ in (1.6). The estimate (1.6) holds whenever $J_\lambda^m x$ and $J_\mu^n x$ are defined and <u>that part of</u> A involved in the definition of $J_\lambda^m x$ and $J_\mu^n x$ is in $G(\omega)$. Hence there are obvious "localized" analogues of Theorem 1.4.

One expects that if $-A$ generates $S$, then $\frac{d}{dt} S(t)x$ should be related to A for $x \in D(A)$. To some extent this is true. Given $S$, define $A_S$ by the equation

$$A_S x = \lim_{t \downarrow 0} \frac{x - S(t)x}{t}$$

where $D(A_S)$ consists of those $x \in C$ for which the limit exists. It is easy to see $S \in Q_\omega(C)$ implies $A_S \in G(\omega)$.

<u>Theorem 1.7.</u>  <u>Let the assumptions of Theorem 1.4 be satisfied and</u> $S \in Q_\omega(\overline{D(A)})$ <u>be generated by</u> $-A$. <u>Then</u> $A_S \cup A \in G(\omega)$. <u>If, moreover, A is a closed subset of</u> $X \times X$, $x \in \overline{D(A)}$ <u>and</u> $t > 0$, <u>then</u>

$$-\frac{d}{dt} S(t)x \in AS(t)x$$

<u>whenever the derivative</u> $\frac{d}{dt} S(t)x$ <u>exists</u>.

The first part of Theorem 1.7 is proved in the course of proving Theorem II of [14], with the additional assumption that $R(I+\lambda A)$ contains the closed convex hull of $D(A)$. Miyadera removed this restriction in [23]. The second part is a simple corollary of the first. See the proof of Theorem II of [14]. The rather complicated statement of Theorem 1.7 is due, quite roughly, to the fact that known arguments provide estimates on right derivatives of $S(t)x$ and suffice to derive estimates from information about left derivatives of $S(t)x$. There is a gap in our knowledge concerning left derivatives which do not agree with right derivatives and vice-versa.

Theorem 1.7 is unsatisfactory in that it does not assert $A_S$ is "big enough". Indeed, it often isn't. In [14], Section 4, an example is given of an $A \in \mathcal{G}(0)$ such that

$$\overline{D(A)} = X \text{ and}$$

$R(I + \lambda A) = X$ for $\lambda > 0$ but, for the $S \in Q_0(X)$ generated by $-A$, $D(A_S) = \emptyset$. In this example, $X = C([-1,1])$ is not reflexive. If $X$ is reflexive, then $S(t)x$ is Lipschitz continuous on bounded sets for $x \in D(A)$ (a consequence of, e.g., (1.6)) and hence is the Bochner integral of its derivative, $\frac{d}{dt} S(t)x$, which exists almost everywhere. See the appendix of [20]. It follows from Theorem 1.7 that in this case $u(t) = S(t)x$ is a <u>strong solution</u> of the Cauchy problem

(CP)
(i) $\quad \dfrac{du}{dt} + Au \ni 0$

(ii) $\quad u(0) = x$

i.e. $u$ is continuous, $u(0) = x$, $u(t) \in D(A)$ a.e., $u'(t)$ exists a.e.,

$$u(t) - u(\tau) = \int_\tau^t u'(s)ds$$

for $t, \tau > 0$ (with the integral in the sense of Bochner), and (CP)(i) holds a.e.. It would be unfortunate if (CP) had a strong solution which was not $S(t)x$. This possibility is ruled out by:

Theorem 1.8. <u>In addition to the assumptions of Theorem 1.4, let $A$ be closed and $S \in Q_\omega(\overline{D(A)})$ be generated by</u> $-A$. <u>If</u> (CP) <u>has a strong solution</u> $u$ <u>for</u> $x \in D(A)$, <u>then</u> $u(t) = S(t)x$ for $t \geq 0$.

Theorem 1.8 is due to Brezis and Pazy [8] in the case $x \in D(A), \omega = 0$. We give another proof based on (1.6). First, let $u(t)$ be a strong solution of (CP) where $x \in \overline{D(A)}$. Then there is a sequence $\{\delta_n\}$, $\delta_n \downarrow 0$ such that $v_n(t) = u(t + \delta_n)$ is a strong solution of (CP) with $x = u(\delta_n)$, $u(\delta_n) \in D(A)$, $v_n'(t)$ is integrable on $[0,T]$ for $0 < T < \infty$ and $v_n(t) = v_n(0) + \int_0^t v_n'(s)ds$. If we show $S(t)u(\delta_n) = v_n(t)$, letting $n \to \infty$ yields $S(t)x = u(t)$. Hence we can assume $u(t) = u(0) + \int_0^t u'(s)ds$ and $x \in D(A)$. Consider the discretized problem

(1.9) $$\frac{u_\varepsilon(t) - u_\varepsilon(t-\varepsilon)}{\varepsilon} + Au_\varepsilon(t) \ni 0 \quad \text{for } t \geq \varepsilon$$

$$u_\varepsilon(t) = x \quad \text{for } \varepsilon > t.$$

This problem has the solution $(I + \varepsilon A)^{-[t/\varepsilon]}x$, where $[t/\varepsilon]$ is the largest integer in $t/\varepsilon$. One sees, via (1.6), that $\lim_{\varepsilon \downarrow 0} u_\varepsilon(t) = S(t)x$ holds uniformly on compact subsets of $[0,\infty)$. Set $u(t) = x$ for $t \leq 0$. Then $u(t)$ satisfies

$$\frac{u(t) - u(t-\varepsilon)}{\varepsilon} + y(t) = g_\varepsilon(t) \quad \text{a.e. on } t \geq 0$$

where $y(t) = -u'(t) \in Au(t)$, $g_\varepsilon(t) = \frac{u(t) - u(t-\varepsilon)}{\varepsilon} - u'(t)$. If $T > 0$, our assumptions yield

$$\lim_{\varepsilon \downarrow 0} \int_0^T \|g_\varepsilon(t)\| dt = 0.$$

We have $u_\varepsilon(t) = (I + \varepsilon A)^{-1} u_\varepsilon(t-\varepsilon)$ and $u(t) = (I + \varepsilon A)^{-1}(u(t-\varepsilon) + \varepsilon g_\varepsilon(t))$ a.e. Thus

$$\|u_\varepsilon(t) - u(t)\| \leq (1-\varepsilon\omega)^{-1}[\|u_\varepsilon(t-\varepsilon) - u(t-\varepsilon)\| + \varepsilon\|g_\varepsilon(t)\|].$$

Integrating over $0 \leq t \leq T$ and rearranging yields

$$\frac{1}{\varepsilon} \int_{T-\varepsilon}^T \|u_\varepsilon(t) - u(t)\| dt$$

$$\leq \frac{(1-\varepsilon\omega)^{-1} - 1}{\varepsilon} \int_0^{T-\varepsilon} \|u_\varepsilon(t) - u(t)\| dt + (1-\varepsilon\omega)^{-1} \int_0^T \|g_\varepsilon(t)\| dt.$$

## SEMIGROUPS OF NONLINEAR TRANSFORMATIONS

Letting $\varepsilon \downarrow 0$ we find

$$\|S(T)x - u(T)\| \leq \omega \int_0^T \|S(t)x - u(t)\| dt .$$

Since $\|S(t)x - u(t)\|$ is continuous, and the above inequality holds for $T \geq 0$, it follows that $\|S(t)x - u(t)\|$ vanishes identically.

Theorems 1.7 and 1.8 assure that there is no harm in <u>defining</u> $S(t)x$, where $-A$ generates $S$, to be the (automatically unique) "generalized solution" of (CP). We adopt this point of view, although it is not completely satisfactory with respect to applications. Improvements concerning this question would be welcome.

It should be noted that writing $S(t)x$ as the limit of solutions of (1.9) usually allows one to show, in examples from partial differential equations, that $S(t)x$ is a "generalized solution" in the sense of satisfying appropriate integral identities.

It remains to review the reasons for Definition 1.3. As remarked above, there exists an $X$, $A \in G(0)$, $\overline{D(A)} = X$ such that if $S \in Q_0(x)$ is generated by $-A$, then $D(A_S) = \emptyset$. This makes it clear that $X$-valued infinitesimal generators of semigroups do not characterize them. The generality of Theorem 1.4 and the assertion of Theorem 1.7 that Definition 1.3 is consistent with the notion of infinitesimal generator, show the definition is a natural one. It was Y. Kōmura in [20] who first recognized the need for "multivalued generators", and "most" semigroups which are generated by subsets of $X \times X$ in the sense of Definition 1.3 are not generated by functions. Moreover, some applications make strong use of concrete multivalued generators. See [4], [5], [6], [7]. Finally, we recall Yosida's method of obtaining a semigroup from a linear candidate $-A$ for a generator. Yosida solved the approximate problem

(1.10) $\qquad \dfrac{du_\lambda}{dt}(t) + A_\lambda u_\lambda(t) = 0, \quad u_\lambda(0) = x ,$

where $A_\lambda = A(I + \lambda A)^{-1}$, and then set

(1.11) $$S(t)x = \lim_{\lambda \downarrow 0} u_\lambda(t).$$

The nonlinear analogue of $A_\lambda$ is $A_\lambda = \lambda^{-1}(I - J_\lambda)$, and if $A \in G(\omega)$, $\lambda \geq 0$, $\lambda\omega < 1$, then $A_\lambda$ is a Lipschitz continuous function. The nonlinear problem (1.10) cannot be solved without some assumption on $D(A_\lambda) = R(I + \lambda A)$. It has a solution under the assumption of Theorem 1.8 if $\overline{D(A)}$ is convex and $x \in \overline{D(A)}$. See, e.g., [11]. Lemma 2.22 of [14] implies that then (1.11) holds where $S$ is the semigroup generated by $-A$. Thus, Definition 1.3 is consistent with the linear theory in all respects, and extends more familiar notions of generation.

For further information on approximation and convergence of semigroups see [9], [24] and their bibliographies.

## 2. Existence of generators.

Perhaps the most difficult problem in the study of nonlinear semigroups has been that of determining whether a given $S \in Q_\omega(C)$ has a generator. The question was first treated by Kōmura [20], [21] in the form of studying whether or not $D(A_S)$ is dense in $C$. Kōmura worked in a Hilbert space, and in this special case (due to Minty's result [22]) the density of $D(A_S)$ in $C$ is equivalent to the existence of an $A \in G(\omega)$ such that $-A$ generates $S$ (if $C$ is closed and convex). Of course, this is false in general $X$. Kōmura obtained a positive answer in [21]. Webb [30] gave the first example where Kōmura's result failed. He exhibited an $X$ and an $S \in Q_0(X)$ such that $D(A_S)$ was not dense in $X$. Kato, in [19], sharpened Kōmura's result and his proof. It is obvious from the discussion in Section 1 that this is not the right question to ask in general spaces, but the methods of Kōmura and Kato have been the only ones used in attacking the problem so far.

We restrict attention to $\omega = 0$. Kōmura's idea was to set $J_{\lambda,t} = (I + (\lambda/t)(I - S(t)))^{-1}$ for $\lambda, t > 0$ and study $\lim_{t \downarrow 0} J_{\lambda,t}$. It is easy to show $J_{\lambda,t}: C \to C$, provided that $S \in Q_0(C)$ and $C$ is closed and convex. We assume $C$ closed and convex in this section. Theorem III of [14] implies:

**Theorem 2.1.** *Let* $S \in Q_0(C)$. *Suppose* $\{t_n\}$ *is a sequence of positive numbers convergent to zero such that*

$$J_\lambda x = \lim_{n \to \infty} J_{\lambda, t_n} x$$

*exists for* $x \in C$ *and* $\lambda > 0$. *Then*

$$A = \bigcup_{\lambda > 0} \{[J_\lambda x, \frac{x - J_\lambda x}{\lambda}] : x \in C\}$$

*is accretive*, $R(I + \lambda A) = C$ *for* $\lambda > 0$ *and* $\overline{D(A)} = C$. *Moreover* $S(t)x$ *is Lipschitz continuous in* $t$ *for* $x \in D(A)$, *and* $-A$ *generates* $S$.

Kato and Kōmura had established the existence of $\lim_{t \downarrow 0} J_{\lambda, t} x$ for $X$ a Hilbert space. It was proved in [14] that the assumptions of Theorem 2.1 are satisfied if $X$ is finite dimensional. (A simpler proof is possible if Lemma 1.1 of [15] is used.) In [15] it was shown that $\lim_{t \downarrow 0} J_{\lambda, t} x$ exists if $X$ is two-dimensional and a counterexample presented in which $X$ is three-dimensional (however, a sequence $\{t_n\}$, etc., exists). Liggett has verified the assumptions of Theorem 2.1 for $X = c_0$ (sequences convergent to zero with the maximum norm), in an unpublished work. Finally, if $X^*$ is uniformly convex and $S$ is uniformly continuous on bounded sets, $\lim_{t \downarrow 0} J_{\lambda, t} x$ exists. But, in general, the question is open. (However, the examples of [14] and [15] are not encouraging.) Perhaps it is best to pose the problem in its simplest form, and in two parts:

**Problem a.** If $S \in Q_0(X)$, is $\{x : S(t)x \text{ is Lipschitz continuous in } t\}$ dense in $X$?

**Problem b.** If $S \in Q_0(X)$ and $S(t)x$ is Lipschitz continuous in $t$ for a dense set of $x$, does there exist $A \in \mathcal{A}(0)$ such that $-A$ generates $S$?

Of course, the property questioned in Problem a is possessed by $S$ if it is generated by $-A$, $A \in \mathcal{A}(0)$, $\overline{D(A)} = X$.

We mention a corollary of Theorem 2.1 which has not been previously pointed out:

**Corollary 2.2.** <u>Let</u> $S, T \in Q_0(C)$, <u>and</u> $R(I+\lambda A_S) \cap R(I+\lambda A_T) \supseteq C$ <u>for</u> $0 < \lambda < \lambda_0$. <u>If</u> $(I+\lambda A_S)^{-1}x = (I+\lambda A_T)^{-1}x$ <u>for</u> $x \in C$ <u>and</u> $0 < \lambda < \lambda_0$, <u>then</u> $S = T$.

**Proof.** One checks that

$$\lim_{t \downarrow 0} J_{\lambda, t} x = (I + \lambda A_S)^{-1} x$$

for $x \in R(I+\lambda A_S)$ and the result follows from Theorem 2.1.

A question related to existence of generators is that of uniqueness. That is, does

$$\lim_{n \to \infty} (I + \frac{t}{n} A)^{-n} = \lim_{n \to \infty} (I + \frac{t}{n} B)^{-n}$$

imply $A = B$? This question was shown to have an affirmative answer if $X$ is a Hilbert space, $A, B \in G(0)$ and $R(I + A) = R(I + B) = X$ in [12] and [13]. The results of Brezis [3] extend this to the case of uniformly convex $X^*$ by establishing an important inequality. Unfortunately, it is not true in general, or even if $X$ is two-dimensional. See [14] Section 4.

**Remark.** In general spaces, the problem may need more careful posing. If the limit is assumed equal on $C$, only that part of $A, B$ involved in $(I + \lambda A)^{-1}C$, $(I + \lambda B)^{-1}C$ may be relevant in general.

Finally, we want to point out a generalized notion of an infinitesimal generator. Let $S \in Q_\omega(C)$ and $\{t_n\}$ be a sequence of positive numbers convergent to $0$. Define $A_S(\{t_n\})$ by $[x;y] \in A_S(\{t_n\})$ if there is a sequence $\{x_n\} \subseteq C$ such that $x_n \to x$ and $t_n^{-1}(x_n - S(t_n)x_n) \to y$. It is easy to show $A_S(\{t_n\}) \in G(\omega)$. In general, $A_S(\{t_n\})$ is multi-valued even though it is defined by a kind of differentiation. Using the definitions, one can see that $\lim_{n \to \infty} J_{\lambda, t_n} x$ exists for a given $x \in C$ if and only if $x \in R(I + \lambda A_S(\{t_n\}))$ and, in this case,

$$\lim_{k \to \infty} J_{\lambda, t_k} x = (I + \lambda A_S(\{t_n\}))^{-1} x .$$

Theorem 2.1 now clarifies the nature of $A_S(\{t_n\})$. In Hilbert spaces $A_S(\{t_n\})$ is independent of $\{t_n\}$ (since $\lim_{t \downarrow 0} J_{\lambda, t}$ exists) but, in general, this is not the case. Moreover, the example of [15] implies $A_S(\{t_n\}) \cup A_S(\{s_m\})$ is not necessarily in $G(\omega)$ if $\{t_n\} \neq \{s_m\}$.

## 3. Results in nice spaces.

The results described in Sections 1 and 2 can be improved under stronger assumptions on $X$. In particular, if $X$ is a Hilbert space the situation is perfect. Some highlights of the theory are outlined below.

**Theorem 3.1.** Let $X$ and $X^*$ be uniformly convex and $A$ satisfy the assumptions of Theorem 1.7. Let $S \in Q_\omega(\overline{D(A)})$ be generated by $-A$. Then

(i)   If $x \in D(A)$, $Ax$ contains a unique element of minimal norm.
(ii)  $S(t)$ leaves $D(A)$ invariant for $t \geq 0$.
(iii) If $x \in D(A)$, then $S(t)x$ is differentiable from the right for $t \geq 0$ and $D_r S(t)x + A^\circ S(t)x = 0$ for $t \geq 0$, where $A^\circ z$ denotes the element of $Az$ of minimal norm and $D_r$ denotes differentiation from the right.
(iv)  $D_r S(t)x$ is continuous from the right for $x \in D(A)$, and is continuous except at a countable number of points. The derivative $\frac{d}{dt} S(t)x$ exists at points of continuity of $D_r S(t)x$.

Thus the relationship of the semigroup generated by $-A$ to (CP) is very clear under the restrictions of Theorem 3.1. For variations on the theme of Theorem 3.1 see [18], [8], [25] and [23]. The assumption that $X^*$ is uniformly convex implies that if $A \in G(\omega)$, $\{[x_n, y_n]\} \subseteq A$, $\{x_n\}$ converges strongly to $x_0$ and $\{y_n\}$ converges weakly to $y_0$, then $A \cup \{[x_0, y_0]\} \in G(\omega)$. If $X^*$ is strictly convex, then $A'$ defined by

$A'x$ = convex hull of $Ax$

will belong to $\mathcal{G}(\omega)$ if $A$ does. These properties were much used in the early works on semigroups. See, e.g., [12], [16], [18].

The situation in Hilbert spaces is even better.

<u>Theorem 3.2.</u>  <u>Let $X$ be a Hilbert space and $C$ be a closed, nonempty, convex subset of $X$. Then</u>

(i)  <u>If $S \in Q_\omega(C)$ there is exactly one $A \in \mathcal{G}(\omega)$ such that $\overline{D(A)} = C$, $R(I + \lambda A) = X$ for $\lambda > 0$, $\lambda\omega < 1$ and $-A$ generates $S$.</u>

(ii) <u>If $A \in \mathcal{G}(\omega)$ and $R(I + \lambda A) = X$ for $\lambda > 0$, $\lambda\omega < 1$, then $\overline{D(A)}$ is convex.</u>

Theorem 3.2(i) is a restatement of the results of the appendix of [12] if $\omega = 0$. See also [27]. For proofs of Theorem 3.2(ii) see [28], [21], [12] or [7]. The condition $R(I + \lambda A) = X$ (rather than $R(I + \lambda A) \supseteq \overline{D(A)}$) is simply a maximality condition if $X$ is a Hilbert space due to Minty's result in [22]. Theorems 3.2 and 1.8 imply that, if $X$ is a Hilbert space, then there is a natural biunique correspondence between the semigroups of contractions on closed convex subsets of $X$ and the maximal monotone (equivalently, accretive) subsets of $X \times X$. If Theorem 3.1 is also used, this correspondence may be described as in [12]. (It was shown in [15] that the analogue of Minty's theorem for accretive subsets of $X \times X$ fails in some Banach spaces $X$.)

**4. Examples.** Here we discuss two examples of operators $A$ arising from concrete problems in partial differential equations for which the assumptions of Theorem 1.4 may be verified. Examples and applications are plentiful in Hilbert spaces ([7], [6], [4]) and part of the point here is the natural use of nonreflexive Banach spaces. The first example is the hyperbolic initial-boundary value problem:

(4.1)
$$\begin{cases} u_t + u_x + u^2 - v^2 = 0 \\ v_t + v_y + v^2 - u^2 = 0 \end{cases}$$

(4.2) $\quad u(0,x,y) = u_0(x,y), \; v(0,x,y) = v_0(x,y)$

and

(4.3) $\quad u(t,0,y) = v(t,x,0) = 0$.

The functions $(u,v)$ are defined on $R^+ \times R^{+2} = \{(t,(x,y)): t,x,y \geq 0\}$. This system was studied by Temam [29] in a Hilbert space setting. T. Liggett pointed out the relationship of this problem to the $L^1$ space described below and has generously allowed his observations to be developed here. From the point of view of the semigroup theory, we study (4.1), (4.2), (4.3) via the study of $A(u,v) = (u_x + u^2 - v^2, v_y + v^2 - u^2)$, where $u,v$ are now functions of $(x,y)$ only. The data $u_0, v_0$ of (4.2) are assumed non-negative (see [29]), and one seeks a nonnegative solution of the problem. We first obtain information about $R(I + \lambda A)$.

<u>Lemma 4.4.</u> <u>Let $f,g$ be continuous on $R^{+2}$ and $0 \leq f, g \leq M$. Then there is a unique classical solution of the problem</u>

(4.5)
$$\begin{aligned} u + \lambda(u_x + u^2 - v^2) &= f, \quad u(0,y) = 0 \\ v + \lambda(v_y + v^2 - u^2) &= g, \quad v(x,0) = 0 \end{aligned}$$

<u>on</u> $x,y \geq 0$ <u>provided</u> $0 < \lambda < 1/4(M+1)$. <u>The solution</u> $(u,v)$ <u>satisfies</u> $0 \leq u, v \leq M$.

Proof. Let $\varphi_M(s) = s^2$ for $|s| < M+1$ and $\varphi_M(s) = (M+1)^2$ for $|s| \geq M+1$. Set $\kappa = 1/\lambda$ and consider the problem:

(4.6)
$$\begin{aligned} u(x,y) &= \int_0^x e^{\kappa(s-x)}(\kappa f(x,y) + \varphi_M(v(s,y)) - \varphi_M(u(s,y)))ds \\ v(x,y) &= \int_0^y e^{\kappa(s-y)}(\kappa g(x,s) + \varphi_M(u(x,s)) - \varphi_M(v(x,s)))ds. \end{aligned}$$

For continuous functions $(u,v)$ (4.6) is equivalent to (4.5) with $u^2$ replaced by $\varphi_M(u)$, etc. Let $w = (u,v)$, and denote the right-hand side of (4.6) by $Tw$, so that (4.6) becomes $w = Tw$. Let $X = C(R^{+2}) \times C(R^{+2})$, where $C(R^{+2})$ is the Banach space of real-valued, continuous, bounded functions on $R^{+2}$. Since $\varphi_M$ is bounded and has $2(M+1)$ as a Lipschitz constant, it is not hard to show $T: X \to X$ and

$$\|Tw - Tz\| \leq 4(M+1)\lambda \|w - z\| \text{ for } z, w \in X.$$

Hence $T$ has a unique fixed point $w_1 = (u_1, v_1) \in X$ if $0 \leq \lambda < 1/4(M+1)$. Next one shows, by standard arguments, that $0 \leq f, g \leq M$ implies $0 \leq u_1, v_1 \leq M$ (and similarly for classical solutions of (4.5)), so that $\varphi_M(u_1) = u_1^2$, $\varphi_M(v_1) = v_1^2$ and $(u_1, v_1)$ is the unique classical solution of (4.5). The restriction on $\lambda$ will be removed later.

Accretive properties of $A$ are checked next. Let

$$\|(u,v)\|_{p,T} = \left( \int_0^T \int_0^T (|u|^p + |v|^p) dx, dy \right)^{1/p}$$

for $0 < T < \infty$.

<u>Lemma 4.7.</u> <u>Let</u> $D_M(A) = \{w = (u,v): u, v, u_x, v_y$ <u>are continuous on</u> $R^{+2}$, $u(0,y) = v(x,0) = 0$ <u>for</u> $x, y \geq 0$ <u>and</u> $0 \leq u, v \leq M\}$. <u>Then</u> $\|(w_1 + \lambda A w_1) - (w_2 + \lambda A w_2)\|_{p,T}$
$\geq (1 - \lambda \omega(M,p)) \|w_1 - w_2\|_{p,T}$ <u>for</u> $w_1, w_2 \in D_M(A)$, $\lambda \geq 0$ <u>and</u>

$$\omega(M, p) = \begin{cases} 4M, & 1 < p < \infty \\ 0, & p = 1. \end{cases}$$

The point of the lemma is that the linear part of $A$ is accretive with respect to $\|\,\|_{p,T}$, while the nonlinear part is Lipschitz on bounded functions. The Lipschitz constant appears as the $\omega$, <u>except</u> for $p = 1$. We verify the case $p = 1$. Let $w_1 = (u_1, v_1)$, $w_2 = (u_2, v_2)$, $\alpha = u_1 - u_2$, and (fixing $y$) $I = \{x: \alpha(x,y) > 0\} \cap [0,T]$, $J = \{x: \alpha(x,y) < 0\} \cap [0,T]$. For $p = 1$, the first component is estimated as follows: Form

$$\int_0^T \int_0^T |\alpha+\lambda(\alpha_x+u_1^2-u_2^2+v_2^2-v_1^2)|\,dxdy\,,$$

and estimate the integral with respect to x first. One has

$$\int_0^T |\alpha+\lambda(\alpha_x+u_1^2-u_2^2+v_2^2-v_1^2)|\,dx \geq$$

$$\geq \int_I (\alpha+\lambda(\alpha_x+u_1^2-u_2^2+v_2^2-v_1^2))\,dx$$

$$+ \int_J (-1)(\alpha+\lambda(\alpha_x+u_1^2-u_2^2+v_2^2-v_1^2))\,dx \geq$$

$$\geq \int_0^T (|\alpha|+\lambda(|u_1^2-u_2^2|-|v_2^2-v_1^2|))\,dx\,,$$

since $\int_I \alpha_x dx \geq 0$, $\int_J \alpha_x dx \leq 0$ and $u_1^2-u_2^2$ has the same sign as $\alpha$. The result follows from integrating with respect to y, estimating the second component in the same way, and adding. The contributions from the nonlinear terms cancel.

We can now complete the example.

**Theorem 4.8.** <u>Let</u> $D(A_0) = \cup_{M>0} D(A_M)) \cap \{w:w \in L^1(R^{+2})$ <u>and</u> $Aw \in L^1(R^{+2})\}$, <u>and</u> $A_0 w = Aw$ <u>for</u> $w \in D(A_0)$. <u>Then the closure</u> $\bar{A}_0$ <u>of</u> $A_0$ <u>in</u> $L^1(R^{+2}) \times L^1(R^{+2})$ <u>has the properties:</u>

(i) $\bar{A}_0$ is accretive in $L^1(R^{+2})$.
(ii) $R(I + \lambda \bar{A}_0) \supseteq \{(f,g): 0 \leq f, g$ <u>and</u> $(f,g) \in L^1(R^{+2})\}$ for $\lambda > 0$.
(iii) $D(\bar{A}_0)$ <u>is dense in</u> $\{(f,g): 0 \leq f, g$ <u>and</u> $(f,g) \in L^1(R^{+2})\}$.

Proof: We use the following simple lemma concerning accretive operators:

**Lemma 4.9.** <u>Let</u> $B \subseteq X \times X$, $B \in G(\omega)$. <u>Then</u>

(i) $\bar{B}$ (<u>the closure of</u> B <u>in</u> $X \times X$) <u>belongs to</u> $G(\omega)$.

(ii) If B is closed, so is $R(I + \lambda B)$ for $\lambda > 0$, $\lambda \omega < 1$.
(iii) If C is closed and convex, $\lambda > \mu > 0$, $\lambda \omega < 1$, then $R(I + \mu B) \supseteq C$ and $(I + \mu B)^{-1} C \subseteq C$ implies $R(I + \lambda B) \supseteq C$ and $(I + \lambda B)^{-1} C \subseteq C$.

<u>Proof.</u> (i) follows at once from the definition. Property (ii) follows from the fact that $B \in \mathcal{G}(\omega)$, $[x_n, y_n] \in B$ and $x_n + \lambda y_n \to z$ as $n \to \infty$ implies $x_n = (I + \lambda B)^{-1}(x_n + \lambda y_n)$ has a limit as $n \to \infty$. For (iii), let $J_\lambda = (I + \lambda B)^{-1}$. It follows from the definitions that

$$J_\lambda x = J_\mu(\frac{\mu}{\lambda} x + \frac{\lambda - \mu}{\lambda} J_\lambda x)$$

whenever $x \in D(J_\lambda)$. Let $x \in C$ be given. The map $z \to J_\mu(\frac{\mu}{\lambda} x + \frac{\lambda - \mu}{\lambda} z)$ is a strict contraction from C into C and has a fixed point, which is easily seen to be $J_\lambda x$. In particular, $x \in R(I + \lambda B)$.

To continue with the proof of Theorem 4.8, it follows from letting $T \to \infty$ in Lemma 4.7 that $A_0$ is accretive, and from Lemma 4.9(i) that $\bar{A}_0$ is accretive. Let $M > 0$ and

$$C_M = \{(f, g) : 0 \leq f, g \leq M \text{ and } (f, g) \in L^1(R^{+2})\}.$$

If $(f, g) \in C_M$, then the $(u, v)$ provided by Lemma (4.4) for $\lambda < 1/(M+1)$ is in $L^1(R^{+2})$ by Lemma 4.7, and thus in $D(A_0)$. Hence $R(I + \lambda A_0)$ is dense in $C_M$ for $\lambda < 1/4(M+1)$, $R(I + \lambda \bar{A}_0) \supseteq C_M$ for $\lambda < 1/4(M+1)$ (by Lemma 4.9(ii)) and $R(I + \lambda \bar{A}_0) \supseteq C_M$ for $\lambda > 0$ by Lemma 4.9(iii). Finally, $R(I + \lambda \bar{A}_0) \supseteq \cup_{M > 0} C_M$, which is dense in the set of non-negative functions in $L^1(R^{+2})$, and $D(A_0)$ contains all non-negative smooth functions with compact support in the interior of $R^{+2}$. Thus (ii) and (iii) of Theorem 4.8 are clear, and the proof is complete.

<u>Remarks.</u> There are several standard ways to generalize the system (4.1) without changing the essentials of the above analysis. For example, $u^2, v^2$ can be replaced by $\varphi(u), \varphi(v)$

for monotone increasing $\varphi$, etc. Indeed, rather than restrict $D(A_0)$ to nonnegative functions we could have used $\varphi(u) = |u|u$ and noted that the convex set of nonnegative functions was then invariant under the semigroup.

As a second example we consider the problem

(4.10)
$$\begin{cases} \dfrac{\partial u}{\partial t} - \Delta \varphi(u) = 0 & \text{in } (0,\infty) \times \Omega \\ u\big|_{t=0} = u_0, \ \varphi(u)\big|_{\partial \Omega} = 0 \end{cases}$$

where $\Omega$ is an open subset of $R^n$ with a smooth boundary. In (4.10), $\varphi$ is a continuous strictly monotone increasing function on R with $\varphi(0) = 0$. The problem (4.10) has been studied under various assumptions on $\varphi$. See, e.g., [1], [2], [17], [26] and the bibliographies of these works. Brezis [5] remarks that (4.10) is associated with a semi-group in $H^{-1}(\Omega)$ with nice properties. Here we point out that (4.10) may be naturally considered in the space $L^1(\Omega)$, and that the generator $-Au = \Delta\varphi(u)$ in this space has the properties of Theorem 1.4. Indeed, in physical situations corresponding to (4.10) u is a density and $\int_\Omega |u(t,x)|\,dx$ has immediate physical significance. We turn now to the stationary questions concerning the associated generator.

<u>Definition 4.11.</u> $D(A_0)$ is the set of continuous functions $u: \bar\Omega \to R$ satisfying

(i) $\qquad \varphi(u) \in C^2(\bar\Omega)$

(ii) $\qquad u\big|_{\partial \Omega} = 0$

(iii) $\qquad$ u has compact support in $\bar\Omega$,

and $A_0 u = -\Delta \varphi(u)$ for $u \in D(A_0)$.

<u>Theorem 4.12.</u> $\underline{A_0 \text{ is accretive in } L^1(\Omega).\ \text{Moreover, if } \Omega}$ $\underline{\text{is bounded and } \lambda > 0,\ \text{then } R(I + \lambda A_0) \text{ is dense in } L^1(\Omega)}$.

Of course, Theorem 4.12 implies that the closure $\bar A_0$ of $A_0$ in $L^1(\Omega) \times L^1(\Omega)$ satisfies the conditions of Theorem 1.4 if $\Omega$ is bounded.

**Proof:** Let $u, v \in D(A_0)$ and define

$$\Omega_+ = \{x \in \Omega : u(x) > v(x)\}, \quad \Omega_- = \{x \in \Omega : u(x) < v(x)\}.$$

For $\lambda > 0$ one has

(4.13) $\int_\Omega |u-v+\lambda(A_0 u - A_0 v)|\, dx \geq \int_{\Omega_+} (u-v+\lambda(A_0 u - A_0 v))\, dx +$

$$+ \int_{\Omega_-} (v-u+\lambda(A_0 v - A_0 u))\, dx =$$

$$= \int_\Omega |u-v| + \lambda \left( \int_{\Omega_+} (A_0 u - A_0 v)\, dx + \int_{\Omega_-} (A_0 v - A_0 u)\, dx \right).$$

Consider, e.g., the term:

(4.14) $\int_{\Omega_+} \frac{\partial^2}{\partial x_1^2} (\varphi(v(x)) - \varphi(u(x)))\, dx.$

Let $h = \varphi(v) - \varphi(u)$, $x = (x_1, x^*)$, where $x^* \in \mathbb{R}^{n-1}$, and

$$\ell(x^*) = \{x_1 : (x_1, x^*) \in \Omega_+\}.$$

If $(a, b)$ is a component of $\ell(x^*)$, then $h(a, x^*) = h(b, x^*) = 0$, while $h(s, x^*) \leq 0$ for $a \leq s \leq b$. Hence $\frac{\partial}{\partial x_1} h(b, x^*) \geq 0$ and $\frac{\partial}{\partial x_1} h(a, x^*) \leq 0$. Thus (4.14) satisfies

$$\int_{\Omega_+} \frac{\partial^2}{\partial x_1^2} h(x)\, dx = \int_{\mathbb{R}^{n-1}} \left( \int_{\ell(x^*)} \frac{\partial^2}{\partial x_1^2} h(x_1, x^*)\, dx_1 \right) dx^* \geq 0.$$

Estimating the last $2n$ terms in (4.13) in this way yields the accretiveness of $A_0$ in $L^1(\Omega)$.

In order to show the problem

(4.15) $\quad u - \lambda \Delta \varphi(u) = f, \quad f \in L^1(\Omega), \quad u \in D(A_0)$

has a solution $u$ for a dense set of $f$ in $L^1(\Omega)$, we first rewrite it as

(4.16) $\quad \beta(w) - \lambda\Delta w = f, \ w \in C^2(\bar{\Omega}), w|_{\partial\Omega} = 0$,

where $\beta = \varphi^{-1}$, $w = \varphi(u)$. This problem has a solution for any $f \in L^2(\Omega)$ by the results of [7], provided that $\Omega$ is bounded so that $-\Delta$ is coercive, and that $w \in C^2(\bar{\Omega}), w|_{\partial\Omega}=0$ is amended to $w \in H^2(\Omega) \cap H_0^1(\Omega)$. To approximate these solutions appropriately we replace $\beta$ by

$$\beta_\eta = \beta((I + \eta\beta)^{-1})$$

for $\eta > 0$. The functions $\beta_\eta$ are Lipschitz continuous, monotone and $|\beta_\eta(x)|$ increases monotonically to $|\beta(x)|$ for each $x$ (to $\infty$ if $x \notin D(\beta)$) as $\eta \downarrow 0$. Let $w_\eta \in H^2(\Omega) \cap H_0^1(\Omega)$ satisfy:

(4.17) $\quad \beta_\eta(w_\eta) - \lambda\Delta w_\eta = f$.

Then, by the arguments of [7], $w_\eta$, $\Delta w_\eta$ and $\beta_\eta(w_\eta)$ converge to $w$, $\Delta w, \beta(w)$, respectively, in $L^2(\Omega)$, where $w$ is the solution of (4.16). Since $\beta_\eta$ is Lipschitz, $w_\eta \in C^2(\bar{\Omega})$ (for $f \in \mathcal{D}(\Omega)$ say) and (4.17) implies

(4.18) $\quad \max_{x \in \Omega} |\beta_\eta(w_\eta)| \leq \max_{x \in \Omega} |f|$.

(Since $\beta_\eta(w_\eta)$ and $w_\eta$ have the same relative maxima and minima.) This last relation implies

(4.19) There is an $\eta_0 > 0$ such that $|\beta(w_\eta(x))| \leq \max_\Omega |f| + 1$
for $0 < \eta < \eta_0$.

Hence $\lim_{\eta \downarrow 0} \beta(w_\eta(x)) = \beta(w(x))$ holds in $L^p(\Omega)$ for $1 \leq p < \infty$. The argument is nearly complete. Setting $u_\eta = \beta(w_\eta), \varphi(u_\eta) = w_\eta$ we have $u_\eta \in D(A_0)$ and

$$u_\eta - \lambda\Delta\varphi(u_\eta) = f - \beta_\eta(w_\eta) + \beta(w_\eta)$$

while $\lim_{\eta \downarrow 0} (\beta(w_\eta) - \beta_\eta(w_\eta)) = 0$ in $L^2(\Omega)$ (and hence in $L^1(\Omega)$). This completes the proof that $R(I + \lambda A_0)$ is dense in $L^1(\Omega)$ when $\Omega$ is bounded.

A more interesting problem is encountered in the case $\Omega = R^n$. Here $A_0$ is still accretive, but the density of $R(I + \lambda A_0)$ has been established only for $n = 1$ by this writer. This problem, and other questions concerning this example, will be studied further elsewhere.

Remark. The author has learned of unpublished work of H. Brezis and W. Strauss which is related to Theorem 4.12. They study the Dirichlet problem for $\beta(w) - \Delta w = f$ in $L^p(\Omega)$, $1 \leq p < \infty$, $\Omega$ bounded, and obtain quite sharp information under more general hypotheses than we have just sketched.

One problem which we have not treated here is that of regularity. I.e., in each of the two examples just presented we did not show to what degree the functions provided by the exponential formula could be differentiated or that the original differential equations and side conditions were satisfied in any usual sense. There are interesting problems here, even in these cases. A recent example of Webb [31] seems to indicate that these questions will have to be studied case by case when using nonreflexive Banach spaces. Webb gives an example of a semilinear generator $A + B$ (where $A$ is a linear generator and $-B$ is continuous and accretive) such that the semigroup $S$ generated by $A + B$ does not leave $D(A)$ invariant.

## REFERENCES

1. D. G. Aronson, Regularity properties of flows through porous media: The interface, Arch. Rat. Mech. and Anal. 37 (1970), 1-10.

2. _____, Regularity properties of flows through porous media: A counterexample, SIAM J. Applied Math. 19 (1970), 299-307.

3. H. Brezis, On a problem of T. Kato, Comm. Pure Appl. Math. 24(1971), 1-6.

4. _____, Semi-groupes non linéaires et applications, Symposium sur les problèmes d'évolution, Instituto Naxionale di Alta Mathematica, Rome, 1970.

5. _____, Propriétes régularisantes de certains semigroups non lineaires, Israel J. of Math., to appear.

6. _____, Problems unilateraux, J. Math. Pures et Appliquées, to appear.

7. H. Brezis, M. Crandall and A. Pazy, Perturbations of nonlinear maximal monotone sets in Banach space, Comm. Pure Appl. Math. 23 (1970), 123-144.

8. H. Brezis and A. Pazy, Accretive sets and differential equations in Banach spaces, Israel J. of Math. 8 (1970), 367-383.

9. _____, Convergence and approximation of semigroups of nonlinear operators in Banach spaces, to appear.

10. F. E. Browder, Nonlinear operators and nonlinear equations of evolution in Banach spaces, Proc. Symp. in Pure Math. 18, Part II, Amer. Math. Soc., to appear.

11. M. G. Crandall, Differential equations on convex sets, J. Math. Soc. Japan 22 (1970), 443-455.

12. M. G. Crandall and A. Pazy, Semi-groups of nonlinear contractions and dissipative sets, J. Functional Analysis 3 (1969), 376-418.

13. _____, On accretive sets in Banach spaces, J. Funct. Anal. 5 (1970), 204-217.

14. M. G. Crandall and T. M. Liggett, *Generation of semi-groups of nonlinear transformations on general Banach spaces*, Amer. J. Math., to appear.

15. _____, *A theorem and a counterexample in the theory of semigroups of nonlinear transformations*, Trans. Amer. Math. Soc., to appear.

16. J. R. Dorroh, *A nonlinear Hille-Yosida-Phillips theorem*, J. Funct. Anal. 3 (1969), 345-353.

17. S. I. Hudjaev and A. I. Vol'pert, *Cauchy's problem for degenerate second order parabolic equations*, Mat. Sbornik 78 (1969); Math. USSR Sbornik 7, (1969), 365-387.

18. T. Kato, *Accretive operators and nonlinear evolution equations in Banach spaces*, Proc. Symp. in Pure Math. 18, Part I, Amer. Math. Soc., Providence, Rhode Island, 138-161.

19. T. Kato, *Note on the differentiability of nonlinear semigroups*, Proc. Symp. Pure Math. 16, Amer. Math. Soc., Providence, Rhode Island, 91-94.

20. Y. Kōmura, *Nonlinear semigroups in Hilbert space*, J. Math. Soc. Japan 19, (1967) 493-507.

21. _____, *Differentiability of nonlinear semigroups*, J. Math. Soc. Japan 21 (1969), 375-402,

22. G. Minty, *Monotone (nonlinear) operators in Hilbert space*, Duke Math. J. 29 (1962), 341-346.

23. I. Miyadera, *Some remarks on semigroups of nonlinear operators*, to appear.

24. I. Miyadera and S. Ôharu, Approximation of semigroups of nonlinear operators, Tôhoku Math. J. 22 (1970), 24-47.

25. S. Oharu, On the generation of semigroups of nonlinear contractions, J. Math. Soc. Japan 22 (1970), 526-550.

26. O. A. Oleinik, On some degenerate quasilinear parabolic equations, Seminari dell Istituto Nazionale di Alta Math. 1962-63, Oderisi, 1964, 355-371.

27. A. Pazy, Semigroups of nonlinear operators in Hilbert space, Proc. of Summer School in Nonlinear Analysis, Varenna, C. I. M. E., (1970), to appear.

28. R. T. Rockafellar, On the virtual convexity of the domain and range of a nonlinear maximal monotone operator, Math. Annalen 185 (1970), 81-90.

29. R. Temam, Sur la résolution exacte et approcheé d'un probleme hyperbolique non linéaire de T. Carleman, Arch. Rat. Math. Mech. 35 (1969), 351-362.

30. G. Webb, Representation of semi-groups of nonlinear nonexpansive transformations in Banach spaces, J. Math. Mech. 19 (1969), 159-170.

31. _____, Continuous nonlinear perturbations of linear dissipative operators in Banach spaces, to appear.

The preparation of this paper was sponsored in part by the Office of Naval Research under Contract N000-14-69-A-0200-4022. Reproduction in whole or in part is permitted for any purpose of the United States Government

>Department of Mathematics
>University of California, Los Angeles
>Los Angeles, California

>Received April 22, 1971.

# Weak and Strong Solutions of Dual Problems

*J. J. MOREAU*

## §1. Introduction.

Many problems arising from various domains of physics consist in the investigation of functions defined, for instance, on a subset S of $\mathbb{R}^n$, fulfilling certain requirements at each interior point of S and other requirements at each boundary point. Such conditions, expressing physical laws of local character, generally involve continuity of the functions and their partial derivatives up to a certain order. At this stage of the problem, physicists frequently characterize the solution, <u>if it exists</u>, by variational properties or even extremal properties: it is that element of a certain class of functions where a certain functional attains its minimum.

These variational characterizations of solutions have suggested to mathematicians the idea of shifting from the so-called <u>strong</u> formulation of problems - i.e. the naive formulation provided by physics - to milder systems of requirements, yielding solutions denoted as <u>weak</u> and whose existence is more easily established. For instance, the function space to which the expected solution of the physical problem belongs is endowed with a topology (and a structure of uniform space); a minimizing sequence of the functional in question does not necessarily converge relatively to this topology, but in some typical cases one proves that convergence is achieved by imbedding the function space in its completion: the element of this completion to which the sequence converges is called a weak solution of the problem, and it is proved that it coincides with the so-called <u>strong solution</u> - i.e. the solution agreeing with the strong formulation- whenever the latter exists.

Another common feature in this connection is the <u>duality</u> between variational properties of solutions. In the classical case where the conventional machinery of the calculus of variations can be applied (this means that the considered functionals are, in some sense, differentiable on the function spaces where they are defined) this duality is well known: see Courant, Hilbert [1]. Dirichlet's problem gives the primary example of such a situation: Let $\Omega$ be a bounded open subset of $\mathbb{R}^n$ ; one looks for a numerical function $u$ , continuous on the closure $\bar{\Omega}$ , harmonic on $\Omega$ and coinciding with a prescribed continuous function on the boundary $\partial\Omega$. Let us suppose here that the boundary data can be extended as a function $f$ , continuous on $\bar{\Omega}$ , possessing a square-integrable gradient on $\Omega$ . A first extremal characterization of the solution $u$ (if it exists) is that, in the set of the functions $v$ , continuous on $\bar{\Omega}$ , agreeing with $f$ on $\partial\Omega$ , possessing a square-integrable gradient on $\Omega$ , this solution is the (unique) element that realizes the minimum of the functional

$$v \to \int_\Omega |\text{grad } v|^2 \, d\tau$$

($d\tau$: volume or Lebesgue measure on $\Omega$). As a <u>dual</u> characterization of $u$ , or rather of the vector field grad $u$ , we have the following: in the set of the square-integrable vector fields on $\Omega$ which are the gradients of harmonic functions, grad $u$ is the (unique) element for which the functional

$$w \to \int_\Omega |w - \text{grad } f|^2 \, d\tau$$

achieves its minimum (for more sophisticated versions of so-called Dirichlet's principle see e.g.: Brelot [1], [2]; Deny, Lions [1]). The underlying geometric pattern is that of projecting a known point of a certain pre-Hilbert space onto two orthogonal affine manifolds. These two projections are known to be complementary to each other by their common equivalence to a third operation of a different sort, the decomposition of a given element of the pre-Hilbert space into the sum of two elements belonging respectively to a certain pair of orthogonal vector subspaces.

In what follows we present a rather general model involving such triplets of equivalent ways of characterizing elements in a function space which is a priori not complete: two minimization properties, said to be dual to each other, and a decomposition property. This is done for <u>nonlinear</u> problems, <u>without differentiability</u> being considered for the functionals in question, but under certain convexity hypotheses. It is stressed that when imbedding the functional space initially considered into its completion, each of the three properties induces, in a natural way, a concept of weak solution, but these extensions do not in general agree with each other; in other words, the weakening procedure may not preserve duality. <u>The core of this report consists in formulating a series of necessary and sufficient conditions for the consistency of the weakening procedures.</u>

For sake of clarity we restrict ourselves to the case where the function space under consideration is endowed with a pre-Hilbert structure, as in the above example of Dirichlet's problem. Such a space making with itself a dual pair, the elements involved in dual minimization properties all belong to the same space, allowing a simple form for the associated decomposition pattern. Such a pre-Hilbert structure is usual in physics: the quadratic form whose polar form provides the scalar product will represent a kinetic energy, an electromagnetic energy, the potential energy of a linear elasticity law, the dissipation function of a linear viscosity law, etc...

First, in Section 2, we recall the necessary basic elements of the duality theory of convex functions defined on vector spaces. This theory, which originated in an idea of W. Fenchel [1], [2] has been extensively developed in recent years as a systematic tool for bringing out duality in convex extremal problems, without differentiability assumptions.

For detailed accounts and bibliography the reader might refer to R. T. Rockafellar's book [3], devoted to finite-dimensional spaces but containing many guiding ideas for the general case, or, for a systematic study of convex functions defined on topological vector spaces, to J. J. Moreau [8]. More recent developments appear in J. L. Joly [1], M. Valadier [1], [2]. For applications to optimization problems in general or to the theory of approximation see e. g. Rockafellar [2], Laurent-Joly [1], Moreau [12]. Following Rockafellar [1] and Ioffe-Tihomirov [1], the theory has been applied to functionals defined by integrals, thus to the calculus of variations and optimal control: see Rockafellar's contribution to this volume. In the same connection these methods have recently proved useful in the study of certain partial differential equations: see Temam [1], [2], [3], [4]. Subdifferential mappings, as defined in paragraph 2.d, provided the primitive example of monotone mappings (G. J. Minty [1]).

The author's interest in this theory was motivated by the mechanics of continua (see: Moreau [4], [6], [10], [14] and Nayroles [1]).

From Section 3 onward, we restrict ourselves to pre-Hilbert or Hilbert spaces. Proximation mappings defined in this context generalize projections on convex sets; thus the present paper is connected to E. H. Zarantonello's article in this volume.

## §2. Elements of the Duality of Convex Functions.

2.a. Polar functions.

Let $X$ and $Y$ be a dual pair of a real vector spaces; we denote by $\langle .,. \rangle$ the bilinear form which "puts these two spaces in duality". For instance $X$ may be a given topological vector space and $Y$ its topological dual, but in what follows $X$ and $Y$ will play perfectly symmetric roles. Each of the two spaces may be endowed with various topologies (actually equivalent when the dimension is finite) said to be compatible with the duality: on the space $X$, for instance, these are the locally convex topologies relative to which the continuous linear forms consist exactly of the functions expressed as $x \to \langle x, y \rangle$ with $y \in Y$ (and each continuous linear form on $X$ corresponds to only one $y \in Y$ if the usual separation condition is fulfilled). Among these topologies, the <u>Mackey topology</u> $\tau(X, Y)$ is the finest, the <u>weak topology</u> $\sigma(X, Y)$ is the coarsest.

An affine function on $X$ (one should understand: continuous for topologies compatible with the duality) is written as

(2.1) $\qquad x \to \langle x, y \rangle - \rho$

with $y \in Y$ (called the <u>slope</u> of the function) and $\rho \in \mathbb{R}$.

Let $f \in \bar{\mathbb{R}}^X$ (i.e. a function defined on $X$ with values in $\bar{\mathbb{R}} = [-\infty, +\infty]$). The affine function (2.1) is, all over $X$, a minorant of $f$ if and only if

(2.2) $\qquad \rho \geq \sup \{\langle x, y \rangle - f(x) : x \in X\} = f^*(y)$.

The function $f^*$ defined thereby on $Y$ is called the <u>polar function</u> of $f$ (or <u>conjugate</u>; in Rockafellar's book [3] the word "polar function" is used in another sense).

<u>Example:</u> Take as $f$ the <u>indicator</u> function $\psi_S$ of a subset $S$ of $X$, i.e.

$$\psi_S(x) = \begin{cases} 0 & \text{if } x \in S \\ +\infty & \text{if } x \notin S \end{cases}$$

The polar function

$$\psi_S^*(y) = \sup \{\langle x, y \rangle - \psi_S(x) : x \in X\}$$

$$= \sup \{\langle x, y \rangle : x \in S\}$$

is conventionally called the <u>support function</u> of the set $S$, with respect to the duality $(X, Y)$.

## 2. b. Closed proper convex functions.

Using the asterisk indifferently to denote the polarity from $\bar{\mathbb{R}}^X$ to $\bar{\mathbb{R}}^Y$ or from $\bar{\mathbb{R}}^Y$ to $\bar{\mathbb{R}}^X$, the polar $f^{**}$ of $f^*$ is
$$f^{**}(x) = \sup \{\langle x, y \rangle - f^*(y) : y \in Y\}.$$

From Condition (2.2) it is easily found that $f^{**}$ equals the pointwise supremum of the affine functions such as (2.1) which minorize $f$.

Denote by $\Gamma(X, Y)$ the set of the numerical functions defined on $X$ which are pointwise suprema of families of such affine functions. By standard separation arguments (i.e. the Hahn-Banach theorem), it may be proved that $\Gamma(X, Y)$ consists of:
1. The constant function $+\infty$.
2. The constant function $-\infty$ (it is the supremum of an empty family of affine functions!)
3. The set of all functions called <u>closed proper convex</u> on $X$, namely the functions with values in $]-\infty, +\infty]$, not everywhere $+\infty$, which are convex and lower semi-continuous for one (thus for all) of the (locally convex) topologies compatible with the duality $(X, Y)$. This latter set will be denoted by $\Gamma_0(X, Y)$.

Similar notations when exchanging $X$ and $Y$.

<u>Example.</u> The indicator $\psi_S$ of $S \subset X$ belongs to $\Gamma(X, Y)$ if and only if the set $S$ is convex and closed for one (thus for all) of the above topologies; $\psi_S$ belongs to $\Gamma_0(X, Y)$ if in addition $S$ is not empty.

Clearly $f^* \in \Gamma(Y, X)$ for any $f \in \bar{\mathbb{R}}^X$; furthermore $f^* \in \Gamma_0(Y, X)$ if and only if $f$ is not everywhere $+\infty$ and possesses at least one affine minorant such as (2.1).

The function $f^{**}$ is the greatest element of $\Gamma(X, Y)$ minorizing $f$; thus it is called the $\Gamma$-<u>hull</u> of $f$.

In particular $f = f^{**}$ if and only if $f \in \Gamma(X, Y)$; this shows that polarity defines a one-to-one correspondence between $\Gamma(X,Y)$ and $\Gamma(Y,X)$. As the constants $+\infty$ and $-\infty$ are polar to each other, the correspondance is also one-to-one between $\Gamma_0(X,Y)$ and $\Gamma_0(Y,X)$.

In other words if it is supposed that $f \in \Gamma(X,Y)$ and $g \in \Gamma(Y,X)$, the relation $f = g^*$ is equivalent to $f^* = g$; in such a case $f$ and $g$ will be said to be <u>mutually polar</u>.

<u>Example</u>. Let A and B respectively be nonempty subset of X and Y; it is easily found that the indicators $\psi_A$ and $\psi_B$ are mutually polar functions if and only if A and B are <u>mutually polar cones</u>, i.e. $A = \{x \in X : \langle x,y \rangle \le 0$ for any $y \in B\}$ and symmetrically (such cones are convex, with vertex at the origin, and closed for the topologies compatible with the duality).

## 2.c. Lower semi-continuous hull and $\Gamma$-hull.

Suppose now that the function $f$ is convex on $X$, with values in $]-\infty,+\infty]$, not everywhere $+\infty$; let $\underline{f}$ be its <u>lower semi-continuous hull</u> with regard to some (locally convex) topology compatible with the duality $(X,Y)$, i.e. the greatest l.s.c. numerical function minorizing $f$. It is known that this l.s.c. hull may be constructed pointwise by

(2.3) $\qquad \underline{f}(x) = \lim_{u \to x} \inf f(u)$ .

It is elementary that the convexity of $f$ implies the convexity of $\underline{f}$; thus, as soon as it is checked that $\underline{f}$ takes nowhere the value $-\infty$ (a common method for this is to check that $f$ possesses at least one continuous affine minorant) one concludes that $\underline{f} \in \Gamma_0(X,Y)$ and, consequently, $\underline{f} = f^{**}$.

Observe at this stage that, due to the convexity of $f$, the choice of the topology when writing (2.3) is immaterial, as long as this (locally convex) topology is compatible with the duality $(X,Y)$.

The following consequence of (2.3) will play an essential part in Section 4: <u>Suppose that the restriction of $f$ to its effective domain</u>

$$\mathrm{dom}\, f = \{x \in X : f(x) < +\infty\}$$

is <u>lower semicontinuous</u> (for the topology induced on this domain); <u>then f has the same restriction to this domain as</u> f .

## 2. d. Subgradients and subdifferentials.

Let $f \in \bar{\mathbb{R}}^X$ ; $y \in Y$ is called a <u>subgradient</u> of f at the point x if y is the slope of an affine minorant of f <u>exact</u> at the point x (i.e. taking at this point the same value as f ). This requires that the value $f(x)$ is finite and that the supposed minorant has the form

$$u \to \langle u-x, y \rangle + f(x).$$

Using Condition (2.2) for an affine function to minorize f , one obtains the following representation for the set, denoted by $\partial f(x)$, of the subgradients of f at the point x

$$\partial f(x) = \{y \in Y : f^*(y) - \langle x, y \rangle \leq -f(x)\}.$$

This set is called the <u>subdifferential</u> of f at the point x . Clearly the convexity and the lower semicontinuity of $f^*$ imply that $\partial f(x)$ is a convex (possibly empty) subset of Y , closed for the topologies compatible with the duality (Y, X). If $\partial f(x)$ is not empty the function f is said to be <u>subdifferentiable</u> at the point x .

Suppose for instance that the function f is convex and <u>weakly differentiable</u> at the point x, with respect to the duality (X, Y) (such is a fortiori the case when X is a normed space, Y its topological dual with f Fréchet-differentiable at the point x ) with the element $y \in Y$ as differential or "gradient"; then it is easily found that $\partial f(x)$ consists of the single element y (About relations between differentials and subdifferentials, see Asplund-Rockafellar [1]).

An evident use of the concept of subdifferential is the characterization of a minimum: the function f attains its infimum at the point x if and only if the set $\partial f(x)$ contains the origin of Y. There thus arises a need for a "subdifferential calculus"; in addition to Rockafellar's book [3] (for general bibliography and finite dimensional cases) the reader might refer to Moreau [7], [8], Valadier [1], [2]. The simplest rules of this calculus concern <u>addition</u> : if $f_1$ and $f_2$ are numerical functions on X, one trivially has, for any $x \in X$,

(2.4) $\qquad \partial f_1(x) + \partial f_2(x) \subset \partial(f_1 + f_2)(x)$

and it is important to formulate sufficient conditions for this inclusion to be an equality of sets. We shall only make use in this paper of this simple one (see Moreau [11]): <u>If $f_1$ and $f_2$ are convex, one of them weakly differentiable at the point x, then the inclusion (2.4) is an equality of sets.</u>

2.e. <u>Conjugate points.</u>

The preceding facts find their clearest setting when one starts with a pair of <u>mutually polar functions</u> $f \in \Gamma_0(X,Y)$, $g \in \Gamma_0(Y,X)$. Then, for $x \in X$ and $y \in Y$, the three following properties are equivalent:

(2.5) $\qquad y \in \partial f(x)$

(2.6) $\qquad x \in \partial g(y)$

(2.7) $\qquad f(x) + g(y) - \langle x, y \rangle = 0$.

(Note that, by the definition of polarity, the left member of this latter equality is essentially nonnegative, whichever are $x \in X$ and $y \in Y$). In such a case the points x and y are said to be <u>conjugate</u> relative to the pair of mutually polar functions (f, g).

For instance, if $A \subset X$ and $B \subset Y$ are mutually polar cones, Condition (2.7) for x and y to be conjugate relative to the pair of mutually polar functions $f = \psi_A$ and $g = \psi_B$ becomes:

$\qquad x \in A, \quad y \in B, \quad \langle x,y \rangle = 0$.

In the very special case where A and B are two vector subspaces of X and Y, each being the orthogonal of the other with respect to the duality, this latter condition reduces to: $x \in A$ and $y \in B$.

As a useful example of the equivalence between (2.5) and (2.6) let us note the following: The (possibly empty) set of the points where the function f attains its infimum equals $\partial g(0)$, the subdifferential of the function g at the origin of Y.

## 2.f. Infimal convolution.

Let $f_1$ and $f_2$ be two elements of $\bar{\mathbb{R}}^X$; we denote by $f_1 \nabla f_2$ the element of $\bar{\mathbb{R}}^X$ defined as the function

(2.8) $\quad x \to (f_1 \nabla f_2)(x) = \inf \{f_1(u) + f_2(x-u) : u \in X\}$

(or equivalently $\quad \inf \{f_1(x-v) + f_2(v) : v \in X\}$).

The possible values of u such that the above sum takes the indeterminate form $\infty - \infty$ must be omitted when constructing the "inf". The operation $\nabla$, called <u>infimal convolution</u>, is a commutative and associative composition law in $\bar{\mathbb{R}}^X$. If $f_1$ and $f_2$ are convex, so is $f_1 \nabla f_2$ (at least in an extended sense, since it may take the value $-\infty$ at some points: for the technique of overcoming such difficulties see Moreau [13]).

<u>Example 1.</u> If $f_2$ is the indicator function of a singleton $\{a\}$, then $f_1 \nabla f_2$ is a <u>translate</u> of $f_1$, namely the function

$$x \to f_1(x-a).$$

<u>Example 2.</u> If A is a subset of X and $\|\cdot\|$ a norm on this vector space, then $(\psi_A \nabla \|\cdot\|)(x)$ is the <u>distance</u> of the point x to the set A.

Taking now polar functions, easy computation yields

(2.9) $\quad (f_1 \nabla f_2)^* = f_1^* + f_2^*.$

This formula provides an answer to the following question: observing that <u>addition</u> is a composition law in $\Gamma(Y,X)$ (at least when extended by the convention $\infty-\infty = -\infty$) and that polarity maps $\Gamma(Y,X)$ one-to-one onto $\Gamma(X,Y)$, what direct interpretation can be given of the composition law obtained in $\Gamma(X,Y)$ as the image of addition ? Suppose that $f_1$ and $f_2$ belong to $\Gamma(X,Y)$ and that $g_1$ and $g_2$, in $\Gamma(Y,X)$, are their polar functions. Equation (2.9), now written as

$$(f_1 \nabla f_2)^* = g_1 + g_2$$

implies

$$(g_1 + g_2)^* = (f_1 \nabla f_2)^{**}.$$

Thus the desired operation is the $\Gamma$-hull of infimal convolution.

This gives a great practical importance to cases where $f_1 \nabla f_2$ happens to belong to $\Gamma(X,Y)$ so that the double asterisk may be omitted. As already mentioned, the convexity of $f_1$ and $f_2$ guarantees that of $f_1 \nabla f_2$ ; in view of paragraph 2.c. it just remains to check whether $f_1 \nabla f_2$ is lower semi-continuous (with nowhere the value $-\infty$). This lies beyond the scope of the present paper (see Moreau [1], [8]).

## 3. Dual Extremum Problems in a Pre-Hilbert Space.

### 3.a. Definitions

Let H be a pre-Hilbert opace over the real numbers, i.e. a real linear space on which a certain positive definite quadratric form has been defined once and for all; the square root of this quadratic form provides the norm $\|\cdot\|$ on H ; its polar form gives the scalar product of any two elements x and y of H , denoted by $(x|y)$. This bilinear form puts H in duality with itself; as a topology compatible with this duality, the weak topology $\sigma(H,H)$ will be considered; but it must be stressed that the norm topology is not compatible with the duality (H H) if H is not norm-complete, i.e. if it is not a Hilbert space.

Denote by Q the quadratic form

$$Q(x) = \tfrac{1}{2}\|x\|^2.$$

Easy computation yields that Q is Fréchet-differentiable, and also weakly differentiable relative to the duality (H, H), with

$$\operatorname{grad} Q(x) = x.$$

The function Q is convex; it belongs to $\Gamma_0(H, H)$ and equals its polar function.

Given $z \in H$, we shall denote by $Q_z$ the function $x \to Q(z-x)$, so that

$$\operatorname{grad} Q_z(x) = x - z.$$

Let f be a convex numerical function defined on H, with values in $]-\infty, +\infty]$; let z be an element of H. The strict convexity of Q implies that the function $f + Q_z$ attains its infimum at <u>at most</u> one point (incidentally, this infimum is $(f \nabla Q)(z)$); <u>if such a point exists, it will be denoted by</u> prox (f;z). The mapping $z \to$ prox (f;z) is called a proximation (Moreau [3]).

<u>Example.</u> Take as f the indicator function $\psi_C$ of a nonempty convex subset C of H; the point prox $(\psi_C; z)$, if it exists, is the nearest point to z in C, frequently called the <u>projection</u> of z onto C and denoted by proj(C;z).

## 3.b. Duality properties

<u>Proposition.</u> Let (f, g) be a pair of mutually polar functions on H and let z be an element of H.

If there exists prox(f;z), denoted by x, then there exists prox(g;z), denoted by y; the points x and y are conjugate with respect to (f, g) and $x + y = z$.

SOLUTIONS OF DUAL PROBLEMS

Conversely if there exists a pair of points x and y which are conjugate with respect to (f, g) and such that x+y = z, then prox(f;z) and prox(g;z) exist and equal x and y, respectively.

Proof: The relation x = prox(f;z) means that the origin belongs to the subdifferential of the function $f + Q_z$ at the point x. By the differentiability of the function $Q_z$ and by the fact mentioned at the end of paragraph 2.d, this is equivalent to

$$0 \in x-z + \partial f(x)$$

or, when putting y = z-x, equivalent to $y \in \partial f(x)$, i.e.: the points x and y are conjugate with respect to (f, g). From this equivalence, the proof is completed by exchanging f with g and x with y. □

Corollary 1. Let A and B be a pair of mutually polar cones in H and let $z \in H$.
If there exists proj(A;z), denoted by x, then there exists proj(B;z), denoted by y; one has (x|y) = 0 and x+y = z.
Conversely, if there exists a pair of points $x \in A$, $y \in B$, such that (x|y) = 0 and x+y = z, then proj(A;z) and proj(B;z) exist, respectively equal to x and y.

Proof: Take $f = \psi_A$ and $f = \psi_B$ in above Proposition. □

Particularizing more, we obtain:

Corollary 2. Let F and G be a pair of vector subspaces of H, each being the orthogonal of the other, and let $z \in H$.
If there exists proj(F;z), denoted by x, then there exists proj(G;z), denoted by y, and x+y = z.
Conversely if there exists a pair of points $x \in F$ and $y \in G$ such that x+y = z, then proj(F;z) and proj(G;z) exist and equal x and y, respectively.

Proof: F and G are mutually polar cones, like A and B in the above Corollary; here $(x|y) = 0$ is automatically ensured by $x \in F$ and $y \in G$. □

3. c. A functional example.

The following is a special case of a situation described, in a context of hydrodynamics, in Moreau [4], [6]; such "unilateral" functional problems have been extensively studied in the recent years by various authors (see e. g. Brezis-Stampacchia [1], Lewy-Stampacchia [1], [2]).

Let $\Omega$ be a bounded open subset of $\mathbb{R}^n$; take as H the space of the vector functions defined and continuous on the closure $\bar{\Omega}$ and which are the gradients of numerical functions. The scalar product of two elements u and v of H is defined by

$$(u|v) = \int_\Omega u(\xi) \cdot v(\xi) \, d\tau(\xi)$$

($\xi$: generic point of $\Omega$; $d\tau$: Lebesgue measure; the dot represents the scalar product of the two elements $u(\xi)$ and $v(\xi)$ of $\mathbb{R}^n$)

For the sake of simplicity we make in this paper the assumption that $\Omega$ is "very smooth", in order that:
  i) there exists on $\partial\Omega$ a continuous field of outward normal unit vectors, denoted by $\xi \to \nu(\xi)$.
  ii) The solutions of the conventional Dirichlet and Neumann problems exist, with gradient in H for any "very smooth" data on $\partial\Omega$ (say twice differentiable data, while $\partial\Omega$ is a twice differentiable hypersurface).

We are given:
  1. An element $z = \text{grad } r$ of H (then the numerical function r is finite and continuous on $\bar{\Omega}$).
  2. A nonnegative "very smooth" function b on $\partial\Omega$.

Call C the (convex) set of the elements $u = \text{grad } p \in H$ such that $p \geq 0$ on $\Omega$ and $p = b$ on $\partial\Omega$.

Call D the (convex) set of the elements $v = \operatorname{grad} q \in H$ such that div v (i.e. the laplacian $\Delta q$), understood in the sense of distributions on the open set $\Omega$, is nonnegative, i.e. is a nonnegative measure on $\Omega$; note that this measure is bounded, by virtue of the continuity of v on $\bar{\Omega}$.

Call f the indicator function of C and define another (convex, non finite) numerical function g on H as taking, for any $v = \operatorname{grad} q \in H$, the following values (d$\sigma$ denotes the area measure on $\partial\Omega$):

$$g(v) = \int_{\partial\Omega} b\, v \cdot \operatorname{grad} q \, d\sigma \,, \quad \text{if } v \in D$$
$$g(v) = +\infty \,, \quad \text{if } v \notin D.$$

<u>Decomposition problem.</u> As it naturally arises from hydrodynamics this problem is formulated: to find in H the elements $x = \operatorname{grad} p$ and $y = \operatorname{grad} q$ such that

(3.1)     $x + y = z$

(3.2)     $x \in C, y \in D$

(3.3)     $p\,\Delta q = 0$.

The latter condition means that the product of the nonnegative measure $\Delta q$ and the continuous function p is the zero measure on $\Omega$; in view of (3.2) it may equivalently be written

$$\int_\Omega p\,\Delta q = 0.$$

Using an integration by parts, the joint conditions (3.2) and (3.3) are then found equivalent to

(3.4)     $f(x) + g(y) - (x|y) = 0$.

Thus the present situation will reduce to the pattern described in paragraph 3.b if we prove that f and g are mutually polar (or in other words that the convex set C is

195

closed for the topology $\sigma(H,H)$ and that g is its support function). Actually the same integration by parts as above shows that the left member of (3.4) is non-negative for any x and y in H, i.e. each of the two functions f and g majorizes the polar of the other. Therefore we need only prove the two reverse inequalities:

1. Let us prove that $f \leq g^*$, i.e. for any $u = \operatorname{grad} p \in H$

$$f(u) \leq \sup \{(\operatorname{grad} p | \operatorname{grad} q) - \int_{\partial \Omega} b \, \nu \cdot \operatorname{grad} q \, d\sigma : \operatorname{grad} q \in D \}$$

From g being positively homogeneous it follows that $g^*$ is an indicator. As f is the indicator of C and in view of the same integration by parts, this inequality reduces to the following statement: "if u is such that

(3.5) $\quad \sup \{-\int_\Omega p \Delta q - \int_{\partial\Omega} (b-p) \nu \cdot \operatorname{grad} q \, d\sigma : \operatorname{grad} q \in D\} = 0$

then $u \in C$".

In fact, if u fulfills condition (3.5), one may first choose q harmonic, and thus conclude that the integral

$$\int_{\partial\Omega} (b-p) \nu \cdot \operatorname{grad} q \, d\sigma$$

is nonnegative, and finally zero since q may be changed into -q. Now $\nu \cdot \operatorname{grad} q$ may be identified with any "very smooth" function having a zero integral over $\partial\Omega$ (this consists in solving a Neumann problem); thus b-p is constant on $\partial\Omega$ or, adjusting the arbitrary constant in the determination of p, p = b on $\partial\Omega$. Therefore (3.5) reduces to

$$\int_\Omega p \Delta q \geq 0$$

for any $\operatorname{grad} q \in D$; now $\Delta q$ may be chosen with an arbitrary support in $\Omega$ and this yields the nonnegativity of p all over $\Omega$.

2. Let us prove that $g \leq f^*$, under the definition:

(3.6) $\quad f^*(v) = \sup \{(\operatorname{grad} p | \operatorname{grad} q): \operatorname{grad} p \in C\}$

where $v = \text{grad } q \in H$. First if $v$ does not belong to $D$ there exists an indefinitely differentiable nonnegative function $\varphi$, with compact support in $\Omega$, such that

$$\langle \varphi, \Delta q \rangle = -(\text{grad } \varphi \mid \text{grad } q) < 0.$$

Then, by taking $p = b + \lambda \varphi$ (b denotes here an extension of b to $\bar{\Omega}$, with gradient in H) and giving to the positive number $\lambda$ arbitrarily large values, one proves that the right member of (3.6) is $+\infty$.

Suppose now $v = \text{grad } q \in D$; in view of the definition of $g$, we have to prove that, for any $\varepsilon > 0$, there exists $\text{grad } p \in C$ such that

$$\int_{\partial \Omega} b \, v \cdot \text{grad } q \, d\sigma \leq (\text{grad } p \mid \text{grad } q) + \varepsilon$$

or, using the integration by parts,

$$\int_\Omega p \, \Delta q \leq \varepsilon.$$

In fact, as $\Delta q$ is a nonnegative bounded measure on $\Omega$, there exists $K$, a compact subset of $\Omega$, such that the integral of $\Delta q$ over $\Omega \setminus K$ is less than $\varepsilon/B$, where $B > 0$ denotes a constant majorizing $b$. As the function $p$ may be chosen majorized by $B$ and null over $K$, this completes the proof.

3.d. <u>Infimal convolution with Q</u>.

We take again an arbitrary pre-Hilbert space and a pair $(f, g)$ of mutually polar functions belonging to $\Gamma_0(H, H)$. As $Q = Q^*$ and $f^* = g$, equation (2.9) yields

(3.7)  $(f \, \nabla \, Q)^* = g + Q.$

Since $g+Q$ is not everywhere $+\infty$, this proves that $f \, \nabla \, Q$ possesses affine minorants; thus this function takes nowhere the value $-\infty$; clearly also, by definition (2.8), it nowhere takes the value $+\infty$; the same conclusions apply to $g \, \nabla \, Q$.

Now (3.7) implies

$$(f \, \nabla Q)^{**} = (g + Q)^*$$

where the right member may be computed as follows, for any $z \in H$:

$$(g+Q)^*(z) = \sup\{(z|y) - g(y) - Q(y) : y \in H\}$$

$$= \sup\{Q(z) - g(y) - Q(z-y) : y \in H\}$$

$$= Q(z) - (g \, \nabla Q)(z).$$

As $(f \, \nabla Q)^{**}$ minorizes $f \, \nabla Q$ (see Paragraph 2.b), this yields in particular the following inequality

(3.8) $\qquad (f \, \nabla Q) + (g \, \nabla Q) \geq Q$ .

The case of equality will be studied in Paragraph 5.a.

## 4. Weakening Procedures and Duality.

### 4.a. Strong problems.

Paragraph 3.b described a rather general situation involving a pair of extremum properties, formulated in the same pre-Hilbert space $H$, and said <u>dual</u> to each other: given $z \in H$, to find $x = \text{prox}(f;z)$ or to find $y = \text{prox}(g;z)$ appear as equivalent problems under the relation $y = z-x$. Furthermore, this pair of equivalent minimization problems has been found equivalent to a third problem of a different sort: that of decomposing $z$ into the sum of two elements $x$ and $y$ which are conjugate relative to $(f,g)$. In what follows we shall refer to any of these three problems as a <u>strong problem</u>: such a problem does not necessarily possess a solution.

In the physical context, it is usually the decomposition problem which directly arises from phenomenological laws: the two minimization "principles" are subsequently derived. So, when going to the task of defining weak solutions of the physical problem variationally, no reason appears for preferring one of them to the other. Now the two choices do not in general provide the same concept of weak solution.

Remark 1. In order to obtain a simpler pattern of implications, we restrict ourselves in this paper to problems associated with a pair of mutually polar functions; actually several results may be obtained under the less stringent hypothesis that f and g are only "superpolar", i.e.: for any x and y in H the inequality

$$f(x) + g(y) - (x|y) \geq 0$$

holds (this means that each of the two functions majorizes the polar of the other).

Remark 2. Results generalizing certain basic facts of the present theory may be formulated when considering, instead of pre-Hilbert or Hilbert spaces, some normed or Banach vector spaces (see Lescarret [1]).

4.b. The imbedding of H in its completion.

Let us denote by $\hat{H}$ the Hilbert completion of the pre-Hilbert space H; we shall treat H as imbedded in $\hat{H}$, hence dense in it for the norm topology. $\hat{H}$ may as well be considered as the topological dual of H when this latter space is endowed with the norm topology. The scalar product in $\hat{H}$ is an extension of the scalar product $(\cdot|\cdot)$ in H so that there is no need of a different notation for it. This bilinear form puts H in duality with itself; it puts in the same way $\hat{H}$ in duality with itself; it also puts in duality the two spaces H and $\hat{H}$. To these three dualities correspond distinct classes of compatible topologies.
On $\hat{H}$, for instance one may consider:
(i) the norm topology, which is compatible with the duality $(\hat{H}, \hat{H})$ (this is actually the Mackey topology $\tau(\hat{H}, \hat{H})$),

(ii) the topology $\sigma(\hat{H}, \hat{H})$, coarser than the preceding one but compatible with the same duality, so that the family of the closed convex sets or the family of the l.s.c. convex numerical functions are the same for both,
(iii) the topology $\sigma(\hat{H}, H)$, coarser than $\sigma(\hat{H}, \hat{H})$.

On the subspace H of $\hat{H}$, these three topologies respectively induce:
(i) the norm topology (this is actually the Mackey topology $\tau(H, \hat{H})$) compatible with the duality $(H, \hat{H})$,
(ii) the topology $\sigma(H, \hat{H})$, coarser but compatible with the same duality,
(iii) the topology $\sigma(H, H)$, coarser than $\sigma(H, \hat{H})$.

## 4.c. Some extensions of f and g to $\hat{H}$.

Let us start, as in Section 3, with a pair of numerical functions f and g, belonging to $\Gamma_0(H, H)$, mutually polar for the duality $(H, H)$.

We shall denote by f' and g' their extensions to $\hat{H}$ with the value $+\infty$ outside of H; these extensions are convex on $\hat{H}$.

The function $\bar{g}$ defined for every $y \in \hat{H}$ by

$$\bar{g}(y) = \sup\{(x|y) - f(x) : x \in H\}$$

$$= \sup\{(x|y) - f'(x) : x \in \hat{H}\}$$

is indifferently the polar of f for the duality $(H, \hat{H})$ or the polar of f' for the duality $(\hat{H}, \hat{H})$; clearly it is an extension of g. Similarly f is the polar of g' for the duality $(\hat{H}, H)$; thus $\bar{g}$ is the bipolar of g' for the same duality, i.e., in view of paragraph 2.c, the l.s.c. hull of g' with respect to a topology on $\hat{H}$ compatible with the duality $(\hat{H}, H)$, e.g. the topology $\sigma(\hat{H}, H)$ (in fact g' possesses at least one continuous affine minorant since its polar function f is not everywhere $+\infty$).

The polar of $\bar{g}$ for the duality $(\hat{H}, \hat{H})$, i.e. the bipolar of f' for this duality will be denoted by $\hat{f}$; it is the l.s.c. hull of f' relatively to any topology compatible with this duality, in particular the norm topology. As f' agrees with f over the set (contained in H) of the points where it differs from $+\infty$, and as f is l.s.c. on $\hat{H}$ for $\sigma(H,H)$ which is coarser than the norm topology, $\hat{f}$ is an extension of f (see the end of paragraph 2.c.).

Symmetrically we define on $\hat{H}$ the function $\hat{g}$, namely the norm - l.s.c. hull of g', and $\bar{f}$, namely the $\sigma(\hat{H}, H)$- l.s.c. hull of f', with similar polarity relations. The following inequalities hold:

$$\bar{f} \leq \hat{f} \leq f'$$

$$\bar{g} \leq \hat{g} \leq g'.$$

### 4.d. Proximations in $\hat{H}$.

The properties of proximations presented in Section 3 for an arbitrary pre-Hilbert space are true in the Hilbert space $\hat{H}$ with this additional simplification: <u>for arbitrary $\varphi \in \Gamma_0(\hat{H}, \hat{H})$ and arbitrary $z \in \hat{H}$, the point $\text{prox}(\varphi; z)$ exists (uniquely)</u>. This is easily proved from the weak compactness of closed balls in a Hilbert space (see Moreau [3]) or by using general conditions for the "exactness" of infimal convolution (i.e. conditions for the infimum in (2.8) to be a minimum; see Moreau [1] [8]).

Therefore the existence of the following elements of $\hat{H}$ is ensured

(4.1) $\qquad \hat{x} = \text{prox}(\hat{f}; z)$, $\quad \bar{y} = \text{prox}(\bar{g}; z)$

these being equivalently characterized by

(4.2) $$\hat{x} + \bar{y} = z$$

$$\hat{f}(\hat{x}) + \bar{g}(\bar{y}) - (\hat{x}|\bar{y}) = 0.$$

Symmetrically, the existence is ensured of

(4.3) $\qquad \bar{x} = \text{prox}(\bar{f}; z)$, $\quad \hat{y} = \text{prox}(\hat{g}; z)$

equivalently characterized by

$$\bar{x} + \hat{y} = z$$

$$\bar{f}(\bar{x}) + \hat{g}(\hat{y}) - (\bar{x}|\hat{y}) = 0 \ .$$

<u>Proposition.</u> Suppose $z \in H$; if one of the four points $\hat{x}, \bar{y}, \bar{x}, \hat{y}$ happens to lie in $H$, so do the other three and then

(4.4) $\qquad \bar{x} = \hat{x} = \text{prox}(f; z)$

(4.5) $\qquad \bar{y} = \hat{y} = \text{prox}(g; z)$

(these proximations referring to $f$ and $g$ are evidently to be understood relative to $H$).
 Conversely, if $z \in H$ and if there exists $x = \text{prox}(f; z)$ in $H$ (or equivalently by paragraph 3.b, if there exists $y = \text{prox}(g; z)$), equalities (4.4) and (4.5) hold.

<u>Proof:</u> This immediately follows from the above characterizations of "prox", since $f$ equals the restriction of $\hat{f}$ or $\bar{f}$ to $H$ and $g$ equals the restriction of $\hat{g}$ or $\bar{g}$ to $H$. □

 At this stage, the replacement of $f$ and $g$, mutually polar numerical functions on $H$, by their extensions $\hat{f}$ and $\hat{g}$, mutually polar numerical functions on $\hat{H}$, appears as a weakening procedure for the so-called strong problem of Paragraph 4.a: In fact this replacement yields a similar problem formulated in $\hat{H}$ instead of $H$ and which possesses a solution for any $z$; this solution coincides with the solution of the strong problem whenever the latter exists.
 The same is true for the replacement of $f$ and $g$ by their extensions $\bar{f}$ and $\bar{g}$, which yields another weakening procedure for the initial problem. More generally, one may choose $\tilde{f} \in \Gamma_0(\hat{H}, \hat{H})$ such that

## SOLUTIONS OF DUAL PROBLEMS

$$\bar{f} \leq \tilde{f} \leq \hat{f}$$

The polar function $\tilde{g}$ of $\tilde{f}$ for the duality $(\hat{H}, \hat{H})$ satisfies

$$\bar{g} \leq \tilde{g} \leq \hat{g}$$

and the same arguments as above show that, for a given $z \in H$, the elements $\text{prox}(\tilde{f}; z)$ and $\text{prox}(\tilde{g}; z)$ of $\hat{H}$, respectively, coincide with $\text{prox}(f; z)$ and $\text{prox}(g; z)$ whenever one (then both) of the latter exist.

Thus we are in the presence of an infinity of weakening procedures for our initial "strong" problem.

### 4.e. Use of minimizing sequences.

Let $\varphi \in \Gamma_0(\hat{H}, \hat{H})$; a sequence $(u_n)$ of points of $\hat{H}$ is classically called a <u>minimizing sequence</u> of the function $\varphi + Q_z$ if

$$\lim_{n \to +\infty} [\varphi(u_n) + Q_z(u_n)] = \inf \{\varphi(u) + Q_z(u): u \in \hat{H}\}.$$

Standard arguments, essentially using the convexity of $\varphi$ and the uniform convexity of $Q_z$, yield: Any minimizing sequence of $\varphi + Q_z$ converges to $\text{prox}(\varphi; z)$ relative to the norm topology of $\hat{H}$.

Now we remark that, $\hat{f}$ being the norm-l.s.c. hull of $f'$ and $Q_z$ being norm-continuous, the function $\hat{f} + Q_z$ is the norm-l.s.c. hull of $f' + Q_z$ and one has

$$\inf\{f(u) + Q_z(u): u \in H\} = \inf\{\hat{f}(u) + Q_z(u): u \in \hat{H}\}$$

(the same would not be true in general if we used instead of $\hat{f}$ the function $\bar{f}$, which is also an extension of $f$ to $\hat{H}$ but inferior to $\hat{f}$). Therefore a sequence $(u_n)$ of points of $H$ which in this space is minimizing for the function $f + Q_z$ is also a minimizing sequence in $\hat{H}$ for the function $\hat{f} + Q_z$; thus:

<u>Proposition</u>. Any minimizing sequence of $f + Q_z$ in $H$ converges in the norm topology of $\hat{H}$ to the element $\text{prox}(\hat{f}; z)$, denoted above by $\hat{x}$; symmetrically any minimizing sequence of $g + Q_z$ in $H$ is normwise convergent in $\hat{H}$ to $\hat{y} = \text{prox}(\hat{g}; z)$.

This proposition just describes the conventional weakening procedure for the two "strong" minimization problems in $H$ whose the supposed solutions have been denoted by $\text{prox}(f; z)$ and $\text{prox}(g; z)$ (for a given $z \in H$). Unfortunately the sum of the elements $\hat{x}$ and $\hat{y}$ of $\hat{H}$ does not in general equal $z$; in other words this weakening destroys the pattern of duality between the two minimization problems and also it destroys their equivalence with a decomposition problem.

Our purpose now is to formulate a series of necessary and sufficient conditions concerning $f$ and $g$, in order that the above minimizing sequence device preserves these basic features; at the same time these conditions will be necessary and sufficient for the various weakening procedures described in Paragraph 4.d to coincide.

## 5. The Case of Uniqueness.

### 5.a. Necessary and sufficient conditions.

<u>Proposition</u>. With the notations defined in the previous paragraphs, the following assertions are equivalent:

(i) $\bar{f} = \hat{f}$

(ii) $\bar{g} = \hat{g}$

(iii) Any pair $(\tilde{f}, \tilde{g})$ of mutually polar functions on $\hat{H}$ which respectively minorize $f$ and $g$ on $H$ coincides with $(\hat{f}, \hat{g})$.

(iv) Any $z \in H$ equals the sum of $\hat{x} = \text{prox}(\hat{f}; z)$ and of $\hat{y} = \text{prox}(\hat{g}; z)$ (let us recall here that $\hat{x}$ and $\hat{y}$ have been interpreted in Paragraph 4.e as the limits in $\hat{H}$ of minimizing sequences for the functions $f + Q_z$ and $g + Q_z$ in $H$).

(v) For any $z \in H$ and any $\varepsilon > 0$ there exist $x_\varepsilon$ and $y_\varepsilon$ in $H$ such that

$$\|x_\varepsilon + y_\varepsilon - z\| \leq \varepsilon$$

$$f(x_\varepsilon) + g(y_\varepsilon) - (x_\varepsilon | y_\varepsilon) \leq \varepsilon \ .$$

(vi) Over all of $H$:

(5.1)
$$f \nabla Q + g \nabla Q = Q \ .$$

(vii) The function $f \nabla Q$ (resp. $g \nabla Q$) is l.s.c. in $H$ relative to the topology $\sigma(H, H)$.

It must be stressed that properties (v) to (vii) only involve elements of the pre-Hilbert space $H$ where the so-called strong problem was formulated; the same is true for property (iv) if it is interpreted in terms of minimizing sequences.

Property (v) was published in S. Maury [1]; it says that any $z \in H$ can be "approximately decomposed" into the sum of two elements which are "approximately conjugate" relatively to $(f, g)$ (in other words, $y_\varepsilon$ is an "approximate subgradient" of $f$ at the point $x_\varepsilon$; concerning this concept, see: Broendsted-Rockafellar [1]). It may equivalently be formulated as follows: for any $z \in H$, there exists in the product space $H \times H$ a sequence $(x_n, y_n)$ which is minimizing at the same time for the two functions $(x, y) \to \|x+y-z\|$ and $(x, y) \to f(x) + g(y) - (x|y)$ (in fact both functions are nonnegative and their infima are zero since $f$ and $g$ are mutually polar).

Proof of the proposition: The equivalence between (i), (ii) and (iii) follows immediately from the fact that polarity reverses the ordering of functions.

Property (ii) implies the equality of the elements $\bar{y}$ and $\hat{y}$ as they are defined by (4.1) and (4.3); so property (iv) follows, by equality (4.2). Conversely, in view of (4.2), assertion (iv) implies that, for any $z \in H$,

$$\text{prox } (\hat{g}; z) = \text{prox } (\bar{g}; z).$$

But proximation mappings are norm-continuous from $\hat{H}$ into itself (they are non-expanding: see Moreau [3]) and H is norm-dense in $\hat{H}$. Therefore this equality also holds for any $z \in \hat{H}$, implying that the functions $\hat{g}$ and $\bar{g}$ differ only by a constant (see Moreau [3]) ; this constant is zero since $\hat{g}$ and $\bar{g}$ agree with g on H (and g is not everywhere $+\infty$).

These assertions also imply (v) : in fact consider in H a minimizing sequence $(x_n)$ of the function $f + Q_z$ and a minimizing sequence $(y_n)$ of the function $g + Q_z$. By paragraph 4.e these sequences are respectively normwise convergent in $\hat{H}$ to $\hat{x}$ and $\hat{y}$, whose sum equals z according to assertion (iv); thereby $\|x_n + y_n - z\|$ is made arbitrarily small. Furthermore $(x_n | y_n)$ tends to $(\hat{x} | \hat{y})$ and, as the function $Q_z$ is norm-continuous, $f(x_n)$ tends to $f(\hat{x})$ and $g(y_n)$ tends to $g(\hat{y})$ : then the second part of assertion (v) follows from the fact that $f(\hat{x}) + g(\hat{y}) - (\hat{x}|\hat{y})$ is zero.

That conversely (v) implies (iv) may be deduced from the following identity:

$$f(x) + Q_z(x) + g(y) + Q_z(y) = Q(z) + Q(x+y-z) + f(x) + g(y) - (x|y).$$

In fact property (v) means that, for each $z \in H$, there exists a sequence $(x_n, y_n) \in H \times H$ which is minimizing for the right member of this identity, the infimum of which is $Q(z)$. Now

$$\inf\{f(x) + Q_z(x) : x \in H\} = (f \nabla Q)(z)$$

$$\inf\{g(y) + Q_3(y) : y \in H\} = (g \nabla Q)(z).$$

Comparing this with inequality (3.8), one finds that the sequence $(x_n)$ is minimizing for the function $f + Q_z$ and that the sequence $(y_n)$ is minimizing for the function $g + Q_z$; as the sequence $(x_n, y_n)$ is minimizing for the function

$(x, y) \to Q(x+y-z)$, the sum $x_n + y_n$ is normwise convergent to $z$ and that implies (iv).

To finish the proof the same identity shows immediately that properties (v) and (vi) are equivalent. Furthermore paragraph 3.d yields

$$(f \nabla Q)^{**} + (g \nabla Q) = Q$$

so that (5.1) is equivalent to

$$(f \nabla Q)^{**} = f \nabla Q.$$

In view of paragraph 2.c this is just property (vii). □

The meaning of properties (vi) and (vii) becomes clearer in some special cases:

Special Case 1. Suppose that $f$ is the indicator function of a subset $C$ of $H$ (convex and $\sigma(H, H)$-closed); for this particular form of $f$, property (vii) means that the distance function

$$z \to \text{dist}(C, z)$$

is $\sigma(H, H)$-l.s.c.

In fact $(f \nabla Q)(z)$ equals half the square of this distance.

Another equivalent assertion is: Any ball in $H$, the radius of which is strictly smaller than the distance of its center to $C$, can be separated from $C$ by a $\sigma(H, H)$-closed hyperplane (i.e. a hyperplane possessing a normal vector in $H$).

In fact the distance function is $\sigma(H, H)$-l.s.c. if and only if, for any $\rho \geq 0$, the (convex) set $\{z \in H : \text{dist}(C; z) \leq \rho\}$ is $\sigma(H, H)$-closed. This in turn is equivalent to the separation property for any point $z_0 \in H$ which does not belong to this set and, finally to the asserted separation property between $C$ and a ball with center $z_0$ and radius $\rho$. Actually direct arguments would easily show that this separation property between $C$ and the balls in $H$

is necessary and sufficient for the norm-closure of C in $\hat{H}$ to coincide with the $\sigma(\hat{H}, H)$-closure of C : this is just the form taken in the present case by property (i).

Special Case 2.  Suppose that f and g are respectively the indicator functions of two mutually polar cones A and B. Then property (vi) reduces to the following "Pythagorean" relation

$$[\text{dist}(A\,;\,z)]^2 + [\text{dist}(B\,;\,z)]^2 = \|z\|^2$$

holding for any $z \in H$.

## 5.b. Functional example.

Let us show that in the functional example presented in paragraph 3.c the preceding properties hold, in fact that assertion (v) is verified.

The given continuous function b on the (compact) set $\partial\Omega$ was supposed nonnegative; just to simplify the following suppose here $b > 0$.

Suppose z arbitrarily given in H; thus z can be approximated as nearly as desired, in the sense of the norm, by an element $\text{grad } r \in H$ such that r is a "very smooth" numerical function on $\partial\Omega$ (see the smoothness assumptions made in paragraph 3.c).

First construct $y_0 = \text{grad } q_0 \in D$ with the following properties:

$$q_0 \leq r \quad \text{on } \bar{\Omega}$$

$$q_0 = r - b \quad \text{on } \partial\Omega.$$

For this purpose call $\gamma$ the Green potential of some negative smooth measure with compact support in $\Omega$ and call h the solution of the Dirichlet problem for the data r-b on $\partial\Omega$. In view of our smoothness assumptions, the functions $\gamma$ and h have their gradients in H ; the function $\gamma$ is negative on $\Omega$ ; thus $q_0 = h + \lambda\gamma$ fulfills above requirements when the positive constant $\lambda$ is chosen sufficiently large.

Our idea consists in applying Poincaré's <u>sweeping out process</u> to the function $q_0$, but "conditionally" with regard to the inequality $q \leq r$.

Denote in general by $q$ a function satisfying the same conditions as above for $q_0$, with gradient belonging to $D$; the Laplacian $\Delta q$ is a bounded nonnegative measure in $\Omega$. Let $B \subset \Omega$ be a closed ball; the function $q'$ defined as equal to $q$ in $\bar{\Omega} \setminus B$ and equal in $B$ to the "harmonic interpolate" of $q$ (i.e. the Poisson integral which solves the Dirichlet problem in the ball $B$ with $q$ as data on $\partial B$) is continuous on $\bar{\Omega}$, subharmonic on $\Omega$, greater than or equal to $q$. As $q'$ agrees with $q$ on a neighborhood of $\partial \Omega$, the nonnegative bounded measures $\Delta q$ and $\Delta q'$ have the same total mass on $\Omega$.

Call "conditional harmonization of $q$ for $B$ under $r$" the replacement of $q$ by $q'$ if this latter function satisfies $q' \leq r$, the conservation of $q$ if not.

Choose then a sequence of balls forming a base for the topology of $\Omega$; a classical construction yields another sequence in which each of these balls recurs infinitely. Starting with the function $q_0$ constructed above, iterate conditional harmonization, using successively the balls of the latter sequence. This generates a non-decreasing sequence of continuous subharmonic minorants of $r$, all agreeing with $r-b$ on $\partial \Omega$; arguments in the style of the theory of potential (see Moreau [9]) prove that this sequence converges uniformly on each compact subset of $\Omega$ to a function $\bar{q}$ possessing the following properties: $\bar{q}$ is a continuous subharmonic minorant of $r$, agreeing with $r - b$ on $\partial \Omega$ and harmonic on the (open) subset of $\Omega$ where it differs from $r$ (in other words the nonnegative measure $(r-\bar{q})\Delta \bar{q}$ is null). As we have supposed $b > 0$ on $\partial \Omega$, the set $K$ on which $\bar{q} = r$ is a compact (possibly empty) subset of $\Omega$. Therefore, given $\varepsilon > 0$, one can stop the process of successive harmonizations after a finite number of steps, obtaining a function $q_1$ which fulfills the following requirement: The difference $r - q_1$ is uniformly less than $\varepsilon$ on a compact set $K_1$, with $K \subset K_1 \subset \Omega$. This compact set can be constructed with smooth boundary; perform an ultimate "harmonization" by replacing $q_1$ on $\Omega \setminus K_1$ by its harmonic interpolate, without alteration on $K_1$. The function $q_2$ constructed thereby is again a continuous subharmonic minorant of $\bar{q}$ (since this latter is harmonic on $\Omega \setminus K_1$) thus of $r$; it agrees with $r-b$ on $\partial \Omega$. The nonnegative measure $\Delta q_2$ has a total mass on $\Omega$ not greater than the total mass $\mu$ of $\Delta q_0$ : in other words our initial measure $\Delta q_0$ has been "swept out" to the boundary of $\Omega$ and to a region where $r - q_2 \leq \varepsilon$. This yields the inequality

$$\int_\Omega (r-q_2) \, \Delta q_2 \leq \mu \varepsilon .$$

But $q_2$ possesses a piecewise continuous gradient, square integrable on $\Omega$; by putting $r - q_2 = p_2$, this allows an integration by parts transforming this inequality into

$$\int_{\partial \Omega} p_2 \, \nu \cdot \operatorname{grad} q_2 \, d\sigma - \int_\Omega \operatorname{grad} p_2 \cdot \operatorname{grad} q_2 \, d\tau \leq \mu \varepsilon$$

while

$$\operatorname{grad} p_2 + \operatorname{grad} q_2 = \operatorname{grad} r .$$

It remains only to approximate, in the sense of the $L^2$ norm, the vector fields $\operatorname{grad} p_2$ and $\operatorname{grad} q_2$ by elements $\operatorname{grad} p$ and $\operatorname{grad} q$ of $H$, respectively belonging to the subsets $C$ and $D$ (alternatively one could have, in the previous process, "smoothed the transitions" after each step of harmonization, so as to deal all the time with functions having their gradients in $H$); $\operatorname{grad} r$ was defined as approximating $z$. In that way one constructs elements $x = \operatorname{grad} p$ and $y = \operatorname{grad} q$ of $H$, the sum of which is approximately $z$, and such that $f(x) + g(y) - (x|y)$ is approximately zero, with $f$ and $g$ defined as in paragraph 3.c. This is precisely property (v) of paragraph 5.a.

This example is obviously connected to the study of the regularity of solutions of certain variational inequalities, as developed in Lewy-Stampacchia [1], [2].

## REFERENCES

E. Asplund, R. T. Rockafellar [1] Gradients of convex functions, Trans. Amer. Math. Soc., 139 (1969), p. 443-467.

M. Brelot [1] Etude et extension du principe de Dirichlet, Ann. Inst. Fourier, 5 (1953-54), p. 371-419.
[2] Eléments de la théorie classique du potentiel, Centre de Documentation Universitaire, Paris.

H. Brezis, G. Stampacchia [1] Sur la régularité de la solution d'inéquations elliptiques, Bull. Soc. Math. France, 96 (1968), p. 153-180.

A. Broendsted, R. T. Rockafellar [1] On the subdifferentiability of convex functions, Proc. Amer. Math. Soc., 16 (1965), p. 605-611.

C. Castaing [1] Proximité et mesurabilité, Seminaire Anal. Unilatérale, Fac. Sci. Montpellier, vol. 1 (1968).

R. Courant, D. Hilbert [1] Methods of mathematical physics, Interscience publ., Vol. 1, Chap. 4, §9 and 11.

J. Deny, J. L. Lions [1] Les espaces du type de Beppo Levi, Ann. Inst. Fourier, 5 (1953-54), p. 305-370.

W. Fenchel [1] On conjugate convex functions, Canad. Journ. Math. 1 (1949), p. 73-77.
[2] Convex cones, sets and functions, mimeographed lecture notes, Princeton University (1951).

A. D. Ioffe, V. M. Tihomirov [1] Duality of convex functions and extremal problems (Russian), Usp. Mat. Nauk 23 (1968), p. 51-116.

J. L. Joly [1] Une famille de topologies et de convergences sur l'ensemble des fonctionnelles convexes, Thèse Sci. Grenoble (1970).

J. L. Joly, P. J. Laurent [1] Stability and duality in convex minimization problems, to appear.

C. Lescarret [1] Applications "prox" dans un espace de Banach, Compt. Rend. Acad. Sci. Paris, Ser. A, 265 (1967), p. 676-678.

H. Lewy, G. Stampacchia [1] On the regularity of the solution of a variational inequality, Comm. Pure Appl. Math., 22 (1969), p. 153 -    .
[2] On the smoothness of superharmonics which solve a minimum problem, Journ. Anal. Math., 23 (1970), 227-236.

J. L. Lions [1] Quelques methodes de résolution de problémes aux limites non linéaires, Dunod et Gauthier-Villars, Paris.

J. L. Lions, G. Stampacchia [1] Variational inequalities, Comm. Pure Appl. Math., 20 (1967), p. 493-519.

S. Maury [1] Proximité dans un espace préhilbertien et complétion, Seminaire Anal. Convexe UER Math. Montpellier (1971).

G. J. Minty [1] On the monotonicity of the gradient of a convex function, Pacific J. Math. 14 (1964), 243-247.

J. J. Moreau [1] Inf-convolution des fonctions numériques sur un espace vectoriel, Compt. Rend. Acad. Sci. Paris, 256 (1963), p. 5047-5049.
[2] Fonctionnelles sous-différentiables, Compt. Rend. Acad. Sci. Paris, 257 (1963), p. 4117-4119.

J. J. Moreau [3] Proximité et dualité dans un espace hilbertien, Bull. Soc. Math. France, 93 (1965), p. 273-299.
[4] One-sided constraints in hydrodynamics, in: J. Abadie, editor, Nonlinear programming, North Holland Pub., Amsterdam (1967), p. 257-279.
[5] Convexity and duality, in E. R. Caianiello, editor, Functional analysis and optimization, Academic Press, New York, 1966, p. 145-169.
[6] Principes extrémaux pour le probléme de la naissance de la cavitation, Journ. de Mécanique, 5 (1966), p. 439-470.
[7] Sous-différentiabilité, in Proceedings of the Colloquium on Convexity, Copenhagen 1965, p. 185-201.
[8] Fonctionnelles convexes, Séminaire sur les équations aux dérivées partielles, Collége de France, Paris 1967 (multigraph 108 p.)
[9] Majorantes sur-harmoniques minimales d'une fonction continue, Seminaire Anal. Unilaterale, Fac. Sci. Montpellier, vol. 1 (1968); an improved version is to appear in: Ann. Inst. Fourier 21, (1971).
[10] La notion de sur-potentiel et les liaisons unilaterales en élastostatique, Compt. Rend. Acad. Sci. Paris, Ser A, 267 (1968), p. 954-957.
[11] Un cas d'addition des sous-différentiels, Séminaire Anal. Unilatérale, Fac. Sci. Montpellier, vol. 2 (1969).
[12] Distance a un convexe d'un espace normé et caractérisation des points proximaux, Séminaire Anal. Unilatérale, Fac. Sci. Montpellier, vol. 2 (1969).
[13] Inf-convolution, sous-additivité, convexité des fonctions numériques, Journ. Math. Pures Appl. 49 (1970), p. 109-154.
[14] Sur les lois de frottement, de plasticité et de viscosité, Compt. Rend. Acad. Sci. Paris, Ser. A, 271 (1970), p. 608-611.

B. Nayroles [1] Essai de théorie fonctionnelle des structures rigides plastiques parfaites, Journ. de Mécanique, 9 (1970), p. 491-506.

R. T. Rockafellar [1] Integrals which are convex functionals, Pacific J. Math. 24 (1968), p. 867-873.
[2] Duality in nonlinear programming, in Mathematics of the Decision Sciences, Part I, Lectures in Applied Mathematics, Vol II, Amer. Math. Soc. (1968), p. 401-422.
[3] Convex Analysis, Princeton Univ. Press (1970).
[4] This volume.

R. Temam [1] Remarques sur la dualité en calcul des variations, Compt. Rend. Acad. Sci. Paris, ser A, 270 (1970) p. 754-757.
[2] Solutions généralisées de certains problémes de calcul des variations, Compt. Rend. Acad. Sci. Paris, ser. A, 271 (1970 p. 1116-1119.
[3] Solutions généralisées de certaines équations du type hypersurfaces minima, to appear.
[4] Solutions généralisées d'équations non lineaires non uniformement elliptiques, Pub. Math. Orsay, UER Math. Univ. Paris II, Centre d'Orsay (1970-71), 71 p.

M. Valadier [1] Sous-différentiels d'une borne supérieure et d'une somme continue de fonctions convexes, Compt. Rend. Acad. Sci. Paris, ser A, 268 (1969), p. 39-42.
[2] Contribution à l'analyse convexe, Thèse Sci. Paris (1970).

E. H. Zarantonello [1]. This volume.

> Faculte des Sciences
> Université de Montpellier
> place Eugene Bataillon
> 34 Montpellier, France
>
> Received March 31, 1971.

# Convex Integral Functionals and Duality

## R. TYRRELL ROCKAFELLAR

## §1. Introduction.

Let $(T, \mathcal{C}, \mu)$ be a measure space, and let $L$ be a linear space of mappings $x : T \to X$, where $X$ is a real vector space. A <u>convex integral functional</u> on $L$ is an extended-real-valued functional of the form

(1) $$I_f(x) = \int_T f(t, x(t)) \mu(dt), \qquad x \in L,$$

where $f(t, \cdot)$ is for each $t \in T$ an extended-real-valued convex functional on $X$. Such a functional is, as the name implies, convex on $L$, if in a rather general sense it is well-defined.

The most familiar functionals of the form $I_f$ are undoubtedly the ones occurring in the classical theory of $L^p$ spaces and Orlicz spaces:

(2) $$I_f(x) = \frac{1}{p} \int_T |x(t)|^p \mu(dt), \qquad 1 \le p < +\infty,$$

or more generally

(3) $$I_f(x) = \int_T N(|x(t)|) \mu(dt),$$

where $N$ is a nondecreasing convex function on $[0, +\infty)$. Here $X = R^1$. Of course, the theory of duality in $L^p$ spaces and Orlicz spaces also involves the study of other, simpler, convex integral functionals, such as the linear functionals

$$(4) \qquad I_f(x) = \int_T x(t)y(t)\mu(dt) .$$

These classes of functionals can be generalized by replacing $X = R^1$ by any normed linear space.

Recent work on convex integral functionals has been motivated not so much by these examples as by broader applications to the extremum problems and variational principles. For instance, many problems in the calculus of variations or optimal control involve extended-real-valued functionals of the form

$$(5) \qquad I(z) = \int_a^b f(t, z(t), \dot{z}(t))dt ,$$

where $z: [a, b] \to Z$ is a "curve" (with derivative $\dot{z}$) in a linear space $Z$. One can view $I$ as a functional (1) in the case where $X$ is $Z \times Z$ and $L$ is the space of functions $(z, w): T \to X$ such that $w = \dot{z}$. The study of $I$ also entails the study, for fixed choices of $z$, of the functional

$$(6) \qquad w \to \int_a^b f(t, z(t), w(t))dt .$$

Even if these integral functionals are not themselves convex, it is often useful to compare them with certain convex integral functionals which they majorize, or to consider their "convexifications".

Here, instead of curves over an interval $[a, b]$, one can investigate functions $z$ defined over a region $\Omega$ of $R_i^m$, the derivative $\dot{z}$ being replaced by a vector of partial derivatives. The generalized gradient operators associated with the functionals $I$ (or closely related functionals) in such cases correspond to variational principles. Many important differential and integral operators, as well as other nonlinear operators on function spaces, belong to this class.

The notion of "continuous addition" of convex functions provides further motivation for a general theory of convex integral functionals. Given a collection of convex functions $f_t = f(t, \cdot)$ on $X$, it is possible to define another convex function

(7) $$F(x) = \int_T f(t,x)\mu(dt),$$

which can be regarded as the integral or "continuous sum" of the functions $f_t$ (with respect to $\mu$). It is an important question whether, or to what extent, certain basic results about the conjugate or subdifferential of a finite sum of convex functions can be extended to such an infinite sum. Observe that $F$ may be identified with the restriction of $I_f$ to the space of all constant functions $x : T \to X$.

Duality has always played a fundamental role in the analysis of convex integral functionals. One may cite in particular the classical results of Luxemburg and Zaanen [21] concerning dual functionals of the form (3). These are examples of convex functionals conjugate to each other in the sense introduced by Fenchel.

General convex integral functionals conjugate to each other were first investigated in the author's paper [26] for $X = R^n$. A concept of "normal convex integrand" was briefly developed there by elementary methods, so as to provide the necessary technical lemmas concerning measurability, measurable selections, etc. Connections between normal convex integrands and the new theory of measurable multifunctions, particularly the results of Castaing [2,3], Debreu [9] and Kuratowski and Ryll-Nardzewski [18], were explored in a separate paper [27]. Taking a different approach to questions of measurability, Ioffe and Tikhomirov [15, 16] studied continuous addition (7) and the dual operation of continuous infimal convolution in a separable Banach space $X$ and its dual. The same operations were treated subsequently by Castaing [4,5,6] in infinite-dimensional space $X$ and by Valadier [38] in $R^n$. Castaing simultaneously extended a sufficient condition of Rockafellar [26] for the weak compactness of the level sets of certain integral functionals (see also Valadier [40]). This compactness condition was also extended by the author [30] in a different manner which made it possible to show necessity, as well as sufficiency, in certain cases [31].

Among other recent work of a more special nature, not to be discussed below, we mention the comprehensive paper of Ioffe and Levin [14] on the subdifferentials of convex integral functionals such as (7). This paper is the latest in a collection including Castaing [6], Gol'shtein [11], Ioffe [12], Levin [19, 20], and Valadier [37, 39]. We mention further the papers of Ioffe [13] and Rockafellar [32, 33] dealing with applications of new results on convex integral functionals to problems in the calculus of variations. The work of Temam [35, 36] should lead to more applications of this sort.

Our aim in this paper is to set forth some of the basic theorems about convex integral functionals in a more general form than has previously appeared in the literature. For the most part, the arguments follow earlier ones, but their extension to a broader context has been made possible by the technical developments cited above, especially in the theory of measurable multifunctions.

## §2. Measurable Multifunctions and Normal Integrands.

We assume henceforth that $\mathcal{A}$ is a $\sigma$-algebra of subsets of T (the measurable sets), and that $\mu$ is a positive, $\sigma$-finite measure on $\mathcal{A}$ which is complete (i.e. every subset of a set of measure zero is measurable). In this section X denotes an arbitrary complete separable metric space with metric d, and $\mathcal{B}$ denotes the $\sigma$-algebra of Borel subsets of X. The $\sigma$-algebra in $T \times X$ generated by the sets $A \times B$, where $A \in \mathcal{A}$ and $B \in \mathcal{B}$, is denoted by $\mathcal{A} \times \mathcal{B}$.

Given a multifunction (set-valued mapping) $\Gamma: T \to X$ and a set $S \subset X$, we denote by $\Gamma^{-1}(S)$ the set of all $t \in T$ such that $\Gamma(t) \cap S \neq \emptyset$. The set

(8) $\qquad D(\Gamma) = \{t \in T \mid \Gamma(t) \neq \emptyset\} = \Gamma^{-1}(X)$

is called the <u>effective domain</u> of $\Gamma$, and the set

(9) $\qquad G(\Gamma) = \{(t, x) \in T \times X \mid x \in \Gamma(t)\}$

the graph of $\Gamma$. We say that $\Gamma$ is measurable if its graph belongs to $\mathcal{G} \times \mathcal{B}$. Other definitions of the measurability of $\Gamma$ are also possible, for example in terms of the measurability of various classes of sets of the form $\Gamma^{-1}(S)$. However, in the case we are really interested in, where $\Gamma$ is closed-valued, it turns out that, under our assumptions on $(T, \mathcal{G}, \mu)$ and $X$, all the reasonable definitions coincide.

**Theorem 1.** Let $\Gamma: T \to X$ be a multifunction such that $\Gamma(t)$ is a closed set for every $t \in T$. Then the following properties of $\Gamma$ are equivalent.

(a) $\Gamma$ is measurable, that is, $G(\Gamma)$ is a measurable set.

(b) $\Gamma^{-1}(C)$ is measurable for every closed set $C \subset X$.

(c) $\Gamma^{-1}(U)$ is measurable for every open set $U \subset X$.

(d) $\Gamma^{-1}(B)$ is measurable for every Borel set $B \subset X$.

(e) $D(\Gamma)$ is measurable, and $d(\Gamma(t), x)$ is a measurable function of $t \in D(\Gamma)$ for each $x \in X$.

(f) $D(\Gamma)$ is measurable, and there exists a countable collection $(x_i, i \in I)$ of measurable functions $x_i: D(\Gamma) \to X$ such that $\Gamma(t)$ is the closure of $\{x_i(t) \mid i \in I\}$ for each $t \in D(\Gamma)$.

(g) There exists a countable collection $(x_i, i \in I)$ of measurable functions $x_i: T \to X$ such that the set $\{x_i(t) \mid i \in I\} \cap \Gamma(t)$ is dense in $\Gamma(t)$ for each $t \in T$, and the set $\{t \in T \mid x_i(t) \in \Gamma(t)\}$ is measurable for each $i \in I$.

This theorem is the key to almost everything involving closed-valued measurable multifunctions. It is due primarily to Castaing, who states it, minus condition (g), as Lemma 2 of [7]; see also [8]. (Castaing also omits condition (b), which he used elsewhere as the definition of measurability, but the implication here from (d) to (b) to (c) is elementary.) A proof has also been furnished by Ioffe and Levin [14, Appendix II]. Most of the implications were established, at least in special cases, in Castaing's dissertation [2] and developed further in a number of papers by that author. The equivalence of (a) and (b), however, was essentially proved earlier by Debreu [9], employing arguments attributed to Freedman and Neveu. The assumptions were weakened, and various implications sharpened, for $X = R^n$ by Rockafellar [27], who introduced condition (g).

In general, the implication from (f) to (g) is trivial, while the implication from (g) to (f) can be proved by the following argument (taking the index set I to be the natural numbers). Let $T_i = \{t \in T \mid x_i(t) \in \Gamma(t)\}$. Define $x_1' : D(\Gamma) \to X$ by

$$x_1'(t) = x_1(t) \text{ for } t \in T_1,$$
$$= x_2(t) \text{ for } t \in T_2 \backslash T_1,$$
$$= x_3(t) \text{ for } t \in T_3 \backslash (T_1 \cup T_2), \text{ etc.},$$

and then for $i = 2, 3, \ldots$, define $x_i' : D(\Gamma) \to X$ by

$$x_i'(t) = x_i(t) \text{ for } t \in T_i,$$
$$= x_1(t) \text{ for } t \in D(\Gamma) \backslash T_i.$$

It is easily checked that the collection $(x_i', i \in I)$ has the properties in condition (f).

Some other recent work on extending Theorem 1 may be found in the dissertation of Valadier [39].

Theorem 1 is indispensible in demonstrating that measurability is achieved or preserved when multifunctions are constructed or manipulated in various ways. In most cases the ultimate purpose of all this is to enable one to invoke the following fact, obtained by specializing condition (f).

Corollary.   If  $\Gamma: T \to X$  is a closed-valued multifunction satisfying any one of the conditions in Theorem 1, then there exists at least one measurable function  $x:T \to X$  such that $x(t) \in \Gamma(t)$ for every  $t \in D(\Gamma)$.

The existence of a measurable selector x when $\Gamma$ satisfies (c), a fact basic to the proof of Theorem 1, was first proved by Rokhlin in 1949 [34, Part I, §2, No. 9, Lemma 2], as Castaing has pointed out. The result was later rediscovered independently by Kuratowski and Ryll-Nardzewski [18] and Castaing [2].

It is convenient in the rest of this paper to refer to a function $f: T \times X \to (-\infty, +\infty]$ as an integrand. For each $t \in T$ we denote by $f_t$ the function $t \to f(t,x)$. The epigraph of $f_t$ is the set

(10) $$\text{epi } f_t = \{(x,\alpha) \in X \times R^1 \mid f_t(x) \leq \alpha\}.$$

Proposition 1.   The following conditions on an integrand f are equivalent:

a) f is  $\alpha \times \beta$ - measurable on  $T \times X$, and for each $t \in T$ the function $f_t$ is lower semicontinuous on X and not identically $+\infty$.

b) The multifunction $t \to \text{epi } f_t$ is measurable, and for each $t \in T$ the set epi $f_t$ is closed and nonempty.

This is easily deduced, arguing by way of the measurability of the function $(t, x, \alpha) \to f(t,x) - \alpha$ and its level sets.

An integrand f satisfying the conditions in Proposition 1 is said to be normal. The normality property can also be expressed in terms of a condition resembling (g) of Theorem 1, and this is a useful approach in dealing with convexity; see [26, 27]. A simple criterion worth mentioning is this: f is a normal integrand if f(t, x) is finite everywhere, measurable in t for fixed x , and continuous in x for fixed t . (Then the functions of the type $t \to (a, f(t, a) + \varepsilon)$ as a ranges over a countable dense subset of X and $\varepsilon$ ranges over the positive rational numbers, form a countable collection having property (f) of Theorem 1 with respect to the multifunction $t \to$ epi $f_t$ , so that this multifunction is measurable.)

Normality ensures in particular that for every measurable function $x: T \to X$ , the function $t \to f(t, x(t))$ is measurable. (The latter function is the composition of f with the measurable mapping $t \to (t, x(t)) \in T \times X$.) If the function $t \to f(t, x(t))$ is summable in the usual sense, or if it majorizes or is majorized by a summable (extended-real-valued) function on T , a natural value (possibly $+\infty$ or $-\infty$) can be assigned to the integral

(11) $$I_f(x) = \int_T f(t, x(t)) \mu(dt).$$

In the remaining case, it has proved useful to adopt the convention that $I_f(x) = +\infty$. In this way, we regard $I_f$ as a well-defined, extended-real-valued functional on the space of all measurable functions $x: T \to X$. The analysis of such a functional depends heavily on the effective use of the equivalences expressed in Theorem 1 and Proposition 1.

## §3. The Conjugate of an Integral Functional.

We assume henceforth that X is a separable reflexive Banach space. The dual of X (which is likewise separable) is denoted by Y , and the natural bilinear pairing between elements $x \in X$ and $y \in Y$ by $\langle x, y \rangle$.

Let f be a normal integrand on $T \times X$ . If $f_t$ is convex on X for every $t \in T$ , we say that f is __convex__. In this event $I_f$ is a convex functional on the linear space $L_0$ consisting of all measurable functions $x: T \to X$ .

The __conjugate__ of the integrand f is the integrand g on $T \times Y$ defined by

(12) $$g(t, y) = \sup \{\langle x, y \rangle - f(t, x) \mid x \in X\}.$$

According to (12), $g_t$ is for each $t \in T$ the function on $Y$ conjugate to $f_t$. The general theory of conjugate functions (see [1, 16, 22, 28]) asserts that in this case $g_t$ is convex and lower semicontinuous. If $f_t$ is itself convex (being already, by virtue of normality, lower semicontinuous and not identically $+\infty$), then $g_t$ is not identically $+\infty$, and $f_t$ is in turn the function on $X$ conjugate to $g_t$:

(13) $\qquad f(t,x) = \sup \{\langle x,y \rangle - g(t,y) \mid y \in Y\}.$

Combining such observations with a measurability argument, we obtain:

Proposition 2. *The integrand* $g$ *conjugate to the normal integrand* $f$ *is a normal convex integrand, provided that for each* $t$ *there is at least one* $y \in Y$ *such that* $g(t,y) < +\infty$. *The latter is true in particular if* $f$ *is convex, and in this event* $f$ *is in turn the integrand conjugate to* $g$.

Proof. We need only show that $g$ is measurable on $T \times Y$. To this end, we choose a countable collection of measurable multifunctions

$$t \to (x_i(t), \alpha_i(t)) \in X \times R^1, \qquad i \in I,$$

such that for each $t \in T$

$$\text{epi } f_t = c\ell \{(x_i(t), \alpha_i(t)) \mid i \in I\}.$$

Such a collection exists by property (f) of Theorem 1, since the multifunction $t \to \text{epi } f_t$ is measurable. For each $i \in I$, let

$$g_i(t,y) = \langle x_i(t), y \rangle - \alpha_i(t).$$

Then $g_i$ is measurable on $T \times Y$, and we have

$$g(t,y) = \sup \{g_i(t,y) \mid i \in I\}.$$

The measurability of $g$ thus follows from the fact that the pointwise supremum of a countable collection of measurable functions is measurable.

The main result of this section concerns the duality between the integral functionals $I_f$ and $I_g$. Let $L$ denote a subspace of the linear space of all measurable functions $x: T \to X$, and let $M$ similarly denote a subspace of the linear space of all measurable functions $y: T \to Y$. We assume that $|\langle x(t), y(t) \rangle|$ is summable in $t$ for each $x \in L$ and $y \in M$, so that the pairing

$$(14) \qquad \langle x, y \rangle_T = \int_T \langle x(t), y(t) \rangle \mu(dt), \qquad x \in L, \qquad y \in M,$$

is well-defined. As an obvious special case, one could take $L = L_X^p$ and $M = L_Y^q$ (the usual Lebesgue spaces of functions on $T$ with values in the Banach spaces $X$ and $Y$) with $1 \leq p \leq \infty$, $(1/p) + (1/q) = 1$.

The space $L$ (or similarly $M$) is said to be <u>decomposable</u> if, whenever $x$ belongs to $L$ and $x_0: S \to X$ is a bounded measurable function on a measurable set $S \subset T$ of finite measure, the function

$$(15) \qquad x'(t) = x_0(t) \text{ for } t \in S$$
$$= x(t) \text{ for } t \in T/S,$$

also belongs to $L$. The Lebesgue spaces, of course, have this property.

The following theorem, first proved by the author in [26] for $X = Y = R^n$, has not previously been stated in such generality. However, certain special infinite-dimensional cases (where $X$ is not necessarily a separable, reflexive Banach space) have been treated by Castaing [4, 5] or are implicit in Ioffe-Tikhomirov [16] and Ioffe-Levin [14]. (Reflexivity is actually used only in the second assertion of the theorem.)

Theorem 2. If L is decomposable and $I_f(x) < +\infty$ for at least one $x \in L$, then the convex integral functional $I_g$ on M is conjugate to the integral functional $I_f$ on L, that is,

(16)     $I_g(y) = \sup\{\langle x, y\rangle_T - I_f(x) | x \in L\}$ for every $y \in M$.

If in addition f is convex, M is decomposable and $I_g(y) < +\infty$ for at least one $y \in M$, then $I_f$ on L is in turn conjugate to $I_g$ on M:

(17)     $I_f(x) = \sup\{\langle x, y\rangle_T - I_g(y) | y \in M\}$ for every $x \in L$.

Proof. It suffices to prove (16), which asserts equivalently that for each $y \in M$ the quantity

$$\int_T g(t, y(t))\mu(dt)$$

is the supremum of

$$\int_T [\langle x(t), y(t)\rangle - f(t, x(t))]\mu(dt)$$

over all functions $x \in L$. Replacing $f_t$ for each t by $f_t - \langle \cdot, y(t)\rangle$ if necessary (this manipulation is normality-preserving), we can reduce the argument to the case where $y(t) \equiv 0$. Thus we need only prove that

(18)     $\inf\{\int_T f(t, x(t))\mu(dt) | x \in L\} = \int_T \varphi(t)\mu(dt)$

where

(19)     $\varphi(t) = \inf\{f(t, x) | x \in X\} = -g(t, 0)$.

Note that $\varphi$ is a measurable function by the argument of Lemma 2. There exists by hypothesis a function $x_1 \in L$ and a summable function $\alpha_1$ such that

(20)     $f(t, x_1(t)) \leq \alpha_1(t)$ for every $t \in T$.

Since $\varphi(t) \leq f(t, x(t))$ for every function $x$ by definition, we see in particular that the integral of $\varphi$ in (18) is in the standard sense well-defined and either finite or $-\infty$, and that the inequality $\geq$ holds in (18). Now let $\beta$ be any real number such that

(21) $$\int_T \varphi(t) \mu(dt) < \beta .$$

We prove the existence of a function $x \in L$ such that

(22) $$\int_T f(t, x(t)) \mu(dt) < \beta ,$$

thereby establishing the theorem. From (21) (and our assumptions on the measure space) there exists a summable function $\alpha_0$, such that $\varphi(t) < \alpha_0(t)$ for every $t$ and

(23) $$\int_T \alpha_0(t) \mu(dt) < \beta .$$

Define the multifunction $\Gamma: T \to X$ by

$$\Gamma(t) = \{x \in X \mid f(t, x) \leq \alpha_0(t)\}.$$

Since the function

$$(t, x) \to f(t, x) - \alpha_0(t)$$

is measurable, the graph of $\Gamma$ is a measurable set, i.e. $\Gamma$ is a measurable multifunction. Moreover, $\Gamma(t)$ is for each $t$ closed (since $f_t$ is lower semicontinuous) and nonempty (by (19) and the fact that $\varphi(t) < \alpha_0(t)$). The corollary of Theorem 1 then implies the existence of a measurable function $x_0$ (not necessarily in $L$) such that $x_0(t) \in \Gamma(t)$ for every $t$. Since (23) holds, it is possible to choose a measurable set $S \subset T$ of finite measure such that

(24) $$\int_S \alpha_0(t) \mu(dt) + \int_{T \setminus S} \alpha_1(t) \mu(dt) < \beta .$$

It can be arranged at the same time that $x_0$ is bounded on $S$. Let

$$x(t) = x_0(t) \quad \text{for } t \in S,$$
$$= x_1(t) \quad \text{for } t \in T \setminus S.$$

Then $x \in L$ by the assumption of decomposability, and we have

$$f(t, x(t)) \leq \alpha_0(t) \quad \text{for } t \in S,$$
$$f(t, x(t)) \leq \alpha_1(t) \quad \text{for } t \in T \setminus S.$$

The latter implies (22), in view of (24), and the proof is complete.

<u>Corollary. Suppose that f is convex, L and M are decomposable, and neither the functional $I_f$ on L nor $I_g$ on M is identically $+\infty$. Then these convex integral functionals are conjugate to each other with respect to the pairing (14), and hence in particular they are lower semi-continuous with respect to any locally convex topologies on L and M compatible with this pairing.</u>

We remark that, in the situation in the corollary, the subdifferential mapping $\partial I_f$, which is a multifunction from L to M, is easily described in terms of the subdifferential mappings $\partial f_t : X \to Y$. Indeed, $\partial I_f(x)$ is for each $x \in L$ the set of all $y \in M$ such that

(25) $\quad\quad\quad y(t) \in \partial f_t(x(t)) \quad \text{for almost every } t.$

As an illustration, let us suppose that $L = L_X^p$ and $M = L_Y^q$, $1 \leq p < \infty$, $(1/p) + (1/q) = 1$. These spaces are decomposable, so that the corollary is applicable if f is any normal convex integrand such that $f(t, x(t))$ is summable in t for at least one function $x \in L_X^p$, and $g(t, y(t))$ is summable in t for at least one function $y \in L_Y^q$. The convex functionals $I_f$ on $L_X^p$ and $I_g$ on $L_Y^q$ are then lower semicontinuous with respect to not only the norm topologies, but also with respect to the weak topologies that

$L_X^p$ and $L_Y^q$ induce on each other. Furthermore, since $L_X^p$ is a Banach space whose dual may be identified with $L_Y^q$, we can conclude that the subdifferential mapping

(26) $$\partial I_f : L_X^p \to L_Y^q ,$$

which as we have seen can be expressed by the relation (25), is a <u>maximal monotone operator</u>, as well as a <u>maximal cyclically monotone operator</u> [29]. Special note should be made of the case $X = Y = R^1$, since then (25) becomes

(27) $$y(t) \in \Gamma(t, x(t)) \quad \text{for almost every } t ,$$

where $\Gamma(t, \cdot)$ is for each $t$ a general maximal monotone operator from $R^1$ to $R^1$. This case is encountered, for example, in the study of the Hammerstein equation and various boundary-value problems.

## 4. A Refinement, With an Application to Weak Compactness.

In the case of Theorem 1 where $L = L_X^1$ and $M = L_Y^\infty$, there is an unanswered question which turns out to be crucial in dealing with many integral functionals that arise in practice. In this case we do have, under the assumptions in the corollary above,

(28) $$I_g(y) = \sup \{ \langle x, y \rangle_T - I_f(x) \mid x \in L_X^1 \} \quad \text{for all } y \in L_Y^\infty ,$$

(29) $$I_f(x) = \sup \{ \langle x, y \rangle_T - I_g(y) \mid y \in L_Y^\infty \} \quad \text{for all } x \in L_X^1 .$$

However, $L_X^1$ cannot be identified with the dual $L_Y^{\infty *}$ of $L_Y^\infty$. On $L_Y^{\infty *}$ we can define another convex functional $I^*$ conjugate to $I_g$,

(30) $$I^*(y^*) = \sup \{ y^*(y) - I_g(y) \mid y \in L_Y^\infty \} ,$$

and the relationship between $I^*$ and $I_f$ requires clarification.

Certainly if the continuous linear functional $y^*$ is of the form

(31) $$y^*(y) = \langle x, y \rangle_T, \qquad x \in L^1_X,$$

then $I^*(y^*) = I_f(x)$ by (29). Thus the restriction of $I^*$ to the copy of $L^1_X$ canonically embedded in $L^{\infty*}_Y$ can be identified with $I_f$. Put another way, $I^*$ can be regarded as a canonical extension of the integral functional $I_f$ to a space more general than $L^1_X$. What is the exact nature of the extension? This question was answered in [30] for $X = R^n$, and the result can now also be formulated for infinite-dimensional $X$.

To do this, we first need to make some observations about the structure of $L^{\infty*}_Y$. A functional $y^* \in L^{\infty*}_Y$ of the type (31) is said to be <u>absolutely continuous</u>. The absolutely continuous functionals thus form a closed subspace of $L^{\infty*}_Y$ isometric to $L^1_X$. What is not so well known, is that this subspace has a natural complement in $L^{\infty*}_Y$, the subspace consisting of the <u>singular</u> functionals. A functional $y^*$ is said to be singular if $T$ can be expressed as the union of an increasing sequence of measurable sets $T_m$ with the property that $y^*(y) = 0$ for all functions $y \in L^\infty_Y$ vanishing everywhere outside of $T_m$. Each $y^* \in L^{\infty*}_Y$ can be expressed uniquely as

(32) $$y^* = y^*_a + y^*_s,$$

where $y^*_a$ is absolutely continuous, $y^*_s$ is singular and

(33) $$\|y^*\| = \|y^*_a\| + \|y^*_s\|.$$

At least for $Y = R^n$, this result can be deduced by representing $L^\infty_Y$ as a space of continuous functions on a compact set and then applying the Lebesgue decomposition theorem to the elements of the dual space, regarded as measures. A more direct proof has been furnished by Dubovitskii and Miliutin [10] for $Y = R^1$, and this has been extended by Ioffe and Levin [14, Appendix I] to an arbitrary, separable Banach space Y. (Some related decomposition theorems may also be found in Ioffe [13] and Rao [24].)

**Theorem 3.** Assume that f is convex, that $I_f(x) < +\infty$ for at least one $x \in L^1_X$, and $I_g(y) < +\infty$ for at least one $y \in L^\infty_Y$. Let $I^*$ be the convex functional on $L^{\infty *}_Y$ defined by (30). Then, with respect to the canonical decomposition (32), one has

(34) $$I^*(y^*) = I_f(x) + J(y^*_s),$$

where x is the element of $L^1_X$ corresponding to the absolutely continuous component $y^*_a$ of $y^*$. Moreover, the functional J is of the special form

(35) $$J(y^*_s) = \sup \{y^*_s(y) | y \in D\},$$

where

(36) $$D = \{y \in L^\infty_Y | I_g(y) < +\infty\}.$$

Proof. The argument given by the author in [30, Theorem 1] extends virtually without change to the present case.

A number of applications of Theorem 3, for example to integral functionals on spaces of continuous functions, have been explored in [30] for $X = Y = R^n$. We limit ourselves here to deducing from Theorem 3 a criterion for weak compactness in $L^1_X$.

**Theorem 4.** Assume that f is convex, and that $f(t, y(t))$ is summable in t for every function $y \in L^\infty_Y$. Then for every real number $\alpha$ the convex set

(37) $$\{x \in L^1_X | I_f(x) \leq \alpha\}$$

is compact in the weak topology induced on $L^1_Y$ by $L^\infty_Y$.

<u>Proof.</u>  We can assume that $I_f(x) < +\infty$ for at least one $x \in L_X^1$, since otherwise the assertion is trivial. Then Theorem 3 is applicable, where $D = L_Y^\infty$ in (35) and (36) by our hypothesis. It follows that the set (37) can be identified with

(38) $$\{y^* \in L_Y^{\infty *} \mid I^*(y^*) \leq \alpha^*\},$$

and we need only show that the latter set is compact in the weak* topology on $L_Y^{\infty *}$. According to a basic theorem about conjugate convex functions proved at the same time by J. J. Moreau [23] and the author [25, Theorem 7A], this is true if $I_g$ is continuous at 0 in the norm topology of $L_Y^\infty$. Certainly $I_g$ is lower semicontinuous throughout $L_Y^\infty$, since $I_g$ is conjugate to $I_f$ (Corollary of Theorem 2). It remains only to recall that a finite, lower semicontinuous, convex function on a Banach space is necessarily continuous (see [1] or [25, Cor. 7C]).

<u>Remark.</u>  If $X = Y = R^n$, the summability hypothesis on g in Theorem 4 can be weakened to the assumption that $g(t,y)$ is summable in t for every $y \in Y$ (or merely for every y in some dense subset of Y); see Rockafellar [30]. More generally, in the infinite-dimensional case it suffices to assume that for each real number r there is a summable function $\alpha_r$ such that

(39) $$g(t,y) \leq \alpha_r(t) \text{ whenever } y \in Y, \|y\| \leq r.$$

Theorem 4 has been proved under somewhat stronger assumptions than this by Castaing [4,5], and the result has been sharpened further by Valadier [40]. The approach of Castaing and Valadier, based on a lemma of Grothendieck, is more direct and has the advantage of avoiding a discussion of the structure of $L_Y^{\infty *}$. On the other hand, the present approach yields additional information about the nature of the sets (38) in situations where they are not weakly compact, as well as a proof of the necessity of the condition in some cases [31].

## REFERENCES

1. A. Brøndsted, "Conjugate convex functions in topological vector spaces", Mat. -Fys. Medd. Danske Vid. Selsk. $\underline{34}$ (1964).

2. C. Castaing, "Sur les multi-applications mesurables", dissertation, Caën, 1967. Partly published in Rev. Franç. Inf. Rech. Op. 1 (1967), 3-34.

3. C. Castaing, "Proximité et mesurabilité," Travaux du Séminaire d'Analyse Unilatérale, Faculté des Sciences, Université de Montpellier, 1968.

4. C. Castaing, "Quelques applications du Théorème de Banach - Dieudonné à l'intégration," Publication No. 67, Secrétariat des Mathématiques, Université de Montpellier, 1970.

5. C. Castaing, "Quelques résultats de compacité liés à l'intégration", C. R. Acad. Sc. Paris $\underline{270}$ (1970), 1732-1735.

6. C. Castaing, "Un théorème de compacité faible dans $L_E^1$, etc.", Publication No. 44, Secrétariat des Mathématiques, Université de Montpellier, 1969.

7. C. Castaing, "Le Théorème de Dunford-Pettis géneralisé", Publication No. 43, Secrétariat des Mathématiques, Université de Montpellier, 1969. See also C. R. Acad. Sc. Paris 268 (1969), 327-329.

8. C. Castaing, "Quelques compléments sur le graphe d'une multi-application mesurable", Travaux du Séminaire d'Analyse Unilatérale, Faculté des Sciences, Université de Montpellier, 1969.

9. G. Debreu, "Integration of correspondences", Proc. Fifth Berkeley Symp. on Statistics and Prob., Vol. II, Part 1, 351-372. Univ. of California Press, Berkeley, 1966.

10. A. Ya. Dubovitskii and A. A. Miliutin, "Necessary conditions for a weak extremum in problems of optimal control with mixed constraints of inequality type", Zh. Vychisl. Mat. i Mat. Fys. $8$ (1968), 725-779. See also USSR Comp. Math. and Math. Phys. 8 (1968).

11. E. G. Gol'shtein, "Problems of best approximation by elements of convex sets and some properties of supporting functionals", Doklady Akad. Nauk SSSR $173$ (1967), 995-998. See also Soviet Math. Dokl. $8$ (1967), 504-507.

12. A. D. Ioffe, "Subdifferentials of restrictions of convex functionals", Uspekhi Mat. Nauk $25$ (1970), 181-182.

13. A. D. Ioffe, "Banach spaces generated by convex integrals and multidimensional variational problems", Doklady Akad. Nauk SSSR $195$ (1970), No. 5.

14. A. D. Ioffe and V. L. Levin, "Subdifferentials of convex functions", Trudy Mosk. Mat. Ob. (to be published early in 1972).

15. A. D. Ioffe and V. M. Tikhomirov, "Duality in problems of the calculus of variations", Doklady Akad. Nauk SSSR 180 (1968), 789-792. See also Soviet Math. Dokl. 9 (1968), 685-688.

16. A. D. Ioffe and V. M. Tikhomirov, "Duality of convex functions and extremum problems", Uspekhi Mat. Nauk $23$ (1968), 51-116. See also Russian Math. Surveys $23$ (1968), 53-124.

17. A. D. Ioffe and V. M. Tikhomirov, "On the minimization of integral functionals", Funkt. Analiz. 3 (1969), 61-70. See also Funct. Anal. Appl. 3 (1969), 218-229.

18. K. Kuratowski and C. Ryll-Nardzewski, "A general theorem on selectors", Bull. Polish Acad. Sci. 13 (1965), 397-411.

19. V. L. Levin, "On some properties of supporting functionals", Mat. Zametki 4 (1968), 685-696.

20. V. L. Levin, "On the subdifferentials of convex functionals", Uspekhi Mat. Nauk 25 (1970), 183-184.

21. W. A. J. Luxemburg and A. C. Zaanen, "Conjugate spaces of Orlicz spaces", Indag. Math. 18 (1956), 217-228.

22. J. J. Moreau, "Fonctionelles Convexes", mimeographed lecture notes, "Séminaire sur les Équations aux Derivées Partielles, Collége de France, 1966-67.

23. J. J. Moreau, "Sur la polaire d'une fonctionelle sémicontinue supérieurment", C. R. Acad. Sc. Paris 258 (1964), 1128-1130.

24. M. M. Rao, "Linear functionals on Orlicz spaces: general theory", Pacific J. Math. 25 (1968), 553-585.

25. R. T. Rockafellar, "Level sets and continuity of conjugate convex functions", Trans. Amer. Math. Soc. 123 (1966), 46-63.

26. R. T. Rockafellar, "Integrals which are convex functionals", Pacific J. Math. 24 (1968), 525-539.

27. R. T. Rockafellar, "Measurable dependence of convex sets and functions on parameters", J. Math. Anal. Appl. 28 (1969), 4-25.

28. R. T. Rockafellar, Convex Analysis, Princeton University Press, 1970.

29. R. T. Rockafellar, "On the maximal monotonicity of subdifferential mappings", Pacific J. Math. 33 (1970), 206-216.

30. R. T. Rockafellar, "Integrals which are convex functionals, II", Pacific J. Math., to appear.

31. R. T. Rockafellar, "Weak compactness of level sets of integral functionals", Proceedings of the Liège Symposium on Functional Analysis (Sept. 1970), H. G. Garnir (editor), to appear.

32. R. T. Rockafellar, "Conjugate convex functions in optimal control and the calculus of variations", J. Math. Anal. Appl. 32 (1970), 174-222.

33. R. T. Rockafellar, "Existence and duality theorems for convex problems of Bolza", Trans. Amer. Math. Soc., September 1971.

34. V. A. Rokhlin, "Selected topics from the metric theory of dynamical systems", Uspekhi Mat. Nauk 4 (1949), 57-128. See also Amer. Math. Soc. Translations, vol. 49 (1966), 171-240.

35. R. Temam, "Solutions généralisées d'équations non linéaires non uniformement elliptiques", Publications Math. d'Orsay, 1970.

36. R. Temam, "Solutions generalisées de certains problémes de calcul de variations", C. R. Acad. Sc. Paris, to appear.

37. M. Valadier, "Sous-différentials d'une borne supérieure et d'une somme continue de fonctions convexes", C. R. Acad. Sc. Paris 268 (1969), 39-42.

38. M. Valadier, "Intégration de convexes fermés, notamment d'epigraphes; inf-convolution continue", Rev. Franç. Inf. Rech. Op. 4 (1970), 57-73.

39. M. Valadier, "Contributions à l'analyse convexe", dissertation, Paris, 1970.

40. M. Valadier, "Un théorème d'inf-compacité", to appear.


The preparation of this report was supported in part by grant AFOSR-71-1994.

> Department of Mathematics
> University of Washington
> Seattle, Washington 98105

> Received April 1, 1971

# Projections on Convex Sets in Hilbert Space and Spectral Theory
## Part I. Projections on Convex Sets
## Part II. Spectral Theory

*EDUARDO H. ZARANTONELLO*

   This paper consists of two separate parts of which the second is an outgrowth of the first and to a certain extent its motivation. The original purpose, when this work began, was to make a general study of projections on convex sets in Hilbert space (a projection being the mapping assigning to any point the nearest point in a closed convex set) without any particular aim in mind. Projections appeared to me as one of the simplest instances of nonlinear mappings that can be defined in abstract terms, and as such worth investigating. Here, but for a few previous timid incursions, I found a large unexplored territory, and as I entered it I saw that I would have to do without the benefit of preestablished paths and visible landmarks. So I began collecting mathematical facts here and there (§1), simple and easy results to be true, but significant. It was not long before I came to the revealing realization that, however different convex sets may look from linear spaces, projections on convex sets do not behave very differently from ordinary linear orthogonal projections. Soon the possibility emerged with Theorem 1.3 of constructing a large class of important nonlinear operators out of projections, in the manner of the Krein-Milman theorem perhaps, or may be in the spirit of the spectral theory of linear selfadjoint operators, I did not know at the time. Since then my work was strongly oriented towards supplementing this idea, and many of my results were just tools and machinery (§§2, 3, 4) for this felt yet unknown task; a few proved themselves useless but, surprisingly, most turned to be precisely that which was needed. It was in the study of the algebra of projections (§5) that the similarities with the linear counterpart became obvious, especially after the specialization to projections on convex cones - which I had resisted all the way. At once all obstacles were removed and the way towards a spectral theory was opened. In fact, the algebra of projections on convex cones retains from the algebra of linear orthogonal projections just those properties required to make such theory possible (§§5,6). The notion of a spectral resolution made up of projections on convex cones makes sense, from it a spectral measure can be built (§7), and a spectral integration theory of real valued functions developed (§8). The resulting integrals are operators in Hilbert space, nonlinear in general, generalizing linear selfadjoint ones whose properties they mimic (§9). These operators, obtained by linear synthesis of projections, are finally characterized in the abstract by a few simple properties (§10).

For the readers mainly interested in the spectral theory, let me indicate a short cut - alas, not too short - across part I: §1, up to Definition 1.2 exclusive; §2 up to Lemma 2.7 exclusive, Definition 2.8 and Lemma 2.16; §3 omit Lemmas 3.4, 3.5; §4, Definition 4.1, Lemmas 4.1, 4.7, Definition 4.3, Theorem 4.1; §5 up to Lemma 5.8 exclusive.

Thanks are due to the Université de Lyon, France, the Universidad de Córdoba, Argentina, the Universidad de Chile, Chile, and the University of Wisconsin, U.S.A. for their generous support while this work was in progress.

Part I.  Projections on Convex Sets in Hilbert Space

§1.  Projections, basic properties.

Let $\mathcal{H}$ be a real Hilbert space and let $\mathcal{K}$ be a closed convex set therein. For each point $x \in \mathcal{H}$ there is a unique closest point in $\mathcal{K}$ called its projection and denoted $P_\mathcal{K} x$. Thus

(1.1) $$\|x - P_\mathcal{K} x\| \leq \|x-y\|, \quad \forall y \in \mathcal{K}$$

is the characteristic property of $P_\mathcal{K} x$. The existence and uniqueness of $P_\mathcal{K} x$ is a well known fact in the geometry of Hilbert space. The convex $\mathcal{K}$ is the set of fixed points of $P_\mathcal{K}$, and so $\mathcal{K}$ and $P_\mathcal{K}$ determine each other. On the other hand, it has been proved that among weakly closed sets convex sets are the only ones admitting projections [27, Theorem 7.8]. Thus closed convex sets and their projections are equivalent classes of objects and the study of one of them amounts to the study of the other. Here we shall view projections as operators in Hilbert space and will study them within the context of operator theory. In so doing a new prospective is gained on convex sets. For the basic facts about convex sets we refer the reader to the books by Bonnesen and Fenchel [2] and Valentine [27].

Lemma 1.1  Condition (1.1) is equivalent to

(1.2) $$\langle x - P_\mathcal{K} x, P_\mathcal{K} x - y \rangle \geq 0, \quad \forall y \in \mathcal{K}^\dagger$$

Proof:  From the identity

(1.3) $$\|x-P_\mathcal{K} x\|^2 - \|x-y\|^2 = \|x-P_\mathcal{K} x\|^2 - \|(x-P_\mathcal{K} x) + (P_\mathcal{K} x - y)\|^2 = -2\langle x - P_\mathcal{K} x, P_\mathcal{K} x - y \rangle - \|P_\mathcal{K} x - y\|^2$$

one sees at once that (1.2) $\Rightarrow$ (1.1). On the other hand, if

---

$^\dagger$Angular parentheses $\langle , \rangle$ denote the scalar product in $\mathcal{H}$; $\|\ \|$ denotes the norm.

(1.3) is written with $y' = ty + (1-t) P_{\mathcal{K}} x$, $0 \leq t \leq 1$, in place of $y$, one obtains

$$\|x - P_{\mathcal{K}} x\|^2 - \|x - y'\|^2 = -2t \langle x - P_{\mathcal{K}} x, P_{\mathcal{K}} x - y \rangle - t^2 \|P_{\mathcal{K}} x - y\|^2 .$$

Thus, if (1.1) holds

$$0 \geq -2t \langle x - P_{\mathcal{K}} x, P_{\mathcal{K}} x - y \rangle - t^2 \|P_{\mathcal{K}} x - y\|^2, \quad 0 < t \leq 1 ,$$

whence, on division by $t$ and by letting $t \to 0$, (1.2) follows.

As a corollary we have the following characterization of projections:

**Theorem 1.1** A mapping $P$ is the restriction of a projection if and only if

(1.4) $\quad \langle x - Px, Px - Py \rangle \geq 0$ , $\forall x, y \in \mathcal{D}(P)$ .

**Proof:** Let $y_i \in \mathcal{D}(P)$ $i = 1, 2, \ldots, n$ and $\alpha_i$ nonnegative numbers with $\sum_1^n \alpha_i = 1$. Then

$$\langle x - Px, Px - \sum_1^n \alpha_i Py_i \rangle = \sum_1^n \alpha_i \langle x - Px, Px - Py_i \rangle \geq 0$$

and hence for any $y \in \mathcal{K}$ = closed convex hull of $\mathcal{D}(P)$, $\langle x - Px, Px - y \rangle \geq 0$, which by Lemma 1.1 says that $Px = P_{\mathcal{K}} x$. This proves sufficiency; necessity is contained in Lemma 1.1.

**Lemma 1.2** Any $P_{\mathcal{K}}$ satisfies the inequality

(1.5) $\quad \langle (I-P)x - (I-P)x', Px - Px' \rangle \geq 0 \quad \forall x, x' \in \mathcal{H}$, [15].

**Proof:** First write down (1.4) with $x'$ in place of $y$:

$$\langle (I - P_{\mathcal{K}})x, P_{\mathcal{K}} x - P_{\mathcal{K}} x' \rangle \geq 0 ;$$

then write the inequality obtained from this by interchanging x and x':

$$\langle (I-P_K)x', P_K x' - P_K x \rangle \geq 0,$$

and finally add the two inequalities. Then (1.5) follows.

Inequality (1.5) is very important and will often make its appearance in the course of this paper. Note that it is invariant under the substitutions of P by I-P, and that if valid for P it is valid for $A^*PA$ for any nonexpansive linear mapping A. The following are alternative ways of writing (1.5):

(1.6) $\quad \|Px-Px'\|^2 \leq \langle x-x', Px-Px' \rangle,$

(1.7) $\quad \|(I-P)x-(I-P)x'\|^2 \leq \langle x-x', (I-P)x-(I-P)x' \rangle.$

It follows that <u>both $P_K$ and $I-P_K$ are monotone operators</u>. Moreover by Schwarz inequality applied to the right hand members of (1.6) and (1.7),

(1.8) $\quad \|Px-Px'\| \leq \|x-x'\|, \quad \|(I-P)x - (I-P)x'\| \leq \|x-x'\|,$

that is, <u>both projections and the complements of projections with respect to the identity are nonexpansive</u> [18], [12]. Thus, <u>projections are continuous mappings</u>. Observe that if we write $y = Px$, $y' = Px'$, and $x = P^{-1}y$, $x' = P^{-1}y'$, (1.6) becomes

$$\|y-y'\|^2 \leq \langle P^{-1}y - P^{-1}y', y-y' \rangle$$

that is, $0 \leq \langle P^{-1}-I)y - (P^{-1}-I)y', y-y' \rangle$, which says, no more and no less, that $P^{-1}-I$ is a monotone (multivalued) mapping; similarly $(I-P)^{-1}-I$ is monotone. Thus, <u>for any closed convex set</u> $K$, $P_K^{-1} - I$ <u>and</u> $(I-P_K)^{-1}-I$ <u>are monotone mappings</u>. In terms of vertices and faces-concepts to be defined in §2 (Defs. 2.4, 2.3)-this result can be stated as saying: <u>The mappings assigning to each point of a convex set its vertex, and to each normal its face are monotone.</u>

**Lemma 1.3** A mapping P satisfies (1.5) if and only if $2P-I$ is nonexpansive.

**Proof:** Set $Q = 2P-I$. The Lemma is a straightforward consequence of the identity,

$$(1.9) \quad \|x-x'\|^2 - \|Qx-Qx'\|^2 = \langle (x+Qx)-(x'+Qx'), (x-Qx)-(x'-Qx') \rangle$$

$$= 4\langle Px-Px', (I-P)x - (I-P)x' \rangle .$$

A number of corollaries follow from this result. Let us first remark that since nonexpansive mappings form a convex family so do the mappings satisfying (1.5). Since by a theorem of Kirzbraun [7] nonexpansive mappings defined in a subset of $\mathcal{H}$ can always be extended as nonexpansive mappings to the whole space, the same holds for mappings satisfying (1.5), that is, <u>any mapping P satisfying (1.5) can be extended to the whole space $\mathcal{H}$ under preservation of (1.5)</u>. For a projection $P_\mathcal{K}$, $2P_\mathcal{K}-I$ is the mapping assigning to each point in space its symmetric with respect to its projection on $\mathcal{K}$; we call this operation **symmetry with respect to $\mathcal{K}$**.

**Corollary 1** Symmetries with respect to convex sets are nonexpansive mappings.

**Corollary 2** The difference of any two projections is nonexpansive.

**Proof:** Let $P_{\mathcal{K}_1}$ and $P_{\mathcal{K}_2}$ be the two projections. By the Lemma above both $2P_{\mathcal{K}_1}-I$ and $I-2P_{\mathcal{K}_2}$ are nonexpansive, and along with them so is their semi-sum $P_{\mathcal{K}_1} - P_{\mathcal{K}_2}$.

**Corollary 3** The complement to the identity of the sum of any two projections is nonexpansive.

**Proof:** As above, $I-2P_{\mathcal{K}_1}$, $I-2P_{\mathcal{K}_2}$ and their semi-sum $I-(P_{\mathcal{K}_1} + P_{\mathcal{K}_2})$ are nonexpansive.

The theorem below tells us that it is (1.5) along with idempotency that characterizes projections:

**Theorem 1.2** Any mapping P from $\mathcal{H}$ into $\mathcal{H}$ satisfying

a. $\mathcal{D}(P) \supset \mathcal{R}(P)$, b. $P^2 = P$, c. $P^{-1} - I$ monotone,

is the restriction of a projection.

**Proof:** Note first that condition c. is just another way of stating (1.5) for x and x' in $\mathcal{D}(P)$. Since $\mathcal{R}(P) \subset \mathcal{D}(P)$ we can set x' = Py, and in doing so we obtain

$$\langle x - Px - Py + P^2 y, Px - P^2 y \rangle \geq 0 \qquad \forall\, x, y \in \mathcal{D}(P).$$

But $P^2 y = Py$ and so,

$$\langle x - Px, Px - Py \rangle \geq 0 \qquad \forall\, x, y \in \mathcal{D}(P),$$

and in consequence by Theorem 1.1, Px is the projection of x onto the closed convex hull of the range of P.

The significance of this theorem lies in the fact that it furnishes a characterization of projections in terms of idempotency and monotonicity only.

Projections and their complements are closed operations when considered as mappings from the weak into the strong topology. More precisely:[†]

**Lemma 1.4**

$$\{x_\alpha \rightharpoonup x,\ \overline{\lim}\, \|x_\alpha\| < \infty,\ P_\mathcal{K} x_\alpha \to y\} \Longrightarrow \{y = P_\mathcal{K} x\}$$

$$\{x_\alpha \rightharpoonup x,\ \overline{\lim}\, \|x_\alpha\| < \infty,\ (I - P_\mathcal{K}) x_\alpha \to y\} \Longrightarrow \{y = (I - P_\mathcal{K}) x\}.$$

---

[†] As usual whole arrows denote strong convergence and half arrows weak convergence.

**Proof:** Let us do the proof for $P_K$ first. By (1.2) we have

$$\langle x_\alpha - P_K x_\alpha, P_K x_\alpha - y' \rangle \geq 0 \qquad \forall\, y' \in K.$$

The left member tends to $\langle x-y, y-y' \rangle$ by virtue of the hypothesis, and so

$$\langle x-y, y-y' \rangle \geq 0, \qquad \forall\, y' \in K$$

which, since $y$ obviously belongs to $K$, implies $y = P_K x$ by Lemma 1.1.. The proof is similar for $I-P$. If $x_\alpha$ is ultimately bounded and converges weakly to $x$ and $(I-P)x_\alpha$ converges strongly to some limit $z$, then $P_K x_\alpha$ converges weakly to an element $y$ in $K$ (for convex sets weak and strong closures coincide). Then

$$x_\alpha \to x, \quad (I-P_K)x_\alpha \to x-y = z,$$

and by passing to the limit in

$$\langle x_\alpha - P_K x_\alpha, P_K x_\alpha - y' \rangle = \langle (I-P_K)x_\alpha, -(I-P_K)x_\alpha + x_\alpha - y' \rangle \geq 0,$$

$$\forall\, y' \in K,$$

one obtains

$$0 \leq \langle x-y, -(x-y) + x-y' \rangle = \langle x-y, y-y' \rangle, \quad \forall\, y' \in K.$$

This means $y = P_K x$, or what is the same $z = x-y = (I-P)x$, and the closedness of $I-P_K$ is established.

Although continuous and demiclosed, projections (and their complements with respect to the identity) do not resemble in general compact operators. In fact, a projection $P_K$ is compact if and only if $K$ is boundedly compact.

In the same vein, examples can be given of projections $P_K$ which are not continuous as mappings of $H$ with the weak topology into $H$ again with the weak topology.

Here is one: let $K$ be the unit ball centered at the origin and let $x_n = \varphi_1 + \varphi_n$, where the $\varphi_i$'s, $i = 1, 2, \ldots$ are orthonormal vectors. Then $P_K x_n = \dfrac{\varphi_1 + \varphi_n}{\sqrt{2}}$, and $x_n \rightharpoonup \varphi_1$, $P_K x_n \rightharpoonup \varphi_1/\sqrt{2}$; yet $P_K \varphi_1 \neq \varphi_1/\sqrt{2}$.

It is known that this last type of continuity is equivalent to ordinary continuity [4, I, Theorem 3.15] in the case of linear operators. For projections it holds when $K$ is either a linear variety or a boundedly compact set, or when it is the direct product of any number, finite or infinite, of such sets. Yet, in spite of this apparent proliferation of examples the impression remains that projections with this type of continuity are rather rare.

On the positive side we have the following lemma:

**Lemma 1.5** The distance from a point to the closed convex set $K$ is weakly lower semicontinuous, that is,

(1.11) $\quad \{x_\alpha \rightharpoonup x\} \Longrightarrow \{\|x - P_K x\| \leq \underline{\lim} \|x_\alpha - P_K x_\alpha\|\}$.

Proof: From

$$\langle x - P_K x, P_K x - P_K x_\alpha \rangle \geq 0$$

one deduces

$$\langle x - P_K x, x_\alpha - P_K x_\alpha \rangle \geq \langle x - P_K x, x_\alpha - P_K x \rangle =$$

$$= \langle x - P_K x, x - P_K x \rangle + \langle x - P_K x, x_\alpha - x \rangle,$$

whence by Schwarz' inequality applied to the left member it follows

$$\|x - P_K x\| \|x_\alpha - P_K x_\alpha\| \geq \|x - P_K x\|^2 + \langle x - P_K x, x_\alpha - x \rangle.$$

Taking the lim inf and noticing that the last term converges to zero, one obtains

$$\|x-P_{\mathcal{K}}x\| \underline{\lim} \|x_\alpha - P_{\mathcal{K}}x_\alpha\| \geq \|x-P_{\mathcal{K}}x\|^2 ,$$

which is equivalent to (1.11).

**Lemma 1.6** The set of fixed points of a mapping P satisfying (1.5) is the closed convex set

$$\mathcal{K} = \{y \mid \langle x-Px, Px-y \rangle \geq 0, \ \forall \ x \in \mathcal{K}\}.$$

**Proof:** Let y be a fixed point and write (1.5) with y in place of x'; then

$$\langle x-Px, Px-y \rangle \geq 0, \ \forall \ x \in \mathcal{K}$$

showing that y belongs to $\mathcal{K}$. Conversely, if y is a point in $\mathcal{K}$, the inequality defining $\mathcal{K}$ should in particular be satisfied by x = y, that is

$$\langle y - Py, Py - y \rangle \geq 0 ,$$

whence Py = y. Hence y is a fixed point.

**Corollary 1** The set of fixed points of a nonexpansive mapping is a closed convex set.

**Proof:** The mapping Q has the same fixed points as the mapping $P = \frac{I+Q}{2}$ which satisfies (1.5) by Lemma 1.3.

**Corollary 2** The inverse image of a point by a projection is either empty or a closed convex cone with vertex at the point [18].

**Proof:** The only points for which $P_{\mathcal{K}}^{-1}y \neq \emptyset$ are the points $y \in \mathcal{K}$. Let y be one such point. Then $P_{\mathcal{K}}^{-1}y$ is the set of x's such that $P_{\mathcal{K}}x = y$, or, which is the same, the set of x's such that $(I-P_{\mathcal{K}})x+y = x$. But this is the set of fixed points for the nonexpansive mapping $x \to (I-P_{\mathcal{K}})x+y$, and as such is closed and convex. By the idempotency of $P_{\mathcal{K}}$, y itself belongs to $P_{\mathcal{K}}^{-1}y$. To conclude the proof we must show that all points $x(t) = P_{\mathcal{K}} x + t(x - P_{\mathcal{K}} x)$, $t \geq 0$, project onto $P_{\mathcal{K}} x$. In fact

$$\langle x(t) - P_{\mathcal{K}} x, P_{\mathcal{K}} x - y \rangle = t \langle x - P_{\mathcal{K}} x, P_{\mathcal{K}} x - y \rangle \geq 0, \quad \forall y \in \mathcal{H}, \forall t \geq 0,$$

and hence $P_{\mathcal{K}} x(t) = P_{\mathcal{K}} x$, by Lemma 1.1.

The following lemma furnishes an estimate for a mapping's deviation from linearity.

<u>Lemma 1.7</u>  If P satisfies (1.5) then

(1.12)
$$\| P(\sum_1^n \alpha_i x_i) - \sum_1^n \alpha_i P x_i \|^2 \leq$$

$$\tfrac{1}{2} \sum_{i,j}^n \alpha_i \alpha_j \langle P x_i - P x_j, (I-P) x_i - (I-P) x_j \rangle$$

for any finite set $\{x_i\}_1^n$ and any sequence $\{\alpha_i\}_1^n$ of non-negative numbers with $\sum_1^n \alpha_i = 1$.

<u>Proof:</u> This is an immediate consequence of the identity

$$\| P(\sum \alpha_i x_i) - \sum \alpha_i P x_i \|^2$$

(1.13)
$$+ \sum \alpha_i \langle P(\sum \alpha_j x_j) - P x_i, (I-P)(\sum \alpha_j x_j) - (I-P) x_i \rangle$$

$$= \tfrac{1}{2} \sum \alpha_i \alpha_j \langle P x_i - P x_j, (I-P) x_i - (I-P) x_j \rangle,$$

whose verification we leave in the reader's hands.

Corollary   If Q is nonexpansive

$$\|Q(\sum_i \alpha_i x_i) - \sum_i \alpha_i Q x_i\|^2 \le$$

(1.14)
$$\le \tfrac{1}{2} \sum_{i,j} \alpha_i \alpha_j (\|x_i - x_j\|^2 - \|Q x_i - Q x_j\|^2)$$

$$\alpha_i \ge 0, \; i = 1, \ldots, n, \; \sum \alpha_i = 1.$$

Proof:   Inequality (1.14) is just (1.12) for $\dfrac{I + Q}{2}$, which by Lemma 1.3 satisfies (1.5).

We shall finish this section with a brief discussion of mappings satisfying (1.5), aimed to determine the particular manner in which projections stand within this class of operators.

Let us briefly go over some standard notions: A set $\Pi$ of the form $\Pi = \{x \mid \langle x, u \rangle = k\}$, $u \ne 0$, is called a hyperplane; it is simply a translation of the subspace of deficiency one perpendicular to $u$. Given a point $x_0$ and a direction $u \ne 0$, there is one and only one hyperplane through $x_0$ perpendicular to $u$, namely $\Pi_u(x_0) = \{x \mid \langle x, u \rangle = \langle x_0, u \rangle\}$. Any hyperplane bounds two closed halfspaces, the sets $\Pi_u^+(x_0) = \{x \mid \langle x, u \rangle \ge \langle x_0, u \rangle\}$ and $\Pi_u^-(x_0) = \{\langle x, u \rangle \le \langle x_0, u \rangle\}$, and two open ones obtained by changing the signs $\le$ and $\ge$ to $<$ and $>$ respectively. Given two <u>closed convex sets a hyperplane is said to separate them if they are contained in closed halfspaces opposite with respect to it.</u>

Definition 1.1   A hyperplane $\Pi$ is said to be a support hyperplane of a closed convex set $\mathcal{K}$ if it bounds a minimal closed halfspace containing $\mathcal{K}$. If $\Pi \cap \mathcal{K} \ne \phi$, $\Pi$ is said to be supporting $\mathcal{K}$ at points of this intersection, whereas if $\Pi \cap \mathcal{K} = \phi$ one says that $\Pi$ supports $\mathcal{K}$ at infinity.

According to this definition $\Pi = \{x \mid \langle x, u \rangle = k\}$ is a support hyperplane of $\mathcal{K}$ if either $k = \sup_{x \in \mathcal{K}} \langle x, u \rangle$ or

$k = \inf_{x \in \mathcal{K}} \langle x, u \rangle$; by an adequate choice of $u$ one may always assume that $k = \sup_{x \in \mathcal{K}} \langle x, u \rangle$.

If $x \in \mathcal{K}$, then by (1.2) $\Pi (P_\mathcal{K} x)$ supports $\mathcal{K}$ at $P_\mathcal{K} x$
$x - P_\mathcal{K} x$
and separates $x$ form $\mathcal{K}$. It is easily seen that any support hyperplane at finite distance is necessarily of this type. It follows that <u>$\mathcal{K}$ is the intersection of all closed halfspaces containing it bounded by supporting hyperplanes.</u>

We also need the classical notion of "pedal" applied to a convex set:

<u>Definition 1.2</u>   If $\mathcal{K}$ is a closed convex set and $x \notin \mathcal{K}$, the set ped $(\mathcal{K}, x) = \{y = P_\Pi x \mid \forall$ supporting $\Pi$'s separating $x$ from $\mathcal{K}\}$ is called the pedal of $\mathcal{K}$ with regard to $x$.†

<u>Lemma 1.8</u>   Let $\mathcal{K}$ be a closed convex set and $x$ a point outside of $\mathcal{K}$; moreover let $\mathfrak{J}(\mathcal{K})$ be the class of all mappings satisfying (1.5) and leaving all points of $\mathcal{K}$ fixed. Then

(1.15)   $\{y \mid y = Px, \forall P \in \mathfrak{J}(\mathcal{K})\}$ = convex hull $\{\{x\} \cup$ ped $(\mathcal{K}, x)\}$

<u>Proof:</u>   We first prove

(1.16)   $\{y \mid y = Px, \forall P \in \mathfrak{J}(\mathcal{K})\} =$

$= \{y \mid y = P_\Pi x, \forall \Pi$'s separating $x$ and $\mathcal{K}\}$.

If $\Pi$ separates $x$ and $\mathcal{K}$, $P_\Pi x$ coincides with the projection of $x$ onto the closed halfspace determined by $\Pi$ that contains $\mathcal{K}$. But the operation of projecting onto such halfspace satisfies (1.5) (Lemma 1.2) and leaves all points of $\mathcal{K}$ invariant, and so $P_\Pi x$ coincides with $Px$ for some $P \in \mathfrak{J}(\mathcal{K})$. Thus the set on the left of (1.16) contains that on the right.

---

†This is not quite the ordinary concept of pedal which does not require that $\Pi$ separate $x$ and $\mathcal{K}$; it rather coincides with the inner loop of the two loops of which the pedal consists, one of which may be at infinity.

Let us prove now that conversely any $Px$ with $P \in \mathfrak{J}(\mathcal{K})$ is a $P_\Pi x$ for a separating $\Pi$. To this effect note that if $z \in \mathcal{K}$, (1.5) with $z$ in place of $x'$ yields

$$\langle (I-P)x, Px-z \rangle \geq 0 ,$$

that is

$$\langle x-Px, Px \rangle \geq \langle x-Px, z \rangle, \quad \forall z \in \mathcal{K} .$$

On the other hand since $\langle x-Px, x-Px \rangle > 0$ (the case $Px = x$ is trivial and can be dealt with separately),

$$\langle x-Px, x \rangle > \langle x-Px, Px \rangle .$$

The two last inequalities simply say that $\Pi = \Pi_{x-Px}(Px)$ separates $x$ and $\mathcal{K}$, and since obviously $Px = P_\Pi x$, (1.16) is proved.

    Next observe that any separating hyperplane is parallel to a separating support hyperplane and that all projections of $x$ onto separating hyperplanes parallel to a fixed separating support hyperplane fill out the segment joining $x$ with its projection onto this hyperplane. Thus either of the members in (1.16) is the set obtained by joining any point in the pedal of $\mathcal{K}$ with regard to $x$ with $x$. But such a set being convex (the class $\mathfrak{J}(\mathcal{K})$ is convex) must also coincide with the convex hull of $x$ and $\text{ped}(\mathcal{K}, x)$. Note that from this argument follows that

$$\text{co}\{\{x\} \cup \text{ped}(\mathcal{K}, x)\} = \{y \mid \langle x-y, y-z \rangle \geq 0, \forall z \in \mathcal{K}\} .$$

<u>Corollary 1</u>    Among all mappings $P$ satisfying (1.5) and leaving all points of a convex set $\mathcal{K}$ fixed the projection $P_\mathcal{K}$ produces the greatest displacement at each point of the space. In other terms

(1.17) $\quad \|x - P_\mathcal{K} x\| \geq \|x - Px\|, \quad \forall x \in \mathcal{H}, \forall P \in \mathfrak{J}(\mathcal{K}) .$

This property characterizes $P_\mathcal{K}$.

**Corollary 2** Among all mappings P satisfying (1.5) and leaving all points of $K$ fixed, $P_K$ is the only one having its range contained in $K$.

**Corollary 3** If P satisfies (1.5) and leaves $K$ fixed pointwise, then

(1.18) $\qquad \langle x-Px, Px-P_K x \rangle \geq 0, \qquad \forall\ x \in H.$

**Proof:** This relation simply states that $P_K x$ lies in the half-space opposite to that containing $x$, with regard to the hyperplane through $Px$ perpendicular to $x-Px$.

**Theorem 1.3** Let $P_i$, $i = 1, 2, \ldots$ be a sequence, finite or infinite, of operators satisfying (1.5), and let $\{\alpha_i\}_1^\infty$ be a sequence of positive numbers with $\sum_1^\infty \alpha_i = 1$. Under these conditions the operator $x \to \sum_1^\infty \alpha_i P_i x$ (the sum being weakly pointwise convergent) is a projection if and only if there is a closed convex set $K$ and vectors $u_i$ with $\sum_1^\infty \alpha_i u$ perpendicular to $K$ such that $P_i x = P_K x + u_i$, $i = 1, 2, \ldots$.

**Proof:** Suppose that there is a $K$ such that $P_K x = \sum_1^\infty \alpha_i P_i x$. Then this $K$ is the range of $P_K$, that is $K = \{z \mid z = \sum_1^\infty \alpha_i P_i x, \forall x \in H\}$. By (1.6) applied to each $P_i$,

$$\|P_i x - P_i x'\|^2 \leq \langle x-x', P_i x - P_i x' \rangle, \quad i = 1, 2, \ldots$$

whence

$$\sum_1^k \alpha_i \|P_i x - P_i x'\|^2 \leq \langle x-x', \sum_1^k \alpha_i P_i x - \sum_1^k \alpha_i P_i x' \rangle.$$

If we let $k \to \infty$, the right hand member converges and hence so does the left, and we have

$$\sum_{1}^{\infty} \alpha_i \|P_i x - P_i x'\|^2 \le \langle x-x', P_\mathcal{K} x - P_\mathcal{K} x' \rangle .$$

For $x' = P_\mathcal{K} x$ the right hand member vanishes, and so along with it all the terms on the left must vanish, that is

(1.19) $\qquad P_i x = P_i P_\mathcal{K} x , \quad i = 1, 2, \ldots .$

Now, pick any two points $z$ and $z'$ in $\mathcal{K}$, and apply (1.7) to each $P_i$,

$$\|(I-P_i)z - (I-P_i)z'\|^2 \le \langle z-z', (I-P_i)z - (I-P_i)z' \rangle ,$$

$$i = 1, 2, \ldots .$$

Multiplying by $\alpha_i$ and adding one obtains as before

$$\sum_{1}^{\infty} \alpha_i \|(I-P_i)z - (I-P_i)z'\|^2 \le$$

$$\le \langle z-z', (z - \sum_{1}^{\infty} \alpha_i P_i z) - (z' - \sum_{1}^{\infty} \alpha_i P_i z') \rangle$$

$$= \langle z-z', (I-P_\mathcal{K})z - (I-P_\mathcal{K})z' \rangle .$$

As $z$ and $z'$ are fixed points of $P$ the right hand member vanishes and therefore

$$(I-P_i)z = (I-P_i)z' , \quad \forall z, z' \in \mathcal{K} , \quad i = 1, 2, \ldots .$$

Thus for each $i$ there is a vector $u_i$ such that

(1.20) $\qquad (I-P_i)z = -u_i , \quad i = 1, 2, \ldots , \quad \forall z \in \mathcal{K} .$

Notice that the sum $\sum \alpha_i u_i$ is weakly convergent to zero. Then with help of (1.19) and (1.20) we deduce

$$P_i x = P_i(P_\mathcal{K} x) = -(I-P_i)P_\mathcal{K} x + P_\mathcal{K} x = P_\mathcal{K} x + u_i, \quad i = 1, 2, \ldots ,$$

and we have proved necessity.

As to the sufficiency it is almost immediate, for if $P_i x = P_K x + u_i$ and if $u = \sum_1^\infty \alpha_i u_i$ is perpendicular to $K$, then

$$\sum_1^\infty \alpha_i P_i x = P_K x + \sum_1^\infty \alpha_i u_i = P_K x + u = P_{K+u} x.$$

__Theorem 1.4__  Let $P_i$, $i = 1, 2, \ldots$ be a sequence, finite or infinite, of operators satisfying (1.5), and let $\{\alpha_i\}_1^\infty$ be a sequence of positive numbers with $\sum_1^\infty \alpha_i = 1$. Under these conditions the operator $x \to \sum_1^\infty \alpha_i P_i x$ is the complement of a projection if and only if there is a closed convex set $K$ and vectors $u_i$ with $\sum_1^\infty \alpha_i u_i$ perpendicular to $K$ such that $P_i x = (I - P_K) x - u_i$, $i = 1, 2, \ldots$.

__Proof:__ Change the equation $(I - P_K) x = \sum_1^\infty \alpha_i P_i x$ to $P_K x = \sum_1^\infty \alpha_i (I - P_i) x$ and apply Theorem 1.3 with $I - P_i$ in place of $P_i$.

Theorems 1.3 and 1.4 state that up to orthogonal translations projections and their complements are extremal points of the convex set of all mappings satisfying (1.5). This brings immediately to mind the Krein-Milman Theorem [4, I. p. 440] and suggests the possibility of expanding a large class of nonlinear operators in terms of projections.

__Corollary__  A sum of the form $\sum_1^\infty \alpha_i P_i$, where the $P_i$'s satisfy (1.5) and the $\alpha_i$'s are positive with $\sum_1^\infty \alpha_i < 1$, is a projection if and only if each $P_i$ maps the whole space into a single point, and it is the complement of a projection if each $P_i$ is a translation.

Proof: Let $P_0$ be the operator mapping the whole space into the origin, and let $\alpha_0 = 1 - \sum_1^\infty \alpha_i$. Then $\sum_1^\infty \alpha_i P_i = \sum_0^\infty \alpha_i P_i$, with $\sum_0^\infty \alpha_i = 1$. An application of the previous theorems yields the result at once.

## §2. Vertices and faces.

We begin this section with a few basic facts concerning dual cones.

**Definition 2.1** For any cone $C$ with vertex at $v$, the cone

(2.1) $$C^\perp = \{x^\perp | \langle x^\perp - v, x - v \rangle \leq 0, \quad \forall \; x \in C \},$$

is called the dual cone of $C$ (often also called "the polar cone", cf. [10]).

$C^\perp$ is a closed convex cone with vertex at $v$; the only point common to $C$ and $C^\perp$ is their vertex. If $C$ is a closed cone Lemma 1.1 tells us that $C^\perp$ is the set of points whose projections on $C$ is $v$. The operation $C \to C^\perp$ is obviously antitonic, that is, the larger $C$ the smaller $C^\perp$; for a linear subspace $C^\perp$ coincides with the orthogonal complement.

For any set $G$ let $\text{co } G$ denote the convex hull of $G$.

**Lemma 2.1** For any cone $C$,

(2.2) $$C^{\perp\perp} = \overline{\text{co } C}.$$

Proof: By the definitions of the perp operation, $C^{\perp\perp} \supset C$, and as $C^{\perp\perp}$ is closed and convex, $C^{\perp\perp} \supset \overline{\text{co } C}$. Let $K = \overline{\text{co } C}$, and for any $z \in C^{\perp\perp}$ let $y = z - P_K z + v$. Since $P_K z \in K$ and $K$ is a convex cone with vertex at $v$, the vector

$w = x+P_\mathcal{K} z-v = v + [(P_\mathcal{K} z-v)+(x-v)]$ belongs to $\mathcal{K}$ for every $x$ in $\mathcal{K}$, and therefore by (1.2),

$$\langle y-v, x-v \rangle = -\langle z-P_\mathcal{K} z, P_\mathcal{K} z -w \rangle \leq 0, \quad \forall\ x \in \mathcal{K},$$

and hence $y \in \mathcal{K}^\perp = (\overline{co\ \mathcal{C}})^\perp \subset \mathcal{C}$. Thus, as $z \in \mathcal{C}^{\perp\perp}$,

$$0 \geq \langle y-v, z-v \rangle = \langle z-P_\mathcal{K} z, z-v \rangle = \|z-P_\mathcal{K} z\|^2 + \langle z-P_\mathcal{K} z, P_\mathcal{K} z-v \rangle.$$

Since $v \in \mathcal{K}$, the last term on the right is nonnegative, and since so is the first, both vanish. Therefore $z = P_\mathcal{K} z \subset \overline{co\ \mathcal{C}}$, and $z$ being arbitrary in $\mathcal{C}^{\perp\perp}$, $\mathcal{C}^{\perp\perp} \subset \overline{co\ \mathcal{C}}$, concluding the proof.

<u>Corollary 1</u>  If $\mathcal{C}$ is a closed convex cone, $\mathcal{C}$ and $\mathcal{C}^\perp$ are the duals of each other.

<u>Corollary 2</u>  If $\{\mathcal{C}_\iota\}$ is a class of closed convex cones with a common vertex $v$, then

(2.3) $$\left(\bigcap \mathcal{C}_\iota \right)^\perp = \overline{co \bigcup \mathcal{C}_\iota^\perp}.$$

Proof: Assuming, as we may, that the vertex is at the origin, we have

$$\bigcap \mathcal{C}_\iota \subset \mathcal{C}_\iota \Rightarrow (\bigcap \mathcal{C}_\iota)^\perp \supset \mathcal{C}_\iota^\perp \Rightarrow (\bigcap \mathcal{C}_\iota)^\perp \supset \overline{co \bigcup \mathcal{C}_\iota^\perp} \Rightarrow$$

$$\Rightarrow \bigcap \mathcal{C}_\iota \subset (\overline{co \bigcup \mathcal{C}_\iota^\perp})^\perp \subset \mathcal{C}_\iota^{\perp\perp} = \mathcal{C}_\iota$$

$$\Rightarrow \bigcap \mathcal{C}_\iota = (\overline{co \bigcup \mathcal{C}_\iota^\perp})^\perp \Rightarrow (\bigcap \mathcal{C}_\iota)^\perp = \overline{co \bigcup \mathcal{C}_\iota^\perp}.$$

<u>Definition 2.2</u>  The dimension and deficiency (or codimension) of a convex set $\mathcal{K}$ is defined as the dimension and deficiency respectively of the closed linear manifold spanned by $\mathcal{K}$.

With this definition it turns out that the deficiency of a cone $\mathcal{C}$ is equal to the dimension of the largest linear manifold contained in $\mathcal{C}^\perp$.

**Lemma 2.2** For any given couple $C_1$ and $C_2$ of dual closed convex cones with vertex at the origin, any point $x \in \mathcal{H}$ can be expressed in one and only one way as the sum of two orthogonal vectors, one in $C_1$ and the other in $C_2$, namely

(2.4) $$x = P_{C_1} x + P_{C_2} x \quad , \quad [10].$$

**Proof:** Let us show uniqueness first. If $x = x_1 + x_2$, $x_1 \in C_1$, $x_2 \in C_2$, $\langle x_1, x_2 \rangle = 0$,

then

$$\langle x - x_1, x_1 - x_1' \rangle = \langle x_2, x_1 - x_1' \rangle = -\langle x_2, x_1' \rangle \geq 0, \forall x_1' \in C_1,$$

and hence by (1.2), $P_{C_1} x = x_1$; similarly $P_{C_2} x = x_2$.

Next let us show that in fact $x = P_{C_1} x + P_{C_2} x$ and that $P_{C_1} x$ and $P_{C_2} x$ are orthogonal. By (1.2),

$$0 \leq \langle x - P_{C_1} x, P_{C_1} x - x_1' \rangle, \quad \forall x_1' \in C_1.$$

Replacing $x_1'$ by $x_1' + P_{C_1} x$, which also belongs to $C_1$, one gets

$$0 \leq -\langle x - P_{C_1} x, x_1' \rangle, \quad \forall x_1' \in C_1,$$

and so $x - P_{C_1} x \in C_1^\perp = C_2$. Moreover, if we set $x_1' = 0$ in the first of these two last inequalities and $x_1' = P_{C_1} x$ in the last, we obtain

$$0 \leq \langle x - P_{C_1} x, P_{C_1} x \rangle, \quad 0 \leq -\langle x - P_{C_1} x, P_{C_1} x \rangle,$$

and so $\langle x - P_{C_1} x, P_{C_1} x \rangle = 0$; similarly $\langle x - P_{C_2} x, P_{C_2} x \rangle = 0$.

Now if $x_1 = P_{C_1}x$, $x_2 = x - P_{C_1}x$, then $x = x_1 + x_2$, $x_1 \in C_1$, $x_2 \in C_2$ and $\langle x_1, x_2 \rangle = 0$. By uniqueness, $x_2 = x - P_{C_1}x = P_{C_2}x$, and the proof is complete.

Corollary 1   For a closed convex set $\mathcal{K}$, $I - P_{\mathcal{K}}$ is a projection if and only if $\mathcal{K}$ is a cone with vertex at the origin.

Proof:   The condition is sufficient by the above Lemma. Necessity follows from the fact that the range of $I - P_{\mathcal{K}}$ is a cone with vertex at $0$, which in turn is a consequence of the identity: $P_{\mathcal{K}}(x + t(x - P_{\mathcal{K}}x)) = P_{\mathcal{K}}x$, $\forall t \geq 0$.

Corollary 2   If $C$ is a closed convex cone with vertex at the the origin $P_C$ is positive homogeneous and

(2.4')  $$\langle x, P_C x \rangle = \|P_C x\|^2.$$

Definition 2.3   The support cone of a closed convex set $\mathcal{K}$ at a point $x$ is the smallest closed convex cone with vertex at the origin containing $\mathcal{K} - x$; it is denoted $S_{\mathcal{K}}(x)$.

This notion is mainly used for points $x$ in $\mathcal{K}$. It is clear that $S_{\mathcal{K}}(x)$ is the closure of the set of all projecting rays:

$$S_{\mathcal{K}}(x) = \overline{\{u \mid u = t(x' - x), \forall x' \in \mathcal{K}, \forall t \geq 0\}}.$$

The following characterization of $S_{\mathcal{K}}(x)$ when $x \in \mathcal{K}$ is often useful.

Lemma 2.3   If $x \in \mathcal{K}$, then $u \in S_{\mathcal{K}}(x)$ if and only if for each $t > 0$ there is an $x(t) \in \mathcal{K}$ such that

(2.5)   $x + tu = x(t) + o(t)$   as $t \to 0$.

Proof:   Any $u$ satisfying (2.5) is the limit of projecting rays and hence belongs to $S_{\mathcal{K}}(x)$. Conversely, if $u \in S_{\mathcal{K}}(x)$,

$u \neq 0$, then there is a sequence $x_n \in \mathcal{K}$ and positive numbers $t_n$ such that $u - (x_n-x)/t_n \to 0$, where by changing the $t_n$ if necessary we may assume that the sequence $\{x_n\}_1^\infty$ is bounded. Since $u \neq 0$, the $t_n$'s must remain bounded; if the $t_n$'s have a limit point $t_0 > 0$, a subsequence of $\{x_n\}$ converges to some $x' \in \mathcal{K}$ and $u = \frac{x'-x}{t_0}$, and $x + tu = (1 - \frac{t}{t_0})x + \frac{t}{t_0}x' = x(t) \in \mathcal{K}$ for $0 \leq t \leq t_0$, in which case (2.5) is satisfied; if $t_n \to 0$ then if we set $x(t) = P_\mathcal{K}(x+tu)$, we have for $0 < t \leq t_n$,

$$\frac{\|x+tu - x(t)\|}{t} \leq \frac{\|x+tu - [(1+t/t_n)x + (t/t_n)x(t_n)]\|}{t} =$$

$$\frac{\|x + t_n u - x(t_n)\|}{t_n} \leq \frac{\|x + t_n u - x_n\|}{t_n}$$

and so,

$$\overline{\lim_{t \to 0}} \frac{\|x+tu - x(t)\|}{t} \leq \frac{\|x + t_n u - x_n\|}{t}.$$

As the right hand member tends to zero as $n \to \infty$,

$$\lim_{t \to 0} \frac{\|x+tu - x(t)\|}{t} = 0.$$

Q. E. D.

Note that if $x \in \mathcal{K}_1 \cap \mathcal{K}_2$, then $\mathcal{S}_{\mathcal{K}_1 \cap \mathcal{K}_2}(x) \subset \mathcal{S}_{\mathcal{K}_1}(x) \cap \mathcal{S}_{\mathcal{K}_2}(x)$; the opposite relation not always holds. The dimension and deficiency of any $\mathcal{S}_\mathcal{K}(x)$ for $x \in \mathcal{K}$ are the same as the dimension and deficiency of $\mathcal{K}$.

Our next lemma simply says that as far as the set of points projecting on a fixed point $x \in \mathcal{K}$ is concerned it does not make any difference if $\mathcal{K}$ is replaced by the smallest closed cone containing $\mathcal{K}$ with vertex at $x$.

## Lemma 2.4
If $x \in K$

(2.6)
$$P_K^{-1} x = S_K^{\perp}(x) + x.$$

Proof: $P_K^{-1} x$ is the set of points in space whose closest point in $K$ is $x$, whereas $S_K^{\perp}(x) + x$ is the set of points whose closest point in $S_K(x) + x$ is $x$. Naturally, since $K \subset S_K(x) + x$, $S_K^{\perp}(x) + x \subset P_K^{-1} x$. On the other hand if $P_K x_0 = x$, $K$ is contained in the halfspace
$\Pi_{x_0 - x}^{-}(x) = \{y \mid \langle y-x, x_0 - x \rangle \leq 0\}$, which in turn contains $S_K(x) + x$. But as $x$ is the point in $\Pi_{x_0 - x}^{-}(x)$ closest to $x_0$, it is also the point in $S_K(x) + x$ closest to $x_0$, and in consequence $x_0 \in (S_K(x) + x)^{\perp} = S_K^{\perp}(x) + x$. By the arbitrariness of $x_0$ in $P_K^{-1} x$, $P_K^{-1} x \subset S_K^{\perp}(x) + x$.

## Definition 2.4
For any point $x$ of a closed convex set $K$, the set

(2.7)
$$U_K(x) = P_K^{-1} x - x = S_K^{\perp}(x)$$

is called the vertex of $K$ at $x$; any $u \in U_K(x)$ is said to be a normal to $K$ at $x$.

(2.8)
$$\{u \in U_K(x)\} \iff \{x = P_K(x+u)\}.$$

A few comments are in order to clarify the above definition and nomenclature. The "size" of the support cone $S_K(x)$ indicates how fully $K$ spreads itself around the point $x$. If $S_K(x)$ is the whole space then the projection cone is dense in $H$ and $x$ appear as an "interior" point, and, as $U_K(x)$ reduces to the zero vector, no point other than $x$ itself projects onto $x$. If $S_K(x)$ is a halfspace $U_K(x)$ is one dimensional and there is only one normal direction at $x$; thus $K$ has a tangent plane there, the point is "smooth" or if one prefers, "flat". On the other hand if $S_K(x)$ is neither a whole or a halfspace then $K$ has a "peak" or a "vertex" at $x$, the sharper the smaller $\mathcal{S}_\mathcal{K}(x)$, from which more than one normal issue filling a larger and larger cone $\mathcal{V}_\mathcal{K}(x)$. Thus $\mathcal{V}_\mathcal{K}(x)$ describes the nature of the vertex whereas $x$ only indicates its position. The distinction between these two elements is very essential for our purposes, for often we will be comparing vertices at different points.

<u>Lemma 2.5</u>  The family of sets $\{x + \mathcal{V}_\mathcal{K}(x)\}_{x \in \mathcal{K}}$ is a collection of closed convex cones covering $\mathcal{H}$, having the properties:

a.   $x + \mathcal{V}_\mathcal{K}(x)$ is a closed convex cone for $\forall x \in \mathcal{K}$,

b.   the sets $x + \mathcal{V}_\mathcal{K}(x)$ are disjoint sets,

c.   $\overline{\lim_{x \to y}} (x + \mathcal{V}_\mathcal{K}(x)) \subset y + \mathcal{V}_\mathcal{K}(y)$,

d.   $\bigcup_{x \in \mathcal{K}} (x + \mathcal{V}_\mathcal{K}(x)) = \mathcal{H}$.

<u>Proof:</u>  All four properties are immediate consequences of the foregoing through the fact that $x + \mathcal{V}_\mathcal{K}(x)$ is the inverse image of $x$ by $P_\mathcal{K}$. We recall that

$$\overline{\lim_{x \to y}} (x + \mathcal{V}_\mathcal{K}(x)) = \bigcap_{\varepsilon > 0} \overline{\left( \bigcup_{0 < \|x-y\| \leq \varepsilon} (x + \mathcal{V}_\mathcal{K}(x)) \right)}.$$

But for the fact that the sets are not compact, $\{x + \mathcal{V}_\mathcal{K}(x)\}_{x \in \mathcal{K}}$ would be "an upper semicontinuous collection" [28, Ch. VII].

<u>Lemma 2.6</u>  For any sequence $\{x_i\}_1^\infty$ in $\mathcal{K}$ and any sequence $\{\alpha_i\}_1^\infty$ of positive real numbers with $\sum_1^\infty \alpha_i = 1$ such that $\sum_1^\infty \alpha_i x_i$ converges weakly,

(2.9) $\qquad \mathcal{V}_\mathcal{K}(\sum_1^\infty \alpha_i x_i) = \bigcap_{i=1}^\infty \mathcal{V}_\mathcal{K}(x_i)$.

Any vector $u$ in the above set is perpendicular to the convex hull of $\{x_i\}_1^\infty$.

PROJECTIONS ON CONVEX SETS

Proof: The set of points in $\mathcal{K}$ admitting $u$ as a normal is the set of x's such that $x = P_{\mathcal{K}}(x+u)$, which, as the set of fixed points of a mapping satisfying (1.5), is closed and convex by Lemma 1.6. Hence, if $u \in \mathcal{V}_{\mathcal{K}}(x_i)$, $i = 1, 2, \ldots$ then $u \in \mathcal{V}_{\mathcal{K}}(\sum_1^\infty \alpha_i x_i)$, and $\mathcal{V}_{\mathcal{K}}(\sum_1^\infty \alpha_i x_i) \supset \bigcap_1^\infty \mathcal{V}_{\mathcal{K}}(x_i)$. Now let us prove the converse, that is, if $u \in \mathcal{V}_{\mathcal{K}}(\sum_1^\infty \alpha_i x_i)$ then $u \in \mathcal{V}_{\mathcal{K}}(x_i)$, $i = 1, 2, \ldots$. To this end for each $j$ pick a number $t_j > 1$ such that $(1-t_j) + t_j \alpha_j > 0$; then the numbers

$$\alpha_i^{(j)} = \begin{cases} t_j \alpha_i, & i \neq j, \\ (1-t_j) + t_j \alpha_j, & i = j, \end{cases}$$

are all positive, $\sum_{i=1}^\infty \alpha_i^{(j)} = 1$, and the point $x_j' = \sum_{i=1}^\infty \alpha_i^{(j)} x_i$ belongs to $\mathcal{K}$; note that $\sum_{i=1}^\infty \alpha_i x_i - x_j' = (1-t_j)(\sum_1^\infty \alpha_i x_i - x_j)$. Since $u \in \mathcal{V}_{\mathcal{K}}(\sum_1^\infty \alpha_i x_i)$ we have

$$\langle u, \sum_1^\infty \alpha_i x_i - x \rangle \geq 0, \quad \forall x \in \mathcal{K}.$$

In particular, for $x = x_j$,

$$0 \leq \langle u, \sum_1^\infty \alpha_i x_i - x_j \rangle,$$

whereas, for $x = x_j'$,

$$0 \leq \langle u, \sum_1^\infty \alpha_i x_i - x_j' \rangle = (1-t_j) \langle u, \sum_1^\infty \alpha_i x_i - x_j \rangle.$$

Since $1-t_j < 0$, it follows that $\langle u, \sum_1^\infty \alpha_i x_i - x_j \rangle = 0$, and then

$$\langle u, x_j - x \rangle = \langle u, x_j - \sum_1^\infty \alpha_i x_i \rangle + \langle u, \sum_1^\infty \alpha_i x_i - x \rangle =$$

$$= \langle u, \sum_1^\infty \alpha_i x_i - x \rangle \geq 0, \quad \forall \ x \in \mathcal{K},$$

which says that $u \in \mathcal{U}_\mathcal{K}(x_j)$. This being true for every $j$, (2.9) is proved. The last part of the lemma follows from the proven relations

$$\langle u, \sum_1^\infty \alpha_i x_i - x_j \rangle = 0, \quad j = 1, 2, \ldots \quad .$$

<u>Definition 2.5</u>  A point $x$ of a closed convex $\mathcal{K}$ is said to be an "inner" point if $x + \mathcal{S}_\mathcal{K}(x)$ coincides with the closed linear manifold spanned by $\mathcal{K}$. The set of all inner points of $\mathcal{K}$ is called the "core" and is denoted $\mathcal{K}^o$; all other points form the shell of $\mathcal{K}$.

The term "inner" has been used in the literature to designate points of a convex set with the property that any ray out of them meets the set in more than one point. This is a purely algebraic notion considerably more restrictive than the one used here (cf [27, Def. 1.9]).

It is clear from this definition that the notion of inner point and hence also that of shell point is at once intrinsic (i.e. depends on $\mathcal{K}$ alone and not on the way $\mathcal{K}$ is embedded in space) and independent of the particular metric defining the topology in $\mathcal{H}$.

To see what the above means in terms of the $\mathcal{U}_\mathcal{K}(x)$'s, remark that the largest linear space contained in $\mathcal{U}_\mathcal{K}(x)$ is the subspace $\mathcal{S}$ orthogonal to the linear manifold spanned by $\mathcal{K}$, and hence is independent of $x$. In fact, $\mathcal{U}_\mathcal{K}(x) = \mathcal{S} + \tilde{\mathcal{U}}_\mathcal{K}(x)$, where $\tilde{\mathcal{U}}_\mathcal{K}(x)$ is the vertex of $\mathcal{K}$ at $x$ relatively to the closed linear manifold spanned by $\mathcal{K}$. Thus at an inner point $\tilde{\mathcal{U}}_\mathcal{K}(x) = 0$ and $\mathcal{U}_\mathcal{K}(x)$ is linear. The above definition can be rephrased by saying that, <u>relatively to the closed manifold spanned by $\mathcal{K}$, shell points are projections of points different from themselves whereas inner points are not.</u>

**Lemma 2.7**   For any closed convex set $K$, $K^o$ is a convex set; if $K$ is separable $K^o$ is dense in $K$. If the dimension of $K$ is finite, then $K^o$ coincides with the interior of $K$ relatively to the linear manifold spanned by $K$.

**Proof:**   It is obviously sufficient to prove the Lemma under the assumption that $K$ spans the whole space $H$. If $x_1$ and $x_2$ are two inner points, then $U_K(x_1) = \{0\}$ and $U_K(x_2) = \{0\}$, and by (2.9) $U_K(\alpha_1 x_1 + \alpha_2 x_2) = \{0\}$, and all points in the segment joining $x_1$ and $x_2$ are inner points. Therefore $K^o$ is convex.

If, $K$ being separable, $K^o$ were not dense in $K$, there would be a point $x_0 \in K$ and a closed ball $B_\rho(x_0)$ around it such that $K \cap B_\rho(x_0)$ has no points in $K^o$. Let $\{x_i\}_1^\infty \subset K \cap B_\rho(x_0)$ be dense in $K \cap B_\rho(x_0)$. Then for any choice of the positive sequence $\{\alpha_i\}_1^\infty$ with $\sum_1^\infty \alpha_i = 1$, the point $\sum_1^\infty \alpha_i x_i$ (this sum is convergent) belongs to $K \cap B_\rho(x_0)$, and since $\sum_1^\infty \alpha_i x_i \in K^o$ there is a nonvanishing $u \in U_K(\sum_1^\infty \alpha_i x_i)$. By Lemma 2.6 this $u$ is perpendicular to the convex hull of $\{x_i\}_1^\infty$ and hence to $K \cap B_\rho(x_0)$, which therefore lies on a hyperplane perpendicular to $u$. But if $K \cap B_\rho(x_0)$ lies on a hyperplane, so does $K$, which is impossible by hypotheses. Hence $K^o$ is dense in $K$. Interior points are inner points regardless of the dimension, so to prove the last part of the lemma it must be shown that if the dimension of $K$ is finite, say $n$, then any inner point $x_0$ is an interior point. To do this we begin by picking $n+1$ points in space $y_1, y_2, \ldots, y_{n+1}$ such that $x_0$ is interior to the $n$-simplex having those points as vertices. Then we remark that since $x_0$ is an inner point and the projecting cone from $x_0$ is dense in $H$, there are, for any given $\varepsilon > 0$, $n+1$ points $x_1', x_2', \ldots, x_{n+1}'$ in $K$ and $n+1$ positive numbers $\lambda_1, \lambda_2, \ldots, \lambda_{n+1}$ such that
$\|\lambda_i(x_i' - x_0) - (y_i - x_0)\| \leq \varepsilon$, $i = 1, 2, \ldots, n+1$. If $\lambda = \max(\lambda_1, \lambda_2, \ldots, \lambda_{n+1})$ then the points $x_i = (1 - \lambda_i/\lambda)x_0 + (\lambda_i/\lambda)x_i'$

all belong to $\mathcal{K}$ and $\lambda(x_i - x_0) = \lambda_i(x_i' - x_0)$, $i = 1, 2, \ldots, n+1$. Hence $\|(x_0 + \lambda(x_i - x_0)) - y_i\| = \|\lambda_i(x_i' - x_0) - (y_i - y_0)\| \le \varepsilon$ $i = 1, 2, \ldots, n+1$. Now, for $\varepsilon$ sufficiently small the n-simplex of vertices $x_0 + \lambda(x_i - x_0)$ contains $x_0$ in its interior, and in consequence so does the simplex of vertices $x_1, \ldots, x_{n+1}$, obtained from the previous set by a homothecy with regard to $x_0$. Hence, as this last simplex is fully contained in $\mathcal{K}$, the proof concludes.

<u>Corollary</u> Any separable closed convex set has a nonempty core.

<u>Lemma 2.8</u> If a closed convex set $\mathcal{K}$ has an interior point then all inner points are interior points, and $\mathcal{K}^0 = \text{int } \mathcal{K}$.

<u>Proof:</u> Let $z$ be an interior point and let $\mathcal{B}_\rho(z)$ be an open ball of radius $\rho$ about it. Now, the cone projecting $\mathcal{K}$ from any inner point $y$ being dense in space, there is an $x \in \mathcal{K}$, $x \ne y$, such that the ray $\{y + t(x-y)\}_{t > 0}$ pierces the open ball $\mathcal{B}_\rho(2y-z) = 2y - \mathcal{B}_\rho(z)$. That is, there is a $z_1 \in \mathcal{B}_\rho(z)$ and a positive $t$ such that $y + t(x-y) = 2y - z_1$, and then

$$y = \frac{x}{1+t} + \frac{t}{1+t} z_1 \in \text{int co}(\{x\} \cup \mathcal{B}_\rho(z)).$$

Hence, since $\text{co}(\{x\} \cup \mathcal{B}_\rho(z)) \subset \mathcal{K}$, $y \in \text{int } \mathcal{K}$.

<u>Definition 2.6</u> A closed convex set in Hilbert space is said to be a convex body if it has a nonempty interior relatively to the smallest closed linear manifold containing it; a total convex body is a convex body spanning the whole space.

In the literature convex bodies are referred to as "convex sets with relative interior", the term convex body being reserved for what we call total convex body [27, Def. 1.10]. In finite dimensions every closed convex set is a convex body by Lemma 2.7. In any case the core of a convex body coincides with its interior relatively to the closed linear manifold it spans.

## PROJECTIONS ON CONVEX SETS

<u>Lemma 2.9</u>  A closed convex set $\mathcal{K}$ with a nonempty core is a convex body if and only if its shell is closed.

<u>Proof:</u>  Necessity is obvious, since for a convex body the shell is the boundary relatively to the smallest closed linear manifold containing it. To prove sufficiency one may assume without loss of generality that $\mathcal{K}$ spans the space $\mathcal{H}$. In such a case, any boundary point being the limit of points outside $\mathcal{K}$ it is also the limit of their projection on $\mathcal{K}$, and hence, if the shell is closed, belongs to the shell, and so the boundary is contained in the shell. Since on the other hand the interior is contained in the core, boundary and interior coincide with shell and core respectively.

<u>Lemma 2.10</u>  The shell of a closed convex set which is not a convex body is dense in the set.

<u>Proof:</u>  The proof follows from the facts: a. The shell is dense in the boundary relatively to the smallest closed linear manifold; b. if a convex set is not a convex body all its points are boundary points.

This lemma holds in normed spaces for $\mathcal{K}$'s admitting a projection [4, II, p. 452].

<u>Lemma 2.11</u>  For any closed convex cone $\mathcal{C}$ with vertex at the origin in a separable Hilbert space, different from a linear subspace,

$$(2.10) \qquad 0 \notin -\mathcal{C}^\circ \cap (\mathcal{C}^\perp)^\circ \neq \emptyset .$$

<u>Proof:</u>  Pick $v^\circ \neq 0$ in $\mathcal{C}^\circ$, and with help of (2.4) write

$$-v^\circ = u + u^\perp, \quad u \in \mathcal{C}, \quad u^\perp \in \mathcal{C}^\perp, \quad \langle u, u^\perp \rangle = 0 .$$

It follows

$$-(v^\circ + u) = u^\perp ,$$

and as the left hand member belongs to $-\mathcal{C}^\circ$ and the right

to $C^\perp$,
$$-C^\circ \cap C^\perp \neq \phi.$$

The origin is not a point of the set above, for if it were it would be an inner point of $C$, and $C^\perp$ — and therefore $C$ — would be a linear space (cf. remarks following Def. 2.5). By symmetry we also have

$$-C \cap (C^\perp)^\circ \neq \phi.$$

Now let us take a vector $w_1 \in -C^\circ \cap C^\perp$ and a vector $w_2 \in -C \cap (C^\perp)^\circ$, and let us consider their sum $w = w_1 + w_2$. From $w_1 \in -C^\circ$, $w_2 \in -C$, one deduces $w = w_1 + w_2 \in -C^\circ$, and from $w_1 \subset C^\perp$, $w_1 \in (C^\perp)^\circ$, $w = w_1 + w_2 \in (C^\perp)^\circ$. Thus the vector $w$ belongs to $-C^\circ \cap (C^\perp)^\circ$ and since it cannot vanish the lemma is proved.

For linear spaces $-m^\circ \cap (m^\perp)^\circ = \{0\}$, so (2.10) characterizes nonlinear cones.

<u>Corollary 1</u>  In a separable space any closed convex cone $C$ with vertex at $0$, different from a linear space, has a vector $u$ such that

$$\langle u, u' \rangle \geq 0, \quad \forall u' \in C,$$

$$\langle u, u' \rangle = 0 \iff -u' \in C,$$

and

$$\langle u, v \rangle \leq 0, \quad \forall v \in C^\perp,$$

$$\langle u, v \rangle = 0 \iff -v \in C^\perp.$$

<u>Corollary 2</u>  Any total convex body in a separable Hilbert space has at each point of its shell a normal making an acute angle with any other normal at the point.

**Lemma 2.12**  Each point $x_0$ in the shell of total convex body $\mathcal{K}$ in $\mathbb{R}^n$ has a neighborhood $\mathcal{U}$ and a support hyperplane $\Pi_{u_0}(x_0)$ such that $(\mathcal{U}_{\mathcal{K}}(x) + x) \cap \Pi_{u_0}(x_0)$ is nonempty and bounded for $\forall x \in \mathcal{U}$.

**Proof:** Take as $u_0$ any normal at $x_0$ making an acute angle with any normal in $\mathcal{U}_{\mathcal{K}}(x_0)$ (Lemma 2.11, Cor. 2). The set $(\mathcal{U}_{\mathcal{K}}(x) + x) \cap \Pi_{u_0}(x_0)$ is empty only if $\langle u, u_0 \rangle \leq 0$, $\forall u \in \mathcal{U}_{\mathcal{K}}(x)$, and, if not empty, is unbounded only if it contains vectors parallel to $\Pi_{u_0}(x_0)$, that is, vectors perpendicular to $u_0$. In either case there is in $\mathcal{U}_{\mathcal{K}}(x)$ a unit vector $u_x$ such that $\langle u_x, u_0 \rangle \leq 0$. Now, points with this property cannot converge to $x_0$ because if they did there would exist a sequence of points $\{x_n\}_1^\infty$ and a sequence of corresponding unit normals $\{u_{x_n}\}_1^\infty$ such that $x_n \to x_0$ and $u_{x_n} \to w$, $\langle u_{x_n}, u_0 \rangle \leq 0$, and $\langle u, u_0 \rangle \leq 0$, which is impossible because by Lemma 2.5 $u \in \mathcal{U}_{\mathcal{K}}(x_0)$. Therefore there is a neighborhood $\mathcal{U}$ of $x_0$ with the mentioned property.

**Remark**  If $\mathcal{U}$ is open then so is the set $\mathfrak{S}$ on $\Pi_{u_0}(x_0)$ projecting on $\mathcal{U}$ and $\{(\mathcal{U}_{\mathcal{K}}(x) + x) \cap \Pi_{u_0}(x_0)\}_{x \in \mathcal{U}}$ is an upper semicontinuous decomposition of $\mathfrak{S}$ into weakly compact convex sets.

The following is a very instructive example:

**Example**  Let $\mathcal{K}$ be the positive cone

$$\mathcal{K} = \{x \mid \langle x, \varphi_\alpha \rangle \geq 0, \ \alpha \in I\}$$

relatively to a orthonormal basis $\{\varphi_\alpha\}_{\alpha \in I}$ in $\mathcal{H}$. Let us see first that $\mathcal{K}^\circ = \{x \mid \langle x, \varphi_\alpha \rangle > 0, \ \alpha \in I\}$; in particular that $\mathcal{K}^\circ$ will be empty if $\mathcal{H}$ is not separable, because a point has at most a countable number of nonvanishing components. In fact if $\langle x, \varphi_\alpha \rangle > 0$, $\alpha \in I$, and if $P_{\mathcal{K}} y = x$ then

$$\langle y-x, x-x' \rangle \geq 0, \qquad \forall x' \in \mathcal{K}.$$

This holds in particular for $x' = x \pm \langle x, \varphi_\alpha \rangle \varphi_\alpha$, and so $\pm \langle x, \varphi_\alpha \rangle \langle y-x, \varphi_\alpha \rangle \geq 0$, $\alpha \in I$, which in turn implies $\langle y-x, \varphi_\alpha \rangle = 0$, $\alpha \in I$. Thus $y = x$, and $x$ is an inner point. On the other hand if for some $\alpha_0$, $\langle x, \varphi_{\alpha_0} \rangle = 0$, then $\langle -\varphi_{\alpha_0}, x-x' \rangle = \langle \varphi_{\alpha_0}, x' \rangle \geq 0$, $\forall x' \in \mathcal{K}$, and $P_\mathcal{K}(x-\varphi_{\alpha_0}) = x$. Hence $x$ is a shell point.

Note the following property of points in $\mathcal{K}^\circ$ in the separable case: If $x \in \mathcal{K}^\circ$ and $\{\xi_i\}_1^\infty$ is a sequence of positive numbers such that

$$\xi_i / \langle x, \varphi_i \rangle = M_i \to +\infty, \quad \sum_1^\infty \xi_i^2 < \infty$$

(in this case $I = \{1, 2, \ldots\}$),

then

$$\langle x-tx_0, \varphi_i \rangle = \langle x, \varphi_i \rangle - t\xi_i = \langle x, \varphi_i \rangle (1-tM_i) \begin{cases} < 0, & t > M_i^{-1}, \\ \geq 0, & t \leq 0, \end{cases}$$

and the open half line $\{x-tx_0\}_{t > 0}$ has no points in $\mathcal{K}$ whereas the closed half line $\{x-tx\}_{t \leq 0}$ is entirely contained in it.

One sees from this that contrary to what happens in finite dimensional case an inner point can be the end point of a maximal segment entirely contained in $\mathcal{H}$. One sees also that the shell might be dense in $\mathcal{K}$ even if $\mathcal{K}^\circ$ is nonempty.

<u>Definition 2.7</u>  A half linear manifold is the intersection of a linear manifold with a halfspace not containing it.

It is easy to see that a closed convex set is a half linear manifold if and only if its shell is a linear manifold.

<u>Lemma 2.13</u>  Unless a closed convex set $\mathcal{K}$ is a half linear manifold or a linear manifold, $\mathcal{K}$ coincides with the closed convex hull of its shell.

We may assume without loss that $\mathcal{K}$ spans the whole space.

**Proof:** Let $\tilde{\mathcal{K}}$ be the closed convex hull of the shell of $\mathcal{K}$; obviously $\tilde{\mathcal{K}} \subset \mathcal{K}$. We shall see that, with the exception of the special cases mentioned in the Lemma, $\mathcal{K}$ and $\tilde{\mathcal{K}}$ have the same support hyperplanes and same normals. Let $\Pi_{u_0}(x_0)$ be the support hyperplane of $\mathcal{K}$ at $x_0 \in \mathcal{K}$ perpendicular to the normal $u_0$. Then $\tilde{\mathcal{K}} \subset \mathcal{K} \subset \Pi_{u_0}(x_0)$, and as $x_0 \in \tilde{\mathcal{K}}$, $\Pi_{u_0}(x_0)$ is a support hyperplane of $\tilde{\mathcal{K}}$ at $x_0$. Conversely, let $\Pi_{\tilde{u}_0}(\tilde{x}_0)$ be a support hyperplane of $\tilde{\mathcal{K}}$ at $\tilde{x}_0$. If $\tilde{x}_0 + t\tilde{u}_0 \in \mathcal{K}$, $\forall t \geq 0$, then for any $x$, and any $\alpha$, $0 < \alpha < 1$,

$$\langle x - P_{\mathcal{K}} x, \; P_{\mathcal{K}} x - (\alpha P_{\mathcal{K}} x + (1-\alpha)(\tilde{x}_0 + t\tilde{u}_0))\rangle \geq 0$$

that is

$$(1-\alpha)\langle x - P_{\mathcal{K}} x, P_{\mathcal{K}} x - \tilde{x}_0 \rangle - (1-\alpha)t\langle x - P_{\mathcal{K}} x, \tilde{u}_0 \rangle \geq 0 \; ,$$

whence dividing by $(1-\alpha)t$ and letting $t \to \infty$ it follows

$$\langle x - P_{\mathcal{K}} x, \tilde{u}_0 \rangle \leq 0, \qquad \forall \; x \in \mathcal{H} \; .$$

Now, if $\mathcal{K} \neq \mathcal{H}$ there are $x$'s not in $\mathcal{K}$ and for any such $x$, $P_{\mathcal{K}} x \in \tilde{\mathcal{K}}$, and in consequence $P_{\mathcal{K}} x = P_{\tilde{\mathcal{K}}} x$. Inequality (1.2) then yields

$$\langle P_{\mathcal{K}} x - \tilde{x}_0, \tilde{u}_0 \rangle = -\langle \tilde{x}_0 + \tilde{u}_0 - P_{\tilde{\mathcal{K}}}(\tilde{x}_0 + \tilde{u}_0), P_{\tilde{\mathcal{K}}}(\tilde{x}_0 + \tilde{u}_0) - P_{\tilde{\mathcal{K}}} x \rangle$$

$$\leq 0 \; ,$$

and adding up these two inequalities,

$$\langle x - \tilde{x}_0, \tilde{u}_0 \rangle \leq 0, \qquad \forall \; x \in \mathcal{K},$$

which says that $\Pi_{\tilde{u}}(\tilde{x}_0) \subset \mathcal{K}$. This means that $\mathcal{K}$ is a half-space, because any convex set different from $\mathcal{H}$ containing a halfspace must necessarily be a halfspace. Thus, if $\mathcal{K}$ is neither the whole space nor a halfspace, for any support hyperplane $\Pi_{\tilde{u}_0}(\tilde{x}_0)$ to $\tilde{\mathcal{K}}$ there is a $t_0 \geq 0$ such that $\tilde{x}_0 + t_0 \tilde{u}_0 \notin \mathcal{K}$. But then $P_{\mathcal{K}}(\tilde{x}_0 + t_0\tilde{u}_0) \in \tilde{\mathcal{K}}$ and

$\tilde{x}_0 = P_{\tilde{\mathcal{K}}}(\tilde{x}_0 + t\,\tilde{u}_0) = P_{\mathcal{K}}(\tilde{x}_0 + t_0\tilde{u}_0)$. Thus $\Pi_{\tilde{u}_0}^-(\tilde{x}_0)$ is a support hyperplane of $\mathcal{K}$ through $\tilde{x}_0$, and $\tilde{u}_0$ is normal to $\mathcal{K}$. In conclusion, the family of halfspaces $\Pi_u^-(x)$ associated with support hyperplanes is the same for both $\mathcal{K}$ and $\tilde{\mathcal{K}}$, and in consequence $\mathcal{K} = \tilde{\mathcal{K}}$.

If $\mathcal{K}$ is bounded the Krein-Milman Theorem says that it is enough to use the extremal points in the shell to generate $\mathcal{K}$; we shall return to this question later.

The following is a rather evident property of the shell:

<u>Lemma 2.14</u>   With exception of the case when a closed convex set $\mathcal{K}$ is the part of a linear manifold between two parallel hyperplanes, the shell of $\mathcal{K}$ is connected.

<u>Proof:</u>   As usual we may work under the assumption that $\mathcal{K}$ generates the whole space. Let $x_1$ and $x_2$ be two points in the shell of $\mathcal{K}$ belonging to different connected components, and let $u_1$ and $u_2$ be two respective unit normals; we shall show that $u_1 + u_2 = 0$. First of all notice that $u_1 \neq u_2$ because if not then by Lemma 2.6 the segment joining $x_1$ and $x_2$ would belong to the shell of $\mathcal{K}$, against the assumption that $x_1$ and $x_2$ belong to different components. Let us assume for contradiction that $u_1 \neq -u_2$. In such a case, since we also must have $u_1 \neq u_2$, $\mathcal{K}$ would be contained in the "wedge" $\{x \mid \langle x, u_1 \rangle \leq \langle x_1, u_1 \rangle, \langle x, u_2 \rangle \leq \langle x_2, u_2 \rangle\}$ and any continuous curve outside this wedge joining $x_1$ and $x_2$ would project onto a continuous curve on the shell joining the two points, which again is impossible. Thus unit normals to $\mathcal{K}$ take one of two opposite values, and the same one for each component. It follows that there are two components in the shell contained in parallel hyperplanes. Each component must fill its hyperplane completely, because any point in a hyperplane outside of the corresponding component but sufficiently close to it must project onto it, generating a normal parallel to the hyperplanes and not perpendicular to them as all normals should be. The proof concludes by an appeal to the previous lemma.

Anderson and Klee [1] have proved an important result

concerning the classification of points of $\mathcal{K}$ by means of their vertices $\mathfrak{h}_\mathcal{K}(x)$, saying essentially that the larger the vertex the rarer the points carrying it. The theorem below is an extension of their result to infinite dimensional Hilbert space; our proof follows the steps of theirs.

**Theorem 2.1** If $\mathfrak{H}$ is separable, then the set of points of a closed convex set $\mathcal{K}$ carrying vertices $\mathfrak{v}_\mathcal{K}(x)$ that are convex bodies of codimension not exceeding $k$ is the union of a countable family of compact sets of finite $k$-dimensional Hausdorff measure, and as such, of dimension not larger than $k$.

This theorem is an almost immediate consequence of the following lemma and of known relations between Hausdorff measures and dimension.

**Lemma 2.15** Let $Z$ be a countable set linear over the rationals and dense in $\mathfrak{H}$. Then any closed linear manifold $\mathfrak{m}^{(k)}$ (affine space) of codimension $k$ is intersected by the class of all $k$-dimensional manifolds determined by $k+1$ points of $Z$ in a set dense in $\mathfrak{m}^{(k)}$.

**Proof:** Let $\mathfrak{m}^{(k)} = x_0 + \mathfrak{v}^{(k)}$, where $\mathfrak{v}^{(k)}$ is a closed subspace of codimension $k$, and let $u^{(k)}$ be a $k$-dimensional subspace spanned by $k$ points in $Z$ such that $\mathfrak{v}^{(k)} \cap u^{(k)} = \{0\}$. The existence of such $u^{(k)}$ results at once from the fact that $Z$ is dense in $\mathfrak{H}$. With this choice $(\mathfrak{v}^{(k)}, u^{(k)})$ is a pair of complementary closed subspaces, and the projection $P$, not necessarily orthogonal, on $\mathfrak{v}^{(k)}$ along $u^{(k)}$, is everywhere defined, linear, and continuous. Now, for any $x$,

$$P(x-x_0) \in \mathfrak{v}^{(k)}, \quad P(x-x_0) - (x-x_0) \in u^{(k)},$$

and so

$$P(x-x_0) + x_0 \in x_0 + \mathfrak{v}^{(k)} = \mathfrak{m}^{(k)}, \quad P(x-x_0) + x_0 \in u^{(k)} + x.$$

Hence, for any $z \in Z$,

$$P(z-x_0) + x_0 \in \mathfrak{m}^{(k)} \cap (\mathfrak{u}^{(k)} + z),$$

and the point $P(z-x_0) + x_0$ belongs to the set along which $\mathfrak{m}^{(k)}$ is intersected by k-dimensional manifolds through k+1 points in $Z$ ($\mathfrak{u}^{(k)}+z$ is one such manifold). Varying $z$ in $Z$, $P(z-x_0) + x_0$ runs over a set dense in $\mathfrak{m}^{(k)}$ because the range of the continuous mapping $x \to P(x-x_0) + x_0$ is $\mathfrak{m}^{(k)}$ and $Z$ is dense in $\mathcal{H}$. The lemma is proved.

Proof of Theorem 2.1: If $\mathfrak{v}_\mathcal{K}(x)$ is a convex body of codimension not exceeding k, $x + \mathfrak{v}_\mathcal{K}(x)$ is intersected by a linear manifold generated by at most k+1 points in $Z$, and $x$ is a point in the projection on $\mathcal{K}$ of such manifold. The manifold being of dimension less than or equal to k, it is the union of countable many compact sets of finite k-dimensional Hausdorff measure, and $x$ belongs to the image by $P_\mathcal{K}$ of one of these sets at least. Moreover, since $P_\mathcal{K}$ is not expansive the images of such sets are again compact and of finite k-Hausdorff measure. Finally, since there is a countable number of manifolds of dimensions not exceeding k determined by points in $Z$, the set of points $x$ for which $\mathfrak{v}_\mathcal{K}(x)$ is a convex body of codimension not greater than k is contained in the union of countably many compact sets of finite k-Hausdorff measure. The assertion about the dimension follows from the fact that the dimension of a set of finite k-measure does not exceed k [cf. 6].

Corollary  For any finite dimensional $\mathcal{K}$ the set of flat points is dense in the shell.

This last result holds for any separable convex body, regardless of the dimension. However in this generality it is not a consequence of our arguments above; for the proof the reader may consult the original article of M. Mazur [9], or the book by Dunford and Schwartz [4, I. p. 450].
Now we turn our attention to the consideration of the faces of a convex set.

Definition 2.8   The face of a closed convex set $\mathcal{K}$ perpendicular to a vector $u$ is the set

(2.11) $$\mathcal{F}_{\mathcal{K}}(u) = \{y \mid y \in \mathcal{K}, \ \langle y, u \rangle = \sup_{x \in \mathcal{K}} \langle x, u \rangle\}.$$

Naturally, a face may be empty; we shall see that in order that $\mathcal{F}_{\mathcal{K}}(u)$ be nonempty it is necessary and sufficient that $u$ belong to the range of $I - P_{\mathcal{K}}$. Any face, being the intersection of $\mathcal{K}$ with a hyperplane, is a closed convex set; the set $\mathcal{K}$ itself can be conceived as the face perpendicular to $0$, or to any vector perpendicular to the manifold spanned by $\mathcal{K}$. It is evident that the same face correspond to all vectors $tu$, $t > 0$, along the same ray out of the origin, for, rather than the individual vectors it is the directions that determine the faces. However, it should also be clear that different directions may determine the same face. We have the following characterization of faces:

Lemma 2.16

(2.12) $$\mathcal{F}_{\mathcal{K}}(u) = \{y \mid y = P_{\mathcal{K}}(y+u)\}.$$

Proof:   If $y \in \mathcal{F}_{\mathcal{K}}(u)$, then by definition

$$\langle u, y-x \rangle \geq 0, \quad \forall \ x \in \mathcal{K},$$

and in consequence

$$\langle y+u-y, y-x \rangle \geq 0, \quad \forall \ x \in \mathcal{K},$$

which by (1.2) says that $y = P_{\mathcal{K}}(y+u)$. Conversely, if $y = P_{\mathcal{K}}(y+u)$ then

$$\langle u, y-x \rangle = \langle y+u-P_{\mathcal{K}}(y+u), P_{\mathcal{K}}(y+u)-x \rangle \geq 0, \quad \forall \ x \in \mathcal{K}$$

and $y \in \mathcal{F}_{\mathcal{K}}(u)$.   Q. E. D.

Thus proper faces ($\neq \mathcal{K}$) are part of the shell; note that $\mathfrak{F}_\mathcal{K}(u) = \mathfrak{F}_\mathcal{K}(u+v)$ for any $v$ perpendicular to the manifold spanned by $\mathcal{K}$, and conclude that the notion of face is intrinsic. We also see that if $\mathfrak{F}_\mathcal{K}(u) \neq \emptyset$, $u = (I-P_\mathcal{K})(y+u) \in \mathfrak{R}(I-P_\mathcal{K})$; as on the other hand for any $u$ the form $x - P_\mathcal{K}x$, $x \in \mathfrak{F}_\mathcal{K}(u)$, we have confirmed the above assertion as to when $\mathfrak{F}_\mathcal{K}(u)$ is nonempty.

Remark that (2.12) is equivalent to

(2.13) $$\mathfrak{F}_\mathcal{K}(u) = (I-P_\mathcal{K})^{-1}u - u.$$

Comparison with (2.7) indicates that vertices and faces are sort of dual notions. <u>If $\mathcal{K}_1$ and $\mathcal{K}_2$ are dual closed cones with vertex at 0, then the faces and vertices of the one are the vertices and faces of the other.</u> We need an auxilliary lemma about cones:

<u>Lemma 2.17</u>  For any vector $u^o$ in the core of a closed convex cone $\mathcal{V}$ with vertex at the origin, the hyperplane through the origin perpendicular to $u^o$ intersects $\mathcal{V}^\perp$ along a linear subspace perpendicular to $\mathcal{V}$.

<u>Proof:</u>  It is obvious that the subspace perpendicular to $\mathcal{V}$ belongs to $\mathcal{V}^\perp$ and is contained in the hyperplane perpendicular to $u^o$. Let us see that conversely any $v_o \in \mathcal{V}^\perp$ such that $\langle v_o, u^o \rangle = 0$ is perpendicular to the whole of $\mathcal{V}$. If $z = u^o + v_o$ then for any $u \in \mathcal{V}$,

$$\langle z-u^o, u^o-u \rangle = \langle v, u^o-u \rangle = -\langle v_o, u \rangle \geq 0,$$

and hence $P_\mathcal{V} z = u^o$, which by the definition of inner point implies that $z - P_\mathcal{V} z = v_o$ is perpendicular to $\mathcal{V}$.  Q.E.D.

<u>Lemma 2.18</u>  In a separable space among all faces of $\mathcal{K}$ containing a point $x$ there is a minimal one, that is, a face contained in all others. This face is $\mathfrak{F}_\mathcal{K}(u)$, for any $u \in (\mathcal{V}_\mathcal{K}(x))^o$.

**Proof:** Let $u^o \in U_K^o(x)$. The face $\mathcal{F}_K(u^o)$ is the intersection of $K$ with the hyperplane $\Pi_{u^o}(x)$, and as $\mathcal{F}_K(u^o) \subset (U_K(x))^\perp$, we have

$$\mathcal{F}_K(u^o) = K \cap (U_K(x))^\perp \cap \Pi_{u^o}(x).$$

By Lemma 2.17, $(U_K(x))^\perp \cap \Pi_{u^o}(x)$ is the linear manifold through $x$ perpendicular to $U_K(x)$, and therefore is contained in any hyperplane $\Pi_u(x)$ perpendicular to any $u \in U_K(x)$. Thus

$$\mathcal{F}_K(u^o) \subset K \cap \Pi_u(x) = \mathcal{F}_K(u), \quad \forall u \in U_K(x). \qquad \text{Q. E. D.}$$

The minimal face through a point will be called the point's face.

The lemma below assures us that any face is some point's face.

**Lemma 2.19** If the space is separable, then any non-empty face $\mathcal{F}_K(u)$ of a closed convex set is the face of any $x^o \in (\mathcal{F}_K(u))^o$.

**Proof:** If $\mathcal{F}_K(u)$ contains only one point there is nothing to prove, $\mathcal{F}_K(u)$ is just the face of that point. If it contains more than one point it contains an inner point $x^o$ (Lemma 2.7), and the cone projecting $\mathcal{F}_K(u)$, and a fortiori, the cone projecting $K$ from $x^o$, is dense in the closed linear manifold spanned by $\mathcal{F}_K(u)$, that is, $U_K^\perp(x^o) + x^o$ contains this manifold. Hence, any $u' \in U_K(x^o)$ is perpendicular to any vector $x' - x^o$, for $x' \in \mathcal{F}_K(u)$, and in consequence $x' \in \mathcal{F}_K(u')$. Since $x'$ is arbitrary in $\mathcal{F}_K(u)$, $\mathcal{F}_K(u) \subset \mathcal{F}_K(u')$, $\forall u' \in U_K(x^o)$.

**Lemma 2.20** If a separable closed convex set $K$ is neither a whole nor a half linear manifold, then $K$ contains inner points of the form: $\sum_{i=1}^{\infty} \alpha_i x_i$, $x_i \in$ shell $K$, $\alpha_i \geq 0$, $\sum_1^\infty \alpha_i = 1$.

Proof: It suffices to carry the proof under the assumption that $K$ spans $H$. By Lemma 2.7 $K$ has an inner point $x^o$; by Lemma 2.13 this point can be approximated by convex linear combinations $x_n = \sum_{i=1}^{k_n} \alpha_i^{(n)} x_i^{(n)}$ of points $x_i^{(n)}$ in the shell. Now, let us order these points into a single sequence $\{z_i\}_1^\infty$, and form a convergent convex linear combination

$$z = \sum_{i=1}^\infty \alpha_i z_i, \quad \alpha_i > 0, \quad \sum_{i=1}^\infty \alpha_i = 1 \text{ (this is always possible)}.$$

Then this $z$ is an inner point, because if it were not, any normal at $z$ would be normal at all $z_i$'s by Lemma 2.6, and therefore normal at $x^o$.

Lemma 2.21  The vertex is the same at all points in the core of a face of a separable closed convex set $K$, and is equal to the intersection of the vertices at the face's shell, if the face is neither a whole nor a half linear manifold.

Proof:  Let $\mathfrak{F}_K$ be a face; then $\mathfrak{F}_K = \mathfrak{F}_K(u^o)$ for $u^o \in \mathcal{V}_K(z^o)$, $z^o \subset \mathfrak{F}_K^o$, by Lemmas 2.19 and 2.20. Hence $u^o$ is normal at all points of $\mathfrak{F}_K$; in particular it is normal at all inner points, that is, $u^o \in \mathcal{V}_K(z_1^o)$ for $\forall z_1^o \in \mathfrak{F}_K^o$, whence

$$\mathcal{V}_K^o(z^o) \subset \mathcal{V}_K(z_1^o)$$

and taking closures (Lemma 2.7),

$$\mathcal{V}_K(z^o) \subset \mathcal{V}_K(z_1^o).$$

The opposite inclusion follows from interchanging the roles of $z^o$ and $z_1^o$. Hence $\mathcal{V}_K(z^o) = \mathcal{V}_K(z_1^o)$, where $z^o$ and $z_1^o$ are any two points in $\mathfrak{F}_K^o$, completing the proof of the first part of the lemma.

As to the last part let us begin by noting that since any $u \in \mathcal{V}_K^o(z^o), (z^o \in \mathfrak{F}_K^o)$, is normal at all points in the face, it is normal at all points in the shell of the face and hence

$$\mathcal{V}_K^o(z^o) \subset \mathcal{V}_K(x), \quad \forall x \in \text{shell } \mathfrak{F}_K,$$

and

$$U_{\mathcal{K}}(z^0) \subset \cap_{x \in \text{shell } \mathfrak{F}_{\mathcal{K}}} U_{\mathcal{K}}(x) \quad , \quad z^0 \in \overset{\circ}{\mathfrak{F}}_{\mathcal{K}}^{\circ} .$$

If now we pick a $z^0$ of the form $\sum_{i=1}^{\infty} \alpha_i x_i$, with the $x_i$'s in the shell of $\mathfrak{F}_{\mathcal{K}}$, whose existence is assured by Lemma 2.20, equation (2.9) tells us that

$$U_{\mathcal{K}}(z^0) = U_{\mathcal{K}}(\sum_{i=1}^{\infty} \alpha_i x_i) = \cap_{i=1}^{\infty} U_{\mathcal{K}}(x_i) \supset \cap_{x \in \text{shell } \mathfrak{F}_{\mathcal{K}}} U_{\mathcal{K}}(x) \quad ,$$

and so

$$U_{\mathcal{K}}(z^0) = \cap_{x \in \text{shell } \mathfrak{F}_{\mathcal{K}}} U_{\mathcal{K}}(x) . \qquad \text{Q. E. D.}$$

Note the duality: <u>The vertex is the same at all points in the core of a face; the face is the same for all normals in the core of a vertex.</u>

<u>Lemma 2.22</u>  Let $\mathcal{K}_1$ and $\mathcal{K}_2$ be two closed convex sets, one of which at least is a convex body. Then, if shell $\mathcal{K}_1 \subset$ shell $\mathcal{K}_2$, one of the following situations holds:
 a. $\mathcal{K}_1$ is either a linear manifold or a half linear manifold bounded by a linear manifold contained in the shell of $\mathcal{K}_2$,
 b. $\mathcal{K}_1$ is contained in the shell of $\mathcal{K}_2$,
 c. $\mathcal{K}_1$ is a section of $\mathcal{K}_2$, i.e., is the intersection of $\mathcal{K}_2$ with a linear manifold.

<u>Proof:</u> Suppose that neither a. nor b. holds. Then $\mathcal{K}_1$ is the convex hull of its shell (Lemma 2.13) and in consequence is contained in $\mathcal{K}_2$. The closed linear manifold spanned by $\mathcal{K}_1$ cuts $\mathcal{K}_2$ along a closed convex set $\tilde{\mathcal{K}}_2$ containing $\mathcal{K}_1$. Our taks is to show that $\tilde{\mathcal{K}}_2 = \mathcal{K}_1$, that is, that $\tilde{\mathcal{K}}_2 \subset \mathcal{K}_1$. Let then $\tilde{x}_2$ be an arbitrary point in $\tilde{\mathcal{K}}_2$. Since $\mathcal{K}_1$ is not

contained in the shell of $\mathcal{K}_2$ whereas its shell is, there must be an inner point $x_1$ of $\mathcal{K}_1$ which is also an inner point of $\mathcal{K}_2$. Then the semiopen segment $\{tx_1+(1-t)\tilde{x}_2\}_{0<t\leq 1}$ joining $x_1$ and $\tilde{x}_2$ lies entirely in $\mathcal{K}_2^o$ and therefore has no point in the shell of $\mathcal{K}_1$. If $\mathcal{K}_1$ is a convex body, such a segment, having one end in the interior of $\mathcal{K}_1$ (relatively to the manifold spanned by $\mathcal{K}_1$) and not piercing the boundary, must be completely contained in $\mathcal{K}_1$. It follows that $x_2$ also belongs to $\mathcal{K}_1$, because $\mathcal{K}_1$ is closed. If $\mathcal{K}_2$ is a convex body, its shell is closed, and any boundary point of $\mathcal{K}_1$, being a limit point of points in its shell, is a limit point of points in the shell of $\mathcal{K}_2$ and hence belongs to the shell of $\mathcal{K}_2$. It follows that $\mathcal{K}_1$ is a convex body, for otherwise its shell would be dense in $\mathcal{K}_1$ (Lemma 2.10) and $\mathcal{K}_1$ would be part of the shell of $\mathcal{K}_2$, contrary to the assumption that b. does not hold. Hence, we are back to the case where $\mathcal{K}_1$ is a convex body, and the proof concludes.

<u>Corollary 1</u>   If $\mathcal{K}$ is a total convex body different from the whole space

$$\mathcal{K} = \bigcap_{\iota \in I} \Pi^-_{u_\iota}(x_\iota) ,$$

where $\{x_\iota\}_{\iota \in I}$ is any set dense in the shell of $\mathcal{K}$, and $\{u_\iota\}_{\iota \in I}$ is any choice of normals at these points.

<u>Proof:</u>   We may discard the case when $\mathcal{K}$ is a half space because in that case the corollary is obvious. We claim that the shell of $\mathcal{K}$ is contained in the shell of $\tilde{\mathcal{K}} = \bigcap_{\iota \in I} \Pi^-_{u_\iota}(x_\iota)$. In fact, any $x_\iota$ belongs to the shell of $\tilde{\mathcal{K}}$, and the $x_\iota$'s are dense in the shell of $\mathcal{K}$, and so shell $\mathcal{K} \subset$ shell $\tilde{\mathcal{K}}$, since shell $\tilde{\mathcal{K}}$ is closed. Therefore, the lemma applies to $\mathcal{K}$ and $\tilde{\mathcal{K}}$, and since $\mathcal{K}$ is neither the whole nor half space, and cannot be contained in a face of $\tilde{\mathcal{K}}$, $\mathcal{K}$ must be a section of $\tilde{\mathcal{K}}$. But $\mathcal{K}$ spans the whole space, so $\mathcal{K} = \tilde{\mathcal{K}}$.   Q.E.D.

<u>Corollary 2</u>   The dual $C^\perp$ of a total conical convex body $C$ in a separable space is the closed convex hull of its one-dimensional faces.

Proof:   By the corollary above

(2.14) $$C = \bigcap_u \Pi_u^-(0)$$

where u runs over the normals at flat points, these being dense in the shell of $C$ (see comments to Corollary of Theorem 2.1). But the normals at flat points of $C$ generate the one-dimensional faces of $C^\perp$, and the desired result follows from (2.14) by applying the operation $\perp$, and appealing to (2.3).

Example   In this corollary one cannot dispense with the hypothesis that the cone be a convex body, as there are couples of dual cones having neither flat points nor one-dimensional faces. For instance the cones $C^+$ and $C^-$ of a.e. nonnegative and nonpositive functions in $L^2[0,1]$. The faces of $C^+$ are the cones of nonnegative functions vanishing on sets of positive measure; they are obviously infinite dimensional. The same is true, of course, for the faces of $C^-$.

Corollary 3   Every bounded closed convex set $K$ in a separable space is the closed convex hull of its point-faces.

Proof:   By the addition of one dimension to the space and translation of $K$, one may work under the assumption that $K$ span a closed linear manifold of codimension one not containing the origin. Then the dual $C^\perp$ of the cone $C$ projecting $K$ from the origin is a total convex body, and by Corollary 2 $C$ is the closed convex hull of its one-dimensional faces. But every such face is the ray projecting a point-face of $K$ from the origin. Thus any $x$ in $K$, being an element of $C$, is the limit of finite sums of the form

$$\sum_i \alpha_i s_i x_i, \quad 0 \le \alpha_i \le 1, \quad \sum_i \alpha_i = 1, \quad s_i \ge 0,$$

where the $x_i$'s are point faces. The component of this sum in the direction perpendicular to the manifold spanned by $K$ is $(\sum_i \alpha_i s_i)\delta$ ($\delta$ distance from the origin to the manifold), whereas the component of any point in $\mathcal{K}$ is $\delta$. Hence the sums $\sum_i \alpha_i s_i$ approach $1$, and any point in $\mathcal{K}$ is the limit of expressions of the form

$$\frac{\sum_i \alpha_i s_i x_i}{\sum_i \alpha_i s_i},$$

and hence belongs to the closed convex hull of the point faces.

Point faces are also called "exposed points" and there is an extensive literature about them (see [3]).

Flat points being dense in the shell of a separable convex body one may surmise that all normals can be constructed out of normals at these points. The lemma below tells us that this is the case in finite dimensions at least.

**Lemma 2.23**  Let $\mathcal{K}$ be a finite dimensional total convex body. Then, for any $x \in$ shell $\mathcal{K}$, $\mathcal{U}_\mathcal{K}(x)$ is the closed convex hull of all limits of normals at flat points as these points approach $x$.

**Proof:**  $s_\mathcal{K}(x)$ is a conical total convex body and by Lemma 2.22, Corollary 2, $\mathcal{U}_\mathcal{K}(x)$ is the closed convex hull of its one-dimensional faces. To prove the lemma it suffices to see that any such face is the limit of normals at flat points. So let $\{tu\}_{t \geq 0}$ be a one dimensional face of $\mathcal{U}_\mathcal{K}(x)$, $\|u\| = 1$, and let $v$ be any nonvanishing vector in the shell of $s_\mathcal{K}(x)$ such that $\mathcal{F}_{\mathcal{U}_\mathcal{K}(x)}(v) = \{tu\}_{t \geq 0}$; naturally $v$ is perpendicular to $u$. By Lemma 2.3, $x + tv - P_\mathcal{K}(x + tv) = o(t)$ as $t \downarrow 0$. Since flat points are dense, for each $t > 0$ there is a flat point $x(t)$ such that $P_\mathcal{K}(x + tv) - x(t) = o(t)$, and hence such that, $x(t) = x + tv + o(t)$. Let $u(t)$ denote the unit normal at $x(t)$, and let $w$ be any limit of such normals as $t \downarrow 0$. By (1.2), since $x \in \mathcal{K}$,

$$\langle u(t), x(t) - x \rangle \geq 0, \qquad \forall\, t \geq 0,$$

that is

$$\langle u(t), tv + o(t)\rangle \geq 0, \qquad \forall\, t \geq 0,$$

whence dividing by $t$ and letting $t \downarrow 0$, $\langle w,v\rangle \geq 0$. On the other hand by Lemma 2.5, $w \in U_{\mathcal{K}}(x)$ and therefore $\langle w,v\rangle \leq 0$. Hence $\langle w,v\rangle = 0$. This says that $w$ belongs to the face of $U_{\mathcal{K}}(x)$ perpendicular to $v$, that is to $\{tu\}_{t \geq 0}$. Hence there is a $t_0 \geq 0$ such that $w = t_0 u$. But then $t_0 = 1$ because both $w$ and $u$ are unitary, and $w = u$. Thus the lemma is proved.

## §3. The range of $I - P_{\mathcal{K}}$.

The range of the complement of a projection is closely related to the "recession cone of a convex set", a notion that we shall presently define [23, p. 61]:

**Definition 3.1** The recession cone of a closed convex set $\mathcal{K}$ is the set

(3.1) $\qquad \mathcal{K}^{\infty} = \{u \mid x + tu \in \mathcal{K}, \forall x \in \mathcal{K}, \forall t \geq 0\}$.

According to this $u \in \mathcal{K}^{\infty}$ if the ray in the direction of $u$ issuing from any point in $\mathcal{K}$ is fully contained in $\mathcal{K}$. The lemma below tells us that much less is sufficient:

**Lemma 3.1** A vector $u$ belongs to $\mathcal{K}^{\infty}$ if there is a sequence of numbers $t_n \to +\infty$ and a point $x_0 \in \mathcal{H}$ such that

$$x_0 + t_n u \in \mathcal{K}, \quad n = 1, 2, \ldots \quad .$$

**Proof:** In fact, if $x$ is any point in $\mathcal{K}$ and $s$ any non-negative number, then $(1 - \frac{s}{t_n}) x + \frac{s}{t_n}(x_0 + t_n u)$ belongs to $\mathcal{K}$ for $n$ sufficiently large, and therefore so does its limit as $n \to \infty$, that is $x + su \in \mathcal{K}$.

Q.E.D.

It is clear that $K^\infty$ <u>is a closed convex cone with vertex at the origin,</u> and in fact is the largest of such cones that can be placed inside $K$. In finite dimensions $K^\infty = \{0\}$ is equivalent to the boundedness of $K$ but not so in infinite dimension, as it can be seen by considering the set $K = \{x \mid |\langle x, \varphi_i \rangle| \leq 1, \, i = 1, 2, \ldots\}$, where $\{\varphi_i\}_1^\infty$ is a complete orthonormal set; $K$ contains no asymptotic direction and yet it is unbounded. Notice in passing that in this case $\Re(I-P_K)$ is the linear space consisting of all <u>finite</u> linear combinations of the $\varphi_i$'s.

<u>Lemma 3.2</u>  A vector $u \in (K^\infty)^\perp$ if and only if

(3.2) $$\lim_{t \to +\infty} \frac{P_K(x_0 + tu)}{t} = 0, \quad \forall x_0 \in H.$$

<u>Proof:</u> Since $\|P_K(x_0 + tu) - P(x_0' + tu)\| \leq \|x_0 - x_0'\|$, if condition (3.2) is satisfied at one point of $x_0$ it is satisfied at any other. Thus we may assume $x_0 \in K$. By (1.2)

$$\langle x_0 + tu - P_K(x_0 + tu), P_K(x_0 + tu) - x_0 \rangle \geq 0,$$

that is

(3.3) $$\|P_K(x_0 + tu) - x_0\|^2 \leq t \langle P_K(x_0 + tu) - x_0, u \rangle.$$

Let us prove necessity now, and assume $u \in (K^\infty)^\perp$. We start out by taking a sequence $t_n \to +\infty$ such that

$$\lim_{n \to \infty} \frac{\|P_K(x + t_n u) - x_0\|}{t_n} = \overline{\lim_{t \to +\infty}} \frac{\|P_K(x_0 + tu) - x_0\|}{t} = \ell.$$

If $\{\|P_K(x_0 + t_n u) - x_0\|\}$ contains a bounded subsequence then $\ell = 0$ and (3.2) follows; if it does not then $\|P_K(x_0 + t_n u) - x_0\| \to +\infty$, $\|P_K(x_0 + t_n u) - x_0\|$ does not vanish for $n$ sufficiently large, and passing to a subsequence if necessary, we may assume that $P_K(x_0 + t_n u) - x_0) / \|P_K(x_0 + t_n u) - x_0\|$ converges weakly to a vector $v$. But then the point

$$(1 - \frac{s}{\|P_{\mathcal{K}}(x_0+t_n u)-x_0\|})x_0 + (\frac{s}{\|P_{\mathcal{K}}(x_0+t_n u)-x_0\|}) P_{\mathcal{K}}(x_0+t_n u)$$

belongs to $\mathcal{K}$ from some value of n on and hence so does its limit $x_0 + sv$, indicating that $v \in \mathcal{K}^\infty$. Thus $\langle u,v \rangle \leq 0$. Moreover, setting $t = t_n$ in (3.3), and dividing by $t_n \|P_{\mathcal{K}}(x_0+t_n u)-x_0\|$, one obtains

$$\frac{\|P_{\mathcal{K}}(x_0+t_n u)-x_0\|}{t_n} \leq \langle \frac{P_{\mathcal{K}}(x_0+t_n u)-x_0}{\|P_{\mathcal{K}}(x_0+t_n u)-x_0\|}, u \rangle,$$

whence by passing to the limit,

$$\lim_{n \to \infty} \frac{\|P_{\mathcal{K}}(x_0+t_n u)-x_0\|}{t_n} \leq \langle v,u \rangle.$$

Hence, as $\langle u,v \rangle \leq 0$, $\ell = 0$ and as before (3.2) follows.

To prove sufficiency, pick any $v$ in $\mathcal{K}^\infty$, and after remarking that $x_0 + tv \in \mathcal{K}$, $\forall t \geq 0$, write

$$\langle x_0 + tu - P_{\mathcal{K}}(x_0+tu), P_{\mathcal{K}}(x_0+tu)-(x_0+tv) \rangle \geq 0 ;$$

then divide by $t^2$ and let $t \to +\infty$, to obtain

$$\langle u,v \rangle \leq 0, \quad \forall v \in \mathcal{K}^\infty.$$

Thus $u \in (\mathcal{K}^\infty)^\perp$.  Q.E.D.

**Lemma 3.3**  In order that a vector $u$ belong to the range $\mathcal{R}(I-P_{\mathcal{K}})$ of $I-P_{\mathcal{K}}$ it is necessary that $\{P_{\mathcal{K}}(x_0+tu)\}_{t \geq 0}$ be bounded for any choice of $x_0$, and it is sufficient that $\{P_{\mathcal{K}}(x_0+t_n u)\}_1^\infty$, be bounded for some sequence $\{t_n\}_1^\infty$ going to $+\infty$ and some choice $x_0$.

**Proof:** It is obvious that $\{P_{\mathcal{K}}(x_0+tu)\}_{t \geq 0}$ and $\{P_{\mathcal{K}}(x_0'+t_u)\}_{t \geq 0}$ are simultaneously bounded or unbounded.

Now, if $u \in \mathcal{R}(I-P_\mathcal{K})$ then for some $z$, $u = z-P_\mathcal{K} z$ and $P_\mathcal{K}(P_\mathcal{K} z + tu) = P_\mathcal{K} z$. Therefore $\{P_\mathcal{K}(x_0+tu)\}_{t \geq 0}$ is bounded for $x_0 = P_\mathcal{K} x$, and hence for any other value of $x_0$. Conversely, if $\{P_\mathcal{K}(x_0+t_n u)\}_1^\infty$ is bounded, then there is a subsequence $t_{n_k} \to +\infty$, and a point $x \in \mathcal{K}$ such that $P_\mathcal{K}(x_0+t_{n_k}u) \to x$. Hence, if we set

$$x_{n_k} = t_{n_k}^{-1}(x_0+t_{n_k}u - P_\mathcal{K}(x_0+t_{n_k}u)) + P_\mathcal{K}(x_0+t_{n_k}u),$$

$$u_{n_k} = (I-P_\mathcal{K})x_{n_k} = t_{n_k}^{-1}(x_0+t_{n_k}u - P_\mathcal{K}(x_0+t_{n_k}u))$$

then $x_{n_k} \to x+u$, $u_{n_k} \to u$, and by Lemma 1.4, $u = (I-P_\mathcal{K})(x+u)$. Therefore, $x \in \mathcal{F}_\mathcal{K}(u)$, $u \in \mathcal{R}(I-P_\mathcal{K})$, and the proof is complete.

Since $\mathcal{R}(I-P_\mathcal{K}) = \bigcup_{x \in \mathcal{K}} \mathcal{V}_\mathcal{K}(x)$ is the union of cones with vertex at the origin, it is itself a cone with the vertex at the same point. Moreover, in general it is neither closed nor convex. The next theorem tells us what its closure is (this is a property of all maximal monotone mappings (cf. [21, Theorem 7']); as to its convex hull it is not difficult to see that it <u>coincides with the set of positive normals to all support planes, at finite or infinite distance.</u>

**Theorem 3.1**     For any closed convex set $\mathcal{K}$

(3.4) $$\overline{\mathcal{R}(I-P_\mathcal{K})} = (\mathcal{K}^\infty)^\perp.$$

**Proof:**     For any $x \in \mathcal{K}$, $\mathcal{S}_\mathcal{K}(x) \supset \mathcal{K}^\infty$, $\mathcal{V}_\mathcal{K}(x) = \mathcal{S}_\mathcal{K}^\perp(x) \subset (\mathcal{K}^\infty)^\perp$ and

$$\mathcal{R}(I-P_\mathcal{K}) = \bigcup_{x \in \mathcal{K}} \mathcal{V}_\mathcal{K}(x) \subset (\mathcal{K}^\infty)^\perp.$$

Therefore, taking closures,

$$\overline{\mathcal{R}(I-P_\mathcal{K})} \subset (\mathcal{K}^\infty)^\perp.$$

On the other hand, if $u \in (\mathcal{K}^\infty)^\perp$, by Lemma 3.2,

$$u = \lim_{t \to \infty} t^{-1}(x_0+tu-P_\mathcal{K}(x_0+tu)),$$

and as $t^{-1}(x_0+tu-P_\mathcal{K}(x_0+tu)) \in \mathcal{R}(I-P_\mathcal{K})$, $u \in \overline{\mathcal{R}(I-P_\mathcal{K})}$. Hence,

$$\overline{\mathcal{R}(I-P_\mathcal{K})} \supset (\mathcal{K}^\infty)^\perp. \qquad \text{Q. E. D.}$$

Combination with Lemma 3.2 yields:

**Corollary 1** A vector $u$ belongs to $\overline{\mathcal{R}(I-P_\mathcal{K})}$ if and only if $P_\mathcal{K}(x_0+tu)/t \to 0$ for some choice of $x_0$ (and hence for any choice).

**Lemma 3.4** The shell points of $\overline{\mathcal{R}(I-P_\mathcal{K})} = (\mathcal{K}^\infty)^\perp$ are the limits of normals $u_n$ at points $x_n$ going to infinity along receding directions (i.e. $x_n/\|x_n\| \to v$, $v \in \mathcal{K}^\infty$, $-v \notin \mathcal{K}^\infty$).

**Proof:** Let $u \in$ shell $\overline{\mathcal{R}(I-P_\mathcal{K})}$, $\|u\|=1$. Then there exists a unit vector $v$ in $\mathcal{K}^\infty$ orthogonal to $u$. If $x_0$ is any fixed point in $\mathcal{K}$ then $x_0+tv' \in \mathcal{K}$, $\forall v' \in \mathcal{K}$, and one may write

$$\langle t(u+v)-P_\mathcal{K}(t(u+v)), P_\mathcal{K}(t(u+v))-(x_0+tv')\rangle \geq 0,$$

$$\forall v' \in \mathcal{K}^\infty, \forall t \geq 0,$$

and dividing by $t^2$ and transposing some terms,

$$-\langle u+v - \frac{P_\mathcal{K}(t(u+v))}{t}, \frac{x_0}{t}\rangle + \langle u+v, \frac{P_\mathcal{K}(t(u+v))}{t}\rangle \geq \frac{\|P_\mathcal{K}(t(u+v))\|^2}{t^2}.$$

The set of vectors $\{P_\mathcal{K}(t(u+v))|t\}_{t>0}$ being bounded has weak limit points as $t \to 0$, all of which belong to $\mathcal{K}^\infty$. If $v''$ is any of them, passage to the limit in the above inequality yields

$$-\langle u+v-v'', v'\rangle + \langle u+v, v''\rangle \geq \|v''\|^2, \forall v' \in \mathcal{K}^\infty,$$

(recall that the norm of the weak limit is smaller than the

limit of the norms), or

$$\langle u+v-v'', v''-v'\rangle \geq 0, \quad \forall\, v' \in \mathcal{K}^\infty,$$

which is equivalent to $v'' = P_\mathcal{K}(u+v) = v$. Thus, since all the weak limits of $P_\mathcal{K}(t(u+v))/t$ are equal to $v$, $P_\mathcal{K}(t(u+v))/t \rightharpoonup v$. Let us show that this limit is also attained in the strong sense. We have

$$\left\| \frac{P_\mathcal{K}(t(u+v))}{t} \right\| \leq \left\| \frac{P_\mathcal{K}(tu)}{t} \right\| + \left\| \frac{P_\mathcal{K}(t(u+v)) - P_\mathcal{K}(tu)}{t} \right\|,$$

and by the contractivity of $P_\mathcal{K}$,

$$\left\| \frac{P_\mathcal{K}(t(u+v))}{t} \right\| \leq \left\| \frac{P_\mathcal{K}(tu)}{t} \right\| + \left\| \frac{t(u+v) - tu}{t} \right\| = \left\| \frac{P_\mathcal{K}(tu)}{t} \right\| + 1.$$

By Lemma 3.2, $P_\mathcal{K}(tu)/t \to 0$, and so

$$\overline{\lim_{t \to +\infty}} \left\| \frac{P_\mathcal{K}(t(u+v))}{t} \right\| = 1.$$

But then, the norm of the weak limit $v$ of $P_\mathcal{K}(t(u+v))/t$ is equal to the lim sup of the norms, and by a well known theorem $P_\mathcal{K}(t(u+v))/t$ converges strongly to $v$. With this at hand the rest of the proof follows easily. Let us write, for $t$ sufficiently large,

$$x(t) = P_\mathcal{K}(t(u+v)), \quad u(t) = \frac{t(u+v) - P_\mathcal{K}(t(u+v))}{\|t(u+v) - P_\mathcal{K}(t(u+v))\|};$$

$u(t)$ is a unit normal at $x(t)$. According to the above $x(t)/t \to v$ and as $\|v\|=1$, $\|x(t)\| \to +\infty$ and $x(t)/\|x(t)\| \to v$. Moreover

$$u(t) = \frac{u+v-(x(t)/t)}{\|u+v-(x(t)/t)\|} \to \frac{u}{\|u\|} = u.$$

Thus, $u$ is the limit of unit normals $u(t)$ at points $x(t)$ going to infinite along the receding direction $v$.

The proof of the converse is much easier. Assume

that $u_n \to u$, where the $u_n$'s are normals at points $x_n$ going to $\infty$ in such a way that $x_n/\|x_n\| \to v$, and let us show that $u \in$ shell $\mathcal{R}(I-P_\mathcal{K})$. Since $v$ must necessarily belong to $\mathcal{K}^\infty$, it will be enough to show that $\langle u, v \rangle = 0$. If $x_0$ is any point in $\mathcal{K}$, and $v'$ any vector in $\mathcal{K}^\infty$, $x_0 + \|x_n\|v'$ belongs to $\mathcal{K}$, and by (1.2),

$$\langle u_n, x_n - (x + \|x_n\|v') \rangle \geq 0 ,$$

whence dividing by $\|x_n\|$ and passing to the limit it follows

$$\langle u, v-v' \rangle \geq 0, \quad \forall \; v' \in \mathcal{K}^\infty .$$

Replacing here $v'$ by $0$ and by $2v$, one finds

$$\langle u, v \rangle \geq 0, \quad \langle u, v \rangle \leq 0 ,$$

and so $\langle u, v \rangle = 0$, as it was to be proved.

<u>Lemma 3.5</u>   A point $u$ in $\mathcal{R}(I-P_\mathcal{K})$ belongs to the shell of $\overline{\mathcal{R}(I-P_\mathcal{K})}$ if and only if the face $\mathcal{F}_\mathcal{K}(u)$ has one receding direction at least.

<u>Proof:</u>   The condition is necessary: Since $u \in$ shell $\overline{\mathcal{R}(I-P_\mathcal{K})}$ = shell $(\mathcal{K}^\infty)^\perp$, there is a nonvanishing vector $v \in \mathcal{K}^\infty$ orthogonal to $u$. If $x_0 \in \mathcal{F}_\mathcal{K}(u)$, we have for all $t \geq 0$,

$$\langle u, x_0+tv \rangle = \langle u, x_0 \rangle = \sup_{x \in \mathcal{K}} \langle u, x \rangle ,$$

and $x_0 + tv \in \mathcal{F}_\mathcal{K}(u)$, $\forall t \geq 0$. In other terms, $v$ belongs to the recession cone of $\mathcal{F}_\mathcal{K}(u)$. The condition is sufficient. If $\mathcal{F}_\mathcal{K}(u)$ has a receding direction $v$ then $\langle u, v \rangle = 0$, and since $u \in (\mathcal{K}^\infty)^\perp$, $v \in \mathcal{K}^\infty$, which characterizes $u$ as a point in the shell of $(\mathcal{K}^\infty)^\perp$, that is, in the shell of $\overline{\mathcal{R}(I-P_\mathcal{K})}$.

The following theorem is important in that it tells how the projection on a face of $\mathcal{K}$ can be obtained from the projection on $\mathcal{K}$.

**Theorem 3.2**  If $u \in \mathcal{R}(I-P_\mathcal{K})$, then

(3.5) $$P_\mathcal{K}(x+tu) \xrightarrow[t\to+\infty]{} P_{\mathcal{F}_\mathcal{K}(u)}(x), \quad \forall\, x \in \mathcal{K}.$$

**Proof:** We know that $\{P_\mathcal{K}(x+tu)\}_{t \geq 0}$ is bounded (Lemma 3.3), hence all we have to prove is that if $P_\mathcal{K}(x+t_n u) \rightharpoonup z$ for some sequence $t_n \to +\infty$, then $z = P_{\mathcal{F}_\mathcal{K}(u)} x$. To this end we pick any point $z'$ in $\mathcal{F}_\mathcal{K}(u)$, and apply (1.6),

$$\|P_\mathcal{K}(x+t_n u) - P_\mathcal{K}(z'+t_n u)\|^2 \leq \langle x-z', P_\mathcal{K}(x+t_n u) - P_\mathcal{K}(z'+t_n u)\rangle,$$

that is, since $P_\mathcal{K}(z'+t_n u) = z'$,

$$\langle x-z', P_\mathcal{K}(x+t_n u) - z'\rangle \geq \|P_\mathcal{K}(x+t_n u) - z'\|^2.$$

Passing to the limit $n \to +\infty$, and recalling that the norm of the weak limit is smaller than the limit of norms, we obtain

$$\langle x-z', z-z'\rangle \geq \|z-z'\|^2,$$

$$\langle x-z, z-z'\rangle \geq 0, \quad \forall\, z' \in \mathcal{F}_\mathcal{K}(u).$$

Thus, by (1.2), $z = P_{\mathcal{F}_\mathcal{K}(u)} x$, as soon as $z \in \mathcal{F}_\mathcal{K}(u)$. But that any weak limit of $\{P_\mathcal{K}(x+tu)\}_{t \geq 0}$ belongs to $\mathcal{F}_\mathcal{K}(u)$ has been shown in the proof of Lemma 3.3; thus (3.5) has been established.

The following result belonging to the same order of ideas is often useful.

**Lemma 3.6**  In order that a vector $u$ belong to $\mathcal{R}(I-P_\mathcal{K})$ it is necessary that the sequence of iterates $\{x_n\}_1^\infty$, by the operation $x \to P_\mathcal{K}(x+u)$, defined by the recurrence relation $x_n = P_\mathcal{K}(x_{n-1} + tu)$, be weakly convergent to a point in $\mathcal{F}_\mathcal{K}(u)$, for any initial point $x_1$ whatever, and it is sufficient that for some initial value it contain a bounded subsequence.

Proof: For any sequence of iterates $\{x_n\}_1^\infty$, (1.2) yields

$$\langle x_{n-1}+u-P_\mathcal{K}(x_{n-1}+u)-x_{n-1}\rangle \geq 0, \quad n \geq 2,$$

that is,

(3.6) $\qquad \langle u, x_n\rangle - \langle u, x_{n-1}\rangle \geq \|x_n - x_{n-1}\|^2,$

whence we see that the numerical sequence $\{\langle u, x_n\rangle\}_1^\infty$ is nondecreasing and as such converges to a limit, finite or infinite.

Now, if $\{x_n\}_1^\infty$ contains a bounded subsequence it contains a subsequence $\{x_{n_k}\}_1^\infty$, weakly converging to a limit $x_\infty$. Then $\lim_{n\to\infty}\langle u, x_n\rangle = \lim_{k\to\infty}\langle u, x_{n_k}\rangle = \langle u, x_\infty\rangle$, and $x_n - x_{n-1} \to 0$ follows from (3.6). Therefore

$$x_{n_k} + u \to x_\infty + u, \quad x_{n_k} + u - P_\mathcal{K}(x_{n_k}+u) = x_{n_k} - x_{n_k+1} + u \to u,$$

and by the demiclosedness of $I - P_\mathcal{K}$ (Lemma 1.4), $u = (I - P_\mathcal{K})(x_\infty + u)$. Thus, $u \in \mathcal{R}(I - P_\mathcal{K})$ and $x_\infty \in \mathcal{F}_\mathcal{K}(u)$. The sufficiency part of the lemma is proved.

As to necessity note that if $u \in \mathcal{R}(I-P)$ then $\mathcal{F}_\mathcal{K}(u)$ is the set of fixed points of the mapping $x \to P_\mathcal{K}(x+u)$ (Lemma 2.16), as this mapping is nonexpansive the successive iterates get closer and closer to any point in $\mathcal{F}_\mathcal{K}(u)$, thus forming a bounded sequence. We have seen in proving sufficiency that any weak limit point of $\{x_n\}_1^\infty$ belongs to $\mathcal{F}_\mathcal{K}(u)$, thus to prove that $\{x_n\}_1^\infty$ is itself weakly convergent we must show that all such weak limits are the same. This we do following an argument due to Opial [17]. Since fixed points of a mapping are attractive points the sequence $\{\|x_n - x\|\}_1^\infty$ is nonincreasing for all $x \in \mathcal{F}_\mathcal{K}(u)$ and thus converges to a nonnegative limit $d(x)$. For any $d > \inf_{x \in \mathcal{F}_\mathcal{K}(u)} d(x) = d_0$, the set $\{x \mid x \in \mathcal{F}_\mathcal{K}(u), d(x) \leq d\}$ is bounded, closed, convex, and nonempty: it is bounded because any of its points is within a fixed distance from the sequence $\{x_n\}_1^\infty$ which is bounded; it is closed because $d(x') = \lim \|x_n - x'\| \leq \lim(\|x_n - x\| + \|x' - x\|) = d(x) + \|x - x'\|$; it is convex because if $\alpha', \alpha'' \geq 0$, $\alpha' + \alpha'' = 1$, then $d(\alpha'x' + \alpha''x'') = \lim \|x_n - (\alpha'x' + \alpha''x'')\| \leq \lim(\alpha' \|x_n - x'\| + \alpha'' \|x_n - x''\|\} = \alpha'd(x') + \alpha''d(x'')$; finally, it is nonempty by choice of $d$. Therefore, since bounded closed convex sets are weakly compact, the intersection of the above sets $\{x \mid x \in \mathfrak{F}_\mathcal{K}(u), d(x) = d\}$ is a nonempty closed convex set. This set cannot contain more than one point because if it did the middle point of the segment joining any two of its points would yield a smaller value for $d(x)$, which is impossible. Let $x_0$ be this point, and let us show that $x_n \rightharpoonup x_0$. Were this not true there would be a subsequence $x_{n_k} \rightharpoonup x_0' \neq x$, and then

$$d(x_0)^2 = \lim \|x_{n_k} - x_0\|^2 = \lim \|x_{n_k}\|^2 + \|x_0\|^2 - 2\langle x_0, x_0' \rangle$$

$$> \lim \|x_{n_k}\|^2 - \|x_0'\|^2 = \lim \|x_{n_k} - x_0'\|^2 = d(x'),$$

in contradiction with the definition of $x_0$; hence $x_n \rightharpoonup x_0$.

Q. E. D.

Remark: Here we cannot say in general what the limit of $x_n$ will be; this limit depends not only on the initial point and on the face $\mathfrak{F}_\mathcal{K}(u)$ but also on the behaviour of $\mathcal{K}$ at the intermediate points. In any case it is not true in general that $x_n \rightharpoonup P_{\mathfrak{F}_\mathcal{K}(u)} x$ as in Lemma 3.2, for if this were so one should also have $x_n \rightharpoonup P_{\mathfrak{F}_\mathcal{K}(u)} x_{n_0}$, for any $n_0$, and then $P_{\mathfrak{F}_\mathcal{K}(u)} x_n = P_{\mathfrak{F}_\mathcal{K}(u)} x_m$, $m, n = 1, 2, \ldots$, and in particular

$$P_{\mathfrak{F}_\mathcal{K}(u)}(P_\mathcal{K}(x+u)) = P_{\mathfrak{F}_\mathcal{K}(u)} x,$$

relation which does not hold even in the case of a right angle circular cone in $\mathbb{R}^3$.

§4. **Translation sets, parallel convex sets, differentiability.**

Definition 4.1  Given two convex sets $K$ and $K'$, the translation set of $K$ into $K'$ is the set

(4.1) $$\tau(K, K') = \{u \mid K + u \subset K'\}.$$

It is seen at once that $\tau(K, K')$, if not empty, is a convex set; it is bounded, closed or compact whenever $K'$ is bounded, closed or compact respectively. This follows from

(4.2) $$\tau(K, K') = \bigcap_{x \in K} (K' - x).$$

This relation is easily proved. If $u \in \tau(K, K')$ then $K + u \subset K'$, and for each $x \in K$, $u \in K - x$; hence $u \in \bigcap_{x \in K}(K'-x)$ and $\tau(K,K') \subset \bigcap_{x \in K}(K'-x)$. Conversely, if $u$ belongs to the set on the right of (4.2), then $u + x \in K'$, $\forall x \in K$, that is, $u + K \subset K'$ and $u \in \tau(K, K')$; therefore $\tau(K, K') \supset \bigcap_{x \in K}(K'-x)$.

A particular case of translation set is that of the translation set of a convex set $K$ into itself. In this case, if $K$ is closed, the situation is completely cleared by the following lemma:

Lemma 4.1  If $K$ is a closed convex set

(4.3) $$\tau(K, K) = K^{\infty}.$$

Proof:  If $u \in \tau(K, K)$ then $K \supset K + u$, and in consequence

$$K \supset K + u \supset K + 2u \supset \ldots \supset K + nu$$

for any nonnegative integer, and by Lemma 3.1, $u \in K^{\infty}$. Hence $\tau(K,K) \subset K^{\infty}$, and as the opposite relation is obvious the lemma is proved.

For any two convex sets $\mathcal{K}_1$ and $\mathcal{K}_2$, $\mathcal{K}_2$ is a set of translation of $\mathcal{K}_1$ into $\mathcal{K}_1 + \mathcal{K}_2$, and so is $\mathcal{K}_2 + \mathcal{K}_1^\infty$, where $\mathcal{K}_1^\infty$ is the recession cone of $\mathcal{K}_1$. The question of finding all the translations of $\mathcal{K}_1$ into $\mathcal{K}_1 + \mathcal{K}_2$ poses itself naturally. Are there any other translations besides those just described?

<u>Lemma 4.2</u> Let $\mathcal{K}_1$ and $\mathcal{K}_2$ be two closed convex sets and let $\mathcal{K}_2$ be bounded. Then

(4.4) $$\tau(\mathcal{K}_1, \mathcal{K}_1 + \mathcal{K}_2) = \mathcal{K}_1^\infty + \mathcal{K}_2 ,$$

(4.5) $$\tau(\mathcal{K}_2, \mathcal{K}_1 + \mathcal{K}_2) = \mathcal{K}_1 .$$

<u>Proof:</u> If $u \in \mathcal{K}_1^\infty + \mathcal{K}_2$, $u = x_1^\infty + x_2$ with $x_1^\infty \in \mathcal{K}_1^\infty$, $x_2 \in \mathcal{K}_2$ and $\mathcal{K}_1 + u = \mathcal{K}_1 + x_1^\infty + x_2 \subset \mathcal{K}_1 + x_2 \subset \mathcal{K}_1 + \mathcal{K}_2$, which says that $u \in \tau(\mathcal{K}_1, \mathcal{K}_1 + \mathcal{K}_2)$; hence $\mathcal{K}_1^\infty + \mathcal{K}_2 \subset \tau(\mathcal{K}_1, \mathcal{K}_1 + \mathcal{K}_2)$. To prove the opposite inclusion we remark first that the sum of two closed convex sets, one of which is bounded, is closed, and conclude that the sets $\mathcal{K}_1 + \mathcal{K}_2$, $\mathcal{K}_1^\infty + \mathcal{K}_2$ and $\tau(\mathcal{K}_1, \mathcal{K}_1 + \mathcal{K}_2)$ are closed. Let $u$ be a vector in $\tau(\mathcal{K}_1, \mathcal{K}_1 + \mathcal{K}_2)$, amd let $v = P_{\mathcal{K}_1^\infty + \mathcal{K}_2} u = x_{1,0}^\infty + x_{2,0}, x_{1,0}^\infty \in \mathcal{K}_1^\infty, x_{2,0} \in \mathcal{K}_2$. If we write $w = u - v$, we have by (1.2),

$$\langle w, x_{1,0}^\infty + x_{2,0} - x_1^\infty - x_2 \rangle \geq 0 , \forall x_1^\infty \in \mathcal{K}_1^\infty, \forall x_2 \in \mathcal{K}_2 .$$

Setting $x_1^\infty = x_{1,0}^\infty$ we obtain first

(4.6) $$\langle w, x_{2,0} - x_2 \rangle \geq 0 , \quad \forall x_2 \in \mathcal{K}_2 ,$$

and then setting $x_1^\infty = 0$, $x_2 = x_{2,0}$,

(4.7) $$\langle w, x_{1,0}^\infty \rangle \geq 0 .$$

Moreover, if we replace $x_2$ by $x_{2,0}$ and $x_1^\infty$ by $x_{1,0}^\infty + x_1^\infty$, we get

(4.8) $$\langle w, x_1^\infty \rangle \leq 0, \quad \forall\, x_1^\infty \in \mathcal{K}_1^\infty,$$

which says that $w \in (\mathcal{K}_1^\infty)^\perp$. Setting $x_1^\infty = x_{1,0}^\infty$ in (4.8) and comparing with (4.7) we derive

(4.9) $$\langle w, x_{1,0}^\infty \rangle = 0.$$

Since $u \in \tau(\mathcal{K}_1, \mathcal{K}_1 + \mathcal{K}_2)$, $P_{\mathcal{K}_1}(tw) + u$ is a point in $\mathcal{K}_1 + \mathcal{K}_2$ and therefore we can write $P_{\mathcal{K}_1}(tw) + u = x_1(t) + x_2(t)$, $x_1(t) \in \mathcal{K}_1$, $x_2(t) \in \mathcal{K}_2$. Then we have by (1.2)

$$\langle tw - P_{\mathcal{K}_1}(tw),\, P_{\mathcal{K}_1}(tw) - x_1(t) \rangle \geq 0, \quad \forall\, t,$$

whence replacing $P_{\mathcal{K}_1}(tw) - x_1(t)$ by $x_2(t) - u$, and dividing by $t$ supposed positive,

$$\langle w - P_{\mathcal{K}_1}(tw)/t,\, x_2(t) - u \rangle \geq 0$$

$$\langle w,\, x_2(t) - u \rangle \geq \langle t^{-1} P_{\mathcal{K}_1}(tw),\, x_2(t) - u \rangle, \quad \forall\, t > 0.$$

From (4.6) we obtain by setting $x_2 = x_2(t)$,

$$\langle w, x_{2,0} - x_2(t) \rangle \geq 0,$$

which added to the previous inequality yields

$$\langle w, x_{2,0} - u \rangle \geq \langle t^{-1} P_{\mathcal{K}_1}(tw),\, x_2(t) - u \rangle, \quad \forall\, t > 0.$$

We are now ready to pass to the limit $t \to +\infty$; we note that since $w \in (\mathcal{K}_1^\infty)^\perp$, $t^{-1} P_{\mathcal{K}_1}(tw) \to 0$ (Lemma 3.2), and that $x_2(t) - u \in \mathcal{K}_2 - u$ remains bounded. Thus

$$\langle w, x_{2,0} - u \rangle \geq 0.$$

But, $x_{2,0} - u = x_{1,0}^\infty + x_{2,0} - u - x_{1,0}^\infty = v - u - x_{1,0}^\infty = -(w + x_{1,0}^\infty)$, so on account of (4.9),

$$-\|w\|^2 = \langle w, -w\rangle = \langle w, -(w+x_{1,0}^\infty)\rangle = \langle w, x_{2,0}-u\rangle \geq 0,$$

that is, $w = 0$. Thus, $u = v \in K_1^\infty + K_2$. Since $u$ was an arbitrary element in $\tau(K_1, K_1+K_2)$, $\tau(K_1, K_1+K_2) \subset K_1^\infty + K_2$, and (4.4) is proved.

As to relation (4.5) its proof is much simpler. We know that $K_1 \subset \tau(K_2, K_1+K_2)$, so we only have to prove that $K_1 \supset \tau(K_2, K_1 + K_2)$. Let $u$ be an element of $\tau(K_2, K_1 + K_2)$ and let $x_{1,0} = P_{K_1} u$. Then if $w = u - x_{1,0}$,

$$\langle w, x_{1,0} - x_1\rangle \geq 0, \quad \forall x_1 \in K_1,$$

by (1.2). Since $\Re(I - P_{K_2}) = \mathcal{H}$, $w$ is normal to $K_2$ at some point $x_{2,0}$ and so

$$\langle w, x_{2,0} - x_2\rangle \geq 0, \quad \forall x_2 \in K_2,$$

which in combination with the previous inequality yields

(4.10) $\langle w, x_{1,0} + x_{2,0} - (x_1 + x_2)\rangle \geq 0, \quad \forall x_1 \in K_1, \forall x_2 \in K_2.$

The vector $u$ being in $\tau(K_2, K_1+K_2)$, $x_{2,0} + u = x_{1,0} + x_{2,0} + w$ lies in $K_1 + K_2$ and so can be identified with $x_1 + x_2$ in (4.10). Doing this we obtain $\|w\|^2 \leq 0$, that is, $w = 0$. It follows that $u \in K_1$, and hence that $K_1 \supset \tau(K_2, K_1+K_2)$.

We do not know precisely when relation (4.4) holds, but we know that it does not hold in general, even when both $K_1$ and $K_2$ are closed. This is an example: In the $\xi_1\xi_2$-plane let $K_1$ be the area enclosed by the parabola $\xi_1 = \xi_2^2$, and $K_2$ the negative $\xi_1$-axis. Then $K_1 + K_2$ and $\tau(K_1, K_1+K_2)$ are the whole plane, $K_1^\infty$ is the positive $\xi_1$-axis and $K_1^\infty + K_2$ is the whole $\xi_1$-axis; clearly $\tau(K_1, K_1+K_2) \neq K_1^\infty + K_2$.

Next we discuss briefly the notion of parallel convex bodies.

**Definition 4.2**  For any closed convex set $\mathcal{K}$, the sets

(4.11) $$\mathcal{K}_t = \begin{cases} \{x \mid \text{dist}(x,\mathcal{K}) \leq t\}, & t \geq 0, \\ \{x \mid \text{dist}(x, \text{complement of } \mathcal{K}) \geq -t\}, & t < 0, \end{cases}$$

are the outer and inner parallel sets to $\mathcal{K}$ according as $t \gtrless 0$ respectively.

As presented here this notion is not intrinsic, it depends on the surrounding space; to make it intrinsic one has to refer it to the space spanned by $\mathcal{K}$. Parallel sets can be obtained by parallel displacement of support hyperplanes as the lemma below shows; it follows that they are closed convex sets.

**Lemma 4.3**  For any closed convex set $\mathcal{K}$ and any real $t$,

(4.12) $$\mathcal{K}_t = \{x \mid \langle u,x \rangle \leq \sup_{y \in \mathcal{K}} \langle u,y \rangle + t,\ \forall u,\ \|u\| = 1\}.$$

**Proof:**  Let us consider the case of positive $t$ first. Let $x_0$ be any point in space; naturally $\text{dist}(x_0, \mathcal{K}) = \|x_0 - P_\mathcal{K} x_0\|$. Now, if $x_0 \in \mathcal{K}_t$, $\|x_0 - P_\mathcal{K} x_0\| \leq t$ and we have for any unitary $u$,

$$\langle u, x_0 \rangle = (\langle u, P_\mathcal{K} x_0 \rangle + t) - (t - \langle u, x_0 - P_\mathcal{K} x_0 \rangle) \leq \langle y, P_\mathcal{K} x_0 \rangle + t$$

$$\leq \sup_{y \in \mathcal{K}} \langle u, y \rangle + t$$

and so $\mathcal{K}_t$ is contained in the set on the right of (4.12). If $x_0 \notin \mathcal{K}_t$, $\|x_0 - P_\mathcal{K} x_0\| > t$ and setting $u = (x_0 - P_\mathcal{K} x_0)/\|x_0 - P_\mathcal{K} x_0\|$ we obtain for any $y \in \mathcal{K}$,

$$\langle u, x_0 \rangle = \langle u, P_\mathcal{K} x_0 \rangle + \langle u, x_0 - P_\mathcal{K} x_0 \rangle = \langle u, P_\mathcal{K} x_0 \rangle + \|x_0 - P_\mathcal{K} x_0\|$$

$$= \langle u, y \rangle + \langle u, P_\mathcal{K} x_0 - y \rangle + \|x_0 - P_\mathcal{K} x_0\|$$

$$\geq \langle u, y \rangle + \|x_0 - P_\mathcal{K} x_0\|,$$

and taking the sup with regard to y,

$$\langle u, x_0 \rangle \geq \sup_{y \in \mathcal{K}} \langle u, y \rangle + \|x_0 - P_\mathcal{K} x_0\| > \sup_{y \in \mathcal{K}} \langle u, y \rangle + t.$$

In consequence x does not belong to the set on the right of (4.12), and the equality of the two members of (4.12) is proved.

Assume now that t is negative. If $x_0$ is any point in $\mathcal{K}_t$ then for any unit vector u and any $t', t' < -t$, $x_0 + t'u \in \mathcal{K}$, and

$$\langle u, x_0 \rangle = \langle u, x_0 - t'u \rangle + t' \leq \sup_{y \in \mathcal{K}} \langle u, y \rangle + t',$$

whence letting $t' \downarrow t$,

$$\langle u, x_0 \rangle \leq \sup_{y \in \mathcal{K}} \langle u, y \rangle + t,$$

which indicates that x belongs to the right member of (4.12). On the other hand, if $x_0 \notin \mathcal{K}_t$ there is a z outside of $\mathcal{K}$ such that $\|z - x_0\| < -t$, and if $u = (z - P_\mathcal{K} z)/\|z - P_\mathcal{K} z\|$, $y \in \mathcal{K}$,

$$\langle u, x_0 \rangle = \langle u, z - P_\mathcal{K} z \rangle + \langle u, P_\mathcal{K} z - y \rangle + \langle u, y \rangle + \langle u, x_0 - z \rangle$$

$$= \|z - P_\mathcal{K} z\| + \frac{\langle z - P_\mathcal{K} z, P_\mathcal{K} z - y \rangle}{\|z - P_\mathcal{K} z\|} + \langle u, y \rangle + \langle u, x_0 - z \rangle$$

$$\geq \|z - P_\mathcal{K} z\| + \langle u, y \rangle + t, \qquad \forall y \in \mathcal{K},$$

and since $z \notin \mathcal{K}$,

$$\langle u, x_0 \rangle > \sup_{y \in \mathcal{K}} \langle u, y \rangle + t.$$

That is, $x_0$ is not in the right member of (4.12), and the proof is complete.

As a simple application of these ideas we shall prove a result to be needed later; it is a generalization of the well known fact that a closed surface whose normals go through a fixed point is a sphere.

Lemma 4.4    Any total convex body having at each point of its shell a normal going through a fixed point in space is a ball.

Proof:    The lemma is automatically valid if $\mathcal{K}$ is of dimension zero, in which case it reduces to a point; if the dimension is one, then the hypotheses are consistent only if the space $\mathcal{H}$ is also one dimensional and again the lemma holds provided half lines are considered as one dimensional balls. Discarding these cases we may assume that the dimension of $\mathcal{K}$ is two at least. For simplicity we shall suppose that the origin is the point where the normals concur. If x is a shell point of $\mathcal{K}$, then x is a normal to $\mathcal{K}$ at x, and therefore the same holds at any limit of shell points, which because of this fact is again a shell point. Now, any ray out of the origin intersecting $\mathcal{K}$ does so along a closed segment. A nonzero endpoint of this segment is the limit of projections of points in the ray outside of $\mathcal{K}$ as the endpoint is approached, and hence is shell point. It follows that the origin belongs to $\mathcal{K}$, for otherwise all points in the shell of $\mathcal{K}$ visible from the origin will have concurring outer normals which is impossible. Another consequence is that all shell points different from the origin are flat points. In fact, a vector of the form x+u, $u \in \mathbb{1}_{\mathcal{K}}(x)$, being outside $\mathcal{K}$ is an outer normal to $\mathcal{K}$ at a point $\alpha(x+u)$, $0 \leq \alpha \leq 1$ ($\alpha=0$ only if x=0); as it is also a normal at x and outer normals do not intersect, $\alpha(x+u)$ must coincide with x, so $u = (\alpha^{-1}-1)x$, and thus that there is only one direction in $\mathbb{1}_{\mathcal{K}}(x)$. The last part of our argument consists of showing that all shell points lie on a sphere. To this end we pick any two linearly independent vectors $x_1, x_2$ in the shell, and cut $\mathcal{K}$ with the two dimensional plane generated by these two vectors. We obtain a two dimensional convex set $\tilde{\mathcal{K}}$ for which the Lemma's hypotheses are satisfied. We let $\rho = \inf \|x\|$, $x \in$ shell $\tilde{\mathcal{K}}$, and consider $\tilde{\mathcal{K}}_{-\rho}$ relatively to the plane containing $\tilde{\mathcal{K}}$. Again, but possibly for spanning the space, $\mathcal{K}_{-\rho}$ meets the hypotheses, but now the origin belongs to the shell of $\tilde{\mathcal{K}}_{-\rho}$. This requires that $\tilde{\mathcal{K}}_{-\rho}$ be reduced to a single point, for if not, for any $x \in$ shell $\tilde{\mathcal{K}}_{-\rho}$, $x/\|x\| \in \mathcal{V}_{\tilde{\mathcal{K}}_{-\rho}}(x)$, $x/\|x\| \in \mathcal{V}_{\tilde{\mathcal{K}}_{-\rho}}^{\perp}(0)$, and it is enough to choose points $x_n$ in the shell in such a manner that $x_n \to 0$ and that $x_n/\|x_n\|$ converge to produce a contradiction. Hence $\tilde{\mathcal{K}}_{-\rho}$ contains the origin only and $\rho = \|x_1\| = \|x_2\|$. These two points being arbitrary points in the shell, the shell is part of a sphere about the origin. An appeal to Lemma 2.13 together with the fact that $\mathcal{K}$ must span the whole space brings the proof to an end.

At points outside of the range projections are not always differentiable. To have a look at this situation we begin by calculating the gradient (differential) of a projection onto a disc of radius $\rho$ in $\mathbb{R}^2$; we obtain for a point $x$ outside the disc

$$(4.13) \qquad \nabla P_{\mathcal{K}}(x) = \rho(\|x - P_{\mathcal{K}} x\| + \rho)^{-1} P_{s_{\mathcal{K}}(P_{\mathcal{K}} x)}$$

Based on this we then derive

$$(4.14) \qquad \nabla P_{\mathcal{K}}(x) = \rho(P_{\mathcal{K}} x)(\|x - P_{\mathcal{K}} x\| + \rho(P_{\mathcal{K}} x))^{-1} P_{s_{\mathcal{K}}(P_{\mathcal{K}} x)}$$

for the projection onto a closed convex set in $\mathbb{R}^2$ bounded by a $C^2$-curve having the radius of curvature $\rho(P_{\mathcal{K}}(x))$ at $P_{\mathcal{K}} x$.

Thus it is clear that the existence of a linear differential is tied up to the curvature of the boundary, which may not exist for a general convex figure. Yet, a projection, being a Lipschitz mapping, is almost everywhere differentiable. Formula (4.14) seems to make sense even if $P_{\mathcal{K}} x$ is a true vertex, provided that both arcs meeting at $P_{\mathcal{K}}(x)$ have a common limiting value of the curvature. In such a case however $\nabla P_{\mathcal{K}} x$ is not a linear operator, it is only homogeneous for positive factors, but satisfies the relation

$$P_K(x+h) = P_K x + (\nabla P_K x) h + o(h),$$

which characterizes differentials. These conclusions carry over to any finite number of dimensions, and presumably to infinite dimensions. At any rate, even in this last extended sense, a projection is not always differentiable outside of its range. In the same vein it has been shown that even in $\mathbb{R}^3$ projections do not necessarily have one sided directional derivatives [8]. Contrary to this, a projection is always differentiable at points in its range. To see it we need some estimates for the difference between the projection on a convex set and the projection on a support cone.

<u>Lemma 4.5</u>   For any closed convex set $K$

(4.15)    $P_K(x+h) = x+h+o(h), \quad x \in K, \quad h \in S_K(x),$

where $o(h)/\|h\| \to 0$ as $h \to 0$ over any locally compact cone of increments in $S_K(x)$.

<u>Proof:</u>   Equation (4.15) follows at once from (2.5); it only remains to prove the uniformity of the limit when the vectors $h/\|h\|$ are part of a compact set in $S_K(x)$. We proceed by contradiction: If the limit were not uniform, then there would be a positive $\varepsilon$ and a sequence $h_n \in S_K$ converging to zero with $h_n/\|h_n\| \to u \in S_K(x)$ such that $\|P_K(x+h_n) - x - h_n\|/\|h_n\| \geq \varepsilon$. Then noticing that for any positive $t$, $Ptx = tPx$,
$tK \quad K$
we would have

$$\varepsilon \leq \| P_{K-x}(h_n) - h_n \|/\|h_n\| = \| P_{\|h_n\|^{-1}(K-x)}(h_n/\|h_n\|) - (h_n/\|h_n\|) \|,$$

and since $t(K-x)$ grows with $t$,

$$\varepsilon \leq \| P_{\rho(K-x)}(h_n/\|h_n\|) - (h_n/\|h_n\|) \|, \qquad \rho > 0,$$

as soon as n is sufficiently large, whence passing to the limit it follows

$$\varepsilon \leq \| P_{\rho(\mathcal{K}-x)} u - u \|.$$

Thus

$$\frac{\varepsilon}{\rho} \leq \| P_{\mathcal{K}}(x+\frac{u}{\rho}) - (x-\frac{u}{\rho}) \|, \quad \forall \rho > 0,$$

in contradiction with (2.5).

<u>Lemma 4.6</u>   For any $x \in \mathcal{K}$,

(4.16)   $P_{\mathcal{K}}(x+h) = x + P_{S_{\mathcal{K}}(x)} h + o(\|h\|),$

where $o(\|h\|)/\|h\| \to 0$ as $h \to 0$ over any locally compact cone of increments.

<u>Proof:</u>   Clearly,

$$\|x+h - P_{x+S_{\mathcal{K}}(x)}(x+h)\|^2 + \|P_{x+S_{\mathcal{K}}(x)}(x+h) - P_{\mathcal{K}}(x+h)\|^2$$

$$+ 2\langle x+h - P_{x+S_{\mathcal{K}}(x)}(x+h), P_{x+S_{\mathcal{K}}(x)}(x+h) - P_{\mathcal{K}}(x+h)\rangle = \|x+h - P_{\mathcal{K}}(x+h)\|^2.$$

So, since the last term on the left is nonnegative and the right member does not exceed $\|x+h - P_{\mathcal{K}} P_{x+S_{\mathcal{K}}(x)}(x+h)\|^2$,

$$\|x+h - P_{x+S_{\mathcal{K}}(x)}(x+h)\|^2 + \|P_{x+S_{\mathcal{K}}(x)}(x+h) - P_{\mathcal{K}}(x+h)\|^2$$

$$\leq \|x+h - P_{\mathcal{K}} P_{x+S_{\mathcal{K}}(x)}(x+h)\|^2,$$

and noting the identity $P_{x+S_{\mathcal{K}}(x)}(x+h) = x + P_{S_{\mathcal{K}}(x)} h$,

$$\|h - P_{S_{\mathcal{K}}(x)} h\|^2 + \|x + P_{S_{\mathcal{K}}(x)} h - P_{\mathcal{K}}(x+h)\|^2 \le \|x+h - P_{\mathcal{K}}(x + P_{S_{\mathcal{K}}(x)} h)\|^2,$$

that is, since $h = P_{S_{\mathcal{K}}(x)} h + P_{U_{\mathcal{K}}(x)} h$,

$$\|P_{\mathcal{K}}(x+h) - x - P_{S_{\mathcal{K}}(x)} h\|$$

$$\le \|P_{U_{\mathcal{K}}(x)} h + x + P_{S_{\mathcal{K}}(x)} h - P_{\mathcal{K}}(x + P_{S_{\mathcal{K}}(x)} h)\|^2 - \|P_{U_{\mathcal{K}}(x)} h\|^2$$

$$\le \langle 2 P_{U_{\mathcal{K}}(x)} h + [x + P_{S_{\mathcal{K}}(x)} h - P_{\mathcal{K}}(x + P_{S_{\mathcal{K}}(x)} h)], [x + P_{S_{\mathcal{K}}(x)} h - P_{\mathcal{K}}(x + P_{S_{\mathcal{K}}(x)} h)] \rangle.$$

By Lemma 4.5 the expression in square brackets is $o(\|P_{S_{\mathcal{K}}(x)} h\|)$, and so by Schwarz' inequality,

$$\|P_{\mathcal{K}}(x+h) - x - P_{S_{\mathcal{K}}(x)} h\|^2 \le (2\|P_{U_{\mathcal{K}}(x)} h\| + o(\|P_{S_{\mathcal{K}}(x)} h\|)) o(\|P_{S_{\mathcal{K}}(x)} h\|),$$

and the proof concludes with the remark that the right hand member is $(o(\|h\|))^2$.

Equation (4.16) can be viewed as saying that $P_{S_{\mathcal{K}}(x)}$, although nonlinear, is the "differential of $P_{\mathcal{K}}$ at $x$". This nonlinear approximation to $P_{\mathcal{K}}(x+h)$, which may be called "conical differential", is closely related to the notion of "subdifferential" [cf. [13]).

To recognize that projections are gradient mappings we need an estimate of a different sort.

**Lemma 4.7** For any closed convex set $K$ and any couple of points $x, y$,

(4.17) $\quad \langle x - P_K x, P_K x - P_K y \rangle = o(\|P_K x - P_K y\|)$,

where $o(\|P_K x - P_K(y)\|)/\|P_K x - P_K y\| \to 0$ uniformly as $P_K x \to P_K y$ over any compact set of $x$'s.

**Proof:** By contradiction. Thus assume the existence of a sequence $x_n$ such that

$$\frac{\langle x_n - P_K x_n, P_K x_n - P_K y \rangle}{\|P_K x_n - P_K y\|} \geq \rho > 0, \qquad P_K x_n \to P_K y.$$

Passing to a subsequence if necessary we may further assume that $x_n \to x$, $(P_K x_n - P_K y)/\|P_K x_n - P_K y\| \to z$ for some $z \in S_K(P_K y)$. Letting $n \to \infty$, $P_K x_n \to P_K x = P_K y$, and from the above, we deduce

$$\langle x - P_K x, z \rangle \geq \rho > 0$$

which is impossible because $x - P_K x \in \mathcal{V}_K(P_K y)$, $z \in S_K(P_K y)$.

We also need the following definition:

**Definition 4.3** A real valued function $\alpha: H \to \mathbb{R}$ is said to be compact differentiable at a point $x$ interior to its domain of definition if there is a vector $u$ such that

(4.18) $\quad \alpha(x+h) = \alpha(x) + \langle h, u \rangle + o(x; h)$

where

$$o(x; h)/\|h\| \to 0$$

as $h \to 0$ over any compact set of $h$'s. The vector $u$ is called the compact gradient of $\alpha$ at $x$, and is denoted $\nabla \alpha(x)$.

This type of differentiability is related to the notion

of Hadamard differential as given by M. Fréchet [5]. For this and similar questions we refer the reader to M. Z. Nashed's article in the proceedings of the 1970 MRC Advanced Seminar on Nonlinear Functional Analysis [16] (see also [26]).

**Theorem 4.1**  A Lipschitz mapping $P: H \to H$ is the projection on a closed convex set if and only if it satisfies the differential equation

(4.19) $\qquad (I-P)x = \nabla \tfrac{1}{2} \|(I-P)x\|^2, \qquad \forall x \in H,$

where $\nabla$ denotes the compact gradient.

**Proof:**  Necessity is a consequence of (4.17) through the following identity

$$\tfrac{1}{2}\|(I-P_K)(x+h)\|^2 = \tfrac{1}{2}\|(I-P_K)x\|^2 + \langle h, (I-P_K)x \rangle$$
$$+ \tfrac{1}{2}[\|(I-P_K)(x+h) - (I-P_K)x\|^2 - 2\langle (I-P_K)x, P_K x - P_K(x+h) \rangle].$$

The proof of sufficiency is rather elementary. Let us write $Q = I - P$ and observe that since $\|Qx\|^2$ is continuously differentiable so is $\|Qx\|$ at points where it does not vanish, and

(4.20) $\qquad Qx/\|Qx\| = \nabla \|Qx\|, \qquad Qx \neq 0.$

Now, through any point $x$ at which $Qx \neq 0$ goes a unique field line $x(s)$ (i.e. a differentiable curve $s \to x(s)$ defined in some interval in $\mathbb{R}$ satisfying the equation $dx(s)/ds = Qx(s)/\|Qx(s)\|$). Such field line may end only at a point where $Q$ vanishes, and along it

$$\tfrac{d}{ds}\|Qx(s)\| = \langle \nabla \|Qx(s)\|, \tfrac{dx(s)}{dx} \rangle = \langle \tfrac{Qx(s)}{\|Qx(s)\|}, \tfrac{Qx(s)}{\|Qx(s)\|} \rangle = 1,$$

by the chain differentiation rule (cf. [16, Theorem 1.8]). Hence if $s_2 \geq s_1$,

(4.21) $$\|Qx(s_2)\| - \|Qx(s_1)\| = s_2 - s_1 \geq \|x(s_2) - x(s_1)\|.$$

On the other hand, by the mean value theorem, if $Q$ does not vanish on the segment joining $x_1$ and $x_2$,

$$\|Qx_2\| - \|Qx_1\| = \|Q(tx_2 + (1-t)x_1)\|_{t=1} - \|Q(tx_2 + (1-t)x_1)\|_{t=0}$$

$$= \left(\frac{d}{dt}\|Q(tx_2 + (1-t)x_1)\|\right)_{t=\theta}$$

(4.22) $$= \langle \nabla \|Q(\theta x_1 + (1-\theta)x_2)\|, x_2 - x_1 \rangle \leq$$

$$\leq \|\nabla\|Q(\theta x_1 + (1-\theta)x_2)\|\| \|x_2 - x_1\|$$

$$= \|x_2 - x_1\|.$$

In particular, letting $x_i = x(s_i)$, $i = 1, 2$,

$$\|Qx(s_2)\| - \|Qx(s_1)\| \leq \|x(s_2) - x(s_1)\|,$$

for $s_2 - s_1$ sufficiently small, and comparison with (4.21) yields

$$\|Qx(s_2)\| - \|Qx(s_1)\| = \|x(s_2) - x(s_1)\| = s_2 - s_1.$$

It follows that field lines are locally straight lines. Moreover, if a field line $x(s)$ is defined for $s = s_0$, then it is defined for any $s > s_0 - \|Qx(s_0)\|$, because for such $s$'s, $Qx(s) \neq 0$. By continuity $Qx(s_0 - \|Qx(s_0)\|) = 0$, and any field line in fact ends at a point in $\mathcal{K} = \{x \mid Qx = 0\}$. If such points are taken as the origin to measure arc lengths along field lines

$$Qx(s) = x(s) - x(0)$$

or equivalently

$$Qx = x - x_{\mathcal{K}}$$

where

$$x_\mathcal{K} = \begin{cases} x & \text{if } x \in \mathcal{K} \\ \text{end point of field line through } x, & \text{if } x \notin \mathcal{K}. \end{cases}$$

From (4.22) one obtains for any $x \in \mathcal{H}$ and any $x' \in \mathcal{K}$, if $x''$ denotes the point in $\mathcal{K}$ on the closed segment joining $x$ to $x'$ nearest to $x$,

$$\|x-x_\mathcal{K}\| = \|Qx\| = \|Qx-Qx''\| \leq \|x-x''\| \leq \|x-x'\|,$$

which indicates that $x_\mathcal{K}$ is a point in $\mathcal{K}$ minimizing the distance to $x$. The straight line $x(s) = x_\mathcal{K} + s \frac{x-x_\mathcal{K}}{\|x-x_\mathcal{K}\|}$, $s > 0$ is the field line through $x$, and so $x(s)_\mathcal{K} = x_\mathcal{K}$. Hence, for any $x' \in \mathcal{K}$ and $\forall s \geq 0$,

$$0 \leq \|x(s)-x'\|^2 - \|x(s)-x_\mathcal{K}\|^2 = \|x_\mathcal{K}-x'+s\frac{x-x_\mathcal{K}}{\|x-x_\mathcal{K}\|}\|^2 - s^2$$

$$= \frac{2s}{\|x-x_\mathcal{K}\|} \langle x-x_\mathcal{K}, x_\mathcal{K}-x' \rangle + \|x_\mathcal{K}-x'\|^2,$$

whence on division by $2s/\|x-x_\mathcal{K}\|$ and passage to the limit $(s \to +\infty)$,

$$\langle x-x_\mathcal{K}, x_\mathcal{K}-x' \rangle \geq 0, \qquad \forall x' \in \mathcal{K}.$$

Therefore, by Theorem 1.1, $x_\mathcal{K}$ is the projection of $x$ on the closed convex hull of $\mathcal{K}$. But as $\mathcal{K}$ is the set of fixed points of the mapping $x \to x_\mathcal{K}$, $\mathcal{K}$ coincides with its closed convex hull and $x_\mathcal{K} = P_\mathcal{K} x$. In other words, $P = P_\mathcal{K}$.      Q.E.D.

Noting that $P_\mathcal{K} x = x - (I-P_\mathcal{K})x = \nabla \frac{1}{2}[\|x\|^2 - \|(I-P_\mathcal{K})x\|^2]$, we obtain:

<u>Corollary 1</u>   Projections and their complements are gradient mappings.

Moreover, since both $P_\mathcal{K}$ and $I-P_\mathcal{K}$ are monotone mappings:

<u>Corollary 2</u>  Both $\|x-P_\mathcal{K} x\|^2$ and $\|x\|^2 - \|x-P_\mathcal{K} x\|^2$ are convex functions.

It is not unreasonable to conjecture that the compact gradient in (4.19) can be replaced by the Gateaux gradient, as the choice of the former appears as dictated by the requirements of the proof rather than by the nature of things.

A significant consequence to be drawn from the above discussion is that, projections being gradient mappings, anything constructed or represented out of them in an additive fashion must, in one way or another, partake of that character and be a gradient mapping again.

Linear mappings are symmetric if and only if they are gradient mappings, namely the gradients of half their associated quadratic forms, and so the notion of gradient mapping might be thought as an extension to the nonlinear realm of the concept of symmetry for linear mappings. However, this extension appears to be insufficient in most situations. For instance, it is not true that any idempotent, continuous gradient mapping is a projection; simply think of the following mapping in $\mathbb{R}$:

$$Px = \begin{cases} 0, & x \leq 0 \\ x, & 0 < x \leq 1 \\ 2-x, & 1 < x \leq 2 \\ 0, & 2 < x \end{cases}.$$

## §5. The algebra of projections.

The class of orthogonal projections on linear subspaces has simple but very useful algebraic properties that make the spectral theory of various types of linear operators in Hilbert space possible. Therefore it is natural to investigate to which extent projections in general participate in such properties. In this search the findings by far exceed the expectations, specially when restricted to projections on convex cones, where enough remains of the original structure to form still a basis for an extended spectral theory for a a class of operators, not necessarily linear, containing the linear selfadjoint ones. Such a theory is the subject of part II. Here in this section we study the meaning of the relations $P_{K_2}P_{K_1} = P_{K_2}$, $P_{K_1}+P_{K_2} = P_{K_3}$ and $P_{K_1}P_{K_2} = P_{K_2}P_{K_1}$ for general projections on convex sets.

**Lemma 5.1** If $P_{K_2}P_{K_1} = P_{K_2}$ then

(5.1) $$U_{K_1}(x) \subset U_{K_2}(P_{K_2}x),$$

(5.2) $$x+U_{K_1}(x) \subset P_{K_2}x + U_{K_2}(P_{K_2}x), \quad \forall x \in K_1.$$

**Proof:** If $u \in U_{K_1}(x)$, then $P_{K_1}(x+tu)=x$, $\forall t > 0$ and $P_{K_2}(x+tu) = P_{K_2}P_{K_1}(x+tu) = P_{K_2}x$. Thus $x+tu-P_{K_2}x \in U_{K_2}(P_{K_2}x)$, $\forall t \geq 0$, and passage to the limit $t \to \infty$ after division by $t$ yields $u \in U_{K_2}(P_{K_2}x)$. Hence (5.1) is proved. As to (5.2) it follows from (5.1) by adding $x-P_{K_2}x$ after observing that $x-P_{K_2}x \in U_{K_2}(P_{K_2}x)$.

**Lemma 5.2** If $P_{K_2}P_{K_1} = P_{K_2}$, then

(5.3) $$P_{K_2}(\mathfrak{I}_{K_1}(u)) \subset \mathfrak{I}_{K_2}(u), \quad \forall u \in \mathcal{H}.$$

Moreover, if in addition $P_{K_1}P_{K_2} = P_{K_2}$ then there is an $\alpha_0$, $0 \leq \alpha_0 \leq +\infty$ such that

$$\mathfrak{I}_{K_2}(u)+\alpha u \subset \begin{cases} K_1, & 0 \leq \alpha \leq \alpha_0 \\ \mathcal{H} \setminus K_1, & \alpha_0 < \alpha. \end{cases}$$

**Proof:** For any $x_1 \in \mathfrak{I}_{K_1}(u)$, $u \in U_{K_1}(x_1) \subset U_{K_2}(P_{K_2}x_1)$. Hence $P_{K_2}x_1 = P_{K_2}(P_{K_2}x_1 + u)$ and $P_{K_2}x_1 \in \mathfrak{I}_{K_2}(u)$ by (2.12). This proves (5.3).

307

To prove (5.4) begin by assuming that $\mathfrak{F}_{K_2}(u)$ contains more than one point. It suffices to show that the segment joining any two points $x_2'$ and $x_2''$ in $\mathfrak{F}_{K_2}(u)$ can be embedded into the shell of $K_1$ by a translation in the u-direction. The vertices $\mathfrak{v}_{K_2}(x_2(s))$ are the same at all points $x_2(s) = sx_2'$ $+ (1-s)x_2''$, $0 < s < 1$, they all contain $u$ and are perpendicular to $x''-x'$. Let $t(s) = \sup t$ for all nonnegative t's such that $x_2(s)+tu \subset K_1$; since $K_2 \subset K_1$, $0 \leq t(s) \leq +\infty$. The points $x_1(s) = x_2(s)+t(s)u$ belong to the boundary of the intersection $\tilde{K}_1$ of $K_1$ with the two dimensional plane through $x_2'$ and $x_2''$ parallel to u; $t(s) = +\infty$ for all s or for none. Naturally, $P_{K_2}x_1(s) = x_2(s)$ and by (5.1), $\mathfrak{v}_{K_1}(x_1(s)) \subset \mathfrak{v}_{K_2}(x_2(s))$, or what is the same $\mathfrak{s}_{K_1}(x_1(s)) \supset \mathfrak{s}_{K_2}(x_2(s))$, $0 < s < 1$. Since for any such s, $\mathfrak{s}_{K_2}(x_2(s))$ contains the whole straight line parallel to $x_2'-x_2''$, so does $\mathfrak{s}_{K_1}(x_1(s))$. But then this also holds relatively to $\tilde{K}_1$, that is, $\mathfrak{s}_{\tilde{K}_1}(x_1(s))$ contains $\{r(x_2'-x_2'')\}_{-\infty < r < +\infty}$. Thus the $\mathfrak{v}_{\tilde{K}_1}(x_1(s))$, $0 < s < 1$, all contain a direction perpendicular to $x_2'-x_2''$, namely u, and the curve $x_1(s)$, $0 < s < 1$ is a straight segment parallel to $x_2'-x_2''$, resulting by parallel translation in the direction of u of the segment $x_2(s)$, $0 < s < 1$, as we wished to prove. If $\mathfrak{F}_{K_2}(u)$ contains only one point the conclusion is almost immediate and is left to the reader's care.

**Lemma 5.3**    The relation $P_{K_2}P_{K_1} = P_{K_2}$ is equivalent to

(5.5) $$\mathfrak{R}(P_{K_1} - P_{K_2}) + K_2 \subset K_1.$$

**Proof:** To show that the first relation implies the second, we have to prove that $y+x_2 \in K_1$ for $\forall y \in \mathfrak{R}(P_{K_1} - P_{K_2})$ and $\forall x_2 \subset K_2$. To this end starting with $x_2$ we build the sequence $\{x_2^{(n)}\}_1^\infty$ of the iterates by the operation $x \to P_{K_2}(x+y)$. Let us first remark that since $y \in \mathfrak{R}(P_{K_1}-P_{K_2})$ $= \mathfrak{R}((I-P_{K_2})P_{K_1}) \in \mathfrak{R}(I-P_{K_2})$ the face $\mathfrak{F}_{K_2}(y)$ is not empty (Lemma 2.16) and consists of all the fixed points of the above mapping. But then as the mapping is nonexpansive the sequence $\{\|x_2^{(n)}-x_2^{(\infty)}\|\}_1^\infty$ is nonincreasing for any $x_2^\infty \in \mathfrak{Z}_{\mathcal{K}_1}\mathfrak{Z}_{\mathcal{K}_2}(y)$; moreover, as it has been seen in the course of the proof of Lemma 3.6, (equation (3.6) and below) $x_2^{(n)}-x_2^{(n-1)} \to 0$. Now by Lemma 5.1,

$$x_2^{(n-1)}+y-P_{\mathcal{K}_1}(x_2^{(n-1)}+y) \in \mathcal{U}_{\mathcal{K}_1}(P_{\mathcal{K}_1}(x_2^{(n-1)}+y))$$

$$\subset \mathcal{U}_{\mathcal{K}_2}(P_{\mathcal{K}_2}P_{\mathcal{K}_1}(x_2^{(n-1)}+y)) = \mathcal{U}_{\mathcal{K}_2}(x_2^{(n)}),$$

and we can write,

$$\langle x_2^{(n-1)}+y-P_{\mathcal{K}_1}(x_2^{(n-1)}+y), x_2^{(n)}-x_2^{(n-1)}\rangle \geq 0.$$

Adding to this inequality

$$\langle x_2^{n-1}+y-P_{\mathcal{K}_1}(x_2^{(n-1)}+y), P_{\mathcal{K}_1}(x_2^{(n-1)}+y)-P_{\mathcal{K}_1}(x_2^{(n)}+y\rangle \geq 0$$

one obtains

$$\langle x_2^{(n-1)}+y-P_{\mathcal{K}_1}(x_2^{(n-1)}+y),(x_2^{(n)}+y-P_{\mathcal{K}_1}(x_2^{(n)}+y))-(x_2^{(n-1)}+y-P_{\mathcal{K}_1}(x_2^{(n-1)}+y))\rangle \geq 0$$

that is,

$$\langle x_2^{n-1}+y-P_{\mathcal{K}_1}(x_2^{n-1}+y), x_2^{(n)}+y-P_{\mathcal{K}_1}(x_2^{(n)}+y)\rangle$$

$$\geq \|x_2^{(n-1)}+y-P_{\mathcal{K}_1}(x_2^{(n-1)}+y)\|^2.$$

The application of Schwarz inequality to the left member yields

(5.6) $\|x_2^{(n)}+y-P_{\mathcal{K}_1}(x_2^{(n)}+y)\| \geq \|x_2^{(n-1)}+y-P_{\mathcal{K}_1}(x_2^{(n-1)}+u)\|$;

in other words, the sequence $\{\|x_2^{(n)}+y-P_{\mathcal{K}_1}(x_2^{(n)}+y)\|\}_1^\infty$ is nondecreasing. By the choice of $y$ there is an $x$ such that $y = P_{\mathcal{K}_1}x-P_{\mathcal{K}_2}x$, that is, such that $y+P_{\mathcal{K}_2}P_{\mathcal{K}_1}x = P_{\mathcal{K}_1}x$. Hence $P_{\mathcal{K}_2}(P_{\mathcal{K}_2}P_{\mathcal{K}_1}x+y) = P_{\mathcal{K}_2}P_{\mathcal{K}_1}x$, and the point $x_2^\infty = P_{\mathcal{K}_2}P_{\mathcal{K}_1}x$

belongs to $\mathcal{F}_{\mathcal{K}_2}(y)$. Since $x_2^\infty + y = P_{\mathcal{K}_1} x \in \mathcal{K}_1$, we may write for all $n$,

$$\langle x_2^{(n)} + y - P_{\mathcal{K}_1}(x_2^{(n)} + y), P_{\mathcal{K}_1}(x_2^{(n)} + y) - x_2^\infty - y \rangle \geq 0$$

or equivalently,

(5.7) $\langle x_2^{(n)} + y - P_{\mathcal{K}_1}(x_2^{(n)} + y), x_2^{(n)} - x_2^\infty \rangle \geq \| x_2^{(n)} + y - P_{\mathcal{K}_1}(x_2^{(n)} + y) \|^2$.

But we also have (equation (1.5)),

$$\langle (I - P_{\mathcal{K}_2}) P_{\mathcal{K}_1}(x_2^{(n-1)} + y) - (I - P_{\mathcal{K}_2})(x_2^\infty + y), P_{\mathcal{K}_2}(x_2^{(n-1)} + y) - P_{\mathcal{K}_2}(x_2^\infty + y) \rangle \geq 0,$$

which by virtue of the identities

$$P_{\mathcal{K}_2} P_{\mathcal{K}_1}(x_2^{(n-1)} + y) = P_{\mathcal{K}_2}(x_2^{(n-1)} + y) = x_2^{(n)},$$

$$P_{\mathcal{K}_2}(x_2^\infty + y) = x_2^\infty,$$

becomes

(5.8) $\langle P_{\mathcal{K}_1}(x_2^{(n-1)} + y) - x_2^{(n)} - y, x_2^{(n)} - x_2^\infty \rangle \geq 0$.

Adding (5.7) and (5.8) one deduces,

$$\langle P_{\mathcal{K}_1}(x_2^{(n-1)} + y) - P_{\mathcal{K}_1}(x_2^{(n)} + y), x_2^{(n)} - x_2^\infty \rangle \geq \| x_2^{(n)} + y - P_{\mathcal{K}_1}(x_2^{(n)} + y) \|^2,$$

and an appeal to Schwarz inequality and to the contractivity of projections leads to

$$\| x_2^{(n)} + y - P_{\mathcal{K}_1}(x_2^{(n)} + y) \|^2 \leq \| P_{\mathcal{K}_1}(x_2^{(n-1)} + y) - P_{\mathcal{K}_1}(x_2^{(n)} + y) \| \, \| x_2^{(n)} - x_2^\infty \|$$

$$\leq \| x_2^{(n-1)} - x_2^{(n)} \| \, \| x_2^{(n)} - x_2^\infty \|.$$

But $\| x_2^{(n-1)} - x_2^{(n)} \| \to 0$ and $\| x_2^{(n)} - x_2^\infty \|$ is decreasing, so

$$\| x_2^{(n)} + y - P_{\mathcal{K}_1}(x_2^{(n)} + y) \|^2 \to 0,$$

and as $\{\|x_2^{(n)}+y-P_{\mathcal{K}_1}(x_2^{(n)}+y)\|\}_1^\infty$ is nondecreasing, all its terms vanish. In particular, for $n = 1$,

$$x_2 + y = P_{\mathcal{K}_1}(x_2+y),$$

and $x_2+y \in \mathcal{K}_1$, as we set out to prove. The proof that (5.5) implies $P_{\mathcal{K}_2}P_{\mathcal{K}_1} = P_{\mathcal{K}_2}$ is much simpler. Since for any $x$, $z$
$= P_{\mathcal{K}_1}x - P_{\mathcal{K}_2}P_{\mathcal{K}_1}x \in \mathcal{R}(P_{\mathcal{K}_1} - P_{\mathcal{K}_2})$, then by (5.5), $\mathcal{K}_2 + z \in \mathcal{K}_1$.
Therefore since $P_{\mathcal{K}_1}x \in \mathcal{K}_2 + z$, we have $P_{\mathcal{K}_1}x = P_{\mathcal{K}_2+z}x$, whence
$x - P_{\mathcal{K}_1}x = x - P_{\mathcal{K}_2+z}x \in \mathfrak{l}_{\mathcal{K}_2+z}(P_{\mathcal{K}_1}x) = \mathfrak{l}_{\mathcal{K}_2}(P_{\mathcal{K}_1}x-z) = \mathfrak{l}_{\mathcal{K}_2}(P_{\mathcal{K}_2}P_{\mathcal{K}_1}x),\ \forall x \in \mathcal{H}.$
On the other hand $P_{\mathcal{K}_1}x - P_{\mathcal{K}_2}P_{\mathcal{K}_1}x \in \mathfrak{l}_{\mathcal{K}_2}(P_{\mathcal{K}_2}P_{\mathcal{K}_1}x)$, and so

$$x - P_{\mathcal{K}_2}P_{\mathcal{K}_1}x = x - P_{\mathcal{K}_1}x + P_{\mathcal{K}_1}x - P_{\mathcal{K}_2}P_{\mathcal{K}_1}x \subset \mathfrak{l}_{\mathcal{K}_2}(P_{\mathcal{K}_2}P_{\mathcal{K}_1}x)$$
$$+ \mathfrak{l}_{\mathcal{K}_2}(P_{\mathcal{K}_2}P_{\mathcal{K}_1}x) = \mathfrak{l}_{\mathcal{K}_2}(P_{\mathcal{K}_2}P_{\mathcal{K}_1}x),$$

which is equivalent to $P_{\mathcal{K}_2}x = P_{\mathcal{K}_2}P_{\mathcal{K}_1}x$.

<u>Lemma 5.4</u>   $P_{\mathcal{K}_2}P_{\mathcal{K}_1} = P_{\mathcal{K}_2}$ if and only if

(5.9)     $\overline{\mathcal{R}(P_{\mathcal{K}_1}-P_{\mathcal{K}_2})} = \tau(\mathcal{K}_2,\mathcal{K}_1) \cap (\mathcal{K}_2^\infty)^\perp.$

<u>Proof:</u>   Since $\tau(\mathcal{K}_2,\mathcal{K}_1) = \bigcap_{x_2 \in \mathcal{K}_2}(\mathcal{K}_1 - x_2)$ (cf. (4.2)), and since by Lemma 5.3, $\mathcal{K}_1 - x_2 \supset \mathcal{R}(P_{\mathcal{K}_1}-P_{\mathcal{K}_2})$,

(5.10)    $\mathcal{R}(P_{\mathcal{K}_1}-P_{\mathcal{K}_2}) \subset \tau(\mathcal{K}_2,\mathcal{K}_1).$

Moreover, if $u \in \mathcal{R}(P_{\mathcal{K}_1}-P_{\mathcal{K}_2})$, $u = x_1 - P_{\mathcal{K}_2}x_1 \in \mathfrak{l}_{\mathcal{K}_2}(P_{\mathcal{K}_2}x_1) \subset (\mathcal{K}_2^\infty)^\perp$

(Lemma 3.1), and

(5.11) $$\overline{R(P_{K_1} - P_{K_2})} \subset (K_2^\infty)^\perp .$$

Then (5.10) and (5.11) together yield

(5.12) $$\overline{R(P_{K_1} - P_{K_2})} \subset \tau(K_2, K_1) \cap (K_2^\infty)^\perp .$$

To prove the opposite relation construct with a fixed element $x_2^0$ in $K_2$, any $u \in \tau(K_2, K_1) \cap (K_2^\infty)^\perp$, and a positive real $t$, the point $x(t) = P_{K_2}(x_2^0 + tu) + t^{-1}(x_2^0 + tu - P_{K_2}(x_2^0 + tu))$. Since

$$x(t) = u + t^{-1} x_2^0 + (1 - t^{-1}) P_{K_2}(x_2^0 + tu) ,$$

$x(t) \in u + K_2$, and as $u$ is a translation of $K_2$ into $K_1$, $x(t) \in K_1$. On the other hand $P_{K_2}(x(t)) = P_{K_2}(x_2^0 + tu)$, and

$$x(t) - P_{K_2}(x(t)) = u + t^{-1}(x_2^0 - P_{K_2}(x_2^0 + tu)).$$

Since $u \in (K_2^\infty)^\perp$, $\lim_{t \to +\infty} P_{K_2}(x_2^0 + tu)/t = 0$ (Lemma 3.2), and letting $t \to +\infty$ in the equation above, we obtain

$$u = \lim_{t \to +\infty} (x(t) - P_{K_2}(x(t))) ,$$

which, since $x(t) - P_{K_2}(x(t)) = (P_{K_1} - P_{K_2}) x(t) \in R(P_{K_1} - P_{K_2})$, implies $u \in \overline{R(P_{K_1} - P_{K_2})}$. But $u$ was any vector in $\tau(K_2, K_1) \cap (K_2^\infty)^\perp$, so

(5.13) $$\overline{R(P_{K_1} - P_{K_2})} \supset \tau(K_2, K_1) \cap (K_2^\infty)^\perp ,$$

completing the proof of (5.9). In view of Lemma 5.3 the converse is obvious.

**Corollary** If $P_{\mathcal{K}_2}P_{\mathcal{K}_1} = P_{\mathcal{K}_2}$ the closure of the range of $P_{\mathcal{K}_1} - P_{\mathcal{K}_2}$ is convex.

**Theorem 5.1** If $P_{\mathcal{K}_2}P_{\mathcal{K}_1} = P_{\mathcal{K}_2}$ then

(5.14) $\quad P_{\mathcal{K}_1} = P_{\mathcal{K}_2} + P_{\mathcal{K}_{1,2}}(I - P_{\mathcal{K}_2}) = P_{\mathcal{K}_2} + P_{\tau(\mathcal{K}_2, \mathcal{K}_1)}(I - P_{\mathcal{K}_2})$

where

(5.15) $\quad \mathcal{K}_{1,2} = \tau(\mathcal{K}_2, \mathcal{K}_1) \cap (\mathcal{K}_2^\infty)^\perp = \overline{\mathcal{R}(P_{\mathcal{K}_1} - P_{\mathcal{K}_2})}.$

**Proof:** For any point in space, $P_{\mathcal{K}_1}x - P_{\mathcal{K}_2}x \in \mathcal{K}_{1,2}$, $P_{\mathcal{K}_1}x \in P_{\mathcal{K}_2}x + \mathcal{K}_{1,2}$. Therefore since $\mathcal{K}_{1,2} + P_{\mathcal{K}_2}x \in \mathcal{K}_1$ (Lemma 5.3),

$$P_{\mathcal{K}_1}x = P_{\mathcal{K}_{1,2} + P_{\mathcal{K}_2}x}x = P_{\mathcal{K}_{1,2}}(x - P_{\mathcal{K}_2}x) + P_{\mathcal{K}_2}x = P_{\mathcal{K}_2} + P_{\mathcal{K}_{1,2}}(I - P_{\mathcal{K}_2}))x.$$

The proof of the second equation in (5.14) is the same, one only has to replace $\mathcal{K}_{1,2}$ by $\tau(\mathcal{K}_2, \mathcal{K}_1)$.

**Lemma 5.5** If $P_{\mathcal{K}_2}P_{\mathcal{K}_1} = P_{\mathcal{K}_2}$ then

(5.16) $\quad P_{\mathcal{K}_2 + u}P_{\mathcal{K}_1} = P_{\mathcal{K}_1}P_{\mathcal{K}_2 + u} = P_{\mathcal{K}_2 + u}$,

where u is the smallest displacement bringing $\mathcal{K}_2$ in contact with $\mathcal{K}_1$. u is the projection of the origin onto $\mathcal{K}_1 - \mathcal{K}_2$, and it is perpendicular to $\mathcal{K}_2$; $\mathcal{K}_2$ is parallel to $\mathfrak{F}_{\mathcal{K}_1}(-u)$.

**Proof:** If $x_2$ is any point in $\mathcal{K}_2$ and if one writes $x_1 = P_{\mathcal{K}_1}x_2$, then

$$x_2 = P_{K_2} x_2 = P_{K_2} P_{K_1} x = P_{K_2} x_1 .$$

The pair of relations $x_1 = P_{K_1} x_2$, $x_2 = P_{K_2} x_1$ indicate that $x_1$ and $x_2$ realize the distance between $K_1$ and $K_2$. It follows by the arbitrariness of $x_2$ in $K_2$ that all points of $K_2$ are at a fixed distance from $K_1$ and that the vector $v = -x_2 + P_{K_1} x_2$ is independent of $x_2$. Clearly $u$ is the smallest displacement bringing $K_2$ in touch with $K_1$; $-u$ is perpendicular to $K_2$, and it is normal to $K_1$ at all the points which are projections of points in $K_2$. Therefore $K_2$ is parallel to $\mathfrak{J}_{K_1}(-u)$ and the translation $u$ embeds $K_2$ into $\mathfrak{J}_{K_1}(u)$. Hence $K_2 + u \in K_1$, and in consequence

$$P_{K_1} P_{K_2} + u = P_{K_2} + u .$$

Moreover, since $u$ is perpendicular to $K_2$, $P_{K_2 + u} = P_{K_2} + u$, and

$$P_{K_2 + u} P_{K_1} = P_{K_2} P_{K_1} + u = P_{K_2} + u = P_{K_2} + u .$$

Now we apply the results obtained above to closed convex cones.

<u>Theorem 5.2</u>  Let $C_1$ and $C_2$ be closed convex cones with vertex at the origin. Then $P_{C_2} P_{C_1} = P_{C_2}$ implies:

(5.17) $$P_{C_1} P_{C_2} = P_{C_2} P_{C_1} = P_{C_2} ,$$

(5.18) $$P_{C_1} P_{C_2^\perp} = P_{C_2^\perp} P_{C_1} = P_{C_1} - P_{C_2} = P_{C_1 \cap C_2^\perp} ,$$

(5.19) $$P_{C_2} P_{C_1^\perp} = P_{C_1^\perp} P_{C_2} = 0 ,$$

(5.20) $$P_{C_1^\perp} P_{C_2^\perp} = P_{C_2^\perp} P_{C_1^\perp} = P_{C_1^\perp} .$$

Proof: Relations (5.17) are direct consequences of Lemma 5.5, since $C_1$ and $C_2$ have the origin in common. It follows that $C_2 \subset C_1$, and since $\tau(C_2, C_1) = C_1$, by (5.14),

$$P_{C_1} = P_{C_2} + P_{C_1} P_{C_2^\perp},$$

that is

$$P_{C_1} - P_{C_2} = P_{C_1} P_{C_2^\perp}.$$

On the other hand since $P_{C_2} P_{C_1} = P_{C_2}$,

$$P_{C_1} - P_{C_2} = P_{C_1} - P_{C_2} P_{C_1} = P_{C_2^\perp} P_{C_1},$$

and the first two equations (5.18) are proved. By (1.2) applied to $P_{C_1}$ and $P_{C_2^\perp}$ respectively,

$$\langle P_{C_2^\perp} x - P_{C_1} P_{C_2^\perp} x, \; P_{C_1} P_{C_2^\perp} x - y \rangle \geq 0, \quad \forall x \in \mathcal{H}, \; \forall y \in C_1$$

$$\langle P_{C_1} x - P_{C_2^\perp} P_{C_1} x, \; P_{C_2^\perp} P_{C_1} x - y \rangle \geq 0, \quad \forall x \in \mathcal{H}, \; \forall y \in C_2^\perp.$$

Hence if $y \in C_2^\perp \cap C_1$, both inequalities above hold. Adding them up and recalling that $P_{C_1} P_{C_2^\perp} = P_{C_2^\perp} + P_{C_1} = P_{C_1} - P_{C_2}$, one obtains

$$\langle P_{C_2^\perp} x + P_{C_1} x - 2(P_{C_1} x - P_{C_2} x), \; P_{C_1} x - P_{C_2} x - y \rangle \geq 0,$$

$$\forall y \in C_2^\perp \cap C_1,$$

or, since $x = P_{C_2^\perp} x + P_{C_2} x$,

$$\langle x - (P_{C_1} x - P_{C_2} x), \; P_{C_1} x - P_{C_2} x - y \rangle \geq 0, \quad \forall x \in \mathcal{H}, \; \forall y \in C_2^\perp \cap C_1.$$

As $P_{C_1}x - P_{C_2}x \in C_2^\perp \cap C_1$, the above simply says that

$P_{C_1^\perp}x - P_{C_2^\perp}x = P_{C_2^\perp \cap C_1}x$, which is the missing part of (5.18).
As to (5.19), note that $P_{C_1^\perp}P_{C_2} = P_{C_1^\perp}(P_{C_1}P_{C_2}) = P_{C_1^\perp}P_{C_1}P_{C_2} = 0$,
and that $P_{C_2}P_{C_1^\perp} = P_{C_2}P_{C_1}P_{C_1^\perp} = 0$.

Finally, by (5.18),

$$P_{C_1^\perp} = P_{C_2^\perp} - (P_{C_2^\perp} - P_{C_1^\perp}) = P_{C_2^\perp} - (P_{C_1} - P_{C_2}) =$$

$$= P_{C_2^\perp} - P_{C_1}P_{C_2^\perp} = P_{C_1^\perp}P_{C_2^\perp} ,$$

which is one of the equations (5.20); the other follows from this in the manner (5.17) followed from $P_{C_2}P_{C_1} = P_{C_2}$.

This theorem can be summarized as follows: <u>If $P_{C_2}P_{C_1} = P_{C_2}$ then any two of the projections $P_{C_1}, P_{C_2}, P_{C_1^\perp}$ and $P_{C_2^\perp}$ commute and their product is equal to the projection on the intersection of the respective cones. By passing to the duals the relation</u> $P_{C_2}P_{C_1} = P_{C_2}$ <u>is turned into</u> $P_{C_1^\perp}P_{C_2^\perp} = P_{C_1^\perp}$.

<u>Theorem 5.3</u>   $P_{C_2}P_{C_1} = P_{C_2}$   if and only if $P_{C_1} - P_{C_2}$ is a projection.

<u>Proof:</u>   Necessity is contained in Theorem 5.2. To prove sufficiency assume that $P_{C_1} - P_{C_2} = P_{C_3}$, that is, that

$$P_{C_1} = P_{C_2} + P_{C_3} .$$

By (1.6),

$$\|P_{C_2}x - P_{C_2}x'\|^2 \le \langle P_{C_2}x - P_{C_2}x', x-x'\rangle, \quad \forall x, x' \in \mathcal{H}$$

$$\|P_{C_3}x - P_{C_3}x'\|^2 \le \langle P_{C_3}x - P_{C_3}x', x-x'\rangle, \quad \forall x, x' \in \mathcal{H}$$

and adding these up

$$\|P_{C_2}x - P_{C_2}x'\|^2 + \|P_{C_3}x - P_{C_3}x'\|^3 \le \langle P_{C_1}x - P_{C_1}x', x-x'\rangle,$$

$$\forall x, x' \in \mathcal{H}.$$

If one sets $x' = P_{C_1}x$, the right hand member vanishes and on consequence so do each of the terms on the left. In particular, $P_{C_2}x = P_{C_2}P_{C_1}x$, $\forall x \in \mathcal{H}$. Q.E.D.

<u>Lemma 5.6</u>  $\{C_2 \subset C_1\} \iff \{\langle P_{C_2}x, x\rangle \le \langle P_{C_1}x, x\rangle\}$.

Proof: If $C_2 \subset C_1$ then $C_2^\perp \supset C_1^\perp$ and the distance of a point $x$ to $C_2^\perp$ is not larger than its distance to $C_1^\perp$, that is,

$$\|x - P_{C_2^\perp}x\|^2 \le \|x - P_{C_1^\perp}x\|^2$$

or equivalently

$$\|P_{C_2}x\|^2 \le \|P_{C_1}x\|^2.$$

By (2.4') this is

$$\langle P_{C_2}x, x\rangle \le \langle P_{C_1}x, x\rangle.$$

Conversely if this last relation holds, so do the two above, and for any $x_1^\perp \in C_1^\perp$, $P_{C_2}x_1^\perp = 0$, that is $x_1^\perp \in C_2^\perp$. But then $C_1^\perp \subset C_2^\perp$ and $C_1 \supset C_2$. Q.E.D.

**Corollary.** If $P_{C_1}$ and $P_{C_2}$ commute

(5.21) $\quad \{\langle x, P_{C_2} x \rangle \le \langle x, P_{C_1} x \rangle\} \iff \{P_{C_2} P_{C_1} = P_{C_1}\}$.

Next we inspect the equation $P_{\mathcal{K}_1} + P_{\mathcal{K}_2} = P_{\mathcal{K}_3}$. If the projections are linear this is possible only if $P_{\mathcal{K}_1}$ and $P_{\mathcal{K}_2}$ are orthogonal projections. What if they are nonlinear? We know no complete answer to this question except when $\mathcal{K}_1$ and $\mathcal{K}_2$ are cones, yet the picture emerging from what we know bears a strong resemblance to that of the linear case.

**Theorem 5.4** If $P_{\mathcal{K}_1} + P_{\mathcal{K}_2} = P_{\mathcal{K}}$ then $\mathcal{K} = \mathcal{K}_1 + \mathcal{K}_2$, and $\mathcal{K}_1$ and $\mathcal{K}_2$ are parallel to faces $\mathfrak{F}_{\mathcal{K}}(-u_2)$ and $\mathfrak{F}_{\mathcal{K}}(-u_1)$, where $u_1$ and $u_2$ are suitable vectors in $\mathcal{K}_1$ and $\mathcal{K}_2$ perpendicular to $\mathcal{K}_2$ and $\mathcal{K}_1$ respectively. Moreover

(5.22) $\quad \mathcal{K}_i - u_i = \mathcal{V}_{\mathcal{K}_i}(u_{i'}) \cap (\mathcal{K}_1 + \mathcal{K}_2 - u_1 - u_2), \quad i = 1, 2$

$$i' = \begin{cases} 1 & \text{if } i = 2 \\ 2 & \text{if } i = 1 \end{cases}.$$

**Proof:** Adding up the inequalities

$$\| P_{\mathcal{K}_i} x - P_{\mathcal{K}_i} x' \|^2 \le \langle x - x', P_{\mathcal{K}_i} x - P_{\mathcal{K}_i} x' \rangle$$

we obtain

$$\| P_{\mathcal{K}_1} x - P_{\mathcal{K}_1} x' \|^2 + \| P_{\mathcal{K}_2} x - P_{\mathcal{K}_2} x' \|^2$$

$$\le \langle x - x', P_{\mathcal{K}_1} x + P_{\mathcal{K}_2} x - P_{\mathcal{K}_1} x' - P_{\mathcal{K}_2} x' \rangle$$

$$= \langle x - x', P_{\mathcal{K}} x - P_{\mathcal{K}} x' \rangle ,$$

and since the right member vanishes for $x' = P_{\mathcal{K}} x$,

(5.23) $\quad P_{K_i} P_K x = P_{K_i} x$ , $\quad i = 1, 2$.

Hence by Lemma 5.2, $R(P_K - P_{K_i}) + K_i \subset K$, that is

$$K_1 + K_2 = R(P_K - P_{K_i}) + K_i = K_{i'} + K_i \subset K.$$

The opposite relation being obvious, $K_1 + K_2 = K$ is proved.

By Lemma 5.5, if $u_{i'}$ is the smallest translation bringing $K_i$ in touch with $K$, $K_i + u_{i'} \in \mathfrak{F}_K(-u_{i'})$, and so for any $x_i \in K_i$, $P_K x_i = x_i + u_{i'}$. Replacing $P_K$ by $P_{K_1} + P_{K_2}$ and noticing that $P_{K_i} x_i = x_i$, it follows

$$P_{K_{i'}} x_i = u_{i'} , \quad i = 1, 2 ,$$

or otherwise said

(5.24) $\quad K_i \subset u_{i'} + V_{K_{i'}}(u_{i'})$ , $\quad i = 1, 2$.

In particular, $u_i$ belongs to $K_i$, $i = 1, 2$.

Now, from $P_{K_i - u_i} x = P_{K_i}(x + u_i) - u_i$ it follows

$$P_{K_i - u_i} x + P_{K_{i'} - u_{i'}} x = P_{K_i}(x + u_i) + P_{K_{i'}}(x + u_{i'}) - u_1 - u_2 .$$

But as $u_{i'}$ is perpendicular to $K_i$, $P_{K_i}(x + u_i) = P_{K_i}(x + u_1 + u_2)$ and so

$$P_{K_1 - u_1} x + P_{K_2 - u_2} x = P_{K_1}(x + u_1 + u_2) + P_{K_2}(x + u_1 + u_2) - u_1 - u_2$$

$$= P_{K_1 + K_2 - u_1 - u_2} x .$$

This shows that the hypotheses of the lemma hold for $\tilde{K}_i = K_i - u_i$, $i = 1, 2$; in this case the corresponding $\tilde{u}_i$ both vanish. By (5.24) applied to the actual case

$$\tilde{K}_i \subset V_{\tilde{K}_i}(0) .$$

Since $0 \in \tilde{K}_i$ then $\tilde{K}_i \subset \tilde{K}_1 + \tilde{K}_2$, $i = 1, 2$, and

(5.25) $\qquad \tilde{K}_i \subset (\tilde{K}_1 + \tilde{K}_2) \cap \mathcal{U}_{\tilde{K}_{i'}}(0)$, $i = 1, 2$.

To prove the opposite relation we pick a point $x$ in the right member of (5.25) and observe first that since $x \in \tilde{K}_1 + \tilde{K}_2$, $x = P_{\tilde{K}_1 + \tilde{K}_2} x = P_{\tilde{K}_1} x + P_{\tilde{K}_2} x$; as $x \in \mathcal{U}_{\tilde{K}_{i'}}(0)$, $P_{\tilde{K}_{i'}} x = 0$, and $x = P_{\tilde{K}_i} x$. Thus $x \in \tilde{K}_i$ and in consequence

$$\tilde{K}_i \supset (\tilde{K}_1 + \tilde{K}_2) \cap \mathcal{U}_{\tilde{K}_{i'}}(0) \qquad i = 1, 2.$$

Replacing here $\tilde{K}_i$ by $K_i - u$, and noting that $\mathcal{U}_{\tilde{K}_{i'}}(0) =$
$= \mathcal{U}_{K_{i'}}(u_i)$, equation (5.22) follows.

In the case of cones Theorem 5.3 becomes

**Corollary** If $C_1$ and $C_2$ are closed convex cones with vertex at $0$, the relation $P_{C_1} + P_{C_2} = P_K$ implies $K = C_1 + C_2$ and

(5.26) $\qquad C_i = C_{i'}^\perp \cap (C_1 + C_2)$, $i = 1, 2$.

We do not know if (5.26) are sufficient for the sum of two projections on cones to be a projection. A different type of necessary and sufficient condition can be stated in terms of the notion of "orthogonal projections", we now introduce:

**Definition 5.1** Two projections $P_{C_1}$ and $P_{C_2}$ on closed convex cones $C_1$ and $C_2$ with vertex at $0$ are said to be orthogonal if

(5.27) $\qquad \langle P_{C_1} x, P_{C_2} x \rangle = 0$, $\forall x \in H$.

In such a case the cones $C_1$ and $C_2$ are also said to be orthogonal.

Projections onto a pair of dual cones are instances of orthogonal projections. Letting $x = P_{C_1} y$ in (5.27) one deduces, on use of (2.4'),

$$0 = \langle P_{C_1} y, P_{C_2} P_{C_1} y \rangle = \| P_{C_2} P_{C_1} y \|^2, \quad \forall\, y \in \mathcal{H}.$$

Thus, <u>the product of two orthogonal projections vanishes</u>. Otherwise said, <u>orthogonal cones are contained in each other duals</u>. This property, however, does not characterize orthogonality, as easily constructed examples in two dimensions show.

**Theorem 5.5** If for a countable family of projections on closed convex cones with vertex at the origin

$$\sum_{i=1}^{n} P_{C_i} x \to P_C x, \quad \forall\, x \in \mathcal{H}$$

then the $P_{C_i}$'s are pairwise orthogonal, $C = \sum_{1}^{\infty} C_i$, $P_{C_i} P_C = P_{C_i}$, $i = 1, 2, \ldots$, and the convergence holds in the strong sense also. Conversely, the sum of a countable number of orthogonal projections always exist and is a projection.

**Proof:** Assume $\sum_{i=1}^{n} P_{C_i} x \to P_C x$. Apply (1.6) to each $P_{C_i}$ and add the results,

$$\sum_{i=1}^{n} \| P_{C_i} x - P_{C_i} x' \|^2 \leq \sum_{i=1}^{n} \langle x - x', P_{C_i} x - P_{C_i} x' \rangle =$$

$$\langle x - x', \sum_{i=1}^{n} P_{C_i} x - \sum_{1}^{n} P_{C_i} x' \rangle.$$

Letting $n \to \infty$, we get

$$\sum_{i=1}^{\infty} \| P_{C_i} x - P_{C_i} x' \|^2 \leq \langle x - x', P_C x - P_C x' \rangle.$$

Let now $x' = P_C x$,

$$\sum_{i=1}^{\infty} \|P_{C_i} x - P_{C_i} P_C x\|^2 \leq \langle x - P_C x, P_C x - P_C^2 x \rangle = 0.$$

Hence

(5.28) $\qquad P_{C_i} = P_C P_{C_i}, \qquad i = 1, 2, \ldots$ .

Therefore, by (2.4), (5.17), (5.18),

$$P_C = P_{C_i} P_C + P_{C_i^\perp} P_C = P_{C_i} + P_{C \cap C_i^\perp}, \qquad i = 1, 2, \ldots,$$

that is

$$\sum_{j \neq i} P_{C_j} = P_{C \cap C_i^\perp}.$$

Thus the deletion of one, and hence of any number of terms from the sum $\sum_{i=1}^{\infty} P_{C_j}$ still yields a projection. The first conclusion is that $P_{C_j} P_{C \cap C_i^\perp} = P_{C_j}$ for $i \neq j$ (simply apply (5.28) to the new situation). Hence $C_j \subset C \cap C_i^\perp$, $i \neq j$, and

$$C_j \subset C_i^\perp, \qquad i \neq j.$$

So

(5.29) $\qquad \langle P_{C_i} x, P_{C_j} x \rangle \leq 0, \qquad i \neq j, \qquad \forall x, y \in H$

On the other hand, if $i \neq j$

$$\langle P_{C_i} x, P_{C_j} x \rangle = \langle P_{C_i} x, P_C x \rangle - \langle P_{C_i} x, \sum_{h \neq i, j} P_{C_h} x \rangle - \langle P_{C_i} x, P_{C_i} x \rangle$$

$$= \langle P_{C_i} x, P_C x \rangle - \sum_{h \neq i, j} \langle P_{C_i} x, P_{C_h} x \rangle - \langle P_{C_i} x, x \rangle$$

$$= - \langle P_{C_i} x, P_{C^\perp} x \rangle - \sum_{h \neq i, j} \langle P_{C_i} x, P_{C_h} x \rangle$$

Since $C_i \subset C$, the first member on the right is nonnegative, and by (5.29) so is each of the others. Hence

$$\langle P_{C_i} x, P_{C_j} x \rangle \geq 0, \quad i \neq j, \quad \forall x \in \mathcal{H},$$

which in conjunction with (5.29) yields

$$\langle P_{C_i} x, P_{C_j} x \rangle = 0, \quad i \neq j, \quad x \in \mathcal{H}.$$

Therefore, the $P_{C_i}$'s are pairwise orthogonal. The strong convergence of $\sum_{i=1}^{\infty} P_{C_i} x$ is a consequence of the known fact that a weakly convergent sum of orthogonal vectors is automatically strongly convergent.

Now, if $x \in C$ then $x = P_C x = \sum_{i=1}^{\infty} P_{C_i} x$, and $x$ is the sum of elements belonging to $C_i$'s. Conversely, every such sum is contained in $C$ since every $C_i$ is. Hence $C = \sum_{i=1}^{\infty} C_i$ (the essential point here is that the set of convergent sums $\sum_{i=1}^{\infty} x_i$, $x_i \in C_i$, is already a closed convex set).

Finally let us demonstrate that the sum of orthogonal projections is a projection. Since the $P_{C_i}$'s are orthogonal, by Theorem 4.1,

$$\nabla \tfrac{1}{2} \left\| \sum_{i=1}^{n} P_{C_i} x \right\|^2 = \nabla \tfrac{1}{2} \sum_{i=1}^{n} \left\| P_{C_i} x \right\|^2 = \sum_{i=1}^{n} \nabla \tfrac{1}{2} \left\| P_{C_i} x \right\|^2$$

$$= \sum_{i=1}^{n} \nabla \tfrac{1}{2} \left\| x - P_{C_i^\perp} x \right\|^2$$

$$= \sum_{i=1}^{n} (x - P_{C_i^\perp} x) = \sum_{i=1}^{n} P_{C_i} x.$$

So, for every $n$, $I - \sum_{i=1}^{n} P_{C_i}$ is a projection because it is Lipschitzian and satisfies the characteristic differential equation (4.19) for projections; since it is positive homogeneous, it is a projection onto a closed convex cone, and the same goes for its complement $\sum_{i=1}^{n} P_{C_i}$. Then, by (1.4), if $m < n$

$$\langle x - (\sum_{i=1}^{n} P_{C_i})x, (\sum_{i=1}^{n} P_{C_i})x - (\sum_{i=1}^{m} P_{C_i})y \rangle \geq 0, \quad \forall x, y \in H,$$

whence letting first $m$ go to infinity and then $n$,

$$\langle x - (\sum_{i=1}^{\infty} P_{C_i})x, (\sum_{i=1}^{\infty} P_{C_i})x - (\sum_{i=1}^{\infty} P_{C_i})y \rangle \geq 0 \quad \forall x, y \in H,$$

proving that $\sum_{i=1}^{\infty} P_{C_i}$ is a projection (cf. Theorem 1.1).

Q. E. D.

**Lemma 5.7** For any orthogonal family of projections $\{P_{C_\alpha}\}$ and any $x$ the set

(5.30) $\qquad A(x) = \{\alpha \mid P_{C_\alpha} x \neq 0\}$

is at most countable and the series $\sum_{\alpha \in A(x)} P_{C_\alpha} x$ converges strongly.

**Proof:** Let $\alpha_1, \alpha_2, \ldots, \alpha_n$ be such that $\|P_{C_{\alpha_i}} x\| \geq 1/k$. Then

$$n^2/k^2 \leq \sum_{i=1}^{n} \|P_{C_{\alpha_i}} x\|^2 = \|(\sum_{i=1}^{n} P_{C_{\alpha_i}})x\|^2 = \|P_{\sum_{i=1}^{n} C_{\alpha_i}} x\|^2 \leq \|x\|^2,$$

and there are at most $k\|x\|$ $\alpha$'s such that $\|P_{C_\alpha} x\| \geq 1/k$. Hence, $A(x)$ is at most countable. The claim concerning the convergence of $\sum_{\alpha \in A(x)} P_{C_\alpha} x$ follows from Theorem 5.5.

**Definition 5.2** The sum $\sum_{\alpha}^{n} P_{C_\alpha}$ of a family of orthogonal projections is defined as follows

(5.31) $$\sum_{\alpha} P_{C_\alpha} x = \sum_{\alpha \in A(x)} P_{C_\alpha} x ,$$

where $A(x)$ is the set of indices (5.30).

In view of the foregoing it is clear that <u>the sum of orthogonal projections always exists and is a projection.</u> <u>Moreover,</u> $P_{C_\beta} \sum_{\alpha} P_{C_\alpha} = P_{C_\beta}$ .

We shall conclude this section with a discussion of the basic question of when the product, or products, of two projections is a projection again. It is well known that if the projections are linear their products in any order are projections if and only if the projections commute, and in such case the product of the projections is the projection on the intersection of their ranges [20, no. 105], but it is not known what precise form, if any, this theorem takes in the nonlinear case. Although commutativity is most likely sufficient it is not strictly necessary - just think of the projections on a hyperplane and on a point outside of it. Nevertheless it seems as if commutativity were still necessary if one were to work modulo translations. This is just a vague conjecture which we cannot state in precise explicit terms at this stage of our investigations. Instead we shall turn our attention mainly to sufficiency and try to give a positive answer to the conjecture that "the product of two commuting projections is a projection".

**Lemma 5.8** If $P_{K_1} P_{K_2} = P_K$ then

(5.32) $$K = K_1 \cap (K_2 + u) ,$$

where $u$ is the smallest vector such that $K_1 \cap (K_2 + u) \neq \emptyset$.

**Proof:** We have, since $P_{\mathcal{K}_1}$ satisfies (5.1),

$$\langle (P_{\mathcal{K}_2}x - P_{\mathcal{K}_1}P_{\mathcal{K}_2}x) - (P_{\mathcal{K}_2}x' - P_{\mathcal{K}_1}P_{\mathcal{K}_2}x'), P_{\mathcal{K}_1}P_{\mathcal{K}_2}x - P_{\mathcal{K}_1}P_{\mathcal{K}_2}x' \rangle \geq 0,$$

and if both $x$ and $x'$ belong to $\mathcal{K}$,

$$-\langle (I-P_{\mathcal{K}_2})x - (I-P_{\mathcal{K}_2})x', x-x' \rangle \geq 0. \quad \forall x, x' \in \mathcal{K}.$$

Then by (1.7) applied to $P_{\mathcal{K}_2}$,

$$\|(I-P_{\mathcal{K}_2})x - (I-P_{\mathcal{K}_2})x'\|^2 \leq \langle (I-P_{\mathcal{K}_2})x - (I-P_{\mathcal{K}_2})x', x-x' \rangle \leq 0,$$

whence it follows,

$$x - P_{\mathcal{K}_2}x = u = \text{const.}, \quad \forall x \in \mathcal{K}.$$

Up to now we have only made use of the property (1.5). If now we take account of the fact that $P_{\mathcal{K}_1}$ and $P_{\mathcal{K}_2}$ are true projections we can see that if we write $x_2 = P_{\mathcal{K}_2}x$, then $x = P_{\mathcal{K}_1}x_2$, and the points $x$ and $x_2$ realize the distance between $\mathcal{K}_1$ and $\mathcal{K}_2$. This means that $u$ is the smallest vector such that $\mathcal{K}_1 \cap (\mathcal{K}_2 + u) \neq \phi$. On the other hand any point $x$ of $\mathcal{K}_1$ whose distance to $\mathcal{K}_2$ is minimal must coincide with the projection on $\mathcal{K}_1$ of its projection on $\mathcal{K}_2$ and in consequence must belong to $\mathcal{K}$. This completes the proof.

**Lemma 5.9** If $P_1$ and $P_2$ are two mappings satisfying (1.5), then for any $u$,

$$(5.33) \quad (I-P_1P_2)^{-1}u - u = \bigcup_{u_1+u_2=u} \{(I-P_1)^{-1}u_1 - u_1) \cap ((I-P_2)^{-1}u_2 - u_2 - u_1)\}.$$

Moreover, at most one set on the right is nonempty.

**Proof:** Let $x$ belong to $((I-P_1)^{-1}u_1 - u_1) \cap ((I-P_2)^{-1}u_2 - u_2 - u_1)$;

then
$$(I-P_1)(x+u_1) = u_1 , \quad (I-P_2)(x+u_1+u_2) = u_2 .$$
From the last equation above we obtain
$$x+u_1 = P_2(x+u_1+u_2) , \quad P_1(x+u_1) = P_1P_2(x+u_1+u_2) ,$$
and by virtue of the first
$$P_1P_2(x+u_1+u_2) = P_1(x+u_1) = x ,$$
that is
$$(I-P_1P_2)(x+u_1+u_2) = u_1+u_2 ,$$
and so
$$x \in (I-P_1P_2)^{-1}(u_1+u_2) - (u_1+u_2).$$
Thus,
$$(I-P_1P_2)^{-1}(u_1+u_2)-(u_1+u_2) \supset ((I-P_1)^{-1}u_1-u_1) \cap ((I-P_2)^{-1}u_2-u_2-u_1),$$
(5.34)
$$\forall u_1, u_2 .$$

Now we write by (1.5),
$$\langle (I-P_1)P_2x - (I-P_1)P_2x', P_1P_2x-P_1P_2x' \rangle \geq 0 ,$$
so that, if $x, x' \in (I-P_1P_2)^{-1}w$,
$$-\langle (I-P_2)x-(I-P_2)x', x-x' \rangle \geq 0 ,$$
and then by (1.7) applied to $P_2$
$$\|(I-P_2)x - (I-P_2)x'\|^2 \leq \langle (I-P_2)x - (I-P_2)x', x-x' \rangle \leq 0.$$

Hence there is a u independent of x such that

$$(I-P_2)x = u, \quad \forall x \in (I-P_1P_2)^{-1}w,$$

and

(5.35) $\quad (I-P_1P_2)^{-1}w \subset (I-P_2)^{-1}u.$

Moreover, from $x-u=P_2 x$ we obtain $P_1(x-u) = P_1P_2 x = x-w$, and then $x-u-P_1(x-u) = w-u$, that is,

$$x \in (I-P_1)^{-1}(w-u)+u, \quad \forall x \in (I-P_1P_2)^{-1}w.$$

So,

$$(I-P_1P_2)^{-1}w \subset (I-P_1)^{-1}(w-u)+u,$$

which combined with (5.35) yields

(5.36)
$$(I-P_1P_2)^{-1}w \subset ((I-P_2)^{-1}u) \cap ((I-P_1)^{-1}(w-u)+u),$$
$$(I-P_1P_2)^{-1}w-w \subset ((I-P_1)^{-1}(w-u)-(w-u)) \cap ((I-P_2)^{-1}u-w).$$

Since the right member above is a set in the class appearing on the right of (5.33), (5.33) follows from (5.34) and (5.36). Now, the sets on the right of (5.33) are all disjoint because the $(I-P_i)^{-1}u_i$ are disjoint, and at the same time, as proved above, the left member is contained in one of them; hence all but this one are empty. Q. E. D.

<u>Corollary</u>  If $P_{\mathcal{K}_1}P_{\mathcal{K}_2} = P_{\mathcal{K}}$ then for any $u \in \mathcal{H}$ there is a unique pair of vectors $u_1$ and $u_2$ with $u = u_1+u_2$ such that

(5.37) $\quad \mathfrak{F}_{\mathcal{K}}(u) = \mathfrak{F}_{\mathcal{K}_1}(u_1) \cap (\mathfrak{F}_{\mathcal{K}_2}(u_2)-u_1);$

for any other choice of $u_1$ and $u_2$ with the same sum the set on the right of (5.37) is empty.

**Lemma 5.10**  If $P_{K_1} P_{K_2} = P_{K_2} P_{K_1}$ then

(5.38) $\{K_1 \cap K_2 + \Re(P_{K_1} - P_{K_1} P_{K_2}) \subset K_1\} \iff \{P_{K_1} P_{K_2} = P_{K_1 \cap K_2}\}$.

**Proof:** If the relation on the left of (5.38) holds, then for any $x \in H$,

$$P_{K_1} x \in K_1 \cap K_2 + P_{K_1} x - P_{K_1} P_{K_2} x \subset K_1,$$

since $P_{K_1} P_{K_2} x \in K_1 \cap K_2$. Therefore,

$$P_{K_1} x = P_{K_1 \cap K_2 - (P_{K_1} x - P_{K_1} P_{K_2} x)} x = P_{K_1} x - P_{K_1} P_{K_2} x$$

$$+ P_{K_1 \cap K_2}(x - P_{K_1} x + P_{K_1} P_{K_2} x),$$

and

$$P_{K_1} P_{K_2} x = P_{K_1 \cap K_2}(P_{K_1} P_{K_2} x + (x - P_{K_1} x)),$$

that is using (2.8),

$$x - P_{K_1} x \in U_{K_1 \cap K_2}(P_{K_1} P_{K_2} x).$$

On the other hand, since $P_{K_1} P_{K_2} x \in K_1 \cap K_2 \subset K_2$

$$P_{K_1} P_{K_2} x = P_{K_1 \cap K_2} P_{K_1} x$$

and

$$P_{K_1} x - P_{K_1} P_{K_2} x \in U_{K_1 \cap K_2}(P_{K_1} P_{K_2} x).$$

Since $U_{K_1 \cap K_2}(P_{K_1} P_{K_2} x)$ is a convex cone,

$$x - P_{K_1} P_{K_2} x = (x - P_{K_1} x) + (P_{K_1} x - P_{K_1} P_{K_2} x) \subset \mathcal{U}_{K_1 \cap K_2} (P_{K_1} P_{K_2} x)$$

$$+ \mathcal{U}_{K_1 \cap K_2} (P_{K_1} P_{K_2} x) \subset \mathcal{U}_{K_1 \cap K_2} (P_{K_1} P_{K_2} x),$$

which is equivalent to $P_{K_1 \cap K_2} x = P_{K_1} P_{K_2} x$. To prove the implication in the opposite direction deduce from $P_{K_1} P_{K_2}$ $= P_{K_1 \cap K_2}$ that $P_{K_1 \cap K_2} P_{K_1} = P_{K_1 \cap K_2}$ and apply Lemma 5.3 to $P_{K_1 \cap K_2}$ and $P_{K_1}$.

In the case of projections on cones we have a more precise result. (In the sequel the letter C, possibly modified by sub- or superscripts, will always indicate a closed convex cone with vertex at the origin).

**Lemma 5.11** Suppose $P_{C_1} P_{C_2} = P_{C_2} P_{C_1}$. If for some x,

(5.39) $\quad \langle x - P_{C_1} P_{C_2} x, \ P_{C_1} P_{C_2} x \rangle = 0,$

then

(5.40) $\quad P_{C_1} P_{C_2} x = P_{C_1 \cap C_2} x,$

and conversely.

**Proof:** We have,

$$x = P_{C_1} P_{C_2} x + x - P_{C_1} P_{C_2} x = P_{C_1} P_{C_2} x + P_{C_2}^{\perp} x + P_{C_2} x - P_{C_1} P_{C_2} x$$

$$= P_{C_1} P_{C_2} x + [P_{C_2}^{\perp} x + P_{C_1}^{\perp} P_{C_2} x]$$

where $P_{C_1}P_{C_2}x \in C_1 \cap C_2$ and $P_{C_2^\perp}x + P_{C_1^\perp}P_{C_2}x \in C_1^\perp + C_2^\perp$
$\subset (C_1 \cap C_2)^\perp$. Now if (5.39) holds the above is a decomposition of $x$ into the sum of two orthogonal vectors, one in $C_1 \cap C_2$ and the other in $(C_1 \cap C_2)^\perp$. Therefore by (2.4), $P_{C_1 \cap C_2}x = P_{C_1}P_{C_2}x$, and (5.39) implies (5.40). Conversely, if (5.40) holds, then so does (5.39) as a consequence of (2.4') applied to $P_{C_1 \cap C_2}$.

<u>Lemma 5.12</u>  If $P_{C_1}P_{C_2} = P_{C_2}P_{C_1} = P_{C_1 \cap C_2}$, then

(5.41) $$(C_1 \cap C_2)^\perp = C_1^\perp + C_2^\perp,$$

(5.42) $$C_1 \cap (C_1^\perp + C_2^\perp) = C_1 \cap C_2^\perp,$$

(5.43) $$C_2 \cap (C_1^\perp + C_2^\perp) = C_2 \cap C_1^\perp.$$

Moreover, $P_{C_2^\perp}P_{C_1}$, $P_{C_1^\perp}P_{C_2}$, $P_{C_1}P_{C_1^\perp + C_2^\perp}$, $P_{C_2}P_{C_1^\perp + C_2^\perp}$ are all projections and

(5.44) $$P_{C_2^\perp}P_{C_1} = P_{C_1}P_{C_1^\perp + C_2^\perp} = P_{C_1 \cap C_2^\perp},$$

(5.45) $$P_{C_1^\perp}P_{C_2} = P_{C_2}P_{C_1^\perp + C_2^\perp} = P_{C_2 \cap C_1^\perp},$$

(5.46) $$I = P_{C_1^\perp} + P_{C_1 \cap C_2^\perp} + P_{C_1 \cap C_2} = P_{C_2^\perp} + P_{C_2 \cap C_1^\perp} + P_{C_1 \cap C_2}.$$

<u>Proof:</u> Since $P_{C_1 \cap C_2}P_{C_1} = P_{C_1}P_{C_2}P_{C_1} = P_{C_1}P_{C_1}P_{C_2} = P_{C_1}P_{C_2}$
$= P_{C_1 \cap C_2}$, then by (5.18) with $C_1 \cap C_2$ in place of $C_2$,

(5.47) $$P_{C_1} = P_{C_1 \cap C_2} + P_{C_1}P_{(C_1 \cap C_2)^\perp} = P_{C_2}P_{C_1} + P_{C_1}P_{(C_1 \cap C_2)^\perp},$$

so

$$P_{c_2^\perp} P_{c_1} = P_{c_1} P_{(c_1 \cap c_2)^\perp} .$$

Moreover, by Theorem 5.2, $P_{c_1} P_{(c_1 \cap c_2)^\perp} = P_{c_1 \cap (c_1 \cap c_2)^\perp}$, and in consequence $P_{c_2^\perp} P_{c_1}$ is also a projection, namely the projection on $c_2^\perp \cap c_1$, by Lemma 5.5. Hence

$$(5.48) \quad P_{c_2^\perp} P_{c_1} = P_{c_1} P_{(c_1 \cap c_2)^\perp} = P_{c_2^\perp \cap c_1} = P_{c_1 \cap (c_1 \cap c_2)^\perp}$$

and similarly

$$(5.49) \quad P_{c_1^\perp} P_{c_2} = P_{c_2} P_{(c_1 \cap c_2)^\perp} = P_{c_1^\perp \cap c_2} = P_{c_2 \cap (c_1 \cap c_2)^\perp} ,$$

proving (5.44) and (5.45). Moreover, by (5.47),

$$I = P_{c_1^\perp + c_1} = P_{c_1^\perp} + P_{c_1 \cap c_2} + P_{c_1 (c_1 \cap c_2)^\perp} = P_{c_1^\perp} + P_{c_1 \cap c_2} + P_{c_2^\perp \cap c_1},$$

which is (5.46). From (5.46),

$$P_{(c_1 \cap c_2)^\perp} = I - P_{c_1 \cap c_2} = P_{c_1^\perp} + P_{c_1 \cap c_2^\perp} ,$$

so by Theorem 5.5,

$$(c_1 \cap c_2)^\perp = c_1^\perp + c_1 \cap c_2^\perp \subset c_1^\perp + c_2^\perp ,$$

which together with the opposite relation resulting from (2.3) yields (5.41). Substitution in (5.48) and (5.49) leads to (5.42) and (5.43), completing the proof.

We do not know if (5.41), (5.42) and (5.43) imply the hypothesis.

**Corollary 1** If $P_{C_1} P_{C_2} = P_{C_2} P_{C_1} = P_{C_1 \cap C_2}$, then $P_{C_1}$ preserves the shell of $C_2$. More specifically if $x_2 \in C_2$, then $P_{C_1} x_2$ belongs to the $C_2$-face of $x_2$ (cf. Lemma 2.18).

**Proof:** Let $x_2^\perp \in C_2^\perp$ be orthogonal to $x_2$. Then $P_{C_2}(x_2 + x_2^\perp) = x_2$ and $P_{C_2^\perp}(x_2 + x_2^\perp) = x_2^\perp$. But,

$$\langle P_{C_1 \cap C_2} x, P_{C_2^\perp} x \rangle = \langle P_{C_1 \cap C_2} x, x \rangle - \langle P_{C_1 \cap C_2} x, P_{C_2} x \rangle$$

$$= \| P_{C_1 \cap C_2} x \|^2 - \| P_{C_1} P_{C_2} x \|^2, \quad \forall x \in \mathcal{H}.$$

So, since $C_1 \cap C_2$ and $C_2^\perp$ are orthogonal cones (from (5.46) via Theorem 5.5),

$$0 = \langle P_{C_1 \cap C_2}(x_2 + x_2^\perp), P_{C_2^\perp}(x_2 + x_2^\perp) \rangle = \langle P_{C_1} P_{C_2}(x_2 + x_2^\perp), x_2^\perp \rangle$$

$$= \langle P_{C_1} x_2, x_2^\perp \rangle.$$

Thus any normal to $C_2$ at $x_2$ is a normal at $P_{C_1} x_2$, which is just another way of saying that $P_{C_1} x_2$ lies in the $C_2$-face of $x_2$.

**Corollary 2** If $P_{C_1} P_{C_2} = P_{C_2} P_{C_1} = P_{C_1 \cap C_2}$, then

(5.50) $\quad 0 = \langle P_{C_1} P_{C_2} x, P_{C_1} P_{C_2^\perp} x \rangle = \langle P_{C_2} P_{C_1} x, P_{C_2} P_{C_1^\perp} x \rangle,$

$$\forall x \in \mathcal{H}.$$

Proof: We have

$$\langle P_{C_1}P_{C_2}x, P_{C_1}P_{C_2^\perp}x\rangle = \langle P_{C_1\cap C_2}x, P_{C_2^\perp}x\rangle - \langle P_{C_1\cap C_2}x, P_{C_1^\perp}P_{C_2^\perp}x\rangle,$$

and since $P_{C_1\cap C_2}$ and $P_{C_2^\perp}$ are orthogonal projections,

$$\langle P_{C_1}P_{C_2}x, P_{C_1}P_{C_2^\perp}x\rangle = -\langle P_{C_1}P_{C_2}x, P_{C_1^\perp}P_{C_2^\perp}x\rangle \geq 0.$$

On the other hand, if $x_2^\perp \in C_2^\perp$ then $0 = P_{C_1}P_{C_2}x_2^\perp$
$= P_{C_2}P_{C_1}x_2^\perp$, that is, $P_{C_1}x_2^\perp \in C_2^\perp$. In particular,
$P_{C_1}P_{C_2^\perp}x \in C_2^\perp$, $\forall x \in \mathcal{H}$, and since $P_{C_1}P_{C_2}x = P_{C_2}P_{C_1}x \in C_2$,

$$\langle P_{C_1}P_{C_2}x, P_{C_1}P_{C_2^\perp}x\rangle \leq 0, \qquad \forall x \in \mathcal{H}.$$

Thus (5.50) holds. Q.E.D.

The theorem below is our only positive result towards proving the conjecture that the product of commuting projections is a projection.

Theorem 5.6  The product of two commuting projections on finite dimensional closed convex cones with vertex at the origin is a projection.

Proof: If $C_1$ and $C_2$ are the cones, the hypothesis is $P_{C_1}P_{C_2} = P_{C_2}P_{C_1}$, and in view of Lemma 5.11, all we have to prove is that

$$(5.51) \qquad \langle x - P_{C_1}P_{C_2}x, P_{C_1}P_{C_2}x\rangle = 0, \qquad \forall x \in \mathcal{H}.$$

Our first remark is that if (5.51) is valid for a set of x's having a common projection on $C_1$, then it is also valid for any weak limit of linear combinations with nonnegative coefficients of such points. Then we observe that $C_1$

being finite dimensional, any point in space outside $C_1$ is either projected by $P_{C_1}$ into an inner point of $C_1$ or belongs to the closed convex hull of limits of points projecting into flat points (Lemma 2.23). Now, if $P_{C_1} x$ is an inner point of $C_1$, $x - P_{C_1} x$ is perpendicular to any vector in $C_1$ and by (2.4')

$$\langle x - P_{C_1} P_{C_2} x, P_{C_1} P_{C_2} x \rangle = \langle x - P_{C_1} x, P_{C_1} P_{C_2} x \rangle$$

$$+ \langle P_{C_1} x - P_{C_2} P_{C_1} x, P_{C_2} P_{C_1} x \rangle = 0.$$

Hence (5.51) holds at $x$. The same is obviously true for any $x$ in $C_1$. Therefore, it is sufficient to consider points outside of $C_1$ projecting on flat points. Hence from now on $x$ will be one such point. As $s$ runs over all nonnegative numbers $x(s) = P_{C_1} x + s(x - P_{C_1} x)$ describes a normal to $C_1$ at $P_{C_1} x$, and hence is projected by $P_{C_1}$ onto $P_{C_1} x$. From $P_{C_1} P_{C_2} = P_{C_2} P_{C_1}$ it follows that $\{P_{C_2} x(s)\}_{s \geq 0}$ is a curve in $C_2$ projected onto $P_{C_1} P_{C_2} x$ by $P_{C_1}$. If for some $s_0 > 0$, $P_{C_2} x(s_0) \in C_1$, then $P_{C_2} x(s_0) = P_{C_1} P_{C_2} x(s_0) = P_{C_2} P_{C_1} x(s_0)$

$= P_{C_1} P_{C_2} x$, and

$$0 = \langle x(s_0) - P_{C_2} x(s_0), P_{C_2} x(s_0) \rangle$$

(5.52)
$$= \langle (1-s_0) P_{C_1} x + s_0 x - P_{C_1} P_{C_2} x, P_{C_1} P_{C_2} x \rangle$$

$$= (1-s_0) \langle P_{C_1} x - P_{C_1} P_{C_2} x, P_{C_1} P_{C_2} x \rangle + s_0 \langle x - P_{C_1} P_{C_2} x, P_{C_1} P_{C_2} x \rangle$$

$$= s_0 \langle x - P_{C_1} P_{C_2} x, P_{C_1} P_{C_2} x \rangle,$$

and x satisfies (5.51). Hence discarding this case, it may be further assumed that for no positive $s$, $P_{C_2}x(s) \in C_1$. This being so, for no positive $s$ and $t$ in the interval $0 \leq t \leq 1$ the point $x(s,t) = P_{C_2}x(s) + t(x(s) - P_{C_2}x(s))$ may be in $C_1$, for if it were, $P_{C_2}x(s)$, which is the projection on $C_2$ of any $x(s,t)$, would belong to $C_1$, by commutativity. Let us consider now the case when for certain $s_0$ and $t_0$, $(s_0 > 0, 0 \leq t_0 \leq 1)$ $P_{C_1}x(s_0,t_0)$ is an inner point of $C_1$. Since $C_1$ is a convex body $P_{C_1}x(s_0,t_0)$ is an interior point of $C_1$ relatively to the manifold spanned by $C_1$, and any point sufficiently close to $x(s_0,t_0)$ would be projected again into an inner point; in particular that would be the case for $x(s_0,t_1)$ if $t_1$ is sufficiently close to $t_0$. Clearly such points are projected perpendicularly to the whole cone $C_1$, and

$$0 = \langle x(s_0,t_0) - P_{C_1}x(s_0,t_0), P_{C_1}P_{C_2}x(s_0,t_0) \rangle$$

$$= \langle x(s_0,t_0) - P_{C_1}P_{C_2}x(s_0,t_0), P_{C_1}P_{C_2}x(s_0,t_0) \rangle$$

$$- \langle P_{C_1}x(s_0,t_0) - P_{C_2}P_{C_1}x(s_0,t_0), P_{C_2}P_{C_1}x(s_0,t_0) \rangle$$

$$= \langle x(s_0,t_0) - P_{C_1}P_{C_2}x(s_0,t_0), P_{C_1}P_{C_2}x(s_0,t_0) \rangle.$$

Since $P_{C_1}P_{C_2}x(s_0,t_0) = P_{C_1}P_{C_2}x(s)$,

$$0 = \langle x(s_0,t_0) - P_{C_1}P_{C_2}x(s_0), P_{C_1}P_{C_2}x(s_0) \rangle.$$

Similarly

$$0 = \langle x(s_0,t_1) - P_{C_1}P_{C_2}x(s_0), P_{C_1}P_{C_2}x(s_0) \rangle,$$

and subtracting

$$0 = \langle x(s_0,t_1)-x(s_0,t_0), P_{C_1}P_{C_2}x(s_0)\rangle$$

$$= (t_1-t_0)\langle x(s_0)-P_{C_2}x(s_0), P_{C_1}P_{C_2}x(s_0)\rangle$$

whence, choosing $t_1 \neq t_0$, as we may, it follows

$$0 = \langle x(s_0) - P_{C_2}x(s_0), P_{C_1}P_{C_2}x(s_0)\rangle ,$$

and (5.51) results by (5.52).

Having disposed of these cases we are left with the situation where $x(s,t)$ does not belong to $C_1$ and is projected by $P_{C_1}$ into the shell of $C_1$, for all values of $s$ and $t$ ($0 < s$, $0 \leq t \leq 1$). The point $x(0,t) = (1-t)P_{C_1}P_{C_2}x + tP_{C_1}x$, $0 \leq t \leq 1$ is the limit of $x(s,t)$ as $s \to 0$, and since it belongs to $C_1$ it is also the limit of $P_{C_1}x(s,t)$. As the latter is a shell point and the shell of a convex body is closed (Lemma 2.9), $x(0,t)$ is also a shell point of $C_1$. Hence the line segment $\{x(0,t)\}_{0 \leq t \leq 1}$ joining $P_{C_1}P_{C_2}x$ to $P_{C_1}x$ lies entirely in the shell of $C_1$ and so belongs to one of its faces, namely the face perpendicular to $x-P_{C_1}x$, since $P_{C_1}x$ is a flat point. But then the generator of $C_1$ through $P_{C_1}P_{C_2}x$, which belongs to this face, is perpendicular to $x-P_{C_1}x$. In other terms, (5.51) holds and the proof concludes.

Remark 1   Upon inspection of the foregoing proof it appears that no use has been made of the finite dimensionality of $C_2$, nor of any other special property besides that of being a closed convex cone, and that of $C_1$ it has only been required that it be a conical convex body and that Lemma 2.23 apply to it. It is just to secure this last property that finite dimensionality was imposed on $C_1$, since so far the property has been proved for finite dimensional sets only. Thus, any extension of Lemma 2.23 should result in an extension of Theorem 5.5.

**Remark 2**  Contrary to the linear case, the fact that $P_{c_1}$ and $P_{c_2}$ commute does not imply that $P_{c_1}$ and $P_{c_2^\perp}$ or $P_{c_2}$ and $P_{c_1^\perp}$ also commute, nor that $P_{c_1} + P_{c_2} - P_{c_1} P_{c_2}$ and $P_{c_1} + P_{c_2} - 2P_{c_1} P_{c_2}$ are projections. To see this assume a situation such that $P_{c_1} P_{c_2} = P_{c_1 \cap c_2}$. Then,

$$P_{c_1 \cap c_2} P_{c_2} = P_{c_1} P_{c_2} P_{c_2} = P_{c_1} P_{c_2} = P_{c_1 \cap c_2}$$

and by Theorem 5.2 and Lemma 5.12,

$$P_{c_2} = P_{c_1 \cap c_2} + P_{c_2}P_{(c_1 \cap c_2)^\perp} = P_{c_1 \cap c_2} + P_{c_2 \cap c_1^\perp},$$

and similarly

$$P_{c_1} = P_{c_1 \cap c_2} + P_{c_1 \cap c_2^\perp}.$$

Therefore,

(5.53)  $$P_{c_1} + P_{c_2} - P_{c_1} P_{c_2} = P_{c_1 \cap c_2} + P_{c_1 \cap c_2^\perp} + P_{c_2 \cap c_1^\perp}$$

(5.54)  $$P_{c_1} + P_{c_2} - 2P_{c_1} P_{c_2} = P_{c_1 \cap c_2^\perp} + P_{c_2 \cap c_1^\perp}.$$

Now, if the operators on the left are to be projections the various terms on the right hand have to be orthogonal (Theorem 5.2). But $P_{c_1 \cap c_2} = P_{c_1} P_{c_2}$ is orthogonal to $P_{c_1 \cap c_2^\perp} = P_{c_2^\perp} P_{c_1}$ and to $P_{c_1^\perp \cap c_2} = P_{c_1^\perp} P_{c_2}$ (Lemma 5.12), and so the orthogonality of $P_{c_1^\perp \cap c_2}$ and $P_{c_2 \cap c_1^\perp}$ is necessary and sufficient for either $P_{c_1} + P_{c_2} - P_{c_1} P_{c_2}$ or $P_{c_1} + P_{c_2} - 2P_{c_1} P_{c_2}$ to be a projection. Such orthogonality condition amounts to

(5.55)   $\langle P_{C_1}x - P_{C_1}P_{C_2}x, P_{C_2}x - P_{C_1}P_{C_2}x \rangle = 0, \quad \forall x \in \mathcal{H}.$

That this is not necessarily a consequence of $P_{C_1}P_{C_2} = P_{C_2}P_{C_1} = P_{C_1 \cap C_2}$ can be easily verified with almost any two dimensional example. Yet, in all cases

$$P_{C_1 \cap C_2^\perp} P_{C_2 \cap C_1^\perp} = P_{C_2^\perp} P_{C_1} P_{C_1^\perp} P_{C_2} = 0.$$

Expanding (5.55) on use of (2.4'), we obtain

(5.56)   $\| P_{C_1} P_{C_2} x \|^2 = \langle P_{C_1}x, P_{C_2}x \rangle$

as a necessary and sufficient condition for (5.53) and (5.54) to be projections. This is also necessary for $P_{C_1}$ and $P_{C_2^\perp}$ to commute, for if they do, then $P_{C_1}P_{C_2^\perp} = P_{C_1 \cap C_2^\perp}$, and

$$0 = \langle x - P_{C_1}P_{C_2^\perp}x, P_{C_1}P_{C_2^\perp}x \rangle = \langle P_{C_2}x + P_{C_2^\perp}x - P_{C_1}P_{C_2^\perp}x, P_{C_1}P_{C_2^\perp}x \rangle$$

$$= \langle P_{C_2}x, P_{C_2^\perp}P_{C_1}x \rangle + \langle P_{C_2^\perp}x - P_{C_1}P_{C_2^\perp}x, P_{C_1}P_{C_2^\perp}x \rangle = \langle P_{C_2}x, P_{C_2^\perp}P_{C_1}x \rangle$$

$$= \langle P_{C_2}x, P_{C_1}x \rangle - \langle P_{C_2}x, P_{C_2}P_{C_1}x \rangle = \langle P_{C_2}x, P_{C_1}x \rangle - \| P_{C_1}P_{C_2}x \|^2.$$

Hence again, in view of the counterexample mentioned above, the commutativity of $P_{C_1}$ and $P_{C_2^\perp}$ does not follow from that of $P_{C_1}$ and $P_{C_2}$.

The lemma below states the expected fact that if projections commute so do the projections on support cones at a common point of their ranges.

**Lemma 5.13** If $P_{\mathcal{K}_1} P_{\mathcal{K}_2} = P_{\mathcal{K}_2} P_{\mathcal{K}_1}$ then

(5.57)
$$P_{S_{\mathcal{K}_1}}(z) P_{S_{\mathcal{K}_2}}(z) = P_{S_{\mathcal{K}_2}}(z) P_{S_{\mathcal{K}_1}}(z) ,$$

$$\forall z \in \mathcal{K}_1 \cap \mathcal{K}_2 ,$$

(5.58)
$$P_{z+S_{\mathcal{K}_1}}(z) P_{z+S_{\mathcal{K}_2}}(z) = P_{z+S_{\mathcal{K}_2}}(z) P_{z+S_{\mathcal{K}_1}}(z) ,$$

$$\forall z \in \mathcal{K}_1 \cap \mathcal{K}_2 .$$

**Proof:** If $z \in \mathcal{K}_1 \cap \mathcal{K}_2$ and $y$ is any point in space, then

$$P_{\mathcal{K}_2}(z+y) = P_{\mathcal{K}_2} z + P_{S_{\mathcal{K}_2}}(y) z + o(\|y\|)$$

by Lemma 4.6. By the same lemma again,

(5.59) $P_{\mathcal{K}_1} P_{\mathcal{K}_2}(z+y) = P_{\mathcal{K}_1}(z + P_{S_{\mathcal{K}_2}}(z) y + o(\|y\|))$

$$= P_{\mathcal{K}_1} P_{\mathcal{K}_2} z + P_{S_{\mathcal{K}_1}}(z)(P_{S_{\mathcal{K}_2}}(z) y + o(\|y\|))$$

$$+ o(\|P_{S_{\mathcal{K}_2}}(z) y + o(\|y\|)\|) = P_{\mathcal{K}_1} P_{\mathcal{K}_2} z$$

$$+ P_{S_{\mathcal{K}_1}}(z) P_{S_{\mathcal{K}_2}}(z) y + o(\|y\|)$$

and similarly,

$$P_{\mathcal{K}_2} P_{\mathcal{K}_1}(z+y) = P_{\mathcal{K}_2} P_{\mathcal{K}_1} z + P_{S_{\mathcal{K}_2}}(z) P_{S_{\mathcal{K}_1}}(z) y + o(\|y\|) .$$

Therefore

$$P_{S_{\mathcal{K}_1}(z)} P_{S_{\mathcal{K}_2}(z)} y = P_{S_{\mathcal{K}_2}(z)} P_{S_{\mathcal{K}_1}(z)} y + o(\|y\|),$$

which, on replacing $y$ by $ty$ ($t > 0$), dividing by $t$ and letting $t \downarrow 0$ yields (5.57). Equation (5.58) follows from (5.57) through the identities

$$P_{S_{\mathcal{K}_1}(z)}(x-z) = P_{z+S_{\mathcal{K}_1}(z)} x - z$$

$$P_{S_{\mathcal{K}_2}(z)} P_{S_{\mathcal{K}_1}(z)}(x-z) = P_{S_{\mathcal{K}_2}(z)}(P_{z+S_{\mathcal{K}_1}(z)} x - z) =$$

$$= P_{z+S_{\mathcal{K}_2}(z)} P_{z+S_{\mathcal{K}_1}(z)} x - z$$

and those obtained by interchanging $\mathcal{K}_1$ and $\mathcal{K}_2$.

<u>Corollary 1</u>  If $\mathcal{K}_1$ and $\mathcal{K}_2$ are finite dimensional, and $P_{\mathcal{K}_1} P_{\mathcal{K}_2} = P_{\mathcal{K}_2} P_{\mathcal{K}_1}$, then

(5.60) $\qquad P_{\mathcal{K}_1} P_{\mathcal{K}_2} x = P_{\mathcal{K}_1 \cap \mathcal{K}_2} x + o(\|x - P_{\mathcal{K}_1 \cap \mathcal{K}_2} x\|)$.

<u>Proof:</u>  If $z \in \mathcal{K}_1 \cap \mathcal{K}_2$, by (5.59) and Theorem 5.5,

$$P_{\mathcal{K}_1} P_{\mathcal{K}_2}(z+y) = P_{\mathcal{K}_1} P_{\mathcal{K}_2} z + P_{S_{\mathcal{K}_1}(z)} P_{S_{\mathcal{K}_2}(z)} y + o(\|y\|)$$

$$= z + P_{S_{\mathcal{K}_1 \cap \mathcal{K}_2}(z)} y + o(\|y\|).$$

By Lemma 4.6, $P_{\mathcal{K}_1 \cap \mathcal{K}_2}(z+y) = z + P_{S_{\mathcal{K}_1 \cap \mathcal{K}_2}(z)} y + o(\|y\|)$, so

$$P_{\mathcal{K}_1} P_{\mathcal{K}_2}(z+y) = P_{\mathcal{K}_1 \cap \mathcal{K}_2}(z+y) + o(\|y\|).$$

This yields (5.60) by setting $z = P_{\mathcal{K}_1} P_{\mathcal{K}_2} x$ and $x = z+y$.

## Part II. Spectral Theory

### §6. The ordering of projections.

Before entering into the matter proper we shall deduce a few basic inequalities concerning a type of operator of frequent occurence in our theory.

**Lemma 6.1** Any positive homogeneous everywhere defined Gateaux gradient mapping $T: \mathcal{H} \to \mathcal{H}$ satisfies the equation

$$(6.1) \qquad Tx = \nabla \tfrac{1}{2} \langle x, Tx \rangle, \qquad \forall x \in \mathcal{H}.$$

**Proof:** Let $\tau(x)$ be a real valued function such that $Tx = \nabla \tau(x)$. Then since $T$ is positive homogeneous,

$$\frac{d}{dt} \tau(tx) = \langle \nabla \tau(tx), x \rangle = \langle T(tx), x \rangle = t \langle Tx, x \rangle, \quad \text{if } t \geq 0,$$

whence, assuming as we may that $\tau(0) = 0$,

$$\tau(x) = \tau(x) - \tau(0) = \int_0^1 \frac{d\tau(tx)}{dt} dt = \int_0^1 t \langle Tx, x \rangle dt = \tfrac{1}{2} \langle Tx, x \rangle,$$

which is (6.1).

Conversely, if (6.1) holds, then setting $\tau(x) = \tfrac{1}{2} \langle x, Tx \rangle$,

$$\frac{d}{du}[e^{-2u}\tau(e^u x)] = -2e^{-2u}\tau(e^u x) + e^{-2u}\frac{d}{du}\tau(e^u x)$$

$$= -2e^{-2u}\tau(e^u x) + e^{-u}\langle \nabla \tau(e^u x), x \rangle$$

$$= -2e^{-2u}\tau(e^u x) + e^{-u}\langle T(e^u x), x \rangle$$

$$= -2e^{-2u}\tau(e^u x) + e^{-2u}\langle T(e^u x), e^u x \rangle = 0.$$

So $e^{-2u}\tau(e^u x)$ does not depend on $u$, and letting $t = e^u$, we may write

$$\tau(tx) = t^2 \tau(x), \quad \forall x \in \mathcal{H}, \quad t \geq 0.$$

This should yield the homogeneity of T at once. Let $x_0 \in \mathcal{H}$ and $t_0 > 0$ be fixed. Then on one hand,

$$T(t_0 x) - T(t_0 x_0) = t_0^2 (T(x) - T(x_0)) = t_0^2 \langle \nabla T(x_0), x - x_0 \rangle + t_0^2 o(\|x - x_0\|)$$

and on the other,

$$T(t_0 x) - T(t_0 x_0) = \langle \nabla T(t_0 x_0), t_0 (x - x_0) \rangle + o(t_0 \|x - x_0\|),$$

whence by comparison

$$\langle t_0 \nabla T(x_0) - \nabla T(t_0 x_0), x - x_0 \rangle = o(\|x - x_0\|),$$

which by the arbitrariness of $x$ yields $t_0 \nabla T(x_0) = \nabla T(t_0 x_0)$, that is $t_0 T x_0 = T(t x_0)$. Q.E.D.

**Lemma 6.2** Let $T: \mathcal{H} \to \mathcal{H}$ be an everywhere defined, monotone, positive homogeneous Gateaux gradient mapping. Then, the real valued function $\frac{1}{2}\langle x, Tx \rangle$ is convex, $Tx = \nabla \frac{1}{2} \langle x, Tx \rangle$ and

(6.2) $$\frac{1}{2}\langle x, Tx \rangle + \frac{1}{2} \langle y, Ty \rangle \geq \langle x, Ty \rangle, \quad \forall\, x, y \in \mathcal{H}.$$

**Proof:** By Lemma 6.1 we know that $Tx = \nabla \frac{1}{2} \langle x, Tx \rangle$. Therefore, having a monotone gradient, $\frac{1}{2} \langle x, Tx \rangle$ is a convex differentiable function. Now, for fixed $y$, the function $\frac{1}{2} \langle x, Tx \rangle - \langle x, Ty \rangle$ is convex and differentiable; its gradient vanishes at $x = y$, and so its minimum is attained there. Hence,

$$\frac{1}{2}\langle x, Tx \rangle - \langle x, Ty \rangle \geq \frac{1}{2}\langle y, Ty \rangle - \langle y, Ty \rangle = -\frac{1}{2} \langle y, Ty \rangle,$$

which is another form of (6.2).

**Corollary 1** (Schwarz inequality)

(6.3) $$\langle x, Ty \rangle \leq \langle x, Tx \rangle^{\frac{1}{2}} \langle y, Ty \rangle^{\frac{1}{2}}$$

## SPECTRAL THEORY

**Proof:** Since $T0 = 0$, and $T$ is monotone, $\langle x, Tx \rangle \geq 0$ for any $x$. From (6.2) with $tx$ and $t^{-1}y$ in place of $x$ and $y$, $(t > 0)$, we get

$$\langle x, Ty \rangle \leq \tfrac{1}{2}[t^2 \langle x, Tx \rangle + t^{-2} \langle y, Ty \rangle],$$

and

$$\langle x, Ty \rangle \leq \inf_{0 < t < \infty} \tfrac{1}{2}[t^2 \langle x, Tx \rangle + t^{-2} \langle y, Ty \rangle] = \langle x, Tx \rangle^{\frac{1}{2}} \langle y, Ty \rangle^{\frac{1}{2}}.$$

**Corollary 2**

(6.4) $\langle x-y, Tx-Ty \rangle \geq |\langle x, Ty \rangle - \langle Tx, y \rangle|$, $\forall\, x, y \in H$.

**Proof:** Interchange $x$ and $y$ in (6.2), and replace the right hand member by $\max[\langle x, Ty \rangle, \langle Tx, y \rangle]$,

$$\langle x, Tx \rangle + \langle y, Ty \rangle \geq 2\max[\langle x, Ty \rangle, \langle Tx, y \rangle] = |\langle x, Ty \rangle - \langle Tx, y \rangle|$$
$$+ \langle x, Ty \rangle + \langle Tx, y \rangle.$$

Transferring to the left the last two terms one obtains (6.4).

**Corollary 3** (Triangle Inequality)

(6.5) $\langle x, Tx \rangle^{\frac{1}{2}} + \langle y, Ty \rangle^{\frac{1}{2}} \geq \langle x+y, T(x+y) \rangle^{\frac{1}{2}}$, $\forall\, x, y \in H$.

**Proof:** Start from

$$\langle x+y, T(x+y) \rangle = \langle x, T(x+y) \rangle + \langle y, T(x+y) \rangle,$$

apply Schwarz inequality to each term on the right,

$$\langle x+y, T(x+y) \rangle \leq \langle x, Tx \rangle^{\frac{1}{2}} \langle x+y, T(x+y) \rangle^{\frac{1}{2}} + \langle y, Ty \rangle^{\frac{1}{2}} \langle x+y, T(x+y) \rangle^{\frac{1}{2}}$$

and divide by $\langle x+y, T(x+y) \rangle^{\frac{1}{2}}$ assumed different from zero.

<u>Corollary 4</u>  The function $\langle x, Tx\rangle^{\frac{1}{2}}$ is convex, and

(6.6) $$\nabla \langle x, Tx\rangle^{\frac{1}{2}} = \frac{Tx}{\langle x, Tx\rangle^{\frac{1}{2}}} \quad \text{if } \langle x, Tx\rangle \neq 0.$$

<u>Proof:</u>  Convexity is a direct consequence of (6.5), by the positive homogeneity of $T$. Equation (6.6) follows from $Tx = \nabla \frac{1}{2}\langle x, Tx\rangle$.

<u>Corollary 5</u>  The operator $x \to Tx/\langle x, Tx\rangle^{\frac{1}{2}}$ is monotone.

<u>Proof:</u>  $Tx/\langle x, Tx\rangle^{\frac{1}{2}}$ is the gradient of a convex function.

<u>Corollary 6</u>  If $P$ is a projection on a closed convex cone with vertex at $0$,

(6.7) $$|\langle x, Py\rangle| \leq \|x\| \langle y, Py\rangle^{\frac{1}{2}}, \quad \forall\, x, y \in \mathcal{H}.$$

<u>Proof:</u>  Replace $\langle x, Px\rangle^{\frac{1}{2}}$ by $\|x\|$ in (6.3), and compare the resulting inequality with that obtained by replacing $x$ by $-x$.

Now we introduce an order in the class of projections on cones and proceed to discuss some of its properties. <u>In this and all subsequent sections all projections will be projections on closed convex cones with vertex at the origin.</u> Consequently we shall no longer qualify this fact nor shall we indicate in the notation the cone on which the projection is made. Whenever reference to this set appears necessary we shall speak of the range of the projection and denote it $\mathcal{R}(P)$. It is important to bear this in mind because letters denoting projections will be affected by subscripts of various kinds, seldom however indicating the range.

<u>Definition 6.1</u>  A projection $P_1$ is said to be weaker than a projection $P_2$ if

(6.8) $$P_1 P_2 = P_1.$$

This fact is indicated by the notation $P_1 < P_2$.

## SPECTRAL THEORY

The relation $<$ is reflexive and transitive: It is a partial ordering. As for linear orthogonal projections $P_1 < P_2$ implies that the range of the first is contained in the range of the second (Theorem 5.2), but unlike in that case, it is not equivalent to this fact. Recapitulating the usual notions about partially ordered structures we recall that a projection P is a lower or an upper bound of a set of projections if it is weaker or stronger respectively than any projection in the set; the class of lower and upper bounds of a set are never empty because they contain 0 (projection onto the origin) and I (projection on the whole space). If among the lower (upper) bounds there is one, and hence only one, stronger (weaker) than any other such bound we say that it is the infimum (supremum) of the set. If the set is $\{P_\alpha\}_{\alpha \in I}$, then we use the notation

$$(6.9) \qquad \inf_{\alpha \in I} P_\alpha = \bigcap_{\alpha \in I} P_\alpha, \qquad \sup_{\alpha \in I} P_\alpha = \bigcup_{\alpha \in I} P_\alpha.$$

Of course the infimum or the supremum may not exist; we do not even know if they do when the class contains two elements only. Theorems 5.2 and 5.5 and the lemmas following them provide the tools to handle the ordering of projections. The basic properties below can be read out of them:

$$(6.10) \qquad P_1 < P_2 \iff P_2^\perp < P_1^\perp, \qquad (P^\perp = I-P)$$

$$(6.11) \qquad P_1 < P_2 \iff P_2 - P_1 = P,$$

$$(6.12) \qquad P_1 < P_2 \iff \{P_1 P_2 = P_2 P_1, \langle x, P_1 x \rangle \leq \langle x, P_2 x \rangle, \forall x \in \mathcal{H}\},$$

$$(6.13) \qquad P_1 < P_2 \implies \mathcal{R} P_1 \subset \mathcal{R} P_2$$

$(6.15) \qquad P_1 < P_2 \implies \{P_1, P_2, P_1^\perp, P_2^\perp$ commute and their products are the projections on the intersection of their ranges$\}$

$$(6.16) \qquad (\bigcap_\alpha P_\alpha)^\perp = \bigcup_\alpha P_\alpha^\perp$$

(6.17) $$\sum_\alpha P_\alpha = \bigcup_\alpha P_\alpha,$$ for any class of orthogonal projections.

**Definition 6.2** A class $\Pi$ of projections is said to be directed upwards (downwards) if for any couple $P_1, P_2 \in \Pi$ there is a $P_3 \in \Pi$ such that $P_1 < P_3$, $P_2 < P_3$ ($P_1 > P_3, P_2 > P_3$).

Clearly, a class directed upwards is a directed set with regard to the relation $<$, whereas one directed downwards is a directed set with regard to the opposite relation $>$. Out of a directed class $\Pi$ of projections and any $x \in \mathcal{H}$ one obtains a directed set of points (net) $\{x_P = Px\}_{P \in \Pi}$. The following lemma answers the question of the convergence of $\{x_P\}_{P \in \Pi}$.

**Lemma 6.3** If $\Pi = \{P\}$ is a class of commuting projections directed either upwards or downwards, then $\lim_< Px$ or $\lim_> Px$ respectively exists for any $x \in \mathcal{H}$. Correspondingly $\bigcup_{P \in \Pi} P$ or $\bigcap_{P \in \Pi} P$ exists, and

(6.18) $$\lim_< Px = (\bigcup_{P \in \Pi} P) x,$$

(6.19) $$\lim_> Px = (\bigcap_{P \in \Pi} P) x.$$

**Proof:** Assume $\Pi$ directed downwards. Then for any $x \in \mathcal{H}$, $\langle Px, x \rangle$, is a nonnegative number decreasing with $P$, and therefore $\lim_> \langle Px, x \rangle = \inf_{P \in \Pi} \langle Px, x \rangle$, that is, for given $\varepsilon > 0$ there is a $P_\varepsilon \in \Pi$ such that

$$\langle Px, x \rangle \leq \inf_{P \in \Pi} \langle Px, x \rangle + \varepsilon^2/4, \quad \forall P < P_\varepsilon, \quad P \in \Pi.$$

Therefore if $P_1$ and $P_2$ belong to $\Pi$ and are both weaker than $P_\varepsilon$,

$$\inf_{P \in \Pi} \langle Px, x \rangle \leq \langle P_i x, x \rangle \leq \inf_{P \in \Pi} \langle Px, x \rangle + \varepsilon^2/4, \quad i = 1, 2,$$

SPECTRAL THEORY

and for any $P_3 \in \Pi$ weaker than both $P_1$ and $P_2$,

$$\langle(P_i-P_3)x,x\rangle = \langle P_i x,x\rangle - \langle P_3 x,x\rangle \leq \varepsilon^2/4, \quad i = 1,2.$$

Since $P_i - P_3$ is a projection we have by (6.7),

$$|\langle(P_i-P_3)x,z\rangle| \leq \|z\| \langle(P_i-P_3)x,x\rangle^{\frac{1}{2}} \leq (\varepsilon/2)\|z\|, \quad i=1,2;$$

$$\forall z \in H,$$

and so

$$\|(P_i - P_3)x\| \leq \varepsilon/2,$$

whence

$$\|(P_1 - P_2)x\| \leq \varepsilon.$$

Therefore the net $x_P = P x$ is a Cauchy net, and by the completeness of $H$ it converges strongly to a limit. Thus $\lim_{>} P x$ exists for any $x \in H$; let us call it $Qx$. Taking limits in the relation

$$\langle x-Px, Px-Py\rangle \geq 0, \quad P \in \Pi, \quad x,y \in H,$$

one obtains

$$\langle x - Qx, Qx - Qy\rangle \geq 0, \quad x,y \in H,$$

indicating that $Q$ is a projection, namely, a projection on a convex cone since it is obviously positive homogeneous. It remains to check that $Q = \bigcap_{P \in \Pi} P$. From the relation

$$P P_1 x = Px, \quad P, P_1 \in \Pi, \quad P < P_1, \quad \forall x \in H,$$

we deduce by taking limits,

$$Q P_1 x = Q x, \quad P_1 \in \Pi, \quad \forall x \in H,$$

which says that $Q$ is a lower bound for $\Pi$. Let $R$ be any

349

lower bound of $\Pi$, then

$$RPx = Rx, \quad \forall P \in \Pi, \quad \forall x \in \mathcal{H},$$

and taking limits again, since $R$ is continuous,

$$RQx = Rx, \quad \forall x \in \mathcal{H},$$

that is $\mathcal{R} < Q$, completing the proof in the case of a class directed downwards. The case of a class directed upwards can be treated by a similar argument, or, better still, be derived from the above by passing to the complements.

## §7. Partitions of the identity. Spectral measures.

**Definition 7.1** A conical partition of the identity is a family of projections $\{P_\lambda\}_{-\infty}^{+\infty}$ depending on a real parameter $\lambda$, $-\infty \leq \lambda \leq +\infty$, satisfying the following properties:

(7.1) $$\lambda_1 \leq \lambda_2 \implies P_{\lambda_1} < P_{\lambda_2},$$

(7.2) $$P_\lambda = P_{\lambda+0}, \text{ i.e., } P_\lambda x = \lim_{\varepsilon \downarrow 0} P_{\lambda+\varepsilon} x, \quad \forall x \in \mathcal{H}, \forall \lambda.$$

(7.3) $$P_{-\infty} = 0, \ P_{+\infty} = I, \text{ i.e., } \lim_{\lambda \to -\infty} P_\lambda x = 0, \ \lim_{\lambda \to +\infty} P_\lambda x = x,$$

$$\forall x \in \mathcal{H}.$$

Partitions of the identity are also called "spectral resolutions".

Partitions of the identity into linear orthogonal projections are instances of conical partitions. The simplest example of a nonlinear partition is perhaps the following: Let $P$ be a projection whatever and set

$$P_\lambda = \begin{cases} 0, & \lambda < 0, \\ P^\perp, & 0 \leq \lambda < \lambda_0 \\ I, & \lambda_0 \leq \lambda \leq +\infty \end{cases}$$

## SPECTRAL THEORY

More complicated examples can be constructed as follows: Let $\mathcal{H} = \mathfrak{m}_1 \oplus \mathfrak{m}_2 \oplus \ldots$ be a decomposition of the space into the direct sum of closed orthogonal subspaces, and let $C_i'$ and $C_i''$ be cones in $\mathfrak{m}_i$, mutually dual relatively to $\mathfrak{m}_i$. Now order the class of projections $\{P_{C_i'}, P_{C_i''}\}_i^{\infty}$ into a single sequence $\{P_i\}_{-\infty}^{+\infty}$, and for any increasing sequence of real numbers $\{\lambda_i\}_{-\infty}^{+\infty}$ define

$$P_\lambda = \sum_{\lambda_i \leq \lambda} P_i, \qquad -\infty \leq \lambda \leq +\infty.$$

This is a nonlinear partition of the identity; we leave the verification to the reader.

The spectral resolutions just described may be termed "discrete partitions". Continuous nonlinear spectral resolution can be obtained much in the same manner from linear ones. If $C^+[\mu, \lambda]$ and $C^-[\mu, \lambda]$ are the cones of all non-negative and nonpositive functions in $L^2(0, \infty)$ vanishing outside of the interval $[\mu, \lambda]$, then

$$P_\lambda = \begin{cases} P_{C^-[-\lambda, +\infty]}, & \lambda < 0 \\ P_{C^+[0, \lambda]} + P_{C^-[0, +\infty]}, & \lambda \geq 0, \end{cases}$$

is a nonlinear continuous partition of the identity in $L^2(0, \infty)$.

If the dimension of $\mathcal{H}$ is finite, say $n$, then any partition of the identity is of the form

$$P_\lambda = \sum_{\lambda_i \leq \lambda} P_i,$$

where the $P_i$'s are no more than $2n$ pairwise orthogonal projections. The reason for this being that in $\mathbb{R}^n$ there are at most $2n$ nonvanishing vectors making obtuse angles with each other, coupled with the fact to be proved below that if $(\mu_1, \lambda_1]$ and $(\mu_2, \lambda_2]$ are disjoint semiopen intervals, then $P_{\lambda_1} - P_{\mu_1}$ and $P_{\lambda_2} - P_{\mu_2}$ are orthogonal, for any spectral resolution.

**Lemma 7.1** Let $\{P_\lambda\}_{-\infty}^{+\infty}$ be a partition of the identity, and write

(7.4) $$P_{(\mu,\lambda]} = P_\lambda - P_\mu, \quad \text{for any } \lambda \geq \mu.$$

Then

(7.5) $$P_{(\mu_1,\lambda_1]} P_{(\mu_2,\lambda_2]} = P_{(\mu_1,\lambda_1]\cap(\mu_2,\lambda_2]} = P_{(\mu_1,\lambda_1]} \cap P_{(\mu_2,\lambda_2]}.$$

Moreover, if $(\mu_1,\lambda_1]\cap(\mu_2,\lambda_2] = \emptyset$ then $P_{(\mu_1,\lambda_1]}$ and $P_{(\mu_2,\lambda_2]}$ are orthogonal projections.

**Proof:** Note that if $\mu = \lambda$ then $(\mu,\lambda] = \emptyset$, and according to (7.4) $P_\emptyset = 0$. Assuming $\mu_1 \leq \mu_2$ we can distinguish the following three cases according to the position of $\lambda_1$:

a. $\mu_1 \leq \lambda_1 \leq \mu_2 \leq \lambda_2$;

b. $\mu_1 \leq \mu_2 \leq \lambda_1 \leq \lambda_2$;

c. $\mu_1 \leq \mu_2 \leq \lambda_2 \leq \lambda_1$.

In case a. write

$$P_{\lambda_2} = P_{\mu_1} + (P_{\lambda_1} - P_{\mu_1}) + (P_{\mu_2} - P_{\lambda_1}) + (P_{\lambda_2} - P_{\mu_2}),$$

and conclude that the four terms on the right must be orthogonal, by Theorem 5.5. In particular, $P_{(\mu_1,\lambda_1]}$ and $P_{(\mu_2,\lambda_2]}$ are orthogonal and their product vanishes, as (7.5) requires.

In case c. we have

$$P_{(\mu_1,\lambda_1]} = P_{(\mu_1,\mu_2]} + P_{(\mu_2,\lambda_2]} + P_{(\lambda_2,\lambda_1]}$$

and by Theorem 5.5 again,

SPECTRAL THEORY

$$P_{(\mu_1,\lambda_1]}P_{(\mu_2,\lambda_2]} = P_{(\mu_2,\lambda_2]} \;,$$

which is (7.5) for the case under consideration.
Finally in case b.,

$$P_{(\mu_1,\lambda_1]}P_{(\mu_2,\lambda_2]} = (P_{(\mu_1,\mu_2]} + P_{(\mu_2,\lambda_1]})P_{(\mu_2,\lambda_2]}$$

$$= P_{(\mu_1,\mu_2]}P_{(\mu_2,\lambda_2]} + P_{(\mu_2,\lambda_1]}P_{(\mu_2,\lambda_2]}$$

and appeal to the cases already treated leads to (7.5) once more. If $(\mu_1,\lambda_1](\mu_2,\lambda_2] = \phi$ we may assume that $\mu_1 \leq \lambda_1 \leq \mu_2 \leq \lambda_2$ and we are in the situation a., in which case the orthogonality of $P_{(\mu_1,\lambda_1]}$ and $P_{(\mu_2,\lambda_2]}$ has been proved.

The main task ahead of us in this section is to construct a projection valued measure - called a "spectral measure" - out of a partition of the identity $\{P_\lambda\}_{-\infty}^{+\infty}$. This is done in various stages, much in the manner as a positive measure over the reals is constructed out of a non-decreasing real valued functions, by ordinary measure theory technique with only occasional reference to the nature of the measure values. The readers familiar with this area of analysis will see their way to Theorem 7.1, and will not need to go through the lengthy and tedious details below.

<u>Definition 7.2</u>   $\Sigma_0$ is the class of subsets E in the extended real line $[-\infty,+\infty]$, of the form

(7.6) $\qquad E = (\mu_1,\lambda_1] \cup (\mu_2,\lambda_2] \cup \ldots \cup (\mu_n,\lambda_n]\;,$

where the $(\mu_i,\lambda_i]$ are disjoint semiopen intervals.

$\Sigma_0$ <u>is clearly a Boolean algebra, that is, it contains the empty set and the whole line, along with any E it contains its complement E', and for any two of its sets contains</u>

their union and their intersection.

<u>Lemma 7.2</u>   Let $\{P_\lambda\}_{-\infty}^{+\infty}$ be a partition of the identity. Then the operation assigning to each $E \in \Sigma_0$ the projection

(7.7) $$P_E = P_{(\mu_1, \lambda_1]} + P_{(\mu_2, \lambda_2]} + \ldots + P_{(\mu_n, \lambda_n]}$$

is independent of the manner

$$E = (\mu_1, \lambda_1] \cup (\mu_2, \lambda_2] \cup \ldots \cup (\mu_n, \lambda_n],$$
$$(\mu_i, \lambda_i] \cap (\mu_j, \lambda_j] = \emptyset, \quad i \neq j,$$

into which $E$ is expressed and thus is well defined. The class $\Pi_0 = \{P_E\}_{E \in \Sigma_0}$ is a Boolean algebra of commuting projections, the algebra operations being

(7.8) $\quad I - P_E = P_{E'}, \quad E'$ complement of $E$,

(7.9) $\quad P_{E_1} \cap P_{E_2} = P_{E_1 \cap E_2} = P_{E_1} P_{E_2}$

(7.10) $\quad P_{E_1} \cup P_{E_2} = P_{E_1 \cup E_2} = P_{E_1} + P_{E_2} - P_{E_1} P_{E_2}$.

Moreover

(7.11) $\quad P_{(-\infty, \lambda]} = P_\lambda$.

<u>Proof:</u>   It is quite clear that if $E$ itself is a semiopen interval $(\mu, \lambda]$, then $P_E = P_{(\mu, \lambda]}$, for any decomposition of $E$ into a union of disjoint semiopen intervals.

If $E = \bigcup_i^n (\mu_i^{(1)}, \lambda_i^{(1)}] = \bigcup_j^m (\mu_j^{(2)}, \lambda_j^{(2)}]$ and if $(\mu_{ij}, \lambda_{ij}] = (\mu_i^{(1)}, \lambda_i^{(1)}] \cap (\mu_j^{(2)}, \lambda_j^{(2)}]$, then $(\mu_i^{(1)}, \lambda_i^{(1)}] = \bigcup_{j=1}^m (\mu_{ij}, \lambda_{ij}]$, and

$$P_{(\mu_i^{(1)}, \lambda_i^{(1)}]} = \sum_{j=1}^m P_{(\mu_{ij}, \lambda_{ij}]},$$

since the $(\mu_{ij}, \lambda_{ij}]$ are obviously disjoint. Adding these up with regard to $i$,

$$\sum_{i=1}^{n} P_{(\mu_i^{(1)}, \lambda_i^{(1)}]} = \sum_{i=1}^{n}\sum_{j=1}^{m} P_{(\mu_{ij}, \lambda_{ij}]},$$

and from this, by symmetry,

$$\sum_{j=1}^{m} P_{(\mu_j^{(2)}, \lambda_j^{(2)}]} = \sum_{j=1}^{m}\sum_{i=1}^{n} P_{(\mu_{ij}, \lambda_{ij}]}$$

and so

$$\sum_{i=1}^{n} P_{(\mu_i^{(2)}, \lambda_i^{(1)}]} = \sum_{j=1}^{m} P_{(\mu_j^{(2)}, \lambda_j^{(2)}]},$$

proving the legitimacy of the operation $E \to P_E$. If $E = (\mu_1, \lambda_1] \cup (\mu_2, \lambda_2] \cup \ldots \cup (\mu_n, \lambda_n]$, then $E' = (-\infty, \mu_1] \cup (\lambda_1, \mu_2] \cup \ldots \cup (\lambda_{n-1}, \mu_n] \cup (\lambda_n, +\infty]$, where we have of course assumed that $\mu_1 \le \mu_2 \le \ldots \le \mu_n$. It is obvious then that

$$P_E + P_{E'} = P_{+\infty} - P_{-\infty} = I,$$

proving (7.8).

Relation (7.9) is proved along the lines used to demonstrate the independence of $P_E$ on the manner $E$ is split into a sum of intervals. If $E_1 = \bigcup_{i=1}^{n}(\mu_i^{(1)}, \lambda_i^{(1)}]$, $E_2 = \bigcup_{j=1}^{m}(\mu_j^{(2)}, \lambda_j^{(2)}]$, then $E_1 \cap E_2 = \bigcup_{i=1}^{n}\bigcup_{j=1}^{m}(\mu_{ij}, \lambda_{ij}]$, where $(\mu_{ij}, \lambda_{ij}] = (\mu_i, \lambda_i] \cap (\mu_j, \lambda_j]$, and

$$P_{E_1 \cap E_2} = \sum_{i=1}^{n}\sum_{j=1}^{m} P_{(\mu_{ij}, \lambda_{ij}]} = \sum_{i=1}^{n}\sum_{j=1}^{m} P_{(\mu_i^{(1)}, \lambda_i^{(1)}]} P_{(\mu_j^{(2)}, \lambda_j^{(2)}]}.$$

On the other hand, for any $i$,

$$E_2 = \bigcup_{j=1}^{m} (\mu_j^{(2)}, \lambda_j^{(2)}] = \bigcup_{j=1}^{m} \{((\mu_j^{(2)}, \lambda_j^{(2)}] \cap (\mu_i^{(1)}, \lambda_i^{(1)}])$$

$$\cup ((\mu_j^{(2)}, \lambda_j^{(2)}] \cap (-\infty, \mu_i^{(1)}]) \cup ((\mu_j^{(2)}, \lambda_j^{(2)}] \cap (\lambda_i^{(1)}, +\infty])\},$$

and since the intervals on the right are disjoint

$$P_{E_2} x = \sum_{j=1}^{m} P_{(\mu_j^{(2)}, \lambda_j^{(2)}]} x = \sum_{j=1}^{m} P_{(\mu_j^{(2)}, \lambda_j^{(2)}]} P_{(\mu_i^{(1)}, \lambda_i^{(1)}]} x$$

$$+ \sum_{j=1}^{m} P_{(\mu_j^{(2)}, \lambda_j^{(2)}]} P_{(-\infty, \mu_i^{(1)}]} x + \sum_{j=1}^{m} P_{(\mu_j^{(2)}, \lambda_j^{(2)}]} P_{(\lambda_i^{(1)}, +\infty]} x$$

for any $x \in \mathcal{H}$. The three vectors represented by the sums on the right are orthogonal, the first belongs to $\mathcal{R}(P_{(\mu_i^{(1)}, \lambda_i^{(1)}]})$ and the last two to $\mathcal{R}(P_{(\mu_i^{(1)}, \lambda_i^{(1)}]})^\perp$. Therefore (Lemma 2.2),

$$P_{(\mu_i^{(1)}, \lambda_i^{(1)}]} P_{E_2} x = \sum_{j=1}^{m} P_{(\mu_j^{(2)}, \lambda_j^{(2)}]} P_{(\mu_i^{(1)}, \lambda_i^{(1)}]} x ,$$

and adding up with regard to $i$,

$$P_{E_1} P_{E_2} x = \sum_{i=1}^{n} \sum_{j=1}^{m} P_{(\mu_j^{(2)}, \lambda_j^{(2)}]} P_{(\mu_i^{(1)}, \lambda_i^{(1)}]} x .$$

Comparison with the expression previously obtained for $P_{E_1 \cap E_2}$ yields $P_{E_1} P_{E_2} = P_{E_1 \cap E_2}$. That this should also be equal to $P_{E_1} \cap P_{E_2}$ is a consequence of the easily proved fact that if the product of two commuting projections is a projection, then it is their infimum. Equation (7.9) is thus established. As to (7.10) note that by (7.8) and (7.9),

$$P_{E_1 \cup E_2} = I - P_{(E_1 \cup E_2)'} = I - P_{E_1' \cap E_2'} = I - P_{E_1'} P_{E_2'} = I - P_{E_2'} + P_{E_1} P_{E_2'}$$

$$= P_{E_2} + P_{E_2'} P_{E_1} = P_{E_2} + P_{E_1} - P_{E_1} P_{E_2}.$$

Furthermore,

$$(P_{E_1 \cup E_2})^\perp = I - P_{E_1 \cup E_2} = P_{E_1'} P_{E_2'} = P_{E_1'} \cap P_{E_2'} = (P_{E_1})^\perp \cap (P_{E_2})^\perp$$

and by (6.16),

$$P_{E_1 \cup E_2} = ((P_{E_1})^\perp \cap (P_{E_2})^\perp)^\perp = P_{E_1} \cup P_{E_2},$$

completing the proof of (7.10). Finally, (7.11) holds by definition of $P_{(-\infty, \lambda]}$.

A substantial part of this lemma can be expressed by saying that <u>the mapping $E \to P_E$ is a homomorphism of the Boolean algebra $\Sigma_0$ onto the Boolean algebra $\Pi_0$</u>.

The next step consists in extending the definition of $P_E$ to the union of countable many disjoint semiopen intervals. Let us write down the formal definition.

<u>Definition 7.3</u>   $\Sigma_1$ is the class of subsets $E$ of the extended real line $[-\infty, +\infty]$ that can be represented in the form

$$E = \bigcup_i (\mu_i, \lambda_i],$$

where $\{(\mu_i, \lambda_i]\}$ is a at most countable class of disjoint semiopen intervals.

<u>$\Sigma_1$ contains $\Sigma_0$ but is not a Boolean algebra,</u> because the complement of a set in $\Sigma_1$ does not necessarily belong to it. However, <u>it contains all open sets in the real line, the intersection of any finite number, and the union of</u>

any countable number of its sets; all its sets are Borel sets.

In other terms, $\Sigma_1$ is a $\sigma$-field of Borel sets containing all open sets.

Extending Lemma 7.2 we have:

**Lemma 7.3** Let $\{P_\lambda\}_{-\infty}^{+\infty}$ be a partition of the identity. Then the mapping assigning to any set $E = \bigcup_{i=1}^{\infty}(\mu_i, \lambda_i]$ in $\Sigma_1$ the projection

$$(7.12) \qquad P_E = \sum_i P_{(\mu_i, \lambda_i]}$$

is independent of the manner into which $E$ has been decomposed, and thus is well defined. The class of projections $\Pi_1 = \{P_E\}_{E \in \Sigma_1}$ is a $\sigma$-field of commuting projections, and the mapping $E \to P_E$ of $\Sigma_1$ onto $\Pi_1$ has the properties,

$$(7.13) \qquad \bigcap_1^n P_{E_i} = P_{\bigcap_1^n E_i} = P_{E_1} P_{E_2} \cdots P_{E_n} ,$$

$$(7.14) \qquad \bigcup_1^\infty P_{E_i} = P_{\bigcup_1^\infty E_i} = \lim_{n \to \infty} \Big( \sum_{i_1=1}^n P_{E_{i_1}} - \sum_{i_1, i_2 = 1}^n P_{E_{i_1}} P_{E_{i_2}}$$

$$+ \sum_{i_1, i_2, i_3 = 1}^n P_{E_{i_1}} P_{E_{i_2}} P_{E_{i_3}} - \cdots - (-1)^n \sum_{i_1, \ldots, i_n = 1}^n P_{E_{i_1}} P_{E_{i_2}} \cdots P_{E_{i_n}} \Big).$$

**Proof:** The first point is to see that $P_E$ is well defined. As in the previous lemma we begin by considering the case when $E$ is an interval and show that if

$$(\mu, \lambda] = \bigcup_{i=1}^\infty (\mu_i, \lambda_i] , \quad (\mu_i, \lambda_i] \cap (\mu_j, \lambda_j] = \emptyset , \quad i \neq j ,$$

then

$$(7.15) \qquad P_{(\mu, \lambda]} = \sum_{i=1}^\infty P_{(\mu_i, \lambda_i]} .$$

SPECTRAL THEORY

Since the $P_{(\mu_i, \lambda_i]}$'s are orthogonal their sum is a projection (Theorem 5.5) and by (6.17)

$$\sum_{i=1}^{\infty} P_{(\mu_i, \lambda_i]} = \bigcup_{i=1}^{\infty} P_{(\mu_i, \lambda_i]}.$$

Moreover by (7.5),

$$(\sum_{i=1}^{\infty} P_{(\mu_i, \lambda_i]}) P_{(\mu, \lambda]} = \sum_{i=1}^{\infty} P_{(\mu_i, \lambda_i]} P_{(\mu, \lambda]} = \sum_{i=1}^{\infty} P_{(\mu_i, \lambda_i] \cap (\mu, \lambda]}$$

$$= \sum_{i=1}^{\infty} P_{(\mu_i, \lambda_i]}.$$

Hence

(7.16) $$\sum_{i=1}^{\infty} P_{(\mu_i, \lambda_i]} < P_{(\mu, \lambda]}.$$

To prove the opposite relation assume first that $(\lambda, \mu]$ is a bounded interval. Now, for any $x$, $\langle P_\lambda x, x \rangle$ is a nondecreasing function of $\lambda$ continuous on the right, and so for any given $\varepsilon > 0$ it is possible to find a sequence $\varepsilon_i > 0$, $i = 1, 2, \ldots$, such that

$$\sum_{i=1}^{\infty} [\langle P_{\lambda_i + \varepsilon_i} x, x \rangle - \langle P_{\lambda_i} x, x \rangle] \leq \varepsilon.$$

Moreover, for any $\eta$, $0 < \eta \leq \lambda - \mu$ (the case $\lambda = \mu$ can be discarded) $[\mu+\eta, \lambda] \subset (\mu, \lambda] \subset \bigcup_{i=1}^{\infty} (\mu_i, \lambda_i + \varepsilon_i]$, and since $[\mu+\eta, \lambda]$ is compact there is an $n$ such that

$$(\mu+\eta, \lambda] \subset [\mu+\eta, \lambda] \subset \bigcup_{1}^{n} (\mu_i, \lambda_i + \varepsilon_i].$$

Hence the increment of the function $\langle P_\lambda x, x \rangle$ in the interval $(\mu+\eta, \lambda]$ does not exceed the sum of its increments on $(\mu_i, \lambda_i + \varepsilon_i]$, so

$$\langle P_\lambda x, x \rangle - \langle P_{\mu+\eta} x, x \rangle$$

$$\leq \sum_{i=1}^{n} (\langle P_{\lambda_i + \varepsilon_i} x, x \rangle - \langle P_{\mu_i} x, x \rangle)$$

$$= \sum_{i=1}^{n} (\langle P_{\lambda_i} x, x \rangle - \langle P_{\mu_i} x, x \rangle) + \sum_{i=1}^{n} (\langle P_{\lambda_i + \varepsilon_i} x, x \rangle - \langle P_{\lambda_i} x, x \rangle)$$

$$\leq \sum_{i=1}^{\infty} \langle P_{(\mu_i, \lambda_i]} x, x \rangle + \varepsilon ,$$

whence letting $\varepsilon \downarrow 0$ first and then $\eta \downarrow 0$, one obtains,

$$\langle P_{(\mu, \lambda]} x, x \rangle \leq \langle (\sum_{i=1}^{\infty} P_{(\mu_i, \lambda_i]}) x, x \rangle .$$

But $x$ being arbitrary and $P_{(\mu, \lambda]}$ commuting with $\sum_{i=1}^{\infty} P_{(\mu_i, \lambda_i]}$,

$$P_{(\mu, \lambda]} \leq \sum_{i=1}^{\infty} P_{(\mu_i, \lambda_i]} ,$$

by (6.12), and the sought equation (7.15) follows by comparison with (7.16). If $(\mu, \lambda]$ is not bounded, we consider first the case when $\lambda = +\infty$, $-\infty < \mu$. If among the $\lambda_i$'s there is one coinciding with $+\infty$, say $\lambda_{i_0}$, then $(\mu, \mu_{i_0}]$ is bounded, $(\mu, \mu_{i_0}] = \sum_{i \neq i_0} (\mu_i, \lambda_i]$, and by the previous case,

$$P_{(\mu, \lambda]} = P_{(\mu, \mu_{i_0}]} + P_{(\mu_{i_0}, \lambda_{i_0}]} = \sum_{i \neq i_0} P_{(\mu_i, \lambda_i]} + P_{(\mu_{i_0}, \lambda_{i_0}]}$$

$$= \sum_{i=1}^{\infty} P_{(\mu_i, \lambda_i]} .$$

If none of the $\lambda_i$'s is $+\infty$, then there is an increasing sequence $\lambda_{i_n} \to +\infty$ and $P_{(\mu, \lambda_{i_n}]} = \sum_{\lambda_i \leq \lambda_{i_n}} P_{(\mu_i, \lambda_i]}$. A passage to the limit $n \to \infty$, yields (7.15) again. If $\mu = -\infty$,

## SPECTRAL THEORY

$\lambda < +\infty$ the proof follows the same pattern. Finally if $\mu = -\infty$, $\lambda = +\infty$, pick any finite $\lambda_{i_0}$, write $P_{(\mu, \lambda]} = P_{(\mu, \lambda_{i_0}]} + P_{(\lambda_{i_0}, \lambda]}$ and apply the previous results to the individual terms on the right.

With (7.15) at hand we may now proceed to show that $P_E$ is well defined for every $E \in \Sigma_1$. Assume two decompositions for $E$:

$$E = \bigcup_{i=1}^{\infty} (\mu_i^{(1)}, \lambda_i^{(1)}] = \bigcup_{j=1}^{\infty} (\mu_j^{(2)}, \lambda_j^{(2)}],$$

and set $(\mu_{ij}, \lambda_{ij}] = (\mu_i^{(1)}, \lambda_i^{(1)}] \cap (\mu_j^{(2)}, \lambda_j^{(2)}]$. Clearly $(\mu_i^{(1)}, \lambda_i^{(1)}] = \bigcup_{j=1}^{\infty} (\mu_{ij}, \lambda_{ij}]$ and in consequence $P_{(\mu_i^{(1)}, \lambda_i^{(1)}]}$

$= \sum_{j=1}^{\infty} P_{(\mu_{ij}, \lambda_{ij}]}$. Adding these up

$$\sum_{i=1}^{\infty} P_{(\mu_i^{(1)}, \lambda_i^{(1)}]} = \sum_{i=1}^{\infty} \sum_{j=1}^{\infty} P_{(\mu_{ij}, \lambda_{ij}]},$$

and by symmetry,

$$\sum_{j=1}^{\infty} P_{(\mu_j^{(2)}, \lambda_j^{(2)}]} = \sum_{j=1}^{\infty} \sum_{i=1}^{\infty} P_{(\mu_{ij}, \lambda_{ij}]}.$$

As the sum of orthogonal projections is independent of the order of the terms, the right hand members in the last two equations above are equal, and in consequence so are the left ones. Thus $P_E$ is well defined. Properties (7.13) and (7.14) result from (7.9) and (7.10) by a limiting process. In fact, if $E_1, E_2 \in \Sigma_1$ choose two increasing sequences $E_1^{(n)}, E_2^{(n)}$ of elements in $\Sigma_0$ such that $P_{E_i} = \lim P_{E_i^{(n)}}$, $i = 1, 2$. Then

$$P_{E_1}P_{E_2} = \lim_{n\to\infty} P_{E_1^{(n)}} P_{E_2^{(n)}} = \lim_{n\to\infty} P_{E_1^{(n)} \cap E_2^{(n)}} = P_{E_1 \cap E_2},$$

and (7.13) is proved for two, and hence for any number of factors. Similarly

(7.17) $\quad P_{E_1} + P_{E_2} - P_{E_1}P_{E_2} = \lim_{n\to\infty} (P_{E_1^{(n)}} + P_{E_2^{(n)}} - P_{E_1^{(n)}} P_{E_2^{(n)}})$

$$= \lim_{n\to\infty} P_{E_1^{(n)} \cup E_2^{(n)}} = P_{E_1 \cup E_2}.$$

By (7.13), $P_{E_1 \cup E_2}$ is an upper bound for $P_{E_1}$ and $P_{E_2}$, whereas from (7.17) follows that any upper bound is stronger than $P_{E_1 \cup E_2}$. Hence $P_{E_1 \cup E_2} = P_{E_1} \cup P_{E_2}$, from which (7.14) follows for any finite number of terms. To extend this to a countable number we prove first that

(7.18) $\quad \bigcup_{i=1}^{\infty} P_{E_i} = \lim_{n\to\infty} \bigcup_{i=1}^{n} P_{E_i}$

$$= \lim_{n\to\infty} \left[ \sum_{i_1=1}^{n} P_{E_{i_1}} - \sum_{i_1, i_2=1}^{n} P_{E_{i_1}} P_{E_{i_2}} + \ldots -(-1)^n \sum_{i_1, i_2, \ldots, i_n=1}^{n} P_{E_{i_1}} P_{E_{i_2}} \ldots P_{E_{i_n}} \right],$$

the limit being taken in the strong pointwise sense. Clearly $\lim_{n\to\infty} \bigcup_{i=1}^{n} P_{E_i}$ exists because $\bigcup_{i=1}^{n} P_{E_i}$ is an increasing sequence of commuting projections (Lemma 6.3), and it is an upper bound for all the $P_{E_i}$'s. We claim that it is the smallest of all such upper bounds. In fact if $P > P_{E_i}$, $i = 1, 2, \ldots$, then

$$(\lim_{n\to\infty} \bigcup_{i=1}^{n} P_{E_i}) P = \lim_{n\to\infty} [(\bigcup_{i=1}^{n} P_{E_i}) P]$$

$$= \lim_{n\to\infty} \left[ \left( \sum_{i_1=1}^{n} P_{E_{i_1}} - \sum_{i_1, i_2=1}^{n} P_{E_{i_1}} P_{E_{i_2}} + \ldots -(-1)^n \sum_{i_1, i_2, \ldots, i_n=1}^{n} P_{E_{i_1}} P_{E_{i_2}} \ldots P_{E_{i_n}} \right) P \right].$$

SPECTRAL THEORY

But introduction of $P$ into the square brackets on the right alters nothing because $P_{E_i} P = P_{E_i}$ and so, $(\lim_{n\to\infty} \bigcup_{i=1}^{n} P_{E_i}) P = \lim_{n\to\infty} \bigcup_{i=1}^{n} P_{E_i}$. Hence $\lim_{n\to\infty} \bigcup_{i=1}^{n} P_{E_i} < P$, and (7.18) is proved.

This done we remark that (7.14) holds as soon as it is shown to hold for an ascending sequence $\{F_i\}_1^\infty$ in $\Sigma_1$, for then application to $F_i = \bigcup_{j=1}^{i} E_j$, leads to

$$\bigcup_{i=1}^{\infty} P_{E_i} = \bigcup_{i=1}^{\infty} (\bigcup_{j=1}^{i} P_{E_j}) = \bigcup_{i=1}^{\infty} P_{\bigcup_{j=1}^{i} E_j} = P_{\bigcup_{i=1}^{\infty} \bigcup_{j=1}^{i} E_j} = P_{\bigcup_{i=1}^{\infty} E_i},$$

which together with (7.18) yields (7.14). So let $\{F_i\}_1^\infty$ be an ascending sequence of sets in $\Sigma_1$. By (7.18) we know already that

(7.19) $$\bigcup_{i=1}^{\infty} P_{F_i} = \lim_{n\to\infty} P_{F_n},$$

and so it only remains to prove that $\lim_{n\to\infty} P_{F_n} = P_{\lim_{n\to\infty} F_n}$.

A preliminary step is to dispose of the case when $\lim_{n\to\infty} F_n = (\mu_0, \lambda_0]$. We claim that for any $\kappa \in (\mu_0, \lambda_0]$ there is a $\nu < \kappa$ in the same interval such that

(7.20) $$P_{(\nu, \kappa]} \lim_{n\to\infty} P_{F_n} = P_{(\nu, \kappa]}.$$

Since $\kappa \in (\mu_0, \lambda_0] = \lim_{n\to\infty} F_n$, $\kappa$ belongs to some $F_{n_\kappa}$, and since $F_{n_\kappa}$ is the union of semiopen intervals, $\kappa$ must belong to one of these, say $(\nu, \rho]$. Therefore, $\kappa \in (\nu, \kappa] \subset (\nu, \rho] \subset F_{n_\kappa}$, and as $\{F_n\}$ is an increasing sequence, $\kappa \subset (\nu, \kappa] \subset F_n$, $n \geq n_\kappa$. This allows us to write,

$$P_{(\nu, \kappa]} P_{F_n} = P_{(\nu, \kappa]}, \quad n \geq n_\kappa,$$

363

whence (7.20) results by letting $n \to \infty$. In particular, identifying $\kappa$ with $\lambda_0$, we see then that the set $N$ of $\nu$'s in $(\mu_0, \lambda_0]$ such that

$$P_{(\nu, \lambda_0]} \lim_{n \to \infty} P_{F_n} = P_{(\nu, \lambda_0]},$$

is not empty and contains points other than $\lambda_0$. We must show that it coincides with the whole interval $(\mu_0, \lambda_0]$. Since along with any $\nu$ it contains all points in the interval $[\nu, \lambda_0]$, it is all a matter of proving that its infimum $\nu_0$ is $\mu_0$. Let us check that $\nu_0 \in N$. To this end take a sequence $\{\nu_k\} \subset N$ decreasing to $\nu_0$; since $\nu_k \in N$,

$$P_{(\nu_k, \lambda_0]} \lim_{n \to \infty} P_{F_n} = P_{(\nu_k, \lambda_0]},$$

which letting $k \to \infty$ leads to

(7.21) $$P_{(\nu_0, \lambda_0]} \lim_{n \to \infty} P_{F_n} = P_{(\nu_0, \lambda_0]},$$

by the continuity on the right of $P_\lambda$. Thus $\nu_0 \in N$ as claimed. Were $\nu_0 \neq \mu_0$, then by (7.20) with $\nu_0$ in place of $\kappa$, there would be a $\nu_0' \subset (\mu_0, \lambda_0]$, $\nu_0' < \nu_0$, such that

$$P_{(\nu_0', \nu_0]} \lim_{n \to \infty} P_{F_n} = P_{(\nu_0', \nu_0]},$$

an equation which when added to (7.21) would yield

$$P_{(\nu_0', \lambda_0]} \lim_{n \to \infty} P_{F_n} = P_{(\nu_0', \lambda_0]},$$

in contradiction with the minimum property of $\nu_0$. Therefore $\nu_0 = \mu_0$, and by (7.21),

(7.22) $$P_{(\mu_0, \lambda_0]} < \lim_{n \to \infty} P_{F_n}.$$

As the opposite relation holds trivially we have proved

$$\lim_{n\to\infty} P_{F_n} = P_{\lim_{n\to\infty} F_n}, \quad \text{when } \lim_{n\to\infty} F_n = (\mu_0, \lambda_0].$$

In the general case $\lim_{n\to\infty} F_n$ is not necessarily a semiopen interval, but is the union of countably many disjoint such intervals: $\lim_{n\to\infty} F_n = \bigcup_k (\mu_k, \lambda_k]$. Clearly, since $(\mu_k, \lambda_k] \subset \lim_{n\to\infty} F_n$, $\lim_{n\to\infty}[(\mu_k, \lambda_k] \cap F_n] = (\mu_k, \lambda_k]$, and by the above result,

$$\lim_{n\to\infty} P_{(\mu_k, \lambda_k] \cap F_n} = P_{\lim_{n\to\infty}[(\mu_k, \lambda_k] \cap F_n]},$$

that is

$$P_{(\mu_k, \lambda_k]} \lim_{n\to\infty} P_{F_n} = P_{(\mu_k, \lambda_k]}, \quad \forall k.$$

Adding these together,

$$P_{\lim_{n\to\infty} F_n} \lim_{n\to\infty} P_{F_n} = \sum_k (P_{(\mu_k, \lambda_k]} \lim_{n\to\infty} P_{F_n}) = \sum_k P_{(\mu_k, \lambda_k]} = P_{\lim_{n\to\infty} F_n}.$$

Thus $P_{\lim_{n\to\infty} F_n} < \lim_{n\to\infty} P_{F_n}$, and since obviously $\lim_{n\to\infty} P_{F_n} < P_{\lim_{n\to\infty} F_n}$,

$$P_{\lim_{n\to\infty} F_n} = \lim_{n\to\infty} P_{F_n},$$

for any ascending sequence of sets in $\Sigma_1$. But this, we have shown, is equivalent to (7.14), and the proof of the lemma comes to an end.

Going on with the construction of the spectral measure we introduce now the class of sets

(7.23) $\quad\quad\quad \Sigma_1' = \{E'\}_{E \in \Sigma_1}, \quad E' = $ complement of $E$.

As $\Sigma_1$, $\Sigma_1'$ <u>contains</u> $\Sigma_0$, <u>is contained in the Boolean algebra of all Borel sets but itself is not a Boolean algebra; it</u> enjoys properties dual of those had by $\Sigma_1$ : <u>it contains all closed sets, it contains the union of any finite number and the intersection of any countable number of its sets.</u>

The lemma below is the counterpart of Lemma 7.3.

<u>Lemma 7.4</u>   The class of projections

(7.24) $\quad\quad \Pi_1' = \{P_{E'}\}_{E' \in \Sigma_1'}, \quad P_{E'} = I - P_E,$

is commutative, and contains the supremum of any finite number and the infimum of any countable number of its elements. The mapping $E' \to P_{E'}$ of $\Sigma_1'$ onto $\Pi_1'$ has the properties:

$$(7.25) \quad \bigcup_{i=1}^{n} P_{E_i'} = P_{\bigcup_{i=1}^{n} E_i'}$$

$$= \sum_{i_1=1}^{n} P_{E_{i_1}'} - \sum_{i_1, i_2=1}^{n} P_{E_{i_1}'} P_{E_{i_2}'} + \ldots -(-1)^n \sum_{i_1, i_2, \ldots, i_n=1}^{n} P_{E_{i_1}'} P_{E_{i_2}'} \ldots P_{E_{i_n}'},$$

$$(7.26) \quad \bigcap_{i=1}^{\infty} P_{E_i'} = P_{\bigcap_{i=1}^{\infty} E_i'} = \lim_{n \to \infty} P_{\bigcap_{i=1}^{n} E_i'}.$$

Moreover, if $A \in \Sigma_1 \cap \Sigma_1'$ then $P_A$ is the same regardless of whether $A$ is considered as an element of $\Sigma_1$ or an element of $\Sigma_1'$.

## SPECTRAL THEORY

**Proof:** Any $P_{E'} \in \Pi_1'$ can be written as

$$P_{E'} = I - P_E = I - \lim_{n \to \infty} P_{E_n} = \lim_{n \to \infty}(I - P_{E_n}) = \lim_{n \to \infty} P_{E_n'},$$

where the $P_{E_n}$'s, and hence the $P_{E_n'}$'s, belong to $\Pi_0$. Since all projections in $\Pi_0$ commute so do those in $\Pi_1'$; by the same token all projections in $\Pi_1$ commute with all projections in $\Pi_1'$. Equations (7.25) and (7.26) are immediate consequences of (7.13) and (7.14) through the relations,

$$\bigcup_{i=1}^{n} P_{E_i'} = I - \bigcap_{i=1}^{n} P_{E_i} = I - P_{\bigcap_{i=1}^{n} E_i} = P_{(\bigcap_{i=1}^{n} E_i)'} = P_{\bigcup_{i=1}^{n} E_i'} =$$

$$= I - \prod_{i=1}^{n}(I - P_{E_i'}) = \sum_{i_1=1}^{n} P_{E_{i_1}'} - \sum_{i_1,i_2=1}^{n} P_{E_{i_1}'} P_{E_{i_2}'}$$

$$+ \ldots - (-1)^n \sum_{i_1, i_2, \ldots, i_n = 1}^{n} P_{E_{i_1}'} P_{E_{i_2}'} \cdots P_{E_{i_n}'},$$

$$\bigcap_{i=1}^{\infty} P_{E_i'} = I - \bigcup_{i=1}^{\infty} P_{E_i} = I - \lim_{n \to \infty} \bigcup_{i=1}^{n} P_{E_i} = \lim_{n \to \infty}(I - \bigcup_{i=1}^{n} P_{E_i}) = \lim_{n \to \infty} \bigcap_{i=1}^{n} P_{E_i'}$$

$$= I - P_{\bigcup_{i=1}^{\infty} E_i} = P_{(\bigcup_{i=1}^{\infty} E_i)'} = P_{\bigcap_{i=1}^{\infty} E_i'}.$$

Finally, if $A \in \Sigma_1 \cap \Sigma_1'$, then $A$ and $A'$ both belong to $\Sigma_1$ and $P_A$ and $P_{A'}$ as defined in Lemma 7.3 are orthogonal and complementary. Hence $P_A = I - P_{A'}$. But the right member here is just $P_A$ when $A$ is considered as a member of $\Sigma_1'$, and both definitions of $P_A$ coincide.

Now we come to the last step in our construction.

367

**Definition 7.4**  A set $A$ in the extended real line $[-\infty, +\infty]$ is said to be measurable relatively to a partition of the identity $\{P_\lambda\}_{-\infty}^{+\infty}$ if

$$(7.27) \qquad \bigcup_{\substack{E' \subset A \\ E' \in \Sigma'}} P_{E'} = \bigcap_{\substack{E \supset A \\ E \in \Sigma_1}} P_E \quad.$$

The class of all measurable sets is denoted $\Sigma$; the common value of both members of (7.27) is a projection, called the "spectral measure of $A$", and is denoted $P_A$.

The theorem below shows, among other things, that the notation $P_A$ is not inconsistent with that already used for sets in $\Sigma_1 \cup \Sigma_1'$. For any set $A$ the class $\{E' \,|\, E' \in \Sigma_1, E' \subset A\}$ is not empty as it contains the empty set, and is directed upwards by virtue of (7.25). So by Lemma 6.3, the left hand member of (7.27), called the "inner measure of $A$", always exists regardless of the nature of $A$. Likewise, the right hand member exists; it is called the "outer measure of $A$". Since,

$$E_1' \subset A \subset E_2' \Rightarrow E_1' \cap E_2' = \phi \Rightarrow P_{E_1'} + P_{E_2'} = P_{E_1' \cup E_2'} \leq I \Rightarrow P_{E_1'} < P_{E_2'},$$

the inner measure is weaker than the outer measure. We denote the outer and inner measure of $A$ by $\bar{P}_A$ and $\underline{P}_A$ respectively.

**Theorem 7.1**  The class of all measurable sets relatively to a partition of the identity $\{P_\lambda\}_{-\infty}^{+\infty}$ is a $\sigma$-Boolean algebra containing all Borel sets, and the class $\Pi = \{P_A\}_{A \in \Sigma}$ is a $\sigma$-Boolean algebra of commuting projections containing $\Pi_1 \cup \Pi_1'$. The spectral measure has the following properties:

$$(7.28) \qquad P_{A'} = I - P_A \quad,$$

$$(7.29) \qquad \bigcup_{i=1}^{\infty} P_{A_i} = \lim_{n \to \infty} \bigcup_{i=1}^{n} P_{A_i} = P_{\bigcup_{i=1}^{\infty} A_i} \quad,$$

SPECTRAL THEORY

$$(7.30) \quad \bigcap_{i=1}^{\infty} P_{A_i} = \lim_{n \to \infty} \bigcap_{i=1}^{n} P_{A_i} = P_{\bigcap_{i=1}^{\infty} A_i},$$

$$(7.31) \quad P_{(-\infty,\lambda]} = P_\lambda.$$

Proof: We need the following result: If $P_1, P_2, P_1^\perp$ and $P_2^\perp$ commute, and $P_1 P_2$ and $P_1^\perp P_2^\perp$ are projections, then $P_1 \cap P_2$ and $P_1 \cup P_2$ exist and

$$P_1 \cap P_2 + P_1 \cup P_2 = P_1 + P_2.$$

In fact, it is immediate that $P_1 P_2 = P_1 \cap P_2$ and hence that $P_1 \cup P_2 = (P_1^\perp \cap P_2^\perp)^\perp = I - (I-P_1)(I-P_2) = P_1 + P_2 - P_1 P_2$.

Now, if $E_1$ and $E_2$ belong to $\Sigma_1$, the four projections $P_{E_1}, P_{E_2}, P_{E_1'} = I - P_{E_1}$ and $P_{E_2'} = I - P_{E_2}$, being limits in the commutative Boolean algebra $\Pi_0$, commute with each other, and their products are projections. It follows by the above proposition that

$$(7.32) \quad P_{A_1} \cap P_{A_2} + P_{A_1} \cup P_{A_2} = P_{A_1} + P_{A_2}$$

for any couple of sets in $\Sigma_1 \cup \Sigma_1'$.

For any set $A$, by (6.16),

$$(\underline{P}_A)^\perp = (\bigcup_{\substack{E' \subset A \\ E' \in \Sigma_1}} P_{E'})^\perp = \bigcap_{\substack{E \supset A' \\ E \in \Sigma_1}} (P_{E'})^\perp = \bigcap_{\substack{E \supset A' \\ E \in \Sigma_1}} P_E = \bar{P}_{A'}$$

and similarly

$$(\bar{P}_A)^\perp = \underline{P}_{A'}.$$

Therefore, if $A$ is measurable so is $A'$ and $P_A + P_{A'} = I$, which is (7.28). Moreover, for any $x \in \mathcal{H}$,

$$\lim_{<} \langle P_{E'} x, x \rangle = \langle P_A x, x \rangle = \lim_{>} \langle P_E x, x \rangle ,$$
$$E' \subset A , \qquad\qquad\qquad\qquad E \supset A$$
$$E' \subset \Sigma_1 \qquad\qquad\qquad\qquad E \in \Sigma_1$$

by Lemma 6.3. Hence for given $\varepsilon > 0$ there are sets $A^{(\varepsilon)} \supset A$ and $A_{(\varepsilon)} \subset A$ in $\Sigma_1$ and $\Sigma_1'$ respectively such that

(7.33) $\quad \varepsilon/2 + \langle P_{A_{(\varepsilon)}} x, x \rangle \geq \langle P_A x, x \rangle \geq \langle P_{A^{(\varepsilon)}} x, x \rangle - \varepsilon/2 ,$

and replacing A by another measurable set B,

$$\varepsilon/2 + \langle P_{B_{(\varepsilon)}} x, x \rangle \geq \langle P_B x, x \rangle \geq \langle P_{B^{(\varepsilon)}} x, x \rangle - \varepsilon/2 .$$

Adding these two equations,

$$\varepsilon + \langle (P_{A_{(\varepsilon)}} + P_{B_{(\varepsilon)}}) x, x \rangle \geq \langle (P_A + P_B) x, x \rangle \geq \langle (P_{A^{(\varepsilon)}} + P_{B^{(\varepsilon)}}) x, x \rangle - \varepsilon ,$$

whence

$$0 \leq \langle (P_{A^{(\varepsilon)}} + P_{B^{(\varepsilon)}}) x, x \rangle - \langle (P_{A_{(\varepsilon)}} + P_{B_{(\varepsilon)}}) x, x \rangle \leq 2\varepsilon ,$$

and by (7.32),

$$0 \leq \langle (P_{A^{(\varepsilon)}} \cup P_{B^{(\varepsilon)}} - P_{A_{(\varepsilon)}} \cup P_{B_{(\varepsilon)}}) x, x \rangle + \langle (P_{A^{(\varepsilon)}} \cap P_{B^{(\varepsilon)}} - P_{A_{(\varepsilon)}} \cap P_{B_{(\varepsilon)}}) x, x \rangle$$
$$\leq 2\varepsilon .$$

Both, $P_{A^{(\varepsilon)}} \cup P_{B^{(\varepsilon)}} - P_{A_{(\varepsilon)}} \cup P_{B_{(\varepsilon)}}$, and $P_{A^{(\varepsilon)}} \cap P_{B^{(\varepsilon)}} - P_{A_{(\varepsilon)}} \cap P_{B_{(\varepsilon)}}$ are limits of projections and so are projections themselves. Hence both terms above are nonnegative, and we can write,

(7.34) $\quad 0 \leq \langle (P_{A^{(\varepsilon)}} \cup P_{B^{(\varepsilon)}} - P_{A_{(\varepsilon)}} \cup P_{B_{(\varepsilon)}}) x, x \rangle \leq 2\varepsilon .$

SPECTRAL THEORY

But $A \cup B \subset A^{(\varepsilon)} \cup B^{(\varepsilon)} \in \Sigma_1$ and $A \cup B \supset A_{(\varepsilon)} \cup B_{(\varepsilon)} \in \Sigma_1'$, so

(7.35) $\qquad P_{A_{(\varepsilon)}} \cup P_{B_{(\varepsilon)}} < \underline{P}_{A \cup B} < \bar{P}_{A \cup B} < P_{A^{(\varepsilon)}} \cup P_{B^{(\varepsilon)}}$

and from (7.34), a fortiori,

$$0 \leq \langle (\bar{P}_{A \cup B} - \underline{P}_{A \cup B})x, x \rangle \leq 2\varepsilon$$

that is, since $\varepsilon$ is an arbitrary positive number,

$$\langle \bar{P}_{A \cup B} x, x \rangle = \langle \underline{P}_{A \cup B} x, x \rangle.$$

As limits of commuting projections $\bar{P}_{A \cup B}$ and $\underline{P}_{A \cup B}$ commute, and so, $x$ being arbitrary,

$$\bar{P}_{A \cup B} = \underline{P}_{A \cup B},$$

by (6.12). Thus $A \cup B$ is measurable, and hence so is the union of any finite number of measurable set. It follows by complementation that the intersection of a finite number of measurable sets is measurable.

From (7.35) and (7.34), since $P_{A \cup B} = \underline{P}_{A \cup B} = \bar{P}_{A \cup B}$,

$$\langle P_{A_{(\varepsilon)}} \cup P_{B_{(\varepsilon)}} x, x \rangle \geq \langle P_{A \cup B} x, x \rangle - 2\varepsilon$$

and as $P_{A_{(\varepsilon)}} \cup P_{B_{(\varepsilon)}} < P_A \cup P_B$,

$$\langle (P_A \cup P_B)x, x \rangle \geq \langle P_{A \cup B} x, x \rangle - 2\varepsilon,$$

whence, by the arbitrariness of $\varepsilon$ and $x$, and (6.12),

$$P_A \cup P_B > P_{A \cup B}.$$

The opposite relation is an obvious consequence of the fact that the spectral measure of a set grows with the set.

Hence (7.29) and in consequence (7.30) are established for any finite number of sets.

The extension of (7.30) to a countable number of sets requires the previous knowledge that the union of a countable family of measurable sets be measurable. In view of what has already been proved, it is sufficient to suppose that the sets $A_n$ form an ascending sequence. This being so, construct, for a given $x \in \mathcal{H}$, two ascending sequences of sets $\{B_n\}$ and $\{C_n\}$, the first contained in $\Sigma_1'$ and the second in $\Sigma_1$, such that $B_n \subset A_n \subset C_n$, and

$$1/n + \langle P_{B_n} x, x \rangle \geq \langle P_{A_n} x, x \rangle \geq \langle P_{C_n} x, x \rangle - 1/n,$$

and let $n \to \infty$. One obtains

(7.36) $\quad \langle \lim_{n \to \infty} P_{B_n} x, x \rangle \geq \langle \lim_{n \to \infty} P_{A_n} x, x \rangle \geq \langle \lim_{n \to \infty} P_{C_n} x, x \rangle.$

Remark now that $B_n \subset \bigcup_{i=1}^{\infty} A_i = \lim_{n \to \infty} A_n$, and so that $P_{B_n} \leq P_{\lim_{n \to \infty} A_n}$ because $B_n \subset \Sigma_1'$. Therefore,

(7.37) $\quad \lim_{n \to \infty} P_{B_n} = \bigcup_{i=1}^{\infty} P_{B_i} \leq P_{\lim_{n \to \infty} A_n}.$

On the other hand, the $C_n$'s being in $\Sigma_1$, $\lim_{n \to \infty} C_n = \bigcup_{i=1}^{\infty} C_n$ also belongs to $\Sigma_1$, and by (7.14)

$$\lim_{n \to \infty} P_{C_n} = \bigcup_{i=1}^{\infty} P_{C_i} = P_{\bigcup_{i=1}^{\infty} C_i} = P_{\lim_{n \to \infty} C_n},$$

which in turn since $\lim_{n \to \infty} A_n \subset \lim_{n \to \infty} C_n \in \Sigma_1$, yields

$$\lim_{n \to \infty} P_{C_n} \geq \bar{P}_{\lim_{n \to \infty} A_n}.$$

SPECTRAL THEORY

Inserting this and (7.37) in (7.36),

$$\langle \underline{P}_{\lim_{n\to\infty} A_n} x, x \rangle \geq \langle \lim_{n\to\infty} P_{A_n} x, x \rangle \geq \langle \bar{P}_{\lim_{n\to\infty} A_n} x, x \rangle .$$

But $\underline{P}_{\lim_{n\to\infty} A_n}$ and $\bar{P}_{\lim_{n\to\infty} A_n}$ commute, and since $x$ is arbitrary, $\underline{P}_{\lim_{n\to\infty} A_n} > \bar{P}_{\lim_{n\to\infty} A_n}$, that is, $\underline{P}_{\lim_{n\to\infty} A_n} = \bar{P}_{\lim_{n\to\infty} A_n}$ because the outer measure is always stronger than the inner measure. In other terms, $\bigcup_{i=1}^{\infty} A_i = \lim_{n\to\infty} A_n$ is measurable. Moreover, $\lim_{n\to\infty} P_{A_n} = \bigcup_{i=1}^{\infty} P_{A_i}$ and $P_{\bigcup_{i=1}^{\infty} A_i} = P_{\lim_{n\to\infty} A_n}$ are the same because their associated functions $\langle P_{\infty} x, x \rangle$ and $\langle (\bigcup_{i=1}^{\infty} P_{A_i}) x, x \rangle$ coincide. Hence (7.30) is proved in its full strength.

The set $A = (-\infty, \lambda]$ belongs to both $\Sigma_1$ and $\Sigma_1'$ and therefore is at once the smallest set in $\Sigma_1$ containing $A$ and the largest set in $\Sigma_1'$ contained in $A$. Therefore, $P_\lambda$ is equal to both its inner and its outer measure, and $A = (-\infty, \lambda]$ is measurable with measure $P_\lambda$. This is (7.31). It follows then that all intervals $(\mu, \lambda]$ and, along with them, all Borel sets are measurable. In particular, all sets in $\Sigma_1 \cup \Sigma_1'$ are measurable. Finally, since the measure of $(\mu, \lambda]$ is $P_\lambda - P_\mu$, and the measure is $\sigma$-additive, the spectral measure of a set $E$ in $\Sigma_1$, is the projection $P_E$ defined by (7.12). Therefore, for sets in $\Sigma_1$ the spectral measure coincides with the projection $P_E$ previously defined in Lemma 7.3. By complementation the same is true over $\Sigma_1'$. It is clear then that $\Pi \supset \Pi_1 \cup \Pi_1'$. The proof of Theorem 7.1 is thus complete.

373

For any $x \in \mathcal{H}$, $\langle P_\lambda x, x \rangle$ is a nondecreasing function of $\lambda$, $-\infty \leq \lambda \leq +\infty$, continuous on the right, varying from 0 to $\|x\|^2$. Associated with it there is a unique positive regular measure $\mu_x$ - the Lebesgue-Stieljes measure - defined over a class $\Sigma_x$ of "measurable sets with regard to $x$", containing all Borel sets and such that $\mu_\lambda((-\infty, \lambda]) = \langle P_\lambda x, x \rangle$ (cf. [4, I, section III, §5]). We recall that the regularity means that for any $x$-measurable set $A$ and any positive $\varepsilon$ there is an open set $O_\varepsilon$ containing $A$ and a compact set $K_\varepsilon$ contained in $A$ such that $\mu_x(O_\varepsilon - K_\varepsilon) \leq \varepsilon$; moreover, any set $A$ for which such sets $O_\varepsilon$ and $K_\varepsilon$ exist belongs to $\Sigma_x$. The following theorem establishes a connection between such measures and the spectral measure.

**Theorem 7.2** Let $\{P_\lambda\}_{-\infty}^{+\infty}$ be a partition of the identity. Then

(7.38) $$\Sigma = \bigcap_{x \in \mathcal{H}} \Sigma_x$$

(7.39) $$\mu_x(A) = \langle P_A x, x \rangle, \quad \forall A \in \Sigma, \quad \forall x \in \mathcal{H}.$$

**Proof:** Set $\tilde{\mu}_x(A) = \langle P_A x, x \rangle$, $A \in \Sigma$. It is clear by Theorem 7.1 that $\tilde{\mu}_x$ is positive and $\sigma$-additive over $\Sigma$, and satisfies $\tilde{\mu}_x((-\infty, \lambda]) = \langle P_\lambda x, x \rangle$; we shall see that in addition it is a regular measure. By (7.33), given $A \in \Sigma$ and $\varepsilon > 0$, there is an $A^{(\varepsilon)} \in \Sigma_1$ containing $A$ such that

$$\tilde{\mu}_x(A) \leq \tilde{\mu}_x(A^{(\varepsilon)}) + \varepsilon/4.$$

If $A^{(\varepsilon)} = \bigcup_{i=1}^{\infty} (\mu_i, \lambda_i]$, take a sequence $\eta_i > 0$, $i = 1, 2, \ldots$ such that $\sum_{i=1}^{\infty} \tilde{\mu}_x((\lambda_i, \lambda_i + \eta_i]) < \varepsilon/4$, and with it construct the open set $O^{(\varepsilon)} = \bigcup_{i=1}^{\infty} (\mu_i, \lambda_i + \eta_i)$. Clearly, $A \subset O^{(\varepsilon)}$ and

SPECTRAL THEORY

$\tilde{\mu}_x(O^{(\varepsilon)}-A) \le \tilde{\mu}_x(O^{(\varepsilon)}-A^{(\varepsilon)}) + \tilde{\mu}_x(A^{(\varepsilon)}-A) < \varepsilon/2$. This construction applied to A' yields, by passing to the complement, a closed set $F^{(\varepsilon)}$ contained in A such that $\mu_x(A - F^{(\varepsilon)}) < \varepsilon/2$. If A is bounded $F^{(\varepsilon)}$ is compact and setting $K^{(\varepsilon)} = F^{(\varepsilon)}$ we have

$$K^{(\varepsilon)} \subset A \subset O^{(\varepsilon)}, \quad \tilde{\mu}_x(O^{(\varepsilon)} - K^{(\varepsilon)}) \le \varepsilon.$$

If A is not bounded, $F^{(\varepsilon)} = \lim_{n\to\infty} F_n^{(\varepsilon)}, F_n^{(\varepsilon)} = F^{(\varepsilon)} \cap [-n, n]$. The $F_n^{(\varepsilon)}$'s are compact and contained in A, and $\lim_{n\to\infty} \tilde{\mu}_x(A - F_n^{(\varepsilon)}) = \tilde{\mu}_x(A-F^{(\varepsilon)}) < \varepsilon/2$. So there is an $n_0$ such that $\tilde{\mu}_x(A-F_{n_0}^{(\varepsilon)}) < \varepsilon/2$, and setting $K^{(\varepsilon)} = F_{n_0}^{(\varepsilon)}$ we obtain the same conclusion as before. Hence $\tilde{\mu}_x$ is a regular measure.

Now, since open sets belong to $\Sigma$ and to all the $\Sigma_x$ as well, and are the countable unions of overlapping semiclosed intervals - over which $\mu_x$ and $\tilde{\mu}_x$ coincide - $\mu_x$ and $\tilde{\mu}_x$ assigned the same measure to any open set and hence to any Borel set. By regularity then any $A \subset \Sigma$ is x-measurable and $\mu_x(A) = \tilde{\mu}_x(A)$. This proves (7.39). It remains to be seen that a set A, x-measurable for any $x \in \mathcal{H}$, is measurable with regard to the spectral resolutions. If $A \in \bigcap_{x \in \mathcal{H}} \Sigma_x$, then by the regularity of $\mu_x$ there are, for given x, a compact set $K^{(\varepsilon)}$ and an open set $O^{(\varepsilon)}$ such that

$$K^{(\varepsilon)} \subset A \subset O^{(\varepsilon)}, \quad \mu_x(O^{(\varepsilon)} - K^{(\varepsilon)}) \le \varepsilon.$$

So since $\mu_x$ and $\tilde{\mu}_x$ coincide on Borel sets,

$$\langle P_{K^{(\varepsilon)}} x, x\rangle \ge \langle P_{O^{(\varepsilon)}} x, x\rangle - \varepsilon,$$

and, a fortiori,

$$\langle \underline{P}_A x, x\rangle \ge \langle \bar{P}_A x, x\rangle - \varepsilon,$$

which by the arbitrariness of $\varepsilon$ and x amounts to $\underline{P}_A > \bar{P}_A$,

375

that is to $\underline{P}_A = \bar{P}_A$. In consequence $A \in \Sigma$. Q.E.D.

A point is said to belong to the support of a measure if any open set containing it has nonvanishing measure. Therefore, the support is a closed set, empty if and only if the measure vanishes identically. This notion applies in particular to any spectral measure $P_E$, to the vector valued measure $P_E x$ and to the positive valued measure $\langle P_E x, x \rangle$ derived from it. The support of these last two measures is the same for any $x$, and in this case, to simplify the language, we speak of "the support of $x$". The support of a spectral resolution $\{P\}_{-\infty}^{+\infty}$ is that of its induced measure. Note that

(7.40) $$\text{supp } x \subset \text{supp } \{P_\lambda\}_{-\infty}^{+\infty}$$

(7.41) $$\text{supp } \{P_\lambda\}_{-\infty}^{+\infty} = \bigcup_{x \in \mathcal{H}} \text{supp } x$$

(7.42) $$\text{supp}(\alpha x + \beta y) \subset \text{supp } x + \text{supp } y, \quad \alpha, \beta \geq 0.$$

This last relation is a consequence of (6.5) applied to projections.

The following lemma has a significant meaning in operator theory:

<u>Lemma 7.5</u>  The spectral resolution has compact support if and only if each point in space has compact support.

<u>Proof:</u>  It is enough to prove sufficiency, as necessity is obvious and contained in (7.40). We proceed by contradiction, and, under the hypothesis that every point has a compact support, suppose that the spectral resolution had a noncompact one. In such a situation there is a sequence of disjoint open intervals $\{J_n\}_1^\infty$ with nonvanishing spectral measure, at increasing distances from the origin going to infinity. With these, an orthonormal vector sequence $\{x_{n_k}\}_1^\infty$, $x_{n_k} \in \mathcal{R}(P_{J_{n_k}})$ is constructed by induction, as follows:

Let $n_1 = 1$ and let $x_{n_1}$ be any unit vector in $\Re(P_{J_{n_1}})$; having constructed $x_{n_1}, x_{n_2}, \ldots, x_{n_h}$, take $n_{h+1}$ so that $J_{n_{h+1}}$ lies outside a closed symmetric interval containing the supports of $\pm x_{n_1}, \pm x_{n_2}, \ldots, \pm x_{n_h}$, and let $x_{n_{h+1}}$ be any unit vector in $\Re(P_{J_{n_{h+1}}})$. By (7.42) any linear combination of $x_{n_1}, x_{n_2}, \ldots, x_{n_h}$ has a support disjoint with $J_{n_{h+1}}$ and so belongs to $\Re(P^\perp_{J_{n_{h+1}}})$. Then, since any linear space contained in the dual of a cone is perpendicular to every vector in the cone, $x_{n_{h+1}}$ is perpendicular to $x_{n_1}, x_{n_2}, \ldots, x_{n_h}$. Hence, the construction is possible. Now if $\{\alpha_k\}_1^\infty$ is a positive sequence with $\sum_{k=1}^\infty \alpha_k^2 < \infty$, the point $x = \sum_{k=1}^\infty \alpha_k x_{n_k}$ is projected by $P_{J_{n_k}}$ into $\alpha_k x_{n_k} \neq 0$, and so it does not have a compact support, against the assumption that it did. Therefore, $\{P_\lambda\}_{-\infty}^{+\infty}$ must have a compact support. Q. E. D.

If the space is separable, in the linear case there is always a point having the same support as the spectral resolution. This is not true for nonlinear spectral resolutions, even in two dimensions, as the example of projections onto three cones leaving right angles in between shows.

## §8. Spectral synthesis.

Spectral synthesis is the process whereby an operator, nonlinear in general, is constructed by integration of a real valued function with regard to the spectral measure associated to a given partition of the identity, just in the manner a self adjoint linear operator is synthesized from a linear partition of the identity. The whole process altogether follows the path of the standard Lebesgue-Stieljes integration with regard to a monotone function. The functions considered here will always be defined on the real line $\mathbb{R}:(-\infty,+\infty)$, or on a subset of it, with values again on the real line or on any of its usual extensions $[-\infty,+\infty), (-\infty,+\infty], [-\infty,+\infty]$, with the standard topologies. We warn the reader that, in manipulating these functions, special precautions have to be taken when two or all the three values $-\infty, 0, +\infty$ appear simultaneously.

**Definition 8.1** A function $f:E \to [-\infty,+\infty]$, $E \subset \mathbb{R}$, is said to be measurable relatively to a partition of the identity $\{P_\lambda\}_{-\infty}^{+\infty}$ whenever the inverse image $f^{-1}(O)$ of any open set is a measurable set with regard to the spectral measure associated to $\{P_\lambda\}_{-\infty}^{+\infty}$.

It follows that the domain $E$ of a measurable function $f$ is measurable, and that $f$ is measurable if and only if its extension to the whole real line obtained by assigning the value $0$ to any point outside $E$ is measurable. In many questions this remark permits one to assume the functions to be everywhere defined. From (7.38) it follows that <u>a function is measurable with regard to a partition of the identity if and only if it is measurable with regard to each monotone function</u> $\langle x, P_\lambda x \rangle$, $x \in \mathcal{H}$. All through this sections the existence of a fixed partition of the identity will be assumed, and the term <u>measurable function</u>, unless otherwise said, will be taken in the sense of Definition 8.1. Hence, measurable functions being just measurable functions relatively to the $\sigma$-field of sets $\Sigma$, enjoy all ordinary properties of such functions. These, which can be found in any measure theory or integration book, will be assumed as known (cf. [4, I, ch. III]).

A <u>simple function</u> is any finitely valued function $f:\mathbb{R} \to \mathbb{R}$. Note that the values $+\infty$ and $-\infty$ are excluded to them. A simple function is measurable if the sets corresponding to each of its values are measurable; it can be written in the form:

$$s(\lambda) = \sum_{i=1}^{n} s_i \chi_{E_i}(\lambda), \quad E_i \in \Sigma, \quad E_i \cap E_j = \phi, \quad i \neq j,$$

where $\chi_{E_i}$ indicates the characteristic function of $E_i$. All measurable simple functions form a function algebra with identity. Among all simple measurable functions we distinguish those taking nonnegative values only; for simplicity we call them <u>positive</u> simple functions, measurability being always assumed.

In correspondence to positive simple functions we have the positive simple operators:

<u>Definition 8.2</u>   A positive simple operator is a mapping $S: \mathcal{H} \to \mathcal{H}$ of the form

(8.1) $\quad S = \sum_{i=1}^{n} s_i P_{E_i}, \quad 0 < s_i < +\infty, \; E_i \in \Sigma, \; E_i \cap E_j = \phi, \; i \neq j.$

Incorporating if necessary the value $s_0 = 0$ one may further assume that $\mathbb{R} = \bigcup_{i=1}^{n} E_i$.

Any finite nonnegative linear combination of positive simple operators is a positive simple operator, and, as we shall see, so is their product, which is independent of the factor order (Lemma 8.2 below). The correspondence

(8.2) $\quad s = \sum_{i=1}^{n} s_i \chi_{E_i} \longleftrightarrow S = \sum_{i=1}^{n} s_i P_{E_i}$

between positive simple functions and positive simple operators is a bijection preserving linear combinations and products. The operator $S$ is in fact the integral of the function $s$ with regard to the partition $\{P_\lambda\}_{-\infty}^{+\infty}$, and our task here is to extend this <u>integral</u> to the class of all measurable functions, and then obtain a class of operators, neither necessarily continuous or everywhere defined, which can be termed the <u>spectral integrals of measurable functions.</u> As a preliminary step we make a study of positive simple operators.

<u>Lemma 8.1</u>   Every positive simple operator $S = \sum_i s_i P_{E_i}$ has the following properties:

(8.3)       $S$ is everywhere defined, positive homogeneous and monotone,

(8.4)       $S$ satisfies the differential equation
$$Sx = \nabla \tfrac{1}{2} \langle x, Sx \rangle ,$$

(8.5)       $\langle x, Sy \rangle \leq \langle x, Sx \rangle^{\frac{1}{2}} \langle y, Sy \rangle^{\frac{1}{2}}, \quad \forall x, y,$

(8.6)       $|\langle Sx, y \rangle - \langle x, Sy \rangle | \leq \langle x-y, Sx - Sy \rangle, \quad \forall x, y,$

(8.7)       $\langle x, Sx \rangle$ is positive homogeneous of degree 2 and convex,

(8.8)       $\langle x, Sx \rangle^{\frac{1}{2}}$ is positive homogeneous of degree 1 and convex,

(8.9)      $\inf_{x \neq y} \{ \langle x-y, Sx-Sy \rangle / \|x-y\|^2 \} = \inf_{x \neq y} \{ \|Sx - Sy\| / \|x-y\| \} = \min_i s_i$

(8.10)     $\sup_{x \neq y} \{ \langle x-y, Sx-Sy \rangle / \|x-y\|^2 \} = \sup_{x \neq y} \{ \|Sx - Sy\| / \|x-y\| \} = \max_i s_i$

(8.11)     $|\langle x, Sy \rangle | \leq (\sup_i s_i)^{\frac{1}{2}} \|x\| \langle y, Sy \rangle^{\frac{1}{2}}, \quad \forall x, y.$

Proof:   As a positive linear combination of projections, $S$ obviously has property (8.3). Moreover, since any projection satisfies (8.4), and

$$\tfrac{1}{2} \langle x, Sx \rangle = \sum_i s_i \tfrac{1}{2} \langle x, P_{E_i} x \rangle ,$$

$$\nabla \tfrac{1}{2} \langle x, Sx \rangle = \sum_i s_i \nabla \tfrac{1}{2} \langle x, P_{E_i} x \rangle = \sum_i s_i P_{E_i} x = Sx ,$$

$S$ itself satisfies (8.4). But then $S$ satisfies the hypotheses of Lemma 6.2, and (8.5)-(8.8) follow.

To deal with (8.9) and (8.10) write

$$\langle x-y, Sx-Sy \rangle = \sum_i s_i \langle x-y, P_{E_i} x - P_{E_i} y \rangle$$

SPECTRAL THEORY

and remark that $\langle x-y, P_{E_i} x - P_{E_i} y \rangle \geq 0$, since projections are monotone. Replacing then the $s_i$'s by their minimum and their maximum one obtains, if $x \neq y$,

$$\min_i s_i \leq \frac{\langle x-y, Sx-Sy \rangle}{\|x-y\|^2} \leq \max_i s_i .$$

These two bounds are attained, as it is seen by setting $y = 0$ and taking for $x$ any vector in the range of the $P_{E_i}$ for which $s_i$ is either minimum or maximum. Hence, the equality of the first and last members of (8.9) and (8.10) is established. As for the rest write

$$\|Sx - Sy\|^2 = \sum_{i,j} s_i s_j \langle P_{E_i} x - P_{E_i} y, P_{E_j} x - P_{E_j} y \rangle ,$$

and note that $\langle P_{E_i} x - P_{E_i} y, P_{E_j} x - P_{E_j} y \rangle \geq 0$, for all values of $i$ and $j$ (this is obvious if $i = j$, and follows from $\langle P_{E_i} x - P_{E_i} y, P_{E_j} x - P_{E_j} y \rangle = -\langle P_{E_i} x, P_{E_j} y \rangle - \langle P_{E_i} y, P_{E_j} x \rangle$, since $P_{E_i} x$ and $P_{E_j} y$ belong to dual cones if $i \neq j$). Substituting then the $s_i s_j$'s by their minimum and maximum

$$(\min_i s_i)^2 \leq \frac{\|Sx-Sy\|^2}{\|x-y\|^2} \leq (\max_i s_i)^2 , \quad x \neq y .$$

As before, these bounds are attained at points in the range of the $P_{E_i}$ corresponding to the minimum and maximum value of the $s_i$'s. This completes the proof of (8.9) and (8.10).
By (8.10), $\langle x, Sx \rangle^{\frac{1}{2}} \leq (\max s_i)^{\frac{1}{2}} \|x\|$; so replacing in (8.5),

$$\langle x, Sy \rangle \leq (\max_i s_i)^{\frac{1}{2}} \|x\| \langle y, Sy \rangle^{\frac{1}{2}},$$

which, in combination with the inequality obtained by replacing $x$ by $-x$ yields (8.11).

**Lemma 8.2** For any positive simple operator $S = \sum_i s_i P_{E_i}$ and any real polynomial $p(\lambda) = a_0 + a_1\lambda + \ldots + a_n\lambda^n$,

(8.12) $\quad p(S) = a_0 I + a_1 S + \ldots + a_n S^n = \sum_i p(s_i) P_{E_i}$.

In particular, $p(S)$ is again a positive simple operator if $p(\lambda)$ takes nonnegative values for nonnegative $\lambda$'s. Moreover, if $q(\lambda) = b_0 + b_1\lambda + \ldots + b_m\lambda^m$ is another such polynomial ($q(\lambda) \geq 0$ for $\lambda \geq 0$),

(8.13) $\quad\quad p(S) q(S) = q(S) p(S) = (pq)(S)$;

(8.14) $\quad\quad \langle p(S)x, q(S)x \rangle = \langle (pq)(S)x, s \rangle$, $\forall x$.

**Proof:** Let

$$S^{(1)} = \sum_i s_i^{(1)} P_{E_i^{(1)}}, \quad S^{(2)} = \sum_j s_j^{(2)} P_{E_j^{(2)}}$$

be two positive simple operators. The sets $E_{ij} = E_i^{(1)} \cap E_j^{(2)}$ are then disjoint and $E_i^{(1)} = \bigcup_j E_{ij}$, $E_j^{(2)} = \bigcup_i E_{ij}$. So,

$$S^{(1)} = \sum_i s_i^{(1)} \sum_j P_{E_{ij}} = \sum_{i,j} s_i^{(1)} P_{E_{ij}}$$

$$S^{(2)} = \sum_j s_j^{(2)} \sum_i P_{E_{ij}} = \sum_{i,j} s_j^{(2)} P_{E_{ij}}$$

and

$$S^{(1)} S^{(2)} x = \sum_{ij} s_i^{(1)} P_{E_{ij}} \left( \sum_{h,k} s_k^{(2)} P_{E_{h,k}} \right) x.$$

But

$$\sum_{h,k} s_k^{(2)} P_{E_{h,k}} x = s_j^{(2)} P_{E_{ij}} x + \sum_{h \neq i, k \neq j} s_k^{(2)} P_{E_{h,k}} x$$

is a decomposition of the left member into the sum of two

perpendicular vector, one belonging to the range of $P_{E_{ij}} x$ and the other to its dual cone. Hence by (2.4),

$$P_{E_{ij}}(\sum_{h,k} s_k^{(2)} P_{E_{h,k}}) = s_j^{(2)} P_{E_{ij}}$$

and on substitution above,

(8.15) $$s^{(1)} s^{(2)} x = \sum_{i,j} s_i^{(1)} s_j^{(2)} P_{E_{ij}}.$$

This is the already mentioned fact that the product of two simple operators is the positive simple operator corresponding to the product of their respective simple functions. Equating the factors to S one obtains, on iteration,

$$S^k = \overline{S\,S}^k \cdots S = \sum_i s_i^k P_{E_i},$$

whence (8.12) follows by linear combination.

On the other hand if in (8.15), $S^{(1)}$ and $S^{(2)}$ are identified with $p(S)$ and $q(S)$ respectively, then $s_i^{(1)} = p(s_i)$, $s_j^{(2)} = q(s_j)$, and

$$p(S)q(S) = \sum_{i,j} p(s_i) q(s_j) P_{E_i \cap E_j} = \sum_i p(s_i) q(s_i) P_{E_i}$$

$$= \sum_i (pq)(s_i) P_{E_i} = (pq)(S),$$

which is (8.13). Moreover,

$$\langle p(S)x, q(S)x \rangle = \sum_{i,j} p(s_i) q(s_j) \langle P_{E_i} x, P_{E_j} x \rangle$$

$$= \sum_i p(s_i) q(s_i) \langle P_{E_i} x, P_{E_i} x \rangle$$

$$= \sum_i p(s_i) q(s_i) \langle P_{E_i} x, x \rangle = \langle (\sum_i (pq)(s_i) P_{E_i}) x, x \rangle$$

$$= \langle (pq)(S)x, x \rangle,$$

and (8.14) is proved.

**Lemma 8.3** Let $\{s_n(\lambda)\}_1^\infty$ be an increasing sequence of positive simple functions converging to a measurable function $u: (-\infty, +\infty) \to [0, +\infty]$, and let $\{S_n\}_1^\infty$ be the corresponding sequence of simple operators. Then if $C$ is the set of points in $H$ where $S_n x$ has a strong limit as $n \to \infty$,

(8.16) $\quad C = \{x \in H \mid \int_{-\infty}^{+\infty} u^2(\lambda) \, d\langle P_\lambda x, x \rangle < +\infty \} \ ;$

$C$ is a convex cone contained in the range $\mathcal{R}(P_{E_\infty}^\perp)$ of $P_{E_\infty}^\perp$ and dense in it, where $E_\infty = \{\lambda \in |u(\lambda) = +\infty\}$. Moreover, for any $x \in C$

(8.17) $\quad \langle \lim_{n \to \infty} S_n x, x \rangle = \int_{-\infty}^{+\infty} u(\lambda) \, d\langle P_\lambda x, x \rangle.$

Both $C$ and $\lim_{n \to \infty} S_n x$ depend on $u(\lambda)$ but not on the particular approximating sequence $\{s_n(\lambda)\}_1^\infty$.

**Proof:** Let us prove (8.16) first. The operator $S_n^2$ is the simple operator corresponding to the simple function $s_n^2(\lambda)$, and in consequence by (8.9),

(8.18) $\quad \|S_n x\| = \langle S_n x, S_n x \rangle^{\frac{1}{2}} = \langle S_n^2 x, x \rangle^{\frac{1}{2}}$

$$= \{\sum_i s_{n,i}^2 \langle P_{E_{n,i}} x, x \rangle\}^{\frac{1}{2}} = \{\int_{-\infty}^{+\infty} s_n^2(\lambda) \langle dP_\lambda x, x \rangle^{\frac{1}{2}}\}.$$

Since the $s_n^2(\lambda)$'s form an increasing sequence and $\mu_x(E) = \langle P_E x, x \rangle$ is a positive measure, the limit of the integral on the right always exists and is equal to the integral of $\lim_{n \to \infty} s_n^2(\lambda) = u^2(\lambda)$. So,

(8.19) $\quad \lim_{n \to \infty} \|S_n x\| = \{\int_{-\infty}^{+\infty} u(\lambda)^2 \, d\langle P_\lambda x, x \rangle\}^{\frac{1}{2}},$

and if $\lim_{n \to \infty} S_n x$ exists, $\int_{-\infty}^{+\infty} u(\lambda)^2 \, d\langle P_\lambda x, x \rangle < +\infty$. To prove the converse apply (8.18) to the simple function $S_n - S_m$, $(n \geq m)$, and get

(8.20) $\qquad \|S_n x - S_m x\| = \{\int_{-\infty}^{+\infty}(s_n(\lambda) - s_m(\lambda))^2 \, d\langle P_\lambda x, x\rangle\}^{\frac{1}{2}}$,

equation which is valid for any values of $m$ and $n$. Writing then $s_n(\lambda) - s_m(\lambda) = (s_n(\lambda) - u(\lambda)) + (u(\lambda) - s_m(\lambda))$ and using the triangle inequality

$$\|S_n x - S_m x\| \leq \{\int_{-\infty}^{+\infty}(s_n(\lambda) - u(\lambda))^2 \, d\langle P_\lambda x, x\rangle\}^{\frac{1}{2}}$$
$$+ \{\int_{-\infty}^{+\infty}(s_m(\lambda) - u(\lambda))^2 \, d\langle P_\lambda x, x\rangle\}^{\frac{1}{2}}.$$

So if $\int_{-\infty}^{+\infty} u(\lambda)^2 \, d\langle P_\lambda x, x\rangle < \infty$, both integrals on the right tend to zero, and $\{S_n x\}_1^\infty$ is a Cauchy sequence and has a strong limit. Thus (8.16) is proved.

Next observe that if $x \in C$, (8.16) says that the set $E_\infty = \{\lambda \,|\, u(\lambda) = +\infty\}$ has measure zero with regard to $\mu_x$. But then, since $I = P_{E_\infty'} + P_{E_\infty}$,

$$\|x\|^2 = \langle P_{E_\infty'} x, x\rangle + \langle P_{E_\infty} x, x\rangle = \langle P_{E_\infty'} x, x\rangle$$

which implies $x = P_{E_\infty'} x$, that is, $x \in \mathcal{R}(P_{E_\infty}^\perp)$. Therefore $C \subset \mathcal{R}(P_{E_\infty}^\perp)$.

To see that $C$ is a convex cone it is enough to show that it contains the sum $x+y$ of any two of its elements, since obviously along with any $z$ it contains $tz$ for any nonnegative $t$. Assuming for a moment that $n \geq m$, $(S_n - S_m)^2$ is a positive simple operator (Lemma 8.2) and by Lemma 8.1 equation (8.8),

$$\langle (S_n - S_m)^2 (x+y), x+y\rangle^{\frac{1}{2}} \leq \langle (S_n - S_m)^2 x, x\rangle^{\frac{1}{2}} + \langle (S_n - S_m)^2 y, y\rangle^{\frac{1}{2}}$$

or, by (8.14),

$$\|S_n(x+y) - S_m(x+y)\| \leq \|S_n x - S_m x\| + \|S_n y - S_m y\|.$$

At this point the condition $n \geq m$ can be dropped and the conclusion be drawn that if both sequences $\{S_n x\}_1^\infty$ and $\{S_n y\}_1^\infty$ have a strong limit so does $\{S_n(x+y)\}_1^\infty$. Thus the claim concerning the convexity of $C$ is proved.

As to $\bar{C} = \mathcal{R}(P_{E_\infty}^\perp)$, begin by noting that if $E_k = \{\lambda \,|\, k-1 \leq u(\lambda) < k\}$ then $E'_\infty = \bigcup_{k=1}^\infty E_k$ and by (7.29),

$$P_{E'_\infty} = \bigcup_{k=1}^\infty P_{E_k} = \lim_{n \to \infty} (P_{E_1} + P_{E_2} + \ldots + P_{E_n}),$$

since clearly the $E'_k$ are measurable disjoint sets. It follows that if $x \in \mathcal{R}(P_{E'_\infty})$

$$x = P_{E'_\infty} x = \sum_{k=1}^\infty P_{E_k} x,$$

whence the sought conclusion will follow as soon as it is established that $\mathcal{R}(P_{E_k}) \subset C$, $k = 1, 2, \ldots$. This is easily proved, for if $x \in \mathcal{R}(P_{E_k})$, then by (8.20),

$$\|S_n x - S_m x\| = \{\int_{-\infty}^{+\infty} (s_n(\lambda) - s_m(\lambda))^2 d\langle P_\lambda x, x\rangle\}^{\frac{1}{2}}$$

$$= \{\int_{-\infty}^{+\infty} (s_n(\lambda) - s_m(\lambda))^2 d\langle P_\lambda P_{E_k} x, P_{E_k} x\rangle\}^{\frac{1}{2}}$$

$$= \{\int_{E_k} (s_n(\lambda) - s_m(\lambda))^2 d\langle P_\lambda x, x\rangle\}^{\frac{1}{2}},$$

whence by Lebesgue's Dominated Convergence Theorem one concludes that $\{S_n x\}_1^\infty$ is a Cauchy sequence, hence convergent, since on $E_k$ the functions $(s_n(\lambda) - s_m(\lambda))^2$ are uniformly bounded. Therefore $x \in C$.

SPECTRAL THEORY

Equation (8.17) is a consequence of the identity

(8.21) $$\langle S_n x, x \rangle = \int_{-\infty}^{+\infty} s_n(\lambda) \, d \langle P_\lambda x, x \rangle ,$$

whose varification we leave to the reader.

Finally, it is apparent from (8.17) that $C$ depends on $\mu$ but not on the approximating sequence. To see that the same is true for $\lim_{n \to \infty} S_n x$ take two increasing sequences $\{s_n^{(1)}(\lambda)\}_1^\infty$ and $\{s_n^{(2)}(\lambda)\}_1^\infty$ converging to $u(\lambda)$ and set $s_n(\lambda) = \max \{s_n^{(1)}(\lambda), s_n^{(2)}(\lambda)\}$. Then $S_n - S_n^{(i)}$, $i = 1, 2$ are positive simple operators, and by (8.18)

$$\| S_n x - S_n^{(i)} x \| = \{ \int_{-\infty}^{+\infty} (s_n(\lambda) - s_n^{(i)}(\lambda))^2 \, d \langle P_\lambda x, x \rangle \}^{\frac{1}{2}} .$$

So, if $x \in C$, the integral on the right tends to zero as $n \to \infty$ because $s_n(\lambda) - s_n^{(1)}(\lambda) \to u(\lambda) - u(\lambda) = 0$, and $\lim_{n \to \infty} S_n x = \lim_{n \to \infty} S_n^{(1)} x = \lim_{n \to \infty} S_n^{(2)} x$.    Q.E.D.

This lemma provides the necessary foundations for the following definitions:

<u>Definitions 8.3</u>  The integral with regard to a partition of the identity $\{P_\lambda\}_{-\infty}^{+\infty}$ of a positive simple function $s(\lambda) = \sum_i s_i \chi_{E_i}$, where the $E_i$ are disjoint measurable functions, is the operator

(8.22) $$\int_{-\infty}^{+\infty} s(\lambda) \, d P_\lambda = \sum_i s_i P_{E_i} .$$

In other words, the integral of $s(\lambda)$ is the corresponding simple operator $S$. The integrals of simple functions are everywhere defined.

387

**Definition 3.4** The integral with regard to a partition of the identity $\{P_\lambda\}_{-\infty}^{+\infty}$ of a nonnegative measurable function $u:(-\infty,+\infty) \to [0,+\infty]$ is the operator $\int_{-\infty}^{+\infty} u(\lambda)\, dP_\lambda$ defined by

(8.23) $$\left(\int_{-\infty}^{+\infty} u(\lambda)\, dP_\lambda\right)x = \lim_{n\to\infty}\left(\int_{-\infty}^{+\infty} s_n(\lambda)\, dP_\lambda\right)x ,$$

where $\{s_n(\lambda)\}_1^\infty$ is any increasing sequence of positive simple functions converging pointwise to $u(\lambda)$; the definition domain of the integral is the set of points where the limit on the right exists.

It is evident, on account of Lemma 8.3, that Definition 8.3 and 8.4 coincide on simple functions.

Finally, to extend the notion of integral to any measurable function $f$ we recall that $f$ is the difference $f^+ - f^-$ of its positive and negative parts, and set

**Definition 8.5** The integral with regard to a partition of the identity $\{P_\lambda\}_{-\infty}^{+\infty}$ of a measurable function $f:(-\infty,+\infty) \to [-\infty,+\infty]$ is the operator

(8.24) $$\int_{-\infty}^{+\infty} f(\lambda)\, dP_\lambda = \int_{-\infty}^{+\infty} f^+(\lambda)\, dP_\lambda - \int_{-\infty}^{+\infty} f^-(\lambda)\, dP_\lambda ,$$

defined over the set of points in $\mathcal{H}$ belonging simultaneously to the domains of definitions of both integrals on the right (by Lemma 8.3, this is the same as the definition domain of the integral of $|f|$).

Again one should notice that Definition 8.4 and 8.5 give the same notion of integral when applied to nonnegative functions.

§9. **The spectral integral.** This section is devoted to the discussion of some basic properties of the operators defined by spectral integrals of measurable functions. We shall restrict our attention to integrals of nonnegative functions, simply because little can be said of integrals of functions in general. In fact, it looks as if negative valued functions do not fit within the framework of our theory. If a nonnegative function $u(\lambda)$ takes the value $+\infty$ on a set $E_\infty$ of nonvanishing spectral measure the integral

$$T = \int_{-\infty}^{+\infty} u(\lambda) \, d P_\lambda$$

is only defined on a convex cone dense in the dual of the range of $P_{E_\infty}$ (Lemma 8.3). Such an operator admits proper maximal extensions to a dense set in $\mathcal{H}$, in fact many extensions, obtainable by redefining $u$ as an almost everywhere finite nonnegative measurable function in $E_\infty$. This lack of maximality robs $T$ of many properties, and, in a way, forces one to further restrict attention to integrals of nonnegative functions finite almost everywhere. In such a case one has

$$\int_{-\infty}^{+\infty} u(\lambda) \, d P_\lambda = \int_0^\infty \lambda \, d P_{E_\lambda} \, , \quad E_\lambda = \{\mu \,|\, u(\mu) \leq \lambda\},$$

and so by changing the spectral resolution ($\{P_{E_\lambda}\}_{-\infty}^{+\infty}$ is a spectral resolution) $T$ can be expressed in the form

(9.1) $$T = \int_0^\infty \lambda \, d P_\lambda \, ,$$

where now $\{P_\lambda\}_{-\infty}^{+\infty}$ is a spectral resolution with <u>support on the nonnegative real line</u>. It is therefore to operators of this type - called <u>spectral integrals</u> - that this section is addressed. Since we shall have nothing to do here with the interrelation between operators defined by different spectral measures, <u>the spectral resolution</u> $\{P_\lambda\}_{-\infty}^{+\infty}$ <u>will be assumed as fixed once for all through the entire section.</u>

The properties below are almost direct consequences of the corresponding properties for simple operators.

**Theorem 9.1** The spectral integral

$$T = \int_0^\infty \lambda \, dP_\lambda$$

has the following properties

(9.2) $T$ is defined on a convex cone dense in $H$, it is positive homogeneous and monotone; its domain is

$$\mathfrak{D}(T) = \{x \,|\, \int_0^\infty \lambda^2 d \langle P_\lambda x, x\rangle < +\infty\},$$

(9.3) $Tx = \nabla \tfrac{1}{2} \langle x, Tx\rangle$, $\forall x \in \mathfrak{D}(T)$, $\nabla$ = Gateaux gradient relative to $\mathfrak{D}(T)$,

(9.4) $\langle x, Ty\rangle \leq \langle x, Tx\rangle^{\tfrac{1}{2}} \langle y, Ty\rangle^{\tfrac{1}{2}}$, $\forall x, y \in \mathfrak{D}(T)$,

(9.5) $|\langle Ty, x\rangle - \langle y, Tx\rangle| \leq \langle x-y, Tx-Ty\rangle$, $\forall x, y \in \mathfrak{D}(T)$,

(9.6) $\langle x, Tx\rangle$ is positive homogeneous of degree 2 and convex,

(9.7) $\langle x, Tx\rangle^{\tfrac{1}{2}}$ is positive homogeneous of degree 1 and convex,

(9.8) $\inf_{x \neq y} \{\langle x-y, Tx-Ty\rangle / \|x-y\|^2\} = \inf_{x \neq y} \{\|Tx-Ty\| / \|x-y\|\}$

$= \inf \text{ support } \{P_\lambda\}_{-\infty}^{+\infty}$,

(9.9) $\sup_{x \neq y} \{\langle x-y, Tx-Ty\rangle / \|x-y\|^2\} = \sup_{x \neq y} \{\|Tx-Ty\| / \|x-y\|\}$

$= \sup \text{ support } \{P_\lambda\}_{-\infty}^{+\infty}$,

(9.10) $|\langle x, Ty\rangle| \leq \{\sup \text{ support } \{P_\lambda\}_{-\infty}^{+\infty}\}^{\tfrac{1}{2}} \|x\| \langle y, Ty\rangle^{\tfrac{1}{2}}$.

**Proof:** Points outside of the support of the spectral resolution do not make any contribution to the spectral integral, which therefore can be taken over the support itself. Hence, if $\chi(\lambda)$ denotes the characteristic function of the support of $\{P_\lambda\}_{-\infty}^{+\infty}$ we can write

(9.11) $$T = \int_0^\infty \lambda \chi(\lambda) \, dP_\lambda.$$

We carry the proof under the only formally more general assumptions that T is of the form

(9.12) $$T = \int_0^\infty u(\lambda) \, dP_\lambda, \qquad u(\lambda) \geq 0.$$

In this new formulation the end members of (9.8) and (9.9) have to be replaced by the essential infimum and supremum of $u(\lambda)$ respectively.

By the definition of integral

$$Tx = \lim_{n \to \infty} S_n x, \qquad S_n = \int_0^\infty s_n(\lambda) \, dP_\lambda,$$

where $s_n(\lambda)$ is an increasing sequence of positive simple functions converging to $u(\lambda)$. With this, properties (9.2), (9.4), (9.5), (9.6), (9.7) and (9.10) follow from (8.16) and by passage to the limit in the corresponding relations for the $S_n$'s (Lemma 8.1). As to (9.9) we have (Lemma 8.1, (8.10)),

$$\langle x-y, S_n x - S_n y \rangle = (\sup s_n(\lambda)) \|x-y\|^2 \leq (\text{ess supp } u(\lambda)) \|x-y\|^2,$$

$$\|S_n x - S_n y\| = (\sup s_n(\lambda)) \|x-y\| \leq (\text{ess supp } u(\lambda)) \|x-y\|,$$

whence letting $n \to \infty$

$$\langle x-y, Tx-Ty \rangle \leq (\text{ess supp } u(\lambda)) \|x-y\|^2,$$

$$\|Tx - Ty\| \leq (\text{ess supp } u(\lambda)) \|x-y\|.$$

That the ess sup $u(\lambda)$ is the best bound is seen, as in the case of simple operators, by setting $y = 0$ and taking for x points with support concentrated near the essential sup of $u(\lambda)$. We have thus proved (9.9). Relation (9.8) follows by applying (9.9) to the operator

$$MI - T_M = \int_0^\infty (M - u_M(\lambda)) \, dP_\lambda, \quad u_M(\lambda) = \max\{u(\lambda), M\}.$$

Differential equation (9.4) will not be proved here, it will appear as part of a more precise result to be obtained later (Theorem 9.8).

Before proceeding to the next lemma let us remind the reader that the equality of two operators means coincidence of their domains as well as equality of their values; an operator $T_1$ is said to be a restriction of another $T_2$, or equivalently $T_2$ is said to be an extension of $T_1$, if $\mathfrak{D}(T_1) \subset \mathfrak{D}(T_2)$ and $T_1 x = T_2 x$, $\forall x \in \mathfrak{D}(T_1)$; this fact is denoted $T_1 \subset T_2$.

<u>Lemma 9.1</u>  Let $\{P_\lambda\}_{-\infty}^{+\infty}$ be a spectral resolution with support on the nonnegative real line, and $u(\lambda)$ and $v(\lambda)$ nonnegative measurable functions. If

(9.13) $\quad U = \int_0^\infty u(\lambda) \, dP_\lambda, \quad V = \int_0^\infty v(\lambda) \, dP_\lambda, \quad W = \int_0^\infty u(\lambda) v(\lambda) \, dP_\lambda,$

then

(9.14) $\qquad \mathfrak{D}(UV) = \mathfrak{D}(V) \cap \mathfrak{D}(W)$

(9.15) $\qquad UV \subset W$

(9.16) $\quad \langle Vx, Ux \rangle = \langle UVx, x \rangle, \; \forall x \in \mathfrak{D}(U) \cap \mathfrak{D}(V) \cap \mathfrak{D}(UV).$

<u>Proof:</u>  A point $x$ belongs to $\mathfrak{D}(UV)$ if $x \in \mathfrak{D}(V)$ and $Vx \in \mathfrak{D}(U)$, that is if

(9.17) $\quad \int_0^\infty v^2(\lambda) \, d \langle P_\lambda x, x \rangle < \infty, \quad \int_0^\infty u^2(\lambda) \, d \langle P_\lambda Vx, Vx \rangle < \infty$

(Lemma 8.3 (8.16)).  Now we have

$$Vx = \int_0^\infty v(\mu) \, dP_\mu x = \int_{[0,\lambda]} v(\mu) \, dP_\mu x + \int_{(\lambda,\infty]} v(\mu) \, dP_\mu x \,.$$

SPECTRAL THEORY

The integrals on the right, as limits of positive linear combinations of orthogonal vectors in the ranges of $P_\lambda$ and $P_\lambda^\perp$ respectively, are orthogonal and belong to the ranges of these projections. Hence (Lemma 2.2),

$$P_\lambda Vx = \int_{[0,\lambda]} v(\mu) \, dP_\mu x$$

and

$$\langle P_\lambda Vx, Vx \rangle$$

$$= \int_{[0,\lambda]} v(\mu) \, d\langle P_\mu x, \int_0^\infty v(\nu) dP_\nu x\rangle = \int_{[0,\lambda]} v(\mu) \, d\langle P_\mu x, \int_{[0,\mu]} v(\nu) dP_\nu x\rangle$$

$$= \int_{[0,\lambda]} v(\mu) \, d[\int_{[0,\mu]} v(\nu) d\langle P_\nu x, P_\mu x\rangle] = \int_{[0,\lambda]} v(\mu) d[\int_{[0,\mu]} v(\nu) d\langle P_\nu x, x\rangle]$$

$$= \int_{[0,\lambda]} v^2(\mu) \, d\langle P_\mu x, x\rangle.$$

Replacing this in (9.17), one obtains

$$\mathcal{D}(UV) = \{x \mid \int_0^\infty v^2(\lambda) d\langle P_\lambda x, x\rangle < +\infty, \int_0^\infty u^2(\lambda) v^2(\lambda) d\langle P_\lambda x, x\rangle < +\infty\}.$$

On the other hand

$$\mathcal{D}(W) = \{x \mid \int_0^\infty u^2(\lambda) v^2(\lambda) \, d\langle P_\lambda x, x\rangle < +\infty\},$$

and so

$$\mathcal{D}(UV) = \mathcal{D}(V) \cap \mathcal{D}(W).$$

Moreover, if $x \in \mathcal{D}(UV)$,

$$UVx = \int_0^\infty u(\lambda) dP_\lambda Vx = \int_0^\infty u(\lambda) d\int_{[0,\lambda]} v(\mu) dP_\mu x = \int_0^\infty u(\lambda) v(\lambda) dP_\lambda x = Wx,$$

and (9.14) and (9.15) are proved. Finally, if $x \in \mathcal{D}(U) \cap \mathcal{D}(V)$,

393

$$\langle Vx, Ux \rangle =$$

$$= \int_0^\infty u(\lambda) d \langle P_\lambda x, \int_0^\infty v(\mu) d P_\mu x \rangle = \int_0^\infty u(\lambda) d \langle P_\lambda x, \int_{[0,\lambda]} v(\mu) d P_\mu x \rangle$$

$$= \int_0^\infty u(\lambda) d \int_{[0,\lambda]} v(\mu) d \langle P_\mu x, P_\lambda x \rangle = \int_0^\infty u(\lambda) d \int_{[0,\lambda]} v(\mu) d \langle P_\mu x, x \rangle$$

$$= \int_0^\infty u(\lambda) v(\lambda) d \langle P_\lambda x, x \rangle = \langle Wx, x \rangle .$$

Q. E. D.

<u>Corollary 1</u>  If $T = \int_0^\infty \lambda \, d P_\lambda$, and $p(\lambda) = a_0 + a_1\lambda + \ldots + a_n \lambda^n$, then

(9.18) $\quad p(T) = a_0 I + a_1 T + \ldots + a_n T^n = \int_0^\infty p(\lambda) \, d P_\lambda$ .

In particular, if $p(\lambda) \geq 0$ for $\lambda \geq 0$, $p(T)$ is a (nonnegative) spectral integral. Moreover, if $q(\lambda) = b_0 + b_1\lambda + \ldots + b_m \lambda^m$, $q(\lambda) \geq 0$ for $\lambda \geq 0$,

(9.19) $\quad\quad\quad p(T)q(T) = q(T)p(T) = (pq)(T)$

(9.20) $\quad\quad\quad \langle p(T)x, q(T)x \rangle = \langle p(T)q(T)x, x \rangle, \ \forall \ x \in \mathcal{D}(p(T)) \cap \mathcal{D}(q(T))$.

<u>Proof:</u>  It is enough to prove that $T^n = \int_0^\infty \lambda^n d P_\lambda$, $n = 0, 1, 2, \ldots$ . This is best done by induction. The proposition is true for $n = 1, 2$ and if it holds for $n-1$ then by Lemma 9.1, with $U, V$ and $W$ identified with $T^{n-1} = \int_0^\infty \lambda^{n-1} d P_\lambda$, $T = \int_0^\infty \lambda \, d P_\lambda$ and $\int_0^\infty \lambda^n d P_\lambda$ respectively,

$$\mathcal{D}(UV) = \mathcal{D}(V) \cap \mathcal{D}(W) ,$$

but since if $n \geq 1$,

$$\mathcal{D}(V) = \{x \mid \int_0^\infty \lambda^2 d \langle P_\lambda x, x \rangle < \infty \} \supset \mathcal{D}(W) = \{x \mid \int_0^\infty \lambda^{2n} d \langle P_\lambda x, x \rangle < \infty \}$$

then

$$\mathcal{D}(UV) = \mathcal{D}(W)$$

and $UV = W$, which is just $T^n = \int_0^\infty \lambda^n dP_\lambda$. This completes the induction argument.

Corollary 2   If $u(\lambda)$ and $v(\lambda)$ are nonnegative essentially bounded measurable functions on the support of $\{P_\lambda\}_{-\infty}^{+\infty}$, then

(9.21) $\quad \alpha \int_0^\infty u(\lambda) dP_\lambda + \beta \int_0^\infty v(\lambda) dP_\lambda = \int_0^\infty (\alpha u(\lambda) + \beta v(\lambda)) dP_\lambda$,

$$\alpha, \beta \geq 0$$

(9.22) $\quad (\int_0^\infty u(\lambda) dP_\lambda)(\int_0^\infty v(\lambda) dP_\lambda) = \int_0^\infty u(\lambda) v(\lambda) dP_\lambda$,

(9.23) $\quad \sup_{x \neq y} \{\|\int_0^\infty u(\lambda) dP_\lambda x - \int_0^\infty u(\lambda) dP_\lambda y\|/\|x-y\|\} = \text{ess sup } u(\lambda)$.

Proof:   It is only a matter of remarking that the domains of all operators involved coincide with the whole space, and applying the Lemma. Equation (9.23) is a restatement of (9.9)

The last two corollaries prompt the following notation: If $T = \int_0^\infty \lambda \, dP_\lambda$ and $u(\lambda)$ is a measurable nonnegative function,

(9.24) $\quad u(T) = \int_0^\infty u(\lambda) \, dP_\lambda$.

The last corollary above is the basis of a "functional calculus" for spectral integral operators. Mutatis mutandis, this is just the ordinary functional calculus for self-adjoint linear operators. It appears convenient to dress it up in the language of Banach semialgebras.

Definition 9.1   A real Banach semialgebra is a set $G = \{A\}$ in which a sum and a multiplication of any two elements is defined, as well as the multiplication of an element by a nonnegative real number. Both multiplications are associative and distributive and $G$ is an Abelian semigroup with regard to addition. Moreover, there is a "norm" satisfying:

a. $|A| \geq 0$, $|A| = 0 \Leftrightarrow A = 0$;

b. $|\lambda A| = \lambda |A|$, $\forall \lambda \geq 0$;

c. $|A| + |B| \geq |A + B|$;

d. $|A||B| \geq |AB|$

e. Any series $\sum_{k=1}^{\infty} A_k$ such that $|\sum_{k=m}^{n} A_k|_{m,n} \to 0$ is convergent, i.e., there is an $A \in \mathcal{G}$ such that $A - \sum_{k=1}^{n} A_k \in \mathcal{G}$, $n = 1, 2, \ldots$, and $|A - \sum_{k=1}^{n} A_k| \xrightarrow[n \to \infty]{} 0$.

A homomorphism of a semialgebra into another is a mapping preserving the algebra operations; an isomorphism is a bijective homomorphism.

<u>Theorem 9.2</u>    Let $\{P_\lambda\}_{-\infty}^{+\infty}$ be a spectral resolution with support on the nonnegative real line, and let $T = \int_0^\infty \lambda \, dP_\lambda$. The mapping

(9.25)    $u(\lambda) \to u(T) = \int_0^\infty u(\lambda) \, dP_\lambda$

establishes an isometric isomorphism between the Banach semialgebra of all essentially bounded nonnegative measurable function on the support of $\{P_\lambda\}_{-\infty}^{+\infty}$ (with the usual identification of functions differing on a set of measure zero) and a semialgebra of everywhere defined monotone Lipschitz mappings of $\mathcal{H}$ into $\mathcal{H}$, the norms being the essential supremum and the Lipschitz' norm respectively.

This theorem is strongly reminiscent of Gelfand's representation of Banach algebras [14] by function algebras and poses the natural question of whether such a representation is also possible for Banach semialgebras.

SPECTRAL THEORY

<u>Theorem 9.3</u>   Let $T = \int_0^\infty \lambda \, dP_\lambda$, and let $u(\lambda)$ and $v(\lambda)$ be two measurable functions nonnegative on $[0, +\infty]$. Then,

(9.26)   $\|u(T)v(T)x - u(T)v(T)y\|^2 \leq \langle u^2(T)x - u^2(T)y, v^2(T)x - v^2(T)y \rangle$

$$\forall\, x, y \in \mathfrak{D}(u(T)^2) \cap \mathfrak{D}(v(T)^2).$$

<u>Proof</u>:   It suffices to carry the proof under the assumption that u and v are simple functions, or what is the same, that $u(T)$ and $v(T)$ are positive simple operators built on the same spectral resolution. A passage to the limit in the relation for approximating sequences yields them the general case. So let

$$S = \sum_i s_i P_{E_i}, \qquad R = \sum_j r_j P_{F_j}$$

be two simple operators; by partitioning the $E_i$'s and $E_j$'s, if necessary, one may assume that $E_i = F_i$, and so write

$$R = \sum_j r_j P_{E_j}.$$

A direct calculation yields

$$\langle S^2 x - S^2 y, R^2 x - R^2 y \rangle - \|SRx - SRy\|^2$$

$$= \sum_{i,j} [s_i^2 r_j^2 - s_i r_i s_j r_j] \langle P_{E_i} x - P_{E_i} y, P_{E_j} x - P_{E_j} y \rangle$$

$$= \tfrac{1}{2} \sum_{i,j} [s_i^2 r_j^2 + s_j^2 r_i^2 - 2 s_i r_i s_j r_j] \langle P_{E_i} x - P_{E_i} y, P_{E_j} x - P_{E_j} y \rangle$$

$$= \tfrac{1}{2} \sum_{i,j} [s_i r_j - s_j r_i]^2 \langle P_{E_i} x - P_{E_i} y, P_{E_j} x - P_{E_j} y \rangle.$$

Since $\langle P_{E_i} x - P_{E_i} y, P_{E_j} x - P_{E_j} y \rangle \geq 0$, $\forall ij$ (cf. Lemma 8.1, proof of (8.9), (8.10)), the last member on the right is nonnegative and (9.26) is proved in the case in question, and hence in general.

Corollary  If $u(T)$ is idempotent it is a projection.

Proof:   Setting $v(T) = I$, (9.26) becomes

$$\|u(T)x - u(T)y\|^2 \le \langle u(T)x - u(T)y, x-y \rangle,$$

which together with idempotency identifies $u(T)$ as a projection (Theorem 1.2).

Theorem 9.4   For any operator $T = \int_0^\infty \lambda\, dP_\lambda$ the four following conditions are equivalent:

a.   $T$ is a Lipschitz mapping,

b.   $T$ is locally bounded at a point $x_0$,

c.   $T$ is everywhere defined,

d.   The spectral resolution has a compact support.

Proof:   This is the scheme of the proof:

$$a. \Rightarrow b. \Rightarrow c. \Rightarrow d. \Rightarrow a.$$

It is obvious that a. implies b. If b. holds then there is a nonempty open ball $\mathcal{B}_\rho(x_0)$ around $x_0$ such that $T$ is bounded over $\mathcal{D}(T) \cap \mathcal{B}_\rho(x_0)$. In this case, since $T$ is boundedly demiclosed (property proved independently in Theorem 9.6), $\mathcal{B}_\rho(x_0)$ is contained in $\mathcal{D}(T)$ is dense in $\mathcal{H}$, and for any point $x_1$ in it the convex hull of $\{x_1\} \cup \mathcal{B}_\rho(x_0)$ is contained in $\mathcal{D}(T)$ and contains the origin as an interior point. Thus $T$ is defined at all points of an open ball about the origin and as $\mathcal{D}(T)$ is a cone with vertex at the origin, everywhere in $\mathcal{H}$. Hence b. implies c.
By (9.9) d. $\Rightarrow$ a. In consequence it only remains to

prove that c. $\Rightarrow$ d. So, suppose that T is everywhere defined and assume for contradiction that the support of $\{P_\lambda\}_{-\infty}^{+\infty}$ were not compact, or equivalently, that there were a point x with non-compact support (Lemma 7.5). This means that a sequence of nonoverlaping intervals $\{[a_n, b_n]_1^\infty\}$ can be found with $a_n \uparrow \infty$ and such that $x_n = P_{(a_n, b_n]} x \neq 0$, $n = 1, 2, \ldots$ . Now take a subsequence $\{a_{n_k}\}_1^\infty$ such that $\sum_{k=1}^\infty 1/a_{n_k}^2 < +\infty$, and with it construct the point

$$y = \sum_{k=1}^\infty \frac{1}{a_{n_k}} \frac{x_{n_k}}{\|x_{n_k}\|} \ ;$$

the series converges by the orthogonality of $x_{n_k}$'s and by the choice of the $a_{n_k}$'s. The support of y is confined to $\bigcup_{k=1}^\infty [a_{n_k}, b_{n_k}]$, and $P_{(a_{n_k}, b_{n_k}]} y = \frac{1}{a_{n_k}} \frac{x_{n_k}}{\|x_{n_k}\|}$. Moreover, for for any choice of the positive integer h,

$$P_{b_{n_h}} T y = \int_{[0, b_{n_h}]} \lambda \, d P_\lambda y$$

and so

$$\|Ty\|^2 \geq \|P_{b_{n_h}} Ty\|^2 \geq \int_{(a_{n_1}, b_{n_h}]} \lambda^2 \, d\langle P_\lambda y, y\rangle \geq \sum_{k=1}^n \int_{(a_{n_k}, b_{n_k}]} \lambda^2 \, d\langle P_\lambda y, y\rangle$$

$$\geq \sum_{k=1}^h a_{n_k}^2 \int_{(a_{n_k}, b_{n_k}]} d \langle P_\lambda y, y\rangle$$

$$= \sum_{k=1}^h a_{n_k}^2 \langle P_{(a_{n_k}, b_{n_k}]} y, y\rangle = \sum_{k=1}^h a_{n_k}^2 \|P_{(a_{n_k}, b_{n_k}]} y\|^2 = \sum_{k=1}^h a_{n_k}^2 / a_{n_k}^2 = h ,$$

which is impossible. Hence $\{P_\lambda\}_{-\infty}^{+\infty}$ must have a compact support. Q. E. D.

The implication c. ⟹ b. corresponds to the Hellinger and Toplitz' theorem for linear selfadjoint operators.

**Theorem 9.5** The range of $T = \int_0^\infty \lambda \, dP_\lambda$ is the whole space if and only if $\lambda=0$ does not belong to the support of $\{P_\lambda\}_{-\infty}^{+\infty}$.

**Proof:** Set $R = \int_0^\infty \lambda^{-1} dP_\lambda$. Then by Lemma 9.1 with $u(\lambda)$, $v(\lambda)$ and $w(\lambda)$ identified with $\lambda, \lambda^{-1}$ and 1 respectively,

$$\mathcal{D}(TR) = \mathcal{D}(R) \cap \mathcal{D}(I) = \mathcal{D}(R), \qquad TR \subset I.$$

If $\lambda = 0$ does not belong to the support of the spectral measure, $\lambda^{-1}$ is a bounded function on this support and by (9.9) R is Lipschitzian and hence everywhere defined (Theorem 4.4). Therefore $TR = I$, and the range of T is H. Conversely, if the range of T is H, we have, again by Lemma 4.1,

$$\mathcal{D}(RT) = \mathcal{D}(T) \cap \mathcal{D}(I) = \mathcal{D}(T), \qquad RT \subset I,$$

and R is everywhere defined because it is defined on the range of T. But then $\lambda^{-1}$ must be bounded on the support of the spectral measure (Lemmas 9.1, 9.4). This is another way of saying that $\lambda = 0$ does not belong to the support of $\{P_\lambda\}_{-\infty}^{+\infty}$. Notice that in this case $R = T^{-1}$.

**Theorem 9.6** Any spectral integral operator $T = \int_0^\infty \lambda \, dP_\lambda$ is boundedly demiclosed, i.e.,

(9.27) $\{x_\alpha \to x, Tx_\alpha \to y, \overline{\lim} \|Tx_\alpha\| < +\infty\} \Longrightarrow \{x \in \mathcal{D}(T), y=Tx\}$.

**Proof:** For any $\lambda$, the operator $TP_\lambda = \int_0^\lambda \mu \, dP_\mu$, is continuous (Theorem 9.1). So, since $P_\lambda Tx_\alpha = TP_\lambda x_\alpha$,

$$\{x_\alpha \to x\} \Longrightarrow \{P_\lambda x_\alpha \to P_\lambda x\}$$
$$\Longrightarrow \{P_\lambda Tx_\alpha = TP_\lambda x_\alpha = TP_\lambda(P_\lambda x_\alpha) \to TP_\lambda P_\lambda x = TP_\lambda x\}$$

and

$$\{x_\alpha \to x, \; Tx_\alpha \rightharpoonup y, \; \overline{\lim} \|Tx_\alpha\| < \infty\}$$
$$\Rightarrow \{Tx_\alpha \rightharpoonup y, \; \overline{\lim} \|Tx_\alpha\| < \infty, \; P_\lambda Tx_\alpha \to TP_\lambda x\}.$$

Hence by Lemma 1.4, $P_\lambda y = TP_\lambda x$, $\forall \lambda \geq 0$, and

$$\|TP_\lambda x\|^2 = \|\int_0^\lambda \mu \, dP_\mu x\|^2 = \int_0^\lambda \mu^2 \, d\langle P_\mu x, x\rangle = \|P_\lambda y\|^2 \leq \|y\|^2,$$

whence if $\lambda \to +\infty$,

$$\int_0^\infty \mu^2 \, d\langle P_\mu x, x\rangle \leq \|y\|^2.$$

Thus, $x \in \mathfrak{D}(T)$. But then $Tx$ being defined $TP_\lambda x = P_\lambda Tx$, and from the above, $P_\lambda y = P_\lambda Tx$, $\forall \lambda \geq 0$. Letting $\lambda \to \infty$ one gets $y = Tx$.  Q.E.D.

Linear selfadjoint mappings are actually demiclosed. We do not know if in the actual case boundedness can be dropped.

<u>Theorem 9.7</u>   Any spectral integral operator $T = \int_0^\infty \lambda \, dP_\lambda$ is hemicontinuous, i.e., it is demicontinuous (strong-weak continuous) over any closed one dimensional segment in its domain.

<u>Proof:</u>   Let $I = \{tx_1 + (1-t)x_2\}_{0 < t < 1}$ be one such segment. Let us check first that $T$ is bounded over $I$. For any $\lambda \geq 0$, $TP_\lambda$ is a continuous everywhere defined spectral integral operator, whose value at any point in $\mathfrak{D}(T)$ converges to the value of $T$ as $\lambda \to \infty$. So if $x(t) = tx_1 + (1-t)x_2$,

$$\|Tx(t)\|^2 = \lim_{\lambda \to \infty} \langle TP_\lambda x(t), TP_\lambda x(t)\rangle$$

and by (9.16)

$$\|Tx(t)\|^2 = \lim_{\lambda \to \infty} \langle x(t), (TP_\lambda)^2 x(t)\rangle,$$

but the pseudoquadratic form $\langle x, (TP_\lambda)^2 x \rangle$ associated with the spectral integral operator $(TP_\lambda)^2$ is convex by (9.6), and in consequence

$$\|T x(t)\|^2$$

$$= \lim_{\lambda \to \infty} \langle x(t), (TP_\lambda)^2 x(t) \rangle \leq \overline{\lim_{\lambda \to \infty}} [t \langle x_1, (TP_\lambda)^2 x_1 \rangle + (1-t) \langle x_2, (TP_\lambda)^2 x_2 \rangle]$$

$$= \overline{\lim_{\lambda \to \infty}} [t \|T P_\lambda x_1\|^2 + (1-t) \|T P_\lambda x_2\|^2] = t \|T x_1\|^2 + (1-t) \|T x_2\|^2$$

$$\leq \max\{\|T x_1\|^2, \|T x_2\|^2\},$$

proving boundedness. Now if $x_n \to x$, $x_n \in I$, the sequence $\{T x_n\}_1^\infty$ is bounded and by the demiclosedness of $T$ all its weak limits coincide with $Tx$. In other terms $T x_n \rightharpoonup Tx$, and the proof concludes.

The same type of argument allows one to prove that $T$ is demicontinuous over any finite dimensional relatively open convex set in $\mathfrak{D}(T)$.

**Theorem 9.8**  The function $\tau(x) = \frac{1}{2} \int_0^\infty \lambda\, d \langle P_\lambda x, x \rangle$ from $\mathcal{H}$ into the extended nonnegative real line $[0, +\infty]$ is a lower semicontinuous convex function. Its subdifferential $\partial \tau(x)$ at a point $x$ is nonempty if and only if $Tx = \int_0^\infty \lambda\, dP_\lambda x$ exists, and in such case $Tx$ is the only subgradient of $\tau$ at $x$. Moreover, over $\mathfrak{D}(T)$, $\tau(x) = \frac{1}{2}\langle x, Tx \rangle$, $\tau$ is Gateaux differentiable, and $Tx = \nabla \frac{1}{2} \langle x, Tx \rangle$.

**Proof:** Clearly $\tau(x) = \lim_{M \to \infty} \tau_M(x)$, where

$$\tau_M(x) = \frac{1}{2} \langle x, TP_M x \rangle = \frac{1}{2} \int_0^M \lambda\, d \langle P_\lambda x, x \rangle.$$

So since the $\tau_M$'s are everywhere defined, continuous and convex, and converge to $\tau(x)$ nondecreasingly, $\tau(x)$ is a lower semicontinuous convex function.

Let us see first that, if defined, $Tx$ is a subgradient of $\tau(x)$. In such case $\tau(x) = \frac{1}{2}\langle x, Tx\rangle$, and if $y$ is any other point in $\mathfrak{g}(T)$, the identity

(9.28) $\langle y, Ty\rangle - \langle x, Tx\rangle - 2\langle Tx, y-x\rangle = \langle Ty-Tx, y-x\rangle + [\langle Ty, x\rangle - \langle y, Tx\rangle]$,

along with the inequality (cf. (9.5)),

$$|\langle Ty, x\rangle - \langle y, Tx\rangle| \leq \langle Ty-Tx, y-x\rangle$$

yield

(9.29) $|\frac{1}{2}\langle y, Ty\rangle - \frac{1}{2}\langle x, Tx\rangle - \langle Tx, x-y\rangle| \leq \langle Ty - Tx, y-x\rangle$.

Now, if $y$ is restricted to a one dimensional convex set in $\mathfrak{g}(T)$ containing $x$,

$$\langle Ty-Tx, y-x\rangle = o(\|y-x\|)$$

by the hemicontinuity of $T$ (Theorem 9.7). So, relatively to $\mathfrak{g}(T)$, $Tx$ is the Gateaux differential of $\tau$ at $x$. In particular, $Tx$ is a subgradient, that is,

$$0 \leq \frac{1}{2}\langle y, Ty\rangle - \frac{1}{2}\langle x, Tx\rangle - \langle Tx, y-x\rangle,$$

relation which also may be checked directly. Now, if $z$ is any point in space, and we apply the above to $P_\lambda z$, letting then $\lambda \to \infty$, we get

$$0 \leq \tau(z) - \tau(x) - \langle Tx, z-x\rangle, \quad \forall z \in \mathcal{H},$$

which says that $Tx$ is a subgradient of $\tau$ at $x$, namely the only one because any subgradient must be, a fortiori, a subgradient relatively to $\mathfrak{g}(T)$, and as such must coincide with the existing gradient there, since $\mathfrak{g}(T)$ is dense in $\mathcal{H}$.

Now we turn our attention to the converse question, namely, that of showing that whenever $\tau$ has a subgradient $T$ is defined. Let us consider the linear space $\mathfrak{m}$ of all points in $\mathcal{H}$ of the form

$$y = \int_0^\infty f(\lambda) \, dP_\lambda x$$

where $f(\lambda)$ is any measurable function, square integrable with regard to $\langle P_\lambda x, x \rangle$. Letting $f^+$ and $f^-$ denote the positive and negative part of $f$ respectively, one has

$$\|y\|^2 =$$
$$= \left\| \int_0^\infty f^+(\lambda) \, dP_\lambda x \right\|^2 + \left\| \int_0^\infty f^-(\lambda) \, dP_\lambda x \right\|^2$$
$$- 2 \langle \int_0^\infty f^+(\lambda) \, dP_\lambda x , \int_0^\infty f^-(\lambda) \, dP_\lambda x \rangle$$
$$= \int_0^\infty (f^+(\lambda))^2 \, d\langle P_\lambda x, x \rangle + \int_0^\infty (f^-(\lambda))^2 \, d\langle P_\lambda x, x \rangle = \int (f(\lambda))^2 \, d\langle P_\lambda x, x \rangle.$$

This shows that $\mathfrak{m}$ is isometrically isomorphic to $L^2([0, +\infty], \mu_x)$, and as such is a closed subspace in $\mathcal{H}$. Hence any $z \in \mathcal{H}$ can be uniquely written in the form

$$z = \int_0^\infty g(\lambda) \, dP_\lambda x + z^\perp \quad , \quad g \in L^2([0, +\infty], \mu_x), \quad z^\perp \perp \mathfrak{m}.$$

Assume now that $z$ is a subgradient of $\tau$ at $x$, that is, that

$$\tau(x+y) - \tau(x) \geq \langle y, z \rangle , \qquad \forall \, y \in \mathcal{H} .$$

Naturally for this to make sense it is necessary that $\tau(x) < +\infty$. Then take for $y$ any vector of the form

$$y = t \int_0^\infty f(\lambda) \, dP_\lambda x ,$$

with $f(\lambda)$ bounded, and $t$ positive. Clearly

$$x + y = \int_0^\infty (1 + t f(\lambda)) \, dP_\lambda x ,$$

and, since for t sufficiently small $1 + t f(\lambda)$ is a positive function,

$$\tau(x+y) - \tau(x) = \tfrac{1}{2}\int_0^\infty \lambda \, d\,\langle P_\lambda(x+y), x+y\rangle - \tfrac{1}{2}\int_0^\infty \lambda \, d\langle P_\lambda x, x\rangle$$

$$= \tfrac{1}{2}\int_0^\infty \lambda[(1+tf(\lambda))^2 - 1]d\langle P_\lambda x, x\rangle = t\int_0^\infty \lambda f(\lambda) d\langle P_\lambda x, x\rangle + \frac{t^2}{2}\int_0^\infty \lambda f^2(\lambda) d\langle P_\lambda x, x\rangle.$$

On the other hand

$$\langle y, z\rangle = \langle \int_0^\infty t f(\lambda) d P_\lambda x, \int_0^\infty g(\lambda) d P_\lambda x + z^\perp\rangle = t\int_0^\infty f(\lambda)g(\lambda) d\langle P_\lambda x, x\rangle,$$

and therefore, for small t,

$$t\int_0^\infty \lambda f(\lambda) d\langle P_\lambda x, x\rangle + \frac{t^2}{2}\int_0^\infty \lambda f^2(\lambda) d\langle P_\lambda x, x\rangle \geq t\int_0^\infty f(\lambda)g(\lambda) d\langle P_\lambda x, x\rangle,$$

whence dividing by t and letting $t \downarrow 0$,

$$\int_0^\infty \lambda f(\lambda) d\langle P_\lambda x, x\rangle \geq \int_0^\infty f(\lambda)g(\lambda) d\langle P_\lambda x, x\rangle.$$

Since this is valid for any bounded measurable $f(\lambda)$, one must have $g(\lambda) = \lambda$ almost everywhere. But then $\lambda$ is square integrable and Tx exists. Q. E. D.

Recalling that the subdifferential mapping $x \to \partial\tau(x)$ associated to a lower semicontinuous convex function is maximal monotone [21] one obtains the important corollary.

<u>Corollary</u>  Any spectral integral operator $T = \int_0^\infty \lambda \, d P_\lambda$ is maximal monotone.

<u>Theorem 9.9</u>  A spectral integral operator $T = \int_0^{+\infty} \lambda \, d P_\lambda$ is compact if and only if

a. The range of the spectral measure of any point on the positive real axis is locally compact,

b. The support of the spectral resolution is either a finite set or a bounded countable one having 0 as its only limit point.

**Proof:** Let us prove sufficiency first. If a. and b. are satisfied, then T is either a finite linear combination of projections on locally compact convex cones or an infinite one with coefficients converging to zero. In the former case T is compact because each of the projections is compact. To deal with the remaining case let us write

$$T = \sum_{k=1}^{\infty} \lambda_k P_{\{\lambda_k\}}, \quad \lambda_k \downarrow 0, \quad \Re(P_{\{\lambda_k\}}) \text{ locally compact,}$$

and let us show that for any bounded sequence $\{x_n\}_1^{\infty}$ there is a subsequence $\{x_{n_k}\}_1^{\infty}$ such that $\{Tx_{n_k}\}_1^{\infty}$ converges strongly. We pick the subsequence $\{x_{n_k}\}_1^{\infty}$ so that $P_{\{\lambda_k\}} x_{n_h} \xrightarrow[h\to\infty]{} y_k \in \Re(P_{\{\lambda_k\}})$, $k = 1, 2, \ldots$ . The $y_k$'s are mutually orthogonal, and since

$$\|\sum_{k=1}^{n} \lambda_k y_k\|^2 = \sum_{k=1}^{n} \lambda_k^2 \|y_k\|^2 \le \lambda_1^2 \sum_{k=1}^{n} \|y_k\|^2 = \lambda_1^2 \lim_{h\to\infty} \sum_{k=1}^{n} \|P_{\{\lambda_k\}} x_{n_h}\|^2$$

$$\le \lambda_1^2 \overline{\lim_{h\to\infty}} \|x_{n_h}\|^2 \le \lambda_1^2 M^2,$$

the series $\sum_{k=1}^{\infty} \lambda_k y_k$ converges strongly to a limit $y$ in $H$. Now, we write $y - Tx_{n_h} = P_\lambda y - P_\lambda Tx_{n_h} + (I - P_\lambda)y - (I - P_\lambda)Tx_{n_h}$.
Taking absolute values it follows by the triangle inequality

$$\|y - Tx_{n_h}\| \le \|P_\lambda y\| + \|P_\lambda Tx_{n_h}\| + \|(I - P_\lambda)y - (I - P_\lambda)Tx_{n_h}\|,$$

and since $\|P_\lambda Tx_{n_h}\| \le \lambda \|x_{n_h}\| \le \lambda M$, and

$$(I - P_\lambda)y - (I - P_\lambda)Tx_{n_h} = \sum_{\lambda_k > \lambda} (y_k - P_{\{\lambda_k\}} Tx_{n_h}),$$

SPECTRAL THEORY

$$\varlimsup_{h\to\infty} \|y-Tx_{n_h}\| \leq \|P_\lambda y\| + \lambda M + \lim_{h\to\infty} \sum_{\lambda_k > \lambda} \|y_k - P_{\{\lambda_k\}} Tx_{n_h}\| = \|P_\lambda y\| + \lambda M.$$

Finally note that since $\|P_\lambda y\| = \|\sum_{\lambda_k \leq \lambda} \lambda_k y_k\|$, $P_\lambda y \to 0$ as $\lambda \to 0$. Thus the right hand member in the last inequality above can be made arbitrarily small by choice of $\lambda$, and $\lim_{n\to\infty} \|y-Tx_{n_h}\| = 0$, that is, $y = \lim_{h\to\infty} Tx_{n_h}$, concluding the proof of sufficiency.

We prove necessity by showing that if one of the conditions a., b. is not satisfied then $T$ is not compact. If a. does not hold then there is a $\mu > 0$, such that $\Re(P_{\{\mu\}})$ is not locally compact. Hence the bounded set $(\Re(P_{\{\mu\}})) \cap \bar{B}_1(0)$ is not compact, and neither is its image by $T$, which is merely an homothetic image with factor $\mu$. Thus in this case $T$ is not compact.

If b. is not met by $T$ then there is a countable family $\{(a_n, b_n)\}_1^\infty$ of disjoint open intervals in a half line $[\lambda_0, +\infty]$, $\lambda_0 > 0$ of lengths $b_n - a_n$ tending to zero and non-vanishing spectral measures $P_{(a_n, b_n)}$. If a subsequence of $\{a_n\}_1^\infty$, say $\{a_{n_k}\}_1^\infty$, converges to $+\infty$ then any sequence $\{x_{n_k}\}_1^\infty$ obtained by picking a unit vector $x_{n_k}$ in the range of $P_{(a_{n_k}, b_{n_k})}$ for each $k$ is transformed by $T$ into an unbounded sequence (since $\|Tx_{n_k}\| \geq a_{n_k} \|x_{n_k}\| = a_{n_k}$), and hence certainly into a nonconvergent one. Therefore $T$ is not compact. If no such sequence exists, there must exist a subsequence $\{a_{n_k}\}_1^{+\infty}$ either increasing or decreasing converging to a positive number $a$. We claim that, if as before, $x_{n_k}$ is a unit vector in the range of $P_{(a_{n_k}, b_{n_k})}$, $k = 1, 2, \ldots$ then the sequence $\{Tx_{n_k}\}_1^\infty$ is not convergent. Consider the case when $a_{n_k} \uparrow a$. Then for given $\varepsilon > 0$

407

and sufficiently large $k$, $(a_{n_k}, b_{n_k}) \subset (a-\varepsilon, a)$, and for such $k$'s, $x_{n_k} \in \Re(P_{(a-\varepsilon, a)})$, $Tx_{n_k} \in \Re(P_{(a-\varepsilon, a)})$. But then if $\{Tx_{n_k}\}_1^\infty$ had a limit $y$, $y \in \Re(P_{(a-\varepsilon, a)})$, $\forall \varepsilon > 0$, and

$$y \in \bigcap_{\varepsilon > 0} \Re(P_{(a-\varepsilon, a)}) = \Re(\bigcap_{\varepsilon > 0} P_{(a-\varepsilon, a)}) = \Re(P_{\bigcap_{\varepsilon > 0} (a-\varepsilon, a)})$$

$$= \Re(P_\emptyset) = \{0\},$$

that is $y = 0$, in contradiction with

$$\|y\| = \lim_{k \to \infty} \|Tx_{n_k}\| \geq \lim_{k \to \infty} a_{n_k} \|x_{n_k}\| = a > 0.$$

So again $T$ is not compact. The proof for the case $a_{n_k} \downarrow a$ is entirely similar and is left to the reader.

This theorem is analogous to the corresponding one for selfadjoint positive operators. The outstanding new feature is the existence of eigenvalues of infinite multiplicity corresponding to projections on infinite dimensional locally compact cones. Such cones can be generated by projecting from the origin infinite dimensional compact convex sets, such as the Hilbert cube, located on a hyperplane not passing through the origin.

We close this section with a discussion of the "spectrum" of a spectral integral operator. The classification of points of the real line either by means of spectral resolution continuity properties ($P_\lambda$ is locally constant, varies continuously, or jumps) or by the existence and nature of the inverse of $T-\lambda T$ ($(T-\lambda I)^{-1}$ exists and is bounded, exists and is unbounded or does not exist), leads to the same result in the case of linear operators. This is not so for nonlinear ones and to restore the equivalence, in addition to continuity properties, the nature of $P_\lambda$ has to be taken into account, namely whether it is linear or not. It all appears as if the operation of taken the inverse of $T-\lambda I$ is too tied up to linearity and that its consideration for non-linear operators is somewhat artificial as it imposes upon them unnatural linear constraints.

As for linear operators <u>a number $\lambda_0$ is said to be an eigenvalue of the operator</u> $T = \int_0^\infty \lambda \, dP_\lambda$ <u>if there is a vector</u> $x \neq 0$ <u>such that</u>

(9.30) $\qquad T x = \lambda_0 x$ ;

in this case $x$ is said to be an eigenvector corresponding to the eigenvalue $\lambda_0$. Applying $P_{(-\infty, \lambda_0)}$ and $P_{(\lambda_0, +\infty)}$ to (9.30), and multiplying scalarly by $x$,

$$\int_{(-\infty, \lambda_0)} \lambda \, d\langle P_\lambda x, x\rangle = \int_{(-\infty, \lambda_0)} \lambda_0 \, d\langle P_\lambda x, x\rangle = \lambda_0 \| P_{(-\infty, \lambda_0)} x \|^2 \, ;$$

$$\int_{(\lambda_0, +\infty)} \lambda \, d\langle P_\lambda x, x\rangle = \int_{(\lambda_0, +\infty)} \lambda_0 \, d\langle P_\lambda x, x\rangle = \lambda_0 \| P_{(\lambda_0, +\infty)} x \|^2 \, .$$

From the first of these two equations we deduce $P_{(-\infty, \lambda_0)} x = 0$ because $\lambda < \lambda_0$ over $(-\infty, \lambda_0)$, and from the second $P_{(\lambda_0, +\infty)} x = 0$, since $\lambda > \lambda_0$ on $(\lambda_0, +\infty)$. Hence the support of $x$ is confined to the point $\lambda_0$. Since (9.30) clearly holds whenever $\{\lambda_0\}$ is the support of $x$, we have the following equivalence

(9.31) $\qquad \{T x = \lambda_0 x\} \iff \{x \in \mathcal{R}(P_{\lambda_0})\}$ .

In other terms, <u>a point $\lambda_0$ is an eigenvalue if and only if it has a nonvanishing spectral measure</u> $P_{\{\lambda_0\}}$; <u>the corresponding set of eigenvectors is the range</u> $\mathcal{R}(P_{\{\lambda_0\}})$ <u>of its measure.</u>

The situation concerning the relations between the behaviour of the spectral measure and that of $(T-\lambda T)^{-1}$ is not completely clear yet. The chart below summarizes the state of our knowledge. The space $\mathcal{H}$ is assumed to be separable and $T$ denotes an operator of the form $\int_0^\infty \lambda \, dP_\lambda$.

$P_{\{\lambda_0\}} \neq 0 \Longrightarrow \lambda_0$ is an eigenvalue and $(T-\lambda_0 I)^{-1}$ does not exist.

$P_{\{\lambda_0\}} = 0$ 
$\begin{cases} \lambda_0 \in \text{support}\{P_\lambda\}_{-\infty}^{+\infty} \begin{cases} P_{\lambda_0} \text{ nonlinear} \Longrightarrow \text{ existence or nonexistence of } (T-\lambda I)^{-1} \text{ unknown.} \\ P_{\lambda_0} \text{ linear} \Longrightarrow (T-\lambda T)^{-1} \text{ exists, is densely defined, boundedly demiclosed but not Lipschitzian.} \end{cases} \\ \lambda_0 \notin \text{support}\{P_\lambda\}_{-\infty}^{+\infty} \begin{cases} P_{\lambda_0} \text{ nonlinear} \Longrightarrow (T-\lambda I)^{-1} \text{ does not exist.} \\ P_{\lambda_0} \text{ linear} \Longrightarrow (T-\lambda I)^{-1} \text{ exists, is everywhere defined and Lipschitzian.} \end{cases} \end{cases}$

The case $P_{\{\lambda_0\}} \neq 0$ has already been discussed, so let us assume that $P_{\{\lambda_0\}} = 0$. In this case $P_{\lambda_0} = P_{(-\infty, \lambda_0)}$, $P_{\lambda_0}^\perp = I - P_{(\lambda_0, +\infty)}$, and so if $C_{\lambda_0} = \mathcal{R}(P_{(-\infty, \lambda_0)})$ then $C_{\lambda_0}^\perp = \mathcal{R}(P_{(\lambda_0, +\infty)})$. Now if $P_{\lambda_0}$ is nonlinear by Lemma 2.11, there is a nonvanishing vector $x_0 \in (-(C_{\lambda_0})^0) \cap (+(C_{\lambda_0}^\perp)^0)$. If in addition $\lambda_0$ does not belong to the support of the spectral measure, it is at a positive distance from it and the integrals $\int_{(-\infty, \lambda_0)} (\lambda_0 - \lambda)^{-1} dP_\lambda$ and $\int_{(\lambda_0, +\infty)} (\lambda - \lambda_0)^{-1} dP_\lambda$

are everywhere defined and continuous (cf. (9.23). Set

$$y_0 = \int_{(-\infty, \lambda_0)} (\lambda_0 - \lambda)^{-1} dP_\lambda(-x_0),$$

$$z_0 = \int_{(\lambda_0, +\infty)} (\lambda - \lambda_0)^{-1} dP_\lambda(+x_0),$$

and proceed to calculate the value of $T - \lambda_0 I$ at these points. One has, since the support $y_0$ is contained in $(-\infty, \lambda_0)$,

$$(T - \lambda_0 I) y_0 = -\int_{(-\infty, \lambda_0)} (\lambda_0 - \lambda) dP_\lambda y_0 + \int_{(\lambda_0, +\infty)} (\lambda - \lambda_0) dP_\lambda y_0 = -\int_{(-\infty, \lambda_0)} (\lambda_0 - \lambda) dP_\lambda y_0$$

$$= -\int_{(-\infty, \lambda_0)} (\lambda_0 - \lambda) d \int_{(-\infty, \lambda)} (\lambda_0 - \mu)^{-1} dP_\mu(-x_0) = -\int_{(-\infty, \lambda_0)} dP(-x_0) = -P(-x_0) = x_0,$$

and similarly, since support of $z_0 \subset (\lambda_0, +\infty)$,

$$(T - \lambda_0 I) z_0 = \int_{(\lambda_0, +\infty)} (\lambda - \lambda_0) dP_\lambda z_0 = P_{(\lambda_0, +\infty)} x_0 = x_0.$$

Hence, since $y_0 \neq z_0$, $(T - \lambda_0 I)^{-1}$ does not exist. This argument fails if $\lambda_0 \in$ support $\{P_\lambda\}_{-\infty}^{+\infty}$ because the existence of $y_0$ and $z_0$ cannot be assured, as the integrals defining them might not converge. All one can do is to replace $y_0$ and $z_0$ by $y_\varepsilon = \int_{(-\infty, \lambda_0 - \varepsilon)} (\lambda_0 - \lambda)^{-1} dP_\lambda(-x_0)$ and $z_\varepsilon = \int_{(\lambda_0 + \varepsilon, +\infty)} (\lambda - \lambda_0)^{-1} dP_\lambda x_0$ respectively and conclude that

$$\lim_{\varepsilon \to 0} (T - \lambda_0 I) y_\varepsilon = \lim_{\varepsilon \to 0} (T - \lambda_0 I) z_\varepsilon = x_0,$$

$$\lim_{\varepsilon \to 0} \|y_\varepsilon - z_\varepsilon\| \geq \overline{\lim_{\varepsilon \to 0}} (\|y_\varepsilon\|^2 + \|z_\varepsilon\|^2)^{\frac{1}{2}} > 0.$$

This, while not proving nonexistence of $(T - \lambda, I)^{-1}$, shows at least that if $(T - \lambda_0 I)^{-1}$ exists it must be quite discontinuous.

Finally, if $P_{\{\lambda_0\}} = 0$ and $P_{\lambda_0}$ is linear, then $C_{\lambda_0}$ $C_{\lambda_0}^\perp$ are orthogonal subspaces and $\mu$ splits into their direct sum. Moreover, $T-\lambda_0 I$ is reduced by this direct sum, that is, $T-\lambda_0 I$ is equal to the direct sum of its restrictions to $C_{\lambda_0}$ and $C_{\lambda_0}^\perp$. It is evident then that $(T-\lambda_0 I)^{-1}$, if existing, must be the operator

$$x \to \int_{(-\infty, \lambda_0)} (\lambda_0 - \lambda)^{-1} d P_\lambda(-x) + \int_{(\lambda_0, +\infty)} (\lambda - \lambda_0)^{-1} P_\lambda x .$$

That this is so can be checked easily by direct computation. If $\lambda_0$ is not in the support of the spectral measure, then $(T-\lambda_0 I)^{-1}$ is a Lipschitz mapping because both integral operators defining it have this property, whereas if $\lambda_0$ is a point of the support, then $(T-\lambda_0 I)^{-1}$ is not Lipschitzian because at least one of the integrals fails to have this property.

§10. Spectral analysis.

In this section we are confronted with the question, inverse to that dealt with in the last two sections, namely, that of finding for a given operator a spectral resolution out of which it can be reconstructed by synthesis. Naturally, the first step is to determine which operators can be analyzed in this manner, that is, to establish necessary and sufficient conditions for an operator to be expressable as a spectral integral; one may then proceed to decompose the operator into its elementary components, and find its spectral resolution. However, it is in the process of proving sufficiency that the spectral resolution is constructed, and so the second step is in fact contained in the first. Here we shall restrict ourselves to characterizing operators of the form

$$T = \int_0^\infty \lambda \, dP_\lambda ,$$

where $\{P_\lambda\}_{-\infty}^{+\infty}$ is a conical spectral resolution with support in the nonnegative real line $\mathbb{R}^+ = [0, +\infty)$.

As in the linear theory this is done in two steps: First the bounded case is treated (in the present context boundedness meaning that the operator is Lipschitzian), and then using it as a stepping stone one proceeds to the general case.

When dealing with nonlinear operators it is necessary to specify clearly what is meant by "a polynomial p(T) of an operator T", since the result may very well depend on the manner the polynomial $p(\lambda)$ is given. To dispell all possible ambiguities we state explicitly the following definition, already used in §9:

<u>Definition 10.1</u>  If $T: H \to H$ and $p(\lambda) = a_0 + a_1\lambda + \ldots + a_n\lambda^n$ then
$$p(T) = a_0 I + a_1 T + a_2 T^2 + \ldots + a_n T^n .$$

Clearly, $\mathcal{D}(p(T))$ is the domain of the highest power of T effectively present in p(T).

<u>Theorem 10.1</u>  Any everywhere defined monotone Lipschitz mapping $T = H \to H$, satisfying

(10.1)  $\quad p(T) q(T) = q(T) p(T)$,

(10.2)  $\quad \langle p(T)x, q(T)x \rangle = \langle x, p(T) q(T) x \rangle , \quad \forall x \in H$,

(10.3)  $\quad \| p(T)x - p(T)y \|^2 \leq \langle p^2(T)x - p^2(T)y, x-y \rangle , \quad \forall x, y \in H$,

for every couple of polynomials $p(\lambda)$ and $q(\lambda)$ nonnegative on $\mathbb{R}^+$, admits a representation of the form

(10.4)  $\quad T = \int_0^\infty \lambda \, dP_\lambda$,

in terms of a conical partition of the identity $\{P_\lambda\}_{-\infty}^{+\infty}$ with compact support in $\mathbb{R}^+$, and conversely. The spectral resolution is uniquely determined by T.

<u>Proof:</u>  For any $x \in H$, let

$$C^{(x)} = \{y \mid y = p(T)x, \forall p(\lambda), p(\lambda) \geq 0 \text{ for } \lambda \geq 0 \} .$$

This is a convex cone with vertex at the origin invariant under $T$, i.e., $T\, C^{(x)} \subset C^{(x)}$. Moreover, if $y_i \in C^{(x)}$, $i = 1, 2, \ldots, n$ and $\sum_{i=1}^{n} \alpha_i y_i \in C^{(x)}$, then

(10.5) $$T(\sum_{i=1}^{n} \alpha_i y_i) = \sum_{i=1}^{n} \alpha_i T y_i.$$

In other words, $T$ is linear over $C^{(x)}$. This perhaps needs a proof: Assume first that the $\alpha_i$'s are nonnegative. By hypothesis there are $n$ polynomials $p_i(\lambda)$, $i = 1, 2, \ldots, n$, nonnegative on $\mathbb{R}^+$, such that $y_i = p_i(T)x$, $i = 1, 2, \ldots, n$. So, by (10.1),

$$T y_i = T p_i(T)x = p_i(T)Tx$$

and

$$\sum_{i=1}^{n} \alpha_i T y_i = (\sum_{i=1}^{n} \alpha_i p_i(T))Tx = T(\sum_{i=1}^{n} \alpha_i p_i(T))x = T(\sum_{i=1}^{n} \alpha_i y_i),$$

proving our claim in the present case. In the general case, write $y_0 = \sum_{i=1}^{n} \alpha_i y_i$, ($y_0 \in C^{(x)}$), transfer the sum over the negative $\alpha_i$'s to the left hand member and apply $T$ to both sides. By the case just discussed $T$ can be distributed and the sought result follows upon transference back to the right. Moreover, from (10.2) we obtain,

(10.6) $$\langle T y_1, y_2 \rangle = \langle y_1, T y_2 \rangle, \quad \forall y_1, y_2 \in C^{(x)},$$

meaning that $T$ acts on $C^{(x)}$ as a linear symmetric operator. Now, if we set

$$T^{(x)}(z_1 - z_2) = T z_1 - T z_2, \quad z_1, z_2 \in C^{(x)},$$

then $T^{(x)}$ is a well defined symmetric linear extension to the linear subspace

## SPECTRAL THEORY

$$m^{(x)} = c^{(x)} - c^{(x)}$$

of the restriction of $T$ to $c^{(x)}$. In fact, it is easily verified with the assistance of (10.5) that the value assigned by $T^{(x)}$ to a point $z \in m^{(x)}$ is independent of the manner $z$ is expressed as the difference of two points in $c^{(x)}$, and so that $T^{(x)}$ is well defined. Its linearity and symmetry result from the linearity and symmetry of $T$ on $c^{(x)}$. Moreover, $T$ being monotone and Lipschitzian so is $T^{(x)}$. Otherwise said, $T^{(x)}$ is a nonnegative bounded operator. Under these conditions $T(x)$ can be extended by continuity to a selfadjoint mapping defined on $\overline{m^{(x)}}$, having the same bounds. Then, by the spectral theorem for bounded selfadjoint operators [20, no. 107], there is a spectral resolution $\{P_\lambda^{(x)}\}_{-\infty}^{\infty}$ consisting of linear projections in $\overline{m^{(x)}}$ with compact support in $[0, +\infty)$, such that

$$(10.7) \qquad T^{(x)} z = \int_{-\infty}^{+\infty} \lambda \, d P_\lambda^{(x)} z \, , \qquad \forall z \in \overline{m^{(x)}},$$

(for simplicity we denote again by $T^{(x)}$ the extension of $T^{(x)}$ to $\overline{m^{(x)}}$). Now, if $\{p_{\lambda_0}^{(n)}(\lambda)\}$ is a sequence of polynomials, nonnegative for nonnegative $\lambda$'s, converging in decreasing manner over any compact set to the characteristic function $\chi_{(-\infty, \lambda_0]}$ of the half line $(-\infty, \lambda_0]$,

$$p_{\lambda_0}^{(n)}(T^{(x)}) z \to P_{\lambda_0}^{(x)} z \, , \qquad \forall z \in \overline{m^{(x)}},$$

and in particular

$$p_{\lambda_0}^{(n)}(T^{(x)}) x = p_{\lambda_0}^{(n)}(T) x \to P_{\lambda_0}^{(x)} x \, .$$

Thus, the strong limit of $p_{\lambda_0}^{(n)}(T) x$ as $n \to \infty$ exists for every $x \in H$; it depends on $\lambda_0$ but not on the particular sequence of polynomials approximating $\chi_{(-\infty, \lambda_0]}$. We call it $P_{\lambda_0} x$, and shall see that $\{P_{\lambda_0}\}_{-\infty}^{+\infty}$ is a conical spectral resolution with compact support on the nonnegative real axis. We begin by noting that

$$P_\lambda x = P_\lambda^{(x)} x, \qquad \forall x \in H,$$

whence it follows,

$$P_\lambda = 0 \text{ for } \lambda < 0, \quad P_\lambda = I \text{ for } \lambda > M, \quad P_{\lambda+0} = P_\lambda.$$

Moreover,

$$P_{\lambda_1} P_{\lambda_2} x = \lim_{n \to \infty} p_{\lambda_1}^{(n)}(T) P_{\lambda_2}^{(x)} x.$$

But as $P_{\lambda_2}^{(x)} x = P_{\lambda_2} x = \lim_{n \to \infty} p_{\lambda_2}^{(n)}(T) x \in \overline{C^{(x)}}$, and since on $\overline{C^{(x)}}$, $T$ and $T^{(x)}$ coincide because they do on $C^{(x)}$ and are both continuous, $p_{\lambda_1}^{(n)}(T) P_{\lambda_2}^{(x)} x = p_{\lambda_1}^{(n)}(T^{(x)}) P_{\lambda_2}^{(x)} x$, and on substitution above

$$P_{\lambda_1} P_{\lambda_2} x = \lim_{n \to \infty} p_{\lambda_1}^{(n)}(T^{(x)}) P_{\lambda_2}^{(x)} x = P_{\lambda_1}^{(x)} P_{\lambda_2}^{(x)}(x) = P_{\min\{\lambda_1,\lambda_2\}}^{(x)} x = P_{\min\{\lambda_1,\lambda_2\}} x$$

Thus the $P_\lambda$'s, but for being projections, satisfy all requirements to form a spectral family with compact support on the nonnegative real line. But they are indeed projections because they are idempotent and on account of (10.3) satisfy

$$\| P_{\lambda_0} x - P_{\lambda_0} y \|^2 = \lim_{n \to \infty} \| p_{\lambda_0}^{(n)}(T) x - p_{\lambda_0}^{(n)}(T) y \|^2$$

$$\leq \lim_{n \to \infty} \langle (p_{\lambda_0}^{(n)})^2(T) x - (p_{\lambda_0}^{(n)})^2(T) y, x-y \rangle$$

$$\leq \langle P_{\lambda_0} x - P_{\lambda_0} y, x-y \rangle,$$

which are characterizing properties of projections (Theorem 1.2). Since in addition they are positive homogeneous, they are projections on closed convex cones. Finally, if $P_\lambda^{(x)} x$

is replaced by $P_\lambda x$ in the spectral representation (10.7) of $T^{(x)}x = Tx$, one obtains (4), and sufficiency is proved. Necessity of monotonicity, the Lipschitz condition and of (10.1)-(10.3) for the validity of (10.4) with the specified type of spectral resolution is contained in Theorems 9.1, 9.2, 9.3 and Lemma 9.1. The spectral resolution is uniquely determined because for every $T$ of the form (10.4),
$$P_\lambda x = \lim_{n \to \infty} p_\lambda^{(n)}(T)x .$$

<u>Theorem 10.2</u>   Any maximal monotone mapping $T: H \to H$ satisfying

(10.8) $\qquad p(T)q(T) = q(T)p(T)$,

(10.9) $\qquad \langle p(T)x, q(T)x \rangle = \langle x, p(T)q(T)x \rangle, \quad \forall x \in \mathcal{D}(p(T)q(T))$,

(10.10) $\quad \|p(T)q(T)x - p(T)q(T)y\|^2$

$$\leq \langle p^2(T)x - p^2(T)y, q^2(T)x - q^2(T)y \rangle ,$$

$$\forall x, y \in \mathcal{D}(p^2(T)) \cap \mathcal{D}(q^2(T)),$$

for any couple of polynomials $p(\lambda)$ and $q(\lambda)$ nonnegative on $\mathbb{R}^+$, admits a representation of the form

(10.11) $\qquad T = \int_0^\infty \lambda \, dP_\lambda ,$

by means of a conical spectral resolution with support in $\mathbb{R}^+$, and conversely. The spectral resolution is uniquely determined by $T$.

<u>Proof:</u>   We start out by checking that (10.8)-(10.9) are invariant under substitution of $T$ by any polynomial $r(T)$ ($r(\lambda) \geq 0$, $\lambda \geq 0$) of $T$. To do this for (10.8) we show that the common value of its two members is $(pq)(T)$. In fact, if $p(\lambda) = \sum_{k=0}^{n} a_k \lambda^k$, $q(\lambda) = \sum_{h=0}^{m} b_h \lambda^h$, then

$$T^k q(T) = q(T)T^k = \sum_{h=0}^{m} b_h T^{h+k}$$

and

$$p(T)q(T) = \sum_{k=0}^{n} a_k T^k q(T) = \sum_{k=0}^{n} \sum_{h=0}^{m} a_k b_h T^{h+k} = (pq)(T).$$

This can be extended at once to any number of factors; in particular $(r(T))^k = r^k(T)$. It follows then,

$$(10.12) \quad p(r(T)) = \sum_{k=0}^{n} a_k r(T)^k = \sum_{k=0}^{n} a_k r^k(T) = (\sum_{k=0}^{n} a_k r^k)(T) = (p \circ r)(T),$$

and from this

$$p(r(T))q(r(T)) = (p \circ r)(T)(q \circ r)(T) = ((p \circ r)(q \circ r))(T) = (pq \circ r)(T)$$

$$= (pq)(r(T)),$$

which is (10.8) with $r(T)$ in place of $T$.

On use of (10.12) the invariance of (10.9) follows readily. As to that of (10.10),

$$\|p(r(T))q(r(T))x - p(r(T))q(r(T))y\|^2$$

$$= \|(p \circ r)(T)(q \circ r)(T)x - (p \circ r)(T)(q \circ r)(T)y\|^2$$

$$\leq \langle (p \circ r)^2(T)x - (p \circ r)^2(T)y, (q \circ r)^2(T)x - (q \circ r)^2(T)y \rangle$$

$$= \langle p^2(r(T))x - p^2(r(T))y, q^2(r(T))x - q^2(r(T))y \rangle.$$

Equations (10.8) and (10.9) are also invariant under the substitution $T \to T^{-1}$, whenever the latter mapping exists and is everywhere defined. Indeed, using the fact that $p(T)$ commutes with any integer power of $T$, positive or negative, and letting $\tilde{p}(\lambda) = \lambda^n p(\lambda^{-1})$, $\tilde{q}(\lambda) = \lambda^m q(\lambda^{-1})$, one deduces

$$p(T^{-1})q(T^{-1}) = \tilde{p}(T)T^{-n}\tilde{q}(T)T^{-m} = \tilde{p}(T)\tilde{q}(T)T^{-n-m} = (\widetilde{pq})(T)T^{-n-m}$$

$$= (\tilde{q}\tilde{p})(T)T^{-n-m} = q(T^{-1})p(T^{-1}).$$

418

No calculation is needed for (10.9) if one observes that for it to hold for any couple of polynomials it is enough that it holds for the nonnegative integer powers, and then that it holds for negative powers whenever $T^{-1}$ exists.

Relation (10.10) is not invariant under the substitution of $T$ by $T^{-1}$. However it leads to the relation

(10.13) $\quad \|p(T^{-1})x - p(T^{-1})y\|^2 \leq \langle p^2(T^{-1})x - p^2(T^{-1})y, x-y \rangle$,

which we proceed to check. Write (10.10) with $\tilde{p}(\lambda) = \lambda^n p(\lambda^{-1})$ in place of $p(\lambda)$, and $\lambda^n$ in place of $q(\lambda)$,

$$\|\tilde{p}(T)T^n x - \tilde{p}(T)T^n y\|^2 \leq \langle \tilde{p}^2(T)x - \tilde{p}^2(T)y, T^{2n}x - T^{2n}y \rangle,$$

and then replace $x$ and $y$ by $T^{-2n}x$ and $T^{-2n}y$ resp. ,

$$\|\tilde{p}(T)T^{-n}x - \tilde{p}(T)T^{-n}y\| \leq \langle \tilde{p}^2(T)T^{-2n}x - \tilde{p}^2(T)T^{-2n}y, x-y \rangle,$$

which, since $\tilde{p}(T)T^{-n} = p(T^{-1})$ and $\tilde{p}^2(T)T^{-2n} = p^2(T^{-1})$, is just (10.13).

With these preliminaries settled we also approach the end of the proof, which simply consists in verifying that the mapping $(I+T)^{-1}$ meets all hypotheses of Theorem 10.1. Since $I$ and $T$ are maximal operators and the former is everywhere defined, $I + T$ is again a maximal monotone operator, by a theorem of T. R. Rockafellar [21, Theorem 5']. Moreover,

$$\langle (I+T)x - (I+T)y, x-y \rangle = \|x-y\|^2 + \langle Tx - Ty, x-y \rangle \geq \|x-y\|^2,$$

and by Schwarz inequality

$$\|(I+T)x - (I+T)y\| \, \|x-y\| \geq \|x-y\|^2,$$

that is

$$\|(I+T)x - (I+T)y\| \geq \|x-y\|,$$

meaning that $I+T$ is an expansive mapping, and as such

one-to-one and coercive. Another theorem of Rockafellar [25] says that under these conditions the range of $I + T$ is the whole space. Therefore, $(I+T)^{-1}$ is everywhere defined, monotone, contractive and hence Lipschitzian, and by the discussion beginning this proof, satisfies conditions (10.1), (10.2) and (10.3) of Theorem 10.1. Consequently, $(I+T)^{-1}$ admits the representation,

$$(I+T)^{-1} = \int_0^\infty \lambda \, dQ_\lambda ,$$

where $Q_\lambda$ is a spectral resolution with support in $[0,1]$. The inverse of such an integral when existing, as is the case here, must be $\int_0^\infty \lambda^{-1} \, dQ_\lambda$ (Lemma 9.1). Hence

$$I + T = \int_0^\infty \lambda^{-1} \, dQ_\lambda$$

and

$$T = \int_0^\infty (\lambda^{-1} - 1) \, dQ_\lambda = \int_{[0,1]} (\lambda^{-1} - 1) \, dQ_\lambda = \int_0^\infty \lambda \, dP_\lambda ,$$

where $P_\lambda = \begin{cases} 0, & \lambda < 0 \\ I - Q_{(1+\lambda)^{-1} - 0}, & \lambda \geq 0; \end{cases}$ naturally, the $P_\lambda$'s form a conical partition of the identity with support in $\mathbb{R}^+$. This proves the first half of the theorem. As to the second, that is, the necessity of the various hypotheses for the validity of (10.11), it is contained in Corollary to Theorem 9.8, Lemma 9.1, Corollary 1, and Theorem 9.3. The uniqueness of the spectral resolution $\{P_\lambda\}_{-\infty}^{+\infty}$ is a consequence of the uniqueness of $\{Q_\lambda\}_{-\infty}^{+\infty}$, proved in Theorem 10.1.

This theorem, although a legitimate result from the purely theoretical point of view, appears to be of little value when it comes to decide whether a given operator is of type (10.11) or not. Other groups of characterizing properties can be obtained among those exhibited in §9, but they all suffer from the same limitations. An entirely new set of necessary and sufficient conditions is needed in order to put this theory on a more practical basis and to open its way towards the applications.

It is interesting to observe that if in Theorem 10.1, (10.3) is strengthened to (10.10), then the Lipschitz condition can be replaced by sequential demiclosedness. This is the resulting theorem:

Theorem 10.3   Any everywhere defined, sequentially demiclosed monotone mapping $T: \mathcal{H} \to \mathcal{H}$, satisfying (10.1), (10.2) and (10.10) can be represented as a spectral integral with respect to a spectral resolution with compact support on $\mathbb{R}^+$, and conversely.

Proof:   The proof is the same as that of the preceding theorem except at the point where it is proved that the range of I+T is the whole space. There the argument should be replaced by the following: "The mapping I+T is strongly monotone, and as it is everywhere defined, it is locally bounded [22, Theorem 1]. But then, being also sequentially demiclosed, its range is the whole space, by a result of the author [29, Theorem 1]."

## REFERENCES

1. Anderson, R. D. and Klee, V. L., Jr., "Convex functions and upper semicontinuous collections", Duke Math. J. 19(1952), 349-357.

2. Bonnesen, T. and Fenchel, W., Theorie der konvexen Körper, Springer Verlag, Berlin 1934.

3. Choquet, G., Carson, H. and Klee, V. L. , "Exposed points of convex sets", Pac. J. Math. 17(1966), 33-43.

4. Dunford, N. and Schwartz, J., Linear operators, parts I and II, Interscience Publishers, New York, 1956, 1963.

5. Fréchet, M., "Sur la notion de différentielle dans l'analyse générale", J. Math. Pures. Appl. 16(1937), 233-250.

6. Hurewicz, W. and Wallman, H., Dimension Theory, Princeton Univ. Press 1948.

7. Kirzbraun, M. D., Über zusamenziehende und Lipschitzche Transformationen", Fund. Math. 22 (1934), 77-108.

8. Kruskal, J. B., "Two convex counterexamples: a discontinuous envelope function and a nondifferentiable nearest point mapping", Proc. Am. Math. Soc. 23 (1969), 697-703.

9. Mazur, S., "Über convexe Menge in linearen normierte Raümen", Studia Math. 4(1933),128-133.

10. Moreau, J. J., "Décomposition orthogonale dans un espace hilbertien selon deux cônes mutuellement polaires", C. R. Acad. Sci., Paris, 255(1962), 233-240.

11. _____, "Proximité et dualité dans un espace hilbertien", Bull. Soc. Math. France, 93(1965), 273-299.

12. _____, "Convexity and duality", in Functional Analysis and Optimization, edited by E. R. Caianiello Acad. Press., New York 1966, pp. 145-169.

13. _____, "Fonctionelles convexes", Séminaire sur les équations á dérivées partielles", Collége de France, 1966-67.

14. Naimark, M. A., Normed Rings, Noordhoff, Groningen 1959.

15. Nashed, M. Z., "A decomposition relative to convex sets", Proc. Am. Math. Soc., 19 (1968), 782-786.

16. ——————, "Differentiability and related properties of nonlinear operators: Some aspects of the role of differentials in nonlinear functional analysis", Nonlinear Functional Analysis and Applications, edited by L. Rall, Academic Press, New York 1971, pp. 103-309.

17. Opial, Z., "Weak convergence of the sequence of successive approximations for nonexpansive mappings", Bull. Am. Math. Soc. 73(1967), 591-597.

18. Phelps, R. R., "Convex sets and nearest points I", Proc. Am. Math. Soc. 8(1957), 790-797.

19. ——————, "Convex sets and nearest points II", Proc. Am. Math. Soc. 9(1958), 867-873.

20. Riesz, F. and Sz-Nagy, B., "Leçons d'analyse fonctionelle", Akadémiai Kiadó, Budapest 1952.

21. Rockafellar, R. T., "Convex functions, monotone operators and variational inequalities", Proc. NATO Adv. Study Inst., Venice, Italy, June 17-30, 1968.

22. ——————, "Local boundedness of monotone operators", Mich. Math. J. 16(1969), 397-307.

23. ——————, Convex Analysis, Princeton Univ. Press. 1970.

24. ——————, "On the virtual convexity of the domain and range of a nonlinear maximal monotone operator", Math. An. 185(1970), 81-90.

25. ——————, "On the maximality of sums of nonlinear monotone operators", To appear.

26. Sova, M., "General theory of differentiability in linear topological spaces", Czech. Math. J. 14(1964), 485-508.

27. Valentine, F. A., Convex Sets, McGraw-Hill, New York 1964.

28. Whyburn, G. T. , Analytic topology, Am. Math. Soc. Coll. Pub. XXVIII, New York, 1942.

29. Zarantonello, E. H., "The closure of the numerical range contains the spectrum", Bull. Am. Math. Soc. 70(1964), 781-787.

Mathematics Research Center
University of Wisconsin
Madison, Wisconsin

Received March 30, 1971

# Nonlinear Functional Analysis and Nonlinear Integral Equations of Hammerstein and Urysohn Type

*FELIX E. BROWDER*

Introduction.

Let $\Omega$ be a measure space with a $\sigma$-finite measure $dx$. By a nonlinear integral equation of Hammerstein type on $\Omega$, one means an equation of the form

(1) $\quad u(x) + \int_\Omega k(x,y) f(y, u(y)) dy = w(x), \quad (x \in \Omega),$

where both the unknown function $u$ and the given function $w$ lie in some class of r-vector functions on $\Omega$, $(r \geq 1)$, $f$ is a given nonlinear mapping of $\Omega \times R^r$ into $R^r$, and $k$ is a function from $\Omega \times \Omega$ to the linear transformations on $R^r$. Hammerstein equations form an important subclass of the more general class of nonlinear integral equations of Urysohn type, where by the latter, one means an equation of the more general form

(2) $\quad u(x) + \int_\Omega f(x, y, u(y)) dy = w(x), \quad (x \in \Omega),$

and $f$ is now a mapping of $\Omega \times \Omega \times R^r$ into $R^r$.

The theory of nonlinear integral equations of Hammerstein type has been since its inception in the paper of Hammerstein [59] one of the most important domains of application of the ideas and techniques of nonlinear functional analysis, second only to the theory of solutions of boundary value problems for nonlinear partial differential equations.

Hammerstein and his immediate successors applied to this domain the direct method of the calculus of variations, sharpened for this application in the later work of Vainberg [100], Krasnoselski [75], and the Russian school. The development of the fixed point and degree theory for compact nonlinear mappings in Banach spaces by Schauder and Leray [82] was strongly influenced in its form by the theory of nonlinear integral equations and was directly applied to this domain by Niemytski, Rothe [99], Krasnoselski [75], and many others. More recently, the development of the theory of monotone mappings from a reflexive Banach space $X$ to its conjugate space $X^*$ has found important applications in this domain.

It is our object in the present discussion to present a unified development of the recently obtained sharpenings of the theory of the Hammerstein equation using the theory of the topological degree for mappings of the form $I - C$ with $C$ compact as well as the basic theory of monotone nonlinear mappings from $X$ to $X^*$. We present in addition an extension of part of this theory to an interesting class of nonlinear Urysohn integral equations, namely those whose kernels $f(x,y,u)$ are of the form

$$f(x,y,u) = \sum_{j=1}^{n} k_j(x,y) f_j(y,u),$$

or more generally

$$f(x,y,u) = \int_\Lambda k_\alpha(x,y) f_\alpha(y,u) d\xi(\alpha),$$

i.e. sums or integrals of Hammerstein kernels. We leave aside in this paper the applications to the domain of nonlinear integral equations of the generalized theory of the topological degree for mappings more general than the compact displacements ([29], [32], [33], [34], [39], [40], [63], [91], [92]), as well as the applications of the topological theory of variational problems and specifically the Lusternik-Schnirelman theory. (Some applications of the latter to nonlinear integral equations have been given in [45], [46], [75], [100].)

For the reduction of the nonlinear Hammerstein equation to a problem in nonlinear functional analysis, an obvious and fairly standard procedure transforms the integral equation (1) into an equation in the Banach space X in which the given function w and the unknown function u are supposed to lie, by the introduction of two basic operators:

(1) The Niemytski operator F corresponding to the nonlinear term in the integral equation,

$$(F(u))(x) = f(x, u(x)), \quad (x \in \Omega);$$

(2) The linear operator K corresponding to the kernel k(x, y) by

$$(K(u))(x) = \int_\Omega k(x, y) u(y) \, dy, \quad (x \in \Omega).$$

In formal terms, the Hammerstein equation becomes the functional equation asking for an element u in X for which

(3) $\quad u + K(F(u)) = w$.

To make sense of the functional equation (3), one introduces hypotheses upon the kernel k(x, y) and the nonlinear term f(y, u) in order that the operators K and F are well-defined and compose in an appropriate fashion so that the operator KF is well-defined on X and maps the Banach space X into itself.

In the simplest version of the latter process, one takes for the Banach space X one of the spaces $L^p(\Omega)$ for $1 < p < \infty$. If we assume that f satisfies the Caratheodory condition, i.e. f(y, u) is continuous in u for almost all y in $\Omega$ and measurable in y for all u in $R^r$, and assume as well an inequality of the form

$$|f(y, u)| \leq c(1 + |u|^{p-1})$$

for all y in $\Omega$, u in $R^r$, for a fixed constant c (and for simplicity assume that $\Omega$ is a finite measure space), then it follows by standard arguments in measure theory that the nonlinear mapping $F$ is a well-defined continuous mapping of $X = L^p(\Omega)$ into its conjugate space $X^* = L^{p'}(\Omega)$, ($p' = p(p-1)$) ($p' = p(p-1)^{-1}$), and $F$ maps bounded sets of $X$ into bounded sets of $X^*$, (more briefly, $F$ is a bounded mapping of $X$ into $X^*$). Under appropriate measurability and boundedness conditions on the kernel $k(x,y)$, it follows also that the linear mapping $K$ maps $X^*$ into the Banach space $X$ for this case, and therefore the Hammerstein operator $KF$ is a well-defined mapping from $X$ to $X$. More general spaces (Orlicz and Kothe spaces) may be used in extensions of this procedure when the nonlinear term $f(y,u)$ is not subjected to the severe constraint of at most polynomial growth in $u$ imposed in the sketched argument above.

Aside from the choice of the Banach space $X$ which is determined to a large degree by the nature of the nonlinearity in $f(y,u)$ in $u$, the character of the problem posed by a nonlinear Hammerstein operator $KF$ depends upon the interrelation of the assumptions made on the nature of the linear operator $K$ and the nonlinear operator $F$. The sharpest distinction of different cases arises from the contrast of the theory for the linear operators $K$ compact and that for $K$ non-compact. If $K$ is compact, we need only minimal continuity assumptions upon $F$ in order that the Hammerstein operator $T = KF$ should be a compact mapping of the Banach space $X$ and that therefore the topological theory of compact mappings be applicable. In the case in which $K$ is not assumed to be compact, sharper assumptions than mere continuity must be imposed upon $F$ to obtain a significant theory. Of these, the most important is the condition that $F$ be a monotone mapping from $X$ to $X^*$, i.e.

$$(F(u) - F(v), u - v) \geq 0, \quad (u, v \in X),$$

where we use the notation $(y,x)$ to indicate the pairing between the element $y$ of $X^*$ and the element $x$ of $X$. In general, we must also assume that $K$ is monotone and that the underlying space $X$ is reflexive.

A second sharp distinction of cases which lies across the distinction between compact and non-compact $K$ refers to the symmetry properties of $K$. In its classical form in the variational theory, this distinction could be enforced by assuming $K$ to be symmetric, i.e.

$$(K(u), v) = (K(v), u), \quad (u, v \in X).$$

In a much less restrictive form introduced by Amann [1], the symmetry of $K$ is replaced by its <u>angle-boundedness,</u> where $K$ is said to be angle-bounded with constant $a \geq 0$ if

$$|(K(u), v) - (K(v), u)| \leq 2a(Ku, j)^{1/2} (Kv, v)^{1/2},$$

where $K$ is assumed in addition to be monotone (i.e. positive semi-definite, $(Ku, u) \leq 0$ for all $u$ in $X$).

Using these two more or less independent dichotomies, we break up our discussion into four cases, discussed in the following order:

(1) $K$ compact, not angle-bounded.

(2) $K$ compact, angle-bounded.

(3) $K$ noncompact, monotone but not angle-bounded; $F$ monotone.

(4) $K$ noncompact, monotone and angle-bounded; $F$ satisfying a weaker hypothesis than monotonicity.

In addition, we have the treatment of:

(5) Urysohn operators which are sums or integrals of Hammerstein operators.

In the study of abstract Hammerstein operators and equations, i.e. equations of the form $(I + KF)$ where $F$ maps $X$ into $X^*$ and $K$ maps $X^*$ into $X$, from the point of view of the theory of mappings of monotone type from a Banach space $X$ to its dual space $X^*$, results have been established under hypotheses on $F$ more general than monotonicity. Some of these are the following:

$F$ is pseudo-monotone from $X$ to $X^*$ if for any weakly convergent sequence $\{u_j\}$ in $X$ with weak limit $u$, for which

$$\varlimsup (F(u_j), u_j - u) \leq 0,$$

we have $F(u_j)$ converging weakly to $F(u)$ in $X^*$ and $(F(u_j), u_j)$ converging to $(F(u), u)$.

F is of type (M) from X to $X^*$ if for any weakly convergent sequence $\{u_j\}$ in X with weak limit u for which

$$\varlimsup (F(u_j), u_j - u) \leq 0,$$

we have $F(u_j)$ converges weakly to $F(u)$ in $X^*$.

F satisfies condition $(S)_+$ if for any weakly convergent sequence $\{u_j\}$ in X with weak limit u for which

$$\varlimsup (F(u_j), u_j - u) \leq 0,$$

we have $u_j$ converging strongly to u and $F(u_j)$ converges weakly to $F(u)$ in $X^*$. Obviously condition $(S)_+$ implies pseudo-monotonicity, which in turn implies that F is of type (M).

These conditions are of greatest significance in the theory of solutions of nonlinear partial differential equations of elliptic, parabolic, and wave-equation type where they correspond to a weakening of monotonicity hypotheses related to the distinction between the properties of nonlinear partial differential operators with respect to the highest order derivatives involved and with respect to lower order derivatives. No such distinction is meaningful in the context of the theory of nonlinear integral equations, and the applicability of the sharper abstract theorems is open to question on this ground. Let us note explicitly, however, that monotonicity and that condition $(S)_+$ are very easily verifiable under conditions of great generality, as in the following Proposition:

Proposition 1: Let X be the Banach space $L^p(\Omega)$, where $\Omega$ is a $\sigma$-finite measure space and $1 < p < \infty$. Let $f: \Omega \times R^1 \to R^1$ be a function which satisfies the Carathedory condition and satisfies an inequality of the form

$$|f(y,u)| \leq c(y) + c_0 |u|^{p-1}, \quad (y \in \Omega, u \in R^1)$$

where $c$ lies in $L^{p'}(\Omega)$. Then:

(a) The Niemytski operator $F$ given by $F(u)(y) = f(y, u(y))$ is a well-defined bounded continuous mapping of $X = L^p(\Omega)$ into $X^* = L^{p'}(\Omega)$.

(b) If $f(y,u)$ is monotone non-decreasing in $u$ for each fixed $y$, then $F$ is monotone from $X$ to $X^*$.

(c) Suppose that $f(y,u)$ is strictly monotone increasing in $u$ for fixed $y$ (i.e. $f(y,u) > f(y,u_1)$ for $u > u_1$, and all $y$ in $\Omega$) and if there exists a constant $c_2 > 0$ and a function $c_1$ in $L^1(\Omega)$ such that

$$f(y,u)u \geq c_2 |u|^p - c_1(y), \quad (y \text{ in } \Omega, u \text{ in } R^1),$$

then the Niemytski operator $F$ satisfies condition $(S)_+$.

We give the proof of Proposition 1 in the Appendix of the paper.

Let us now survey the detailed results of the following Sections. In Section 1, we consider the case of $K$ compact but not necessarily angle-bounded, $F$ demicontinuous and bounded. In Theorem 1.1, we show the existence of a solution of the equation $(I + KF)(u) = w$ under the assumption that on the boundary of a bounded open set containing $G$, $(F(u), u - w) \geq 0$, (a result essentially due to Amann [8]). In Theorem 1.2, we assume that $(I + KF)$ is locally one-one and that $(I + KF)^{-1}$ is bounded, and conclude that $(I + KF)$ has all of $X$ as its range and is continuously invertible. In Theorem 1.3, we assume that $F(\xi u) = \xi F(u)$ for $u$ outside some ball, $\xi \geq 1$, and that $(I + KF)$ is injective, and conclude that $(I + KF)$ maps onto $X$. In Theorem 1.4, we assume that $F$ is odd outside of some ball and that $(I + KF)^{-1}$

431

is bounded, and conclude that the range of $(I + KF)$ is all of $X$. In Theorem 1.5, we assume that $F = F_0 + F_1$ such that $F_0$ is odd outside of some ball and such that $(I + K(F_0 + tF_1))^{-1}$ is uniformly bounded for $t$ in $[0,1]$, and conclude that $(I + KF)$ maps onto $X$. In Theorem 1.6, we specialize Theorem 1.5 by assuming that $F_0(\xi u) = \xi F_0(u)$ outside of some ball with $\xi \geq 1$, that $(I + KF_0)^{-1}(0)$ is bounded, and that $F_1(u) \|u\|^{-1}$ goes to 0 as $\|u\| \to \infty$, and obtain once more the conclusion that $(I + KF)$ has all of $X$ as its range. All these theorems are consequences of the general theory of compact operators in Banach spaces and serve as models for results for the noncompact case under more stringent hypotheses on $F$.

In Section 2, we begin the discussion of angle-bounded monotone linear mappings $K$. We begin with Propositions 2.1 and 2.2, which give the proof of the Splitting Lemma of Browder-Gupta [32] and of various consequences. In Theorem 2.1, we consider the case of $K$ monotone compact and angle-bounded with constant $a \geq 0$, and prove a variant of Theorem 1.1 where the hypothesis on $F$ on the boundary of $G$ is weakened to the inequality

$$(F(u), u - w) \geq (1 + a^2)^{-1} \|K\|^{-1} \|u - w\|_X^2 .$$

(This result is a sharpened form of recent theorems of Amann [8] and Hess [56].)

In Section 3, we turn to the case in which $K$ is non-compact, monotone, but not necessarily angle-bounded. Here, we apply the theory of maximal monotone mappings from $X$ to $2^{X^*}$ and must assume that the basic Banach space $X$ is reflexive. The arguments applied in this Section develop the ideas presented in Browder-De Figueiredo-Gupta [36]. We consider the case in which $K$ is not necessarily assumed to be linear. In Theorem 3.1, we assume $K$ and $F$ hemi-continuous and monotone, $K(0) = 0$, and for a given $w$ in $X$, we assume that

$$(F(u), u - w) \geq 0$$

for $\|u\| \geq R$. Then, $w$ must lie in $(I + KF)(X)$. In Theorem

3.2, we extend this result to the case in which F is pseudo-monotone and bounded. (In the case of K linear, F monotone, and X a Hilbert space, such a theorem was first proved by Dolph-Minty [48] and extended to reflexive Banach spaces X and F of type (M) by Brezis [10].) In Theorem 3.3, we prove that if K and F are monotone, F bounded, and if $(I + KF)^{-1}$ is locally bounded, then the range of $(I + KF)$ is all of $X^*$. In the later results of Section 3, we assume that K is a monotone linear mapping from $X^*$ to X. In Theorem 3.4, we show that if F is pseudo-monotone and bounded, K monotone and linear, then $(I + KF)$ is demiclosed (i.e. if $u_j$ converges weakly to u in X, $(I + KF)(u_j)$ converges strongly to w, then $(I + KF)(u) = w$). As a consequence, $(I + KF)(C)$ is closed for any bounded closed convex subset C of X. In Theorem 3.5, we assume that F is bounded and satisfies the stronger condition $(S)_+$ and obtain the result that $(I + KF)$ is a proper mapping from bounded closed subsets of X into X and hence maps bounded closed sets of X on closed subsets of X. In Theorem 3.6, we show that if K is monotone and linear and F monotone hemicontinuous (but not necessarily bounded), then it is still true that if $(I + KF)^{-1}$ is locally bounded, then the range of $(I + KF)$ is all of X. (These two Theorems, Theorem 3.3 and 3.6 are based upon and generalize a corresponding result for maximal monotone mappings due independently to Browder [31] and Rockafellar [97].) In Theorem 3.7, we extend part of the preceding result to a mapping F which is bounded, pseudo-monotone and subcoercive (i.e. $(F(u), u) \geq -k\|u\|$ for a fixed constant $k$ and all u in X). For such F, it is shown that if $(I + KF)^{-1}$ is bounded, then $R(I + KF) = X$.

In Theorem 3.8, we consider a pseudo-monotone subcoercive mapping F of X into $X^*$, where $F = F_0 + F_1$ with $F_0$ homogeneous of first order outside of some ball in X and satisfying condition $(S)_+$, $F_1(u)\|u\|^{-1}$ converging to 0 as $\|u\| \to \infty$. Then if $(I + KF_0)^{-1}(0)$ is bounded, it follows that $R(I + KF) = X$. In Theorem 3.9, we consider the structure of the mapping $(I + KF)^{-1}$ for K and F monotone and hemicontinuous. For each w in X, it is shown that $(I + KF)^{-1}(w)$ is a closed convex subset of X, which is at most a single point if $F$ is strictly monotone. If $K$ is linear and $(I + KF)^{-1}$ is bounded, then $(I + KF)^{-1}$ is upper-semicontinuous from $X$ to $2^X$ from the strong topology of $X$ to the weak topology of $X$. If $F$ satisfies condition $(S)_+$, it is upper-semicontinuous from the strong topology of $X$ to the strong topology of $X$. If $F$ is strictly monotone and satisfies condition $(S)_+$, $(I + KF)^{-1}$ is well-defined as a single-valued mapping and continuous. In the remainder of Section 3, we concern ourselves with perturbations of Hammerstein operators by compact mappings of $X$ into $X$. In Theorem 3.10, we give an extension of the Leray-Schauder principle for a compact perturbation of a proper mapping $T$ such that $T^{-1}(y)$ is closed, convex, non-empty for each $y$. In Theorem 3.11, this is applied to give a Leray-Schauder principle for a family of equations of the form $(I + KF + C_t)$ with $K$ monotone linear, $F$ monotone and satisfying condition $(S)_+$, and $\{C_t\}$ a continuous family of compact mappings of $X$ into $X$. This re result is sharpened in Theorem 3.2 by applying more sophisticated technical devices from the theory of monotone mappings. In Theorem 3.13, a corresponding extension of the Borsuk-Ulam theorem is presented.

In Section 4, we turn to the consideration of non-compact $K$ which are monotone and angle-bounded. Here, we may drop the assumption made throughout Section 3 that the Banach space $X$ is reflexive, as well as weaken the assumed inequalities on the mapping $F$. In Theorem 4.1, we consider a general Banach space $X$, $K$ a monotone linear mapping of $X^*$ into $X$ which is angle-bounded with constant $a$, $F$ a demicontinuous mapping of $X$ into $X^*$ such that $F = F_0 + F_1$, where

$$(F_0(u) - F_0(v), u - v) \geq -c\|u - v\|_X^2$$

with $c(1 + a^2)\|K\| < 1$, and $F_1$ a demicontinuous pseudo-monotone mapping of $X$ into $X^*$ such that for each $w$ in $X$, there exists a constant $k_w$ such that

$$(F_1(u), u - w) \geq -k_w\|u - w\|_X .$$

Then for each w in X, there exists u in X such that $(I + KF)(u) = w$. If $F_1 = 0$, the solution u is unique, and if in addition F is continuous, then the solution u depends continuously upon w. In Theorem 4.2, we give a corresponding extension of the Leray-Schauder principle for mappings of the form $(I + KF + C_t)$ with $C_t$ compact from X to X, and in Theorem (4.3) a corresponding extension of the Borsuk-Ulam principle.

In Section 5, we turn to Urysohn operators T from X to X, where $T = \sum_{j=1}^{n} K_j F_j$ is the sum of Hammerstein operators. The basic result, Theorem 5.1 considers a Urysohn operator T where for each j, $K_j$ is a monotone angle-bounded mapping from $X^*$ to X with a constant a, $\|K_j\| \leq k_0$, and $\{F_1, \ldots, F_n\}$ a corresponding family of demicontinuous mappings of X into $X^*$. Generalizing the theory of Hammerstein operators for angle-bounded K given in Section 4, one imposes upon the family $\{F_j\}$ the condition that for any ordered n-tuples of elements of X, $\{u_j\}$ and $\{v_j\}$, we have

$$\sum_{j=1}^{n}(F_j(\sum_{k=1}^{n} u_k) - F(\sum_{k=1}^{n} v_k), u_j - v_j) \geq -c \sum_{j=1}^{n} \|u_j - v_j\|_X^2$$

where $(1 + a^2)\|K\|c < 1$. Then $R(I + T) = X$. We give without proof the statement of Theorem 5.2 giving an extension of Theorem 5.1 to the case of $T = \int K_\alpha F_\alpha \, d\xi(\alpha)$, and the methodological result Proposition 5.1 which is the key step in the proof of Theorem 5.2.

The bibliography contains an extensive set of references to recent work on Hammerstein equations, monotone mappings, the generalized topological degree, and closely related areas of nonlinear functional analysis.

Section 1: Hammerstein operators with K compact.

For a Hammerstein operator $T = KF$ with K compact and linear and the nonlinear mapping F demicontinuous and bounded, it follows immediately that T is a compact mapping of X into X (i.e. T is continuous and maps bounded subsets of X into relatively compact subsets of X ). Indeed, F maps bounded subsets of X into bounded subsets of $X^*$ and by hypothesis, K maps bounded subsets of $X^*$ into relatively compact subsets of X . In addition, F maps strongly convergent sequences in X into weakly convergent subsequences in $X^*$ (by demicontinuity) while since K is compact and linear, K maps weakly convergent sequences in $X^*$ into strongly convergent sequences in X . Thus, T is also continuous in the strong topology of X .

By this simple remark, it follows that for the linear mapping K compact, the theory of Hammerstein operators in a general Banach space X may be subsumed under the topological theory of compact mappings in a Banach space due to Schauder and Leray [82]. In particular, we may apply the Leray-Schauder principle as stated below, as well as Schauder's generalization of the Brouwer principle of invariance of domain, and the corresponding generalization of the antipodal fixed point theorem of Borsuk-Ulam. Using these arguments, we now establish a series of results for the case of K compact:

Theorem 1.1. Let X be a Banach space, w a fixed element of X . Suppose that F is a demicontinuous, bounded mapping of X into $X^*$ , K a compact monotone linear mapping of $X^*$ into X . Suppose further that there exists a bounded open subset G of X such that w lies in G and

$$(F(u), u - w) \geq 0$$

for all u on the boundary of G .
 Then the equation $u + KF(u) = w$ has at least one solution u in $\bar{G}$ .

Theorem 1.1 is essentially due to Amann [8], (see also Amann [4] and Hess [65]). We apply the Leray-Schauder principle: For a compact mapping $T_0$ of the closure of an open subset G of X into X and an element w of X , a sufficient condition that there exist a solution of the equation $u + T_0(u) = w$ in G is that w lies in G while for each t in [0,1], there exists no solution u of the equation

$u + tT_0(u) = w$ on the boundary of $G$.

Proof of Theorem 1.1. Following an argument due to Stanley Weiss which is applied by Amann in [8], there exists a bounded, continuous mapping $S_0$ of $X$ into $X^*$ such that for all $u$ in $X$,

$$\|S_0(u)\|_{X^*} \leq \|u\|_X, \quad (S_0(u), u) \geq \frac{1}{2} \|u\|_X^2.$$

The existence of such a map follows by an elementary argument on partitions of unity.

For each $\varepsilon > 0$, we define a new mapping $F_\varepsilon$ of $X$ into $X^*$ from our original nonlinear mapping $F$ by setting

$$F_\varepsilon(u) = F(u) + \varepsilon S_0(u - w), \quad (u \in X).$$

Let $T_\varepsilon = KF_\varepsilon$, where each $T_\varepsilon$ is a compact mapping of $X$ into $X$. Suppose that we can show for each $\varepsilon > 0$ that there exists a solution $u_\varepsilon$ in the bounded set $G$ of the equation $u_\varepsilon + T_\varepsilon(u_\varepsilon) = w$. For this solution, we should then have

$$(I + T)(u_\varepsilon) = w - \varepsilon KS_0(u_\varepsilon - w) = w_\varepsilon$$

where $w_\varepsilon$ converges strongly to $w$ as $\varepsilon \to 0$, i.e. $w$ lies in the closure of $(I + T)(G)$. Since for $T = KF$ compact, $(I + T)(\bar{G})$, which is our desired conclusion.

Let $\varepsilon > 0$ be fixed. It follows from our preceding remarks that it suffices to prove that $w$ lies in $(I + T_\varepsilon)(G)$ for each such $\varepsilon$. By the Leray-Schauder principle, this will follow if we show that for any $t$ in $[0,1]$, there exists no solution $u$ of $(I + tT_\varepsilon)(u) = w$ with $u$ on the boundary of $G$. Suppose, however, that such a solution $u$ existed for some value of $t$ in $[0,1]$. Then we should have

$$0 = (F_\varepsilon(u), u + tT_\varepsilon(u) - w) = F_\varepsilon(u), u - w) + t(F_\varepsilon(u), KF_\varepsilon(u)).$$

Since $K$ is monotone by hypothesis,

$$(F_\varepsilon(u), KF_\varepsilon(u)) \geq 0.$$

Hence

$$0 \geq (F_\varepsilon(u), u - w) = (F(u), u-w) + \varepsilon(S_0(u-w), u-w)$$
$$\geq \varepsilon(S_0(u-w), u-w) \geq \frac{\varepsilon}{2} \|u-w\|^2 .$$

Hence $u = w$, which contradicts the fact that $w$ lies in $G$ and $u$ on the boundary of $G$. This contradiction establishes the possibility of applying the Leray-Schauder principle, and hence the conclusion of our Theorem. q.e.d.

For the next result, we apply the Schauder principle of invariance of domain for mappings of the form $(I + T)$ with $T$ compact: If $T$ is a compact mapping of the Banach space $X$ into $X$ and $(I + T)$ is locally one-to-one, then $(I + T)(G)$ is open for any open subset $G$ of $X$.

<u>Theorem 1.2.</u> <u>Let $X$ be a Banach space, $F$ a demicontinuous bounded mapping of $X$ into $X^*$, $K$ a compact linear mapping of $X^*$ into $X$. Let $T = KF$, and suppose that $(I + T)$ is locally one-to-one (i.e. there exists a neighborhood $V$ of each point $x_0$ of $X$ such that $(I + T)(u) = (I + T)(v)$ for $u$ and $v$ in $V$ implies that $u = v$). Suppose that there exists a continuous function $k: R^+ \to R^+$ such that if $(I + T)(u) = w$ then</u>

$$\|u\| \leq k(\|w\|) ,$$

(<u>i.e.</u> $(I + T)^{-1}$ <u>is bounded</u>).
<u>Then for each $w$ in $X$, there exists exactly one solution $u$ of the equation</u> $u + KF(u) = w$.

Proof of Theorem 1.2. Under the hypotheses of the Theorem, $T = KF$ is a compact mapping of $X$ into $X$ and $(I + T)$ is locally one-to-one. Hence by the Schauder principle of invariance of domain, $(I + T)$ is an open mapping and hence a local homeomorphism of $X$ into $X$ (i.e. $(I + T)$ maps some neighborhood $V$ of each point $x_0$ of $X$ homeomorphically onto a neighborhood of $(I + T)(x_0)$). To show that $(I + T)$ is a homeomorphism of $X$ onto $X$, it suffices to show that

$(I + T)$ maps closed sets of $X$ on closed sets of $X$. This will follow if $(I + T)$ is proper, i.e. for each compact subset $K$ of $X$, $(I + T)^{-1}(K)$ is compact.

Let $K$ be a compact subset of $X$. Since $K$ is bounded and $(I + T)^{-1}$ maps bounded sets on bounded sets by hypothesis, $(I + T)^{-1}(K)$ is bounded. Let $\{u_j\}$ be an infinite sequence in $(I + T)^{-1}(K)$, and let $v_j = (I + T)(u_j)$. We wish to extract an infinite subsequence of the sequence $\{u_j\}$ which converges strongly in $X$. We may assume without loss of generality since $T$ is compact, $\{u_j\}$ is bounded, and $K$ is compact that $v_j$ converges strongly to some element $w$ of $X$. Then $u_j = v_j - Tu_j$ converges strongly to $v - w = u$, and $(I + T)(u) = v$, i.e. $u$ lies in $(I + T)^{-1}(K)$. Hence $(I + T)$ is proper and the proof of the Theorem is complete. q.e.d.

An interesting application of Theorem 1.2 is the following nonlinear extension of the Fredholm alternative for linear integral equations (which is somewhat closer to the linear case than the variant versions of the Fredholm alternative for nonlinear mappings in the recent literature, many of which are cited in the bibliography below):

<u>Theorem 1.3. Let $X$ be a Banach space, $F$ a demicontinuous bounded mapping of $X$ into $X^*$, $K$ a compact linear mapping of $X^*$ into $X$. Let $T = KF$ and suppose that $(I + T)$ is one-to-one. Suppose further that there exists $R_0 > 0$ such that</u>

$$F(\xi u) = \xi F(u)$$

<u>for</u> $\|u\| \geq R_0$, $\xi \geq 1$.
    <u>Then for each</u> $w$ <u>in</u> $X$, <u>there exists exactly one</u> $u$ <u>in $X$ such that</u>

$$u + KF(u) = w.$$

Proof of Theorem 1.3. Since $(I + T)$ is assumed to be one-to-one, it follows from Theorem 1.2 that we need only show that $(I + T)^{-1}$ is bounded under the hypotheses of the present Theorem.

We choose $R > R_0$ such that if $u_0 = (I + T)^{-1}(0)$, then $\|u_0\| < R$. (This last condition is taken as vacuous if no such element $u_0$ exists.) If $(I + T)^{-1}$ is not a bounded mapping, there exists a sequence $\{u_j\}$ in $X$ such that $\|u_j\| \to \infty$ while $\|(I + T)(u_j)\| \leq M$ for a suitable constant $M$ and all $j \geq 1$. We may assume without loss of generality that $\|u_j\| \geq R$ for all $j$, and we set

$$\xi_j = R^{-1} \|u_j\|, \quad v_j = \xi_j^{-1} u_j.$$

Then $\|v_j\| = R$ for all $j$, $\xi_j \geq 1$, and $u_j = \xi_j v_j$. Since $F$ is homogeneous of degree one as above and $K$ is linear, it follows that

$$(I + T)(u_j) = (I + T)(\xi_j v_j) = \xi_j (I + T)(v_j).$$

Thus

$$\|(I + T)(v_j)\| = \xi_j^{-1} \|(I + T)(u_j)\| \leq M\xi_j^{-1} \to 0, \quad (j \to \infty).$$

Since $T$ is compact, $(I + T)$ maps each bounded closed subset of $X$ on a closed subset of $X$. Since $0$ lies in the closure of the image under $(I + T)$ of the sphere $S_R(0;X)$ of radius $R$ about $0$, there exists $v$ in $X$ with $\|v\| = R$ such that $(I + T)(v) = 0$. This is excluded by our choice of $R$, and this contradiction proves that $(I + T)^{-1}$ is bounded. The conclusion of our Theorem follows. q.e.d.

For the next result, we apply the extended principle of Borsuk-Ulam for compact mappings in the following form: Let $X$ be a Banach space, $T$ a compact mapping of $X$ into $X$. Suppose that $T$ is odd outside some ball in $X$, i.e. there exists $R > 0$ such that $T(-u) = -T(u)$ for $\|u\| \geq R$, and suppose that $(I + T)^{-1}$ maps bounded sets of $X$ on bounded sets of $X$. Then $(I + T)(X) = X$.

More generally, suppose that $T$ is not itself odd, but that $T = T_0 + T_1$ where $T_0$ and $T_1$ are compact, $T_0$ is odd outside of some ball in $X$, and that there exists a continuous function $k: R^+ \to R^+$ such that if

$$u + T_0(u) + tT_1(u) = w$$

for some $t$ in $[0,1]$, then $\|u\| \leq k(\|w\|)$. Then $(I + T)(X) = X$, i.e. there exists $u$ in $X$ for each given $w$ such that $u + T(u) = w$.

Theorem 1.4. <u>Let $X$ be a Banach space, $F$ a demicontinuous bounded mapping of $X$ into $X^*$, $K$ a compact linear mapping of $X^*$ into $X$. Let $T = KF$, and suppose that $F$ is odd outside of some ball in $X$. Suppose further that $(I+T)^{-1}$ is bounded.</u>

<u>Then for each</u> $w$ <u>in</u> $X$, <u>there exists</u> $u$ <u>in</u> $X$ <u>such that</u> $u + KF(u) = w$.

Proof of Theorem 1.4. Since $T$ is compact under our given hypotheses and since $F$ being odd outside of some ball and $K$ being linear implies that $T$ is odd outside of some ball, the result of Theorem 1.4 follows from the simpler form of the general Borsuk-Ulam principle as stated above for compact mappings. q.e.d.

Theorem 1.5. <u>Let $X$ be a Banach space, $F_0$ and $F_1$ two bounded demicontinuous mappings of $X$ into $X^*$, $K$ a compact linear mapping of $X^*$ into $X$. Suppose that $F = F_0 + F_1$, $T = KF$, and that $F_0$ is odd outside of some ball in</u> $X$. <u>Suppose further that there exists a continuous function $k: R^+ \to R^+$ such that if</u> $u + K(F_0 + tF_1)(u) = w$ <u>for any</u> $w$ <u>in</u> $X$ <u>and any</u> $t$ <u>in</u> $[0,1]$, <u>then</u> $\|u\| \leq k(\|w\|)$.

<u>Then for each</u> $w$ <u>in</u> $X$, <u>then equation</u> $u + KF(u) = w$ <u>has a solution</u> $u$ <u>in</u> $X$.

Proof of Theorem 1.4. Since $T$ is compact under our given hypotheses and since $F$ being odd outside of some ball and $K$ being linear implies that $T$ is odd outside of some ball, the result of Theorem 1.4 follows from the simpler form of the general Borsuk-Ulam principle as stated above for compact mappings.    q.e.d.

**Theorem 1.5.**    Let X be a Banach space, $F_0$ and $F_1$ two bounded demicontinuous mappings of X into $X^*$, K a compact linear mapping of $X^*$ into X. Suppose that $F = F_0 + F_1$, $T = KF$, and that $F_0$ is odd outside of some ball in X. Suppose further that there exists a continuous function $k: R^+ \to R^+$ such that if $u + K(F_0 + tF_1)(u) = w$ for any w in X and any t in [0,1], then $\|u\| \le k(\|w\|)$.
    Then for each w in X, then equation $u + KF(u) = w$ has a solution u in X.

**Proof of Theorem 1.5.** Let $T_0 = KF_0$, $T_1 = KF_1$. Then $T_0$ and $T_1$ are compact and $T_0$ is odd outside of some ball in X. The conclusion of Theorem 1.5 then follows from the more general form of the Borsuk-Ulam principle for compact mappings as stated above.    q.e.d.

As an application of Theorem 1.5, we obtain the following simple variant of the generalized Fredholm alternative:

**Theorem 1.6.**    Let X be a Banach space, $F_0$ and $F_1$ two bounded demicontinuous mappings of X into $X^*$, K a compact linear mapping of $X^*$ into X. Suppose that $F_0$ is odd outside of some ball in X, and that

$$F_0(\xi u) = \xi F_0(u)$$

for $\|u\| \ge R$, $\xi \ge 1$. Suppose further that

$$\frac{\|F_1(u)\|}{\|u\|} \to 0 \quad (\|u\| \to \infty).$$

Suppose finally that there exists M such that if $u + KF_0(u) = 0$, then $\|u\| \le M$.

Then for each $w$ in $X$, the equation

$$u + K(F_0 + F_1)(u) = w$$

has at least one solution $u$ in $X$.

Proof of Theorem 1.6. Let $T_0 = KF_0$, $T_1 = KF_1$. Then $T_0$ and $T_1$ are compact mappings of $X$ into $X$. We shall derive the conclusion of Theorem 1.6 by verifying that the hypotheses of Theorem 1.5 are fulfilled.

By our present hypothesis, there exists a function $c: R^+ \to R^+$ with $\lim_{r \to \infty} c(r) = 0$ such that

$$\|F_1(u)\|_{X^*} \le c(\|u\|_X) \|u\|_X, \quad (u \in X).$$

We assert that there exists a constant $k \ge 0$ such that if $R \ge M + 1$ and $\|u\| \ge R$, then

$$\|u\| \le k \|(I + T_0)u\|.$$

We prove this inequality under the assumption as in the hypothesis of our Theorem that $F_0(\xi u) = \xi F_0(u)$ for $\|u\| \ge R$, $\xi \ge 1$. Indeed, suppose that no such inequality holds. Then we may find a sequence $\{u_j\}$ in $X$ with $\|u_j\| \ge R$ for all $j$ such that

$$\|u_j\| \ge j \|(I + T_0)(u_j)\|.$$

For each $j$, we set $\xi_j = R^{-1}\|u_j\|$ and $v_j = \xi_j^{-1} u_j$. Then $\|v_j\| = R$ for each $j$, and $u_j = \xi_j v_j$ with $\xi_j \ge 1$. In particular, it follows from the homogeneity of $F_0$ and the linearity of $K$ that

$$(I + T_0)(u_j) = \xi_j (I + T_0)(v_j).$$

Hence

$$\|(I + T_0)(v_j)\| \le \xi_j^{-1} \|(I + T_0)(u_j)\| \le \xi_j^{-1} j^{-1} \|u_j\| = Rj^{-1} \to 0$$

as $j \to \infty$. Since $T_0$ is compact, $(I + T_0)(S_R(0;X))$ is closed in $X$. Hence, there would exist $v$ in $X$ with $\|v\| = R > M$ such that

$$v + KF_0(v) = 0,$$

which contradicts the choice of the constant $M$. This contradiction establishes the existence of a constant $k$ such that

$$\|u\| \le k \|(I + T_0)(u)\|, \quad (u \in S, \|u\| \ge R).$$

Suppose finally that

$$U + T_0(u) + t\, T_1(u) = w$$

for some $w$ in $X$ and $t$ in $[0,1]$. We assume that $\|u\| \ge R$, and also that $\|u\| \ge R_1$ where $R_1$ is chosen so that $c(s) \|K\| \le \frac{1}{2} k^{-1}$ for $s \ge R_1$. Then we have

$$k^{-1} \|u\| \le \|(I + T_0)(u)\| \le \|w\| + \|KF_1(u)\|$$

$$\le \|w\| + \|K\| c(\|u\|) \|u\| \le \|w\| + \frac{1}{2} k^{-1} \|u\|.$$

Hence $\|u\| \le 2k \|w\|$, the hypothesis of Theorem 1.5 is satisfied, and the conclusion of Theorem 1.6 follows. q.e.d.

## Section 2: Hammerstein operators with $K$ compact and angle-bounded.

In the present Section, we consider the sharpening of some of the results of Section 1, and in particular of Theorem 1.1, that can be obtained if one assumes the compact monotone linear mapping $K$ to be angle-bounded as well. This disucssion presents us with an opportunity, moreover, to develop the simplest part of the machinery that we need to treat the case where $K$ is non-compact but still monotone and angle-bounded.

We recall from the Introduction that a monotone linear mapping K of X into $X^*$ is said to be <u>angle-bounded</u> with constant of angle-boundedness $a \geq 0$ if for each u and v in X,

(2.1) $\qquad |(Ku,v) - (Kv,u)| \leq 2a(Ku,u)^{1/2}(Kv,v)^{1/2}$.

The case of angle-boundedness with $a = 0$ corresponds to K being <u>symmetric</u>, i.e. $(Ku,v) = (Kv,u)$ for all u and v in X.

<u>Proposition 2.1.</u> <u>Let X be a Banach space, K a monotone linear mapping of X into $X^*$ with K angle-bounded with constant of angle-boundedness $a \geq 0$. Then:</u>

(a) <u>There exists a Hilbert space H, a bounded linear mapping S of X into H, and a skew-adjoint bounded linear mapping B of H such that</u>

$$L = S^*(I + B)S .$$

(b) <u>If the range of K lies in a closed subspace Y of $X^*$, S and B can be chosen so that the range of $S^*$ is contained in Y.</u>

(c) <u>For any S and B which satisfy the conditions of part (a), we have</u>

$$\|S\|^2 \leq \|K\|; \|B\| \leq a .$$

(d) <u>If K is compact, then S is compact.</u>

<u>Proof of Proposition 2.1.</u> The assertion of Proposition 2.1 in its general form as well as the proof given below are taken from Browder-Gupta [32].

Since K is monotone and everywhere defined, it is locally bounded. Since K is linear and locally bounded, it is bounded. Let $K^*: X^{**} \to X^*$ be the adjoint of K, and after identifying X with its canonical image in $X^{**}$, let

$$K_1(u) = \frac{1}{2}(K(u) + K^*(u))$$

$$K_2(u) = \frac{1}{2}(K(u) - K^*(u))$$

for each $u$ in $X$. We define a positive semi-definite inner product on $X$ by setting for each $u$ and $v$ in $X$,

$$[u,v] = (K_1(u), v) .$$

Then $[u,u] \geq 0$ and $[u,v] = [v,u]$ for each $u$ and $v$ in $X$, and hence by the generalized Schwarz inequality

$$|[u,v]| \leq [u,u]^{1/2} [v,v]^{1/2} .$$

Let $N = \{u \mid u \in X, [u,u] = 0\}$. Since $K$ is continuous, $N$ is a closed subspace of $X$, while the given inner product on $X$ induces a positive definite inner product on the quotient space $X/N$. With respect to this inner product, $X/N$ is a pre-Hilbert space and we may define $H$ as the completion of $X/N$ to a Hilbert space. We denote by $S$ the composition of the quotient mapping $\beta$ of $X$ into $X/N$ with the injection of $X/N$ into $H$. We denote the inner product on $H$ also by $[\cdot,\cdot]$.

By definition of $S$, for each pair of elements $u$ and $v$ of $X$,

$$[Su, Sv] = (K_1 u, v)$$

$S$ is a bounded linear mapping of $X$ into $H$ since $\|Su\|_H^2 = (K_1 u, v) \leq \|K\| \cdot \|u\|_X^2$, and in particular $\|S\|^2 \leq \|K\|$. Since $K$ is angle-bounded with constant $a$, we know that

$$|(K_2 u, v)| = \frac{1}{2} |(Ku, v) - (Kv, u)| \leq a[u,u]^{1/2} [v,v]^{1/2} =$$

$$= a \|Su\|_H \|Sv\|_H ,$$

for all $u$ and $v$ in $X$, while $(K_2 u, v)$ is skew-symmetric in $u$ and $v$. Hence $(K_2 u, v)$ induces a bounded skew-symmetric bilinear form on $X/N$ as a dense subset of the

Hilbert space $H$. It follows that there exists a bounded skew-adjoint linear mapping $B$ of $H$ such that

$$(K_2 u, v) = [BSu, Sv]$$

for each $u$ and $v$ in $X$.

Finally,

$$(Ku, v) = (K_1 u, v) + (K_2 u, v) = [Su, Sv] + [BSu, Sv] =$$

$$= [(I + B)Su, Sv] = (S^*(I + B)Su, v)$$

where $S^*$ is the linear mapping of $H$ into $X^*$ adjoint to $S$. Hence

$$K = S^*(I + B)S$$

and the proof of part (a) of our Proposition is complete.

For the proof of part (b), we note that since $R(K) \subset Y$, it follows that $S^*$ maps $(I + B)RS$ into $Y$. Since $Y$ is assumed to be a closed subspace of $X^*$, it follows that $S^*$ maps the closure of $(I + B)R(S)$ in $H$ into $Y$. By its construction, $R(S)$ is dense in $H$. Since $B$ is bounded and skew-adjoint, $(I + B)$ is an isomorphism of $H$ with $H$. Hence $(I + B)R(S)$ is dense in $H$, and $S^*$ maps $H$, its closure, into $Y$. Thus, the proof of part (b) is complete.

To prove part (c), we note that if $K = S^*(I + B)S$, then for each $u$ in $X$,

$$(Ku, u) = (S^*(I + B)Su, u) = [(I + B)Su, Su] = [Su, Su] = \|Su\|_H^2,$$

since $B$ is skew-adjoint. Hence $\|S\|^2 \leq \|K\|$. Furthermore, we have

$$[BSu, Sv] = (K_2 u, v) \leq a \|Su\|_H \|Sv\|_H$$

for all $u$ and $v$ in $X$. Hence $\|B\| \leq a$, and the proof of part (c) is complete.

For the proof of part (d), we note that if $\{u_j\}$ is a bounded infinite sequence in $X$ and if $K$ is assumed to be compact, we may find an infinite subsequence (which we denote by the same index set as the original sequence) such that $Ku_j$ converges strongly in $X^*$. For this subsequence, we have

$$\|Su_j - Su_k\|_H^2 = \|S(u_j - u_k)\|_H^2 = (Ku_j - Ku_k, u_j - u_k) \to 0$$

as $j, k \to \infty$. Hence, $S$ is compact.   q.e.d.

**Proposition 2.2.** *Let $K$ be a bounded monotone linear mapping of $X$ into $X^*$ which is angle-bounded with constant $a$. Let the Hilbert space $H$, the bounded linear mapping $S$ of $X$ into $H$, and the skew-adjoing bounded linear mapping $B$ of $H$ satisfy the conditions of part (a) of Proposition 2.1. Then:*

(a) $(I + B)^{-1}$ *is a well-defined bounded linear mapping of $H$ which satisfies the inequality*

$$[(I + B)^{-1} h, h] \geq (1 + a^2)^{-1} \|h\|^2$$

*for all $h$ in $H$.*

(b) *The mapping $K$ satisfies the inequality*

$$(Ku, u) \geq (1 + a^2)^{-1} \|K\|^{-1} \|Ku\|_{X^*}^2 .$$

**Proof of Proposition 2.2.** By hypothesis, $B^* = -B$, $\|B\| \leq a$. Hence $[(I + B)h, h] = \|h\|_H^2$ for each $h$ in $H$. By the Lax-Milgram Lemma, $(I + B)$ is an isomorphism of $H$, and $(I + B)^{-1}$ is a well-defined bounded linear mapping of $H$. For a given $f$ in $H$, let $h = (I + B)^{-1} f$. Then $f = (I + B)h$, and we have

$$[(I + B)^{-1} f, f] = [h, (I + B)h] = \|h\|_H^2 .$$

On the other hand,

$$\|f\|^2 = [(I + B)h, (I + Bh] = \|h\|_H^2 + \|Bh\|_H^2 \leq (1 + \|B\|^2) \|h\|_H^2 .$$

Hence

$$[(I + B)^{-1} f, f] \geq (1 + \|B\|^2)^{-1} \|f\|_H^2 \geq (1 + a^2)^{-1} \|f\|_H^2 ,$$

and the proof of part (a) is complete.

By Proposition 2.1, we assume that $K = S^*(I + B)S$. For $u$ in $X$, we have therefore

$$(Ku, u) = (S^*(I + B)Su, u) = [(I + B)Su, Su] = \|Su\|_H^2.$$

On the other hand,

$$\|Ku\|_{X^*}^2 = \|S^*(I + B)Su\|_{X^*}^2 \leq \|S^*\|^2 \|(I + B)Su\|_H^2,$$

while

$$\|(I + B)Su\|_H^2 = \|Su\|_H^2 + \|BSu\|_H^2 \leq (1 + \|B\|^2)\|Su\|_H^2 \leq (1 + a^2)\|Su\|_H^2.$$

Since

$$\|S^*\|^2 = \|S\|^2 \leq \|K\|,$$

we obtain the inequality

$$(Ku, u) \geq (1 + a^2)^{-1} \|K\|^{-1} \|Ku\|_{X^*}^2.$$

With the proof of the last inequality, the proof of Proposition 2.2 is complete. q.e.d.

We note that the inequality of part (b) of Proposition 2.2 was first explicitly noted by Hess [56].

We now proceed to our principal result for the case of $K$ compact monotone linear and angle-bounded.

<u>Theorem 2.1.</u> <u>Let</u> $X$ <u>be a Banach space</u>, $w$ <u>an element of</u> $X$, $F$ <u>a demicontinuous bounded mapping of</u> $X$ <u>into</u> $X^*$, $K$ <u>a compact monotone linear mapping of</u> $X^*$ <u>into</u> $X$ <u>which is angle-bounded with constant</u> $a$. <u>Let</u> $G$ <u>be a bounded open subset of</u> $X$ <u>which contains the point</u> $w$. <u>Suppose that for each point</u> $u$ <u>on the boundary of</u> $G$, <u>we have</u>

$$(F(u), u - w) \geq -(1 + a^2)^{-1} \|K\|^{-1} \|u - w\|_X^2.$$

Then the equation

$$u + KF(u) = w$$

has at least one solution u in G.

Note that the inequality assumed in the hypothesis of Theorem 2.1 is a weakening of that assumed in Theorem 1.1 with 0 replaced by the negative term $-(1 + a^2)^{-1} \|K\|^{-1} \|u-w\|^2$. Theorem 2.1 is a sharpened form of recent results of Amann [8] and Hess [56] who use somewhat sharper restrictions. In particular, Amann assumes that $(F(u), u-w) \|u-w\|_X^{-2} \to 0$ as $\|u\| \to \infty$ while Hess assumes a linear bound from below.

Proof of Theorem 2.1. We may assume without loss of generality that there exists no solution u of the equation $u + KF(u) = w$ lying on the boundary of G. By the Leray-Schauder principle, to show that a solution u of this equation exists in G, it suffices to show that for each t in [0,1], no solution of the equation

$$u + tKF(u) = w$$

exists on the boundary of G.

Since $t = 1$ has already been treated, we may assume that $0 \leq t < 1$. Suppose that u is a solution of $u + tKF(u) = w$ with u in the boundary of G. Then

$$u - w = -tKF(u),$$

and after we apply $F(u)$ to both sides of the last equality, we obtain:

$$0 = (F(u), u - w) + t(F(u), KF(u)).$$

By part (b) of Proposition 2.2, which we may apply to K considered as a map from $X^*$ to $X^{**}$,

$$(F(u), KF(u)) \geq (1 + a^2)^{-1} \|K\|^{-1} \|KF(u)\|_X^2.$$

By the hypothesis of our Theorem, we know that

$$(F(u), u - w) \geq -(1 + a^2)^{-1} \|K\|^{-1} \|u - w\|_X^2 .$$

Since

$$\|u - w\|_X = t \|KF(u)\|_X ,$$

we have

$$0 \geq (1 + a^2)^{-1} \|K\|^{-1} (1 - t) \|KF(u)\|_X^2 ,$$

and hence $KF(u) = 0$. It follows that $\|u - w\|_X = 0$, $u = w$, and $u$ lies in $G$ contrary to assumption. This contradiction establishes the applicability of the Leray-Schauder principle, and the proof of Theorem 2.1 is thereby complete. q.e.d.

## Section 3: Hammerstein operators with K monotone non-compact but not angle-bounded.

When we turn to the case of Hammerstein operators $T = KF$ with the linear operator $K$ not necessarily compact and without angle-boundedness assumptions, we need more restrictive hypotheses upon the Banach space $X$ in which the Hammerstein operators act. The basic assumption which we apply is that $X$ is reflexive.

We consider the theory of Hammerstein operators $T$ in reflexive Banach spaces as a generalization or a variant of the theory of monotone operators from a Banach space $X$ to its conjugate space $X^*$. Both for the sake of motivation and also because of its essential role in the arguments, we recall at this point some of the basic results of the theory of monotone operators.

If $T$ is a mapping from the Banach space $X$ to $2^{X^*}$, $T$ is said to be monotone if for $y$ in $T(x)$, $w$ in $T(u)$, we have

$$(y - w, x - u) \geq 0 .$$

Such a mapping $T$ is said to be maximal monotone if it is maximal in the family of monotone mappings from $X$ to $2^{X^*}$ in terms of the ordering by inclusion of graphs. The effective domain $D(T)$ of a mapping $T$ from $X$ to $2^{X^*}$ is the set of those $x$ in $X$ for which $T(x)$ is non-empty. The range $R(T)$ of $T$ consists of those $w$ in $X^*$ such that $w$ lies in $T(x)$ for some $x$ in $X$.

A mapping $f$ from $X$ to $X^*$ is said to be demicontinuous if it is continuous from the strong topology of $X$ to the weak topology of $X^*$. The mapping $f$ is said to be hemicontinuous if $f$ is continuous from each line segment in $X$ to the weak topology of $X^*$.

A monotone mapping $f$ from $X$ to $X^*$ which is hemicontinuous is maximal monotone (in the sense of mappings from $X$ to $2^{X^*}$).

The basic result on monotone mappings which we shall apply in the following discussion is the following (proved in Browder [26]): Let $X$ be a reflexive Banach space, $T$ a maximal monotone mapping of $X$ into $2^{X^*}$, with $0 \in D(T)$. Suppose that there exists $R > 0$ such that for $u$ in $X$ with $\|u\| \geq R$, $w \in T(u)$, we have $(w, u) \geq 0$. Then $0$ lies in $R(T)$, and there exists $u_0$ with $\|u_0\| \leq R$ such that $0 \in T(u_0)$.

We apply the preceding result to the first of the Theorems of the present Section:

<u>Theorem 3.1.</u>   <u>Let $X$ be a reflexive Banach space, $w$ an element of $X$. Let $F$ be a hemicontinuous monotone mapping of $X$ into $X^*$, $K$ a hemicontinuous monotone mapping of $X^*$ into $X$ with $K(0) = 0$ (and $K$ need not be linear). Suppose that there exists $R > 0$ such that for $\|u\| \geq R$, we have</u>

$$(F(u), u - w) \geq 0.$$

<u>Then there exists $u$ in $B_R(0; X)$ such that</u>

$$u + KF(u) = w.$$

Proof of Theorem 2.1. The result as stated, together with the proof given below is essentially contained in the paper of Browder-De Figuiredo-Gupta [36]. For K linear it is contained as a special case of a more general result with a more difficult proof given by Brezis [10] described in more detail below.

If we introduce the new variable $v = u - w$, the equation $u + KF(u) = w$ is equivalent to the equation in $v : v + F_w(v) = 0$, where for each $v$ in $X$, $F_w(v) = F(v + w)$. Since

$$(F_w(v), v) = (F(u), u - w) ,$$

we see that there exists $R_1 > 0$ such that for $\|v\| \geq R_1$, we have $(F_w(v), v) \geq 0$. Hence, we may assume without loss of generality that $w = 0$.

Let $K^{-1}$ be the inverse mapping of $K$ considered as a map of $X$ into $2^{X^*}$. Since $K$ is monotone and hemicontinuous from $X^*$ to $X = X^{**}$, $K$ is maximal monotone from $X^*$ to $2^X$. Since the inverse of a maximal monotone mapping is obviously maximal monotone, $K^{-1}$ is maximal monotone from $X$ to $2^{X^*}$. Moreover, the fact that $K(0) = 0$ implies that $0 \in K^{-1}(0)$. Since the single-valued maontone mapping $F$ from $X$ to $X^*$ is hemicontinuous and hence maximal monotone, it follows from a theorem of Rockafellar [98] (see also Brezis-Crandall-Pazy [15] and Browder-Hess [38]) that the mapping $T$ of $X$ into $2^{X^*}$ given by $T(u) = K^{-1}(u) + F(u)$ for each $u$ in $X$ is also maximal monotone from $X$ to $2^{X^*}$.

The equation $u + KF(u) = 0$ is equivalent to the equation:

$$-F(u) \in K^{-1}(u)$$

or

$$0 \in K^{-1}(u) + F(u) = T(u) .$$

If $\|u\| \geq R$, $w_1 \in T(u)$, we have $w_1 = z + F(u)$ for some $z$ in $K^{-1}(u)$, and

$$(w_1, u) = (z, u) + (F(u), u) .$$

By hypothesis, we know that $(F(u), u) \geq 0$ for $\|u\| \geq R_1$. Since $K^{-1}$ is monotone and $0 \in K^{-1}(0)$, it follows that

$$(z, u) = (z - 0, u - 0) \geq 0 .$$

Hence, $(w_1, u) \geq 0$. If we now apply the basic result on maximal monotone mappings in reflexive Banach spaces stated above, it follows that $0 \in T(B_R(0; X))$. This, in turn, implies the conclusion of our Theorem. q. e. d.

We now extend the result of Theorem 3.1 by modifying the hypotheses on the mapping $F$. We recall that a mapping $F$ of $X$ into $X^*$ is said to be **pseudo-monotone** if $F$ is continuous from each finite-dimensional subspace of $X$ to the weak topology of $X^*$ and if for any sequence $\{u_j\}$ in $X$ which converges weakly to $u_0$ in $X$ and for which

$$\overline{\lim}(F(u_j), u_j - u_0) \leq 0 ,$$

we have $F(u_j)$ converging weakly to $F(u_0)$ in $X^*$ and $(F(u_j), u_j)$ converging to $(F(u_0), u_0)$ as $j \to \infty$. (The weaker condition in which we demand only that $F(u_j)$ converge weakly to $F(u_0)$ in $X^*$ defines the mapping of **type (M)** as given by Brezis in [10]. The precise form of the definition we use is that given in Browder [32] and is a weakening of the original definition of Brezis for pseudo-monotonicity in avoiding an explicit condition on filters rather than sequences and dropping a boundedness hypothesis. It has been considered in greater detail in Browder-Hess [38].) We use the following basic result (Browder-Hess [38]): Let $X$ be a reflexive Banach space, $T_0$ a maximal monotone mapping of $X$ into $2^{X^*}$ with $0 \in D(T_0)$, $F$ a pseudo-monotone, bounded mapping of $X$ into $X^*$. Suppose that there exists $R > 0$ such that for $\|u\| \geq R$, $w \in T(u) = T_0(u) + F(u)$,

$$(w, u) \geq 0 .$$

Then $0$ lies in $R(T_0 + F)$.

**Theorem 3.2.** <u>Let X be a reflexive Banach space, w an element of X. Let F be a pseudo-monotone mapping of X into $X^*$ with F bounded (i.e. mapping bounded subsets of X into bounded subsets of $X^*$). Let K be a hemicontinuous monotone mapping of $X^*$ into K with K(0) = 0 (and K need not be linear). Suppose that there exists R > 0 such that for $\|u\| \geq R$, we have</u>

$$(F(u), u - w) \geq 0.$$

<u>Then there exists a solution u in X of the equation</u> u + KF(u) = w.

<u>Proof of Theorem 3.2.</u> This result is also contained in Browder-De Figueiredo-Gupta [36]. For K linear, it was proved for mappings of type (M) by Brazis [10] using a more delicate argument. (For variants of the latter argument, see De Figueiredo-Gupta [51] and Petryshyn-Fitzpatrick [94].)

For the proof of Theorem 3.2, we apply the same argument as in the proof of Theorem 3.1, and note that if F is pseudo-monotone and bounded, the same is true for the mapping $F_w$ given by $F_w(v) = F(v + w)$. The conclusion then follows as in the proof of Theorem 3.1 from the existence theorem for mappings T of the form $T = T_0 + F$ with $T_0$ maximal monotone from X to $2^{X^*}$ and F pseudo-monotone and bounded from X to $X^*$.    q.e.d.

To sharpen our existence theorems, we now apply the following more general theorem for maximal monotone mappings (due independently to Browder [31] and Rockafellar [97]): Let X be a reflexive Banach space, T a maximal monotone mapping of X into $2^{X^*}$. Then a necessary and sufficient condition that $R(T) = X^*$ is that $T^{-1}$ be locally bounded (i.e. for any sequence $\{z_j\}$ in $X^*$ converging strongly to z in $X^*$ and for any sequence $\{u_j\}$ in X such that $z_j \in T(u_j)$ for each j, we must have $\|u_j\|$ uniformly bounded.)

**Theorem 3.3.** <u>Let X be a reflexive Banach space, F a hemicontinuous bounded monotone mapping of X into $X^*$, K a hemicontinuous monotone mapping of $X^*$ into X.</u>

Suppose that the mapping $(I + KF)$ of $X$ into $X$ has a locally bounded inverse $(I + KF)^{-1}$ mapping $X$ into $2^X$. Then for each $w$ in $X$, there exists $u$ in $X$ such that $u + KF(u) = w$.

Proof of Theorem 3.3. The result of the present Theorem and its proof are new, though a variant of the proof has been given independently by Calvert-Gustafson [43] under the stronger hypothesis that $(I + KF)^{-1}$ is single-valued and continuous.

To prove the Theorem, it suffices to show that $0$ lies in $R(I + KF)$ since for any fixed $w$ in $X$, the mapping $F_w$ will satisfy the hypotheses of the Theorem if and only if $F$ does, where $F_w(v) = F(v + w)$ for each $v$ in $X$. (Indeed, suppose that $u_j + KF_w(u_j) = z_j$ where $z_j$ converges strongly to $z$ in $X$. This is equivalent to having

$$(u_j + w) + KF(u_j + w) = z_j + w \to z + w.$$

If $(I + KF)^{-1}$ is locally bounded, it follows that the sequence $\{u_j + w\}$ is bounded, and hence that $\{u_j\}$ is bounded.)

To show that $0$ lies in $R(I + KF)$, it suffices to show that there exists $u$ in $X$ such that

$$u + KF(u) = 0.$$

If $y = F(u)$, then $u \in F^{-1}(y)$, and

$$0 \in F^{-1}(y) + K(y).$$

Conversely, if $0 \in F^{-1}(y) + K(y)$, then there exists $u$ in $F^{-1}(y)$ such that $0$ W $u + K(y)$, i.e. $0 = u + KF(u)$. Hence, it suffices to prove that $0$ lies in $R(F^{-1} + K)$, and *a fortiori* that $R(F^{-1} + K) = X$.

Since $F$ is a hemicontinuous monotone mapping of $X$ into $X^*$, it is maximal monotone. Hence $F^{-1}$ is a maximal monotone mapping of $X^*$ into $2^X$. By hypothesis, $K$ is a hemicontinuous monotone mapping of $X^*$ into $X$. By the theorem of Rockafellar [98], $F^{-1} + K$ is a maximal monotone mapping of $X^*$ into $2^X$. Hence, in order to show that $R(F^{-1} + K)$ is all of $X$, it suffices to show that $(F^{-1} + K)^{-1}$ is locally bounded.

Suppose finally that we are given a sequence $\{v_j\}$ in $X^*$ and a sequence $\{x_j\}$ in $X$ such that $x_j$ converges strongly to $x$ in $X$ and for each $j$, $x_j \in (F^{-1} + K)(v_j)$. We wish to show that the sequence $\{x_j\}$ is bounded, and if we do so, then by the conclusion of the last paragraph the proof of our Theorem will be complete. For each $j$, there exists $u_j$ in $F^{-1}(v_j)$ such that $x_j = u_j + K(v_j)$. It follows that $v_j = F(u_j)$, and we have $x_j = u_j + KF(u_j)$. Since $(I + KF)^{-1}$ is locally bounded, it follows that the sequence $\{u_j\}$ is bounded. Since $F$ is a bounded mapping, $\{v_j\} = \{F(u_j)\}$ is a bounded sequence.     q.e.d.

For the remainder of the Section, we restrict ourselves to the case in which $K$ is linear, which is obviously the interesting case from the point of view of applications to Hammerstein integral equations. A basic result in this case is the following:

<u>Theorem 3.4.</u>   <u>Let</u> $X$ <u>be a reflexive Banach space</u>, $K$ <u>a monotone linear mapping of</u> $X^*$ <u>into</u> $X$, $F$ <u>a bounded mapping of</u> $X$ <u>into</u> $X^*$ <u>with</u> $F$ <u>pseudo-monotone (or more generally of type</u> (M)). <u>Then:</u>

   (a) <u>For any sequence</u> $\{u_j\}$ <u>in</u> $X$ <u>such that</u> $u_j$ <u>converges weakly to</u> $u$ <u>while</u> $(I + KF)(u_j)$ <u>converges strongly to</u> $w$ <u>in</u> $X$, <u>it follows that</u>

$$(I + KF)(u) = w.$$

   (b) <u>For any closed bounded convex subset</u> $C$ <u>of</u> $X$ (<u>or more generally, any sequentially weakly compact subset</u>), $(I + KF)(C)$ <u>is closed in</u> $X$.

<u>Proof of Theorem 3.4.</u>   The conclusion of (b) follows from (a) since for any sequence $\{u_j\}$ in $C$ for which $(I + KF)(u_j)$ converges weakly to an element $u$ of $C$.

To prove (a), let $z_j = F(u_j)$, $w_j = u_j + KF(u_j)$. By hypothesis, $w_j$ converges strongly to $w$. Since $F$ is bounded and $X^*$ is reflexive, we may assume without loss of generality that $z_j$ converges weakly to $z$ in $X^*$. Since $w_j = u_j + Kz_j$ while $K$ being monotone linear is bounded and hence continuous in the weak topology, $Kz_j$ converges weakly to $Kz$, and $w = u + Kz$. Hence, it suffices to prove that $z = F(u)$, i.e. $F(u_j)$ converges weakly to $F(u)$.

For each $j$,

$$(F(u_j), u_j - u) = (F(u_j), -KF(u_j) + w_j - u)$$

$$= -(z_j, Kz_j) + (F(u_j), w_j - u).$$

Since $F(u_j)$ converges weakly to $z$, while $w_j - u$ converges strongly to $w - u = Kz$, it follows that

$$(F(u_j), w_j - u) \to (z, Kz).$$

Hence

$$\overline{\lim} \, (F(u_j), u_j - u) \leq (z, Kz) - \underline{\lim} \, (z_j, Kz_j).$$

Finally, the function $f(z) = (z, Kz)$ on $X^*$ is sequentially weakly lower-semi-continuous on $X^*$ since it is the square of the non-negative convex function $(Kz, Kz)^{1/2}$. Hence

$$(z, Kz) - \underline{\lim} \, (z_j, Kz_j) \leq 0.$$

Since $F$ is pseudo-monotone (and in particular of type (M)), it follows from the fact that $\overline{\lim}(F(u_j), u_j - u) \leq 0$ with $u_j$ converging weakly to $u$ that $F(u_j)$ converges weakly to $F(u)$. q.e.d.

We recall that a mapping $F$ of $X$ into $X^*$ is said to satisfy the condition $(S)_+$ if the following holds: For any sequence $\{u_j\}$ in $X$ with $u_j$ converging weakly to $u$ in $X$ for which $\overline{\lim}(F(u_j), u_j - u) \leq 0$, we have $u_j$ converging strongly to $u$ and $F(u_j)$ converges weakly to $F(u)$. (The second condition will follow from the first if $F$ is demicontinuous; we always assume that $F$ is continuous from finite dimensional subspaces of $X$ to the weak topology of $X^*$.)

**Theorem 3.5.** *Let* $X$ *be a reflexive Banach space,* $K$ *a monotone linear mapping of* $X^*$ *into* $X$, $F$ *a bounded mapping of* $X$ *into* $X^*$ *which satisfies condition* $(S)_+$. *Then:*

(a) $(I + KF)$ *is a proper mapping from bounded closed subsets of* $X$ *into* $X$ (*i.e. for each compact subset* $K_0$ *of* $X$ *and each closed ball* $B$ *of* $X$, $(I + KF)^{-1}(K_0) \cap B$ *is compact*).

(b) *For each bounded closed subset* $A$ *of* $X$, $(I + KF)(A)$ *is closed in* $X$.

**Proof of Theorem 3.5.** The conclusion of part (b) follows from that of part (a). Indeed, let $\{u_j\}$ be a sequence in $A$ such that $(I + KF)(u_j) \to w$. The subset $K_0$ of $X$ which consists of $\{w\} \cup \bigcup_j \{(I + KF)(u_j)\}$ is compact, and if by part (a), $(I + KF)$ is proper, then the sequence $\{u_j\}$ lies in a compact subset of $X$. After we extract a strongly convergent subsequence (which we may identify with the original sequence), we see that $u_j$ converges to some element $u$ of $A$. Since the sequence $\{u_j\}$ eventually lies in $(I + KF)^{-1}(\{w\} \cup \bigcup_{j>k} \{(I + KF)(u_j)\})$, and the latter is compact and hence closed in $X$, it follows that for each $k$

$$(I + KF)(u) \in \bigcup_{j>k} (I + Kf)(u_j) \cup \{w\}.$$

Hence, $(I + KF)(u) = w$, i.e. $w$ lies in $(I + KF)(A)$ and the latter set is strongly closed in $X$.

For the proof of the assertion of part (a), we consider a sequence $\{u_j\}$ which is bounded and such that $I + KF)(u_j)$ lies in the given compact subset $K_0$ of $X$. We wish to show that we may extract an infinite subsequence of this sequence which is strongly convergent in $X$. We may assume without loss of generality by the compactness of $K_0$ that $(I + KF)(u_j) = w_j$ converges strongly to an element $w$ of $K_0$. Since $\{u_j\}$ is bounded and $F$ maps bounded sets into bounded sets, $\{F(u_j)\}$ is bounded in $X^*$. If we pass once more to an infinite subsequence, we may assume that $u_j$ converges weakly to $u$ in $X$ and $F(u_j)$ converges weakly to an element $z$ of $X$. Applying the argument of the proof of Theorem 3.4, it follows that

$$\overline{\lim}_j (F(u_j), u_j - u) \leq 0,$$

while

$$w - u = Kz.$$

Since F satisfies the condition $(S)_+$, it follows that $u_j$ converges strongly to u in X and $F(u_j)$ converges weakly to $F(u)$ in $X^*$. Hence, $z = F(u)$, and we have $u + KF(u) = w$, which implies that u lies in $(I + KF)^{-1}(K_0)$. Thus, $(I + KF)(I + KF)^{-1}(K_0) \cap B$ is compact for each closed ball B in X. q.e.d.

**Theorem 3.6.** *Let* X *be a reflexive Banach space,* K *a monotone linear mapping of* $X^*$ *into* X, F *a monotone hemicontinuous mapping of* X *into* $X^*$. *Suppose that* $(I + KF)^{-1}$ *is a locally bounded mapping of* X *into* $2^X$.

*Then for each* w *in* X, *there exists a solution* u *in* X *of the equation*

$$(I + KF)(u) = w,$$

i.e. $R(I + KF) = X$.

Proof of Theorem 3.6. It suffices as in the proof of Theorem 3.3. to prove that 0 lies in $R(I + KF)$. Applying the argument used in the proof of Theorem 3.1, we may show that $0 \in (I + KF)(u)$ if and only if

$$0 \in K^{-1}(u) + F(u) = T(u),$$

where $T(u) = K^{-1}(u) + F(u)$ defines a maximal monotone mapping T of X into $2^{X^*}$. Such a conclusion will certainly hold if the range of the mapping T is all of $X^*$, and by the previously quoted theorem on maximal monotone mappings in reflexive Banach spaces, this will be the case if $T^{-1}$ is a locally bounded mapping of $X^*$ into $2^X$. Hence, it suffices to prove the local boundedness of $T^{-1}$.

Let $\{u_j\}$ be a sequence in $X$, $\{w_j\}$ a strongly convergent sequence in $X^*$ with limit $w$ such that for each $j$, $w_j \in T(u_j)$. We seek to show that the sequence $\{u_j\}$ is bounded in $X$. By the definition of $T$, for each $j$, there exists an element $v_j$ of $K^{-1}(u_j)$ such that

$$w_j = v_j + F(u_j).$$

Hence

$$Kw_j = Kv_j + KF(u_j) = u_j + KF(u_j).$$

Since $K$ is monotone and linear, $K$ is continuous. Hence $Kw_j$ converges strongly to $Kw$. Since $(I + KF)^{-1}$ is locally bounded, $\{u_j\}$ is bounded. q.e.d.

To extend the last result to the case in which $F$ is pseudo-monotone rather than monotone, we introduce the following condition on a mapping $F$ of $X$ into $X^*$: $F$ is said to be <u>subcoercive</u> if there exists a constant $k$ such that for all $u$ in $X$,

$$(F(u), u) \geq -k\|u\|.$$

(This condition was introduced without the use of the term in Browder-Hess [38].) For pseudo-monotone mappings $F$, the following useful extension of the previously applied theorems on maximal monotone mappings was proved in Browder-Hess [38]: Let $X$ be a reflexive Banach space, $T$ a maximal monotone mapping of $X$ into $2^{X^*}$ with $0 \in D(T_0)$, $F$ a bounded pseudo-monotone mapping of $X$ into $X^*$ which is subcoercive. Let $T = T_0 + F$ and suppose that $T^{-1}$ is bounded. Then $R(T) = X^*$.

<u>Theorem 3.7.</u> <u>Let $X$ be a reflexive Banach space, $K$ a monotone linear mapping of $X^*$ into $X$, $F$ a bounded pseudo-monotone mapping of $X$ into $X^*$ which is subcoercive. Suppose that $(I + KF)^{-1}$ is a bounded mapping of $X$ into $X$, i.e. there exists a continuous function $\xi(r)$ such that if $w = (I + KF)(u)$ for any $u$ in $X$, then $\|u\| \leq \xi(\|w\|)$.</u>

Then R(I + KF) = X , <u>i.e. for each w in X , there exists at least one</u> u <u>in</u> X <u>such that</u> u + KF(u) = w .

<u>Proof of Theorem 3.7.</u>  Let $T(u) = K^{-1}(u) + F(u)$ for each u in X . $K^{-1}$ is a maximal monotone mapping of X into $2^{X^*}$, and following the argument of the proof of Theorem 3.6, it suffices to prove that $R(T) = X^*$ . By the result just quoted from [38], it suffices to show that $T^{-1}$ is a bounded mapping from $X^*$ into $2^X$ . Let $\{u_j\}$ be a sequence in X , $\{w_j\}$ a bounded sequence in $X^*$ such that $w_j \in T(u_j) = K^{-1}(u_j) + F(u_j)$ for each j . There exists $v_j$ in $K^{-1}(u_j)$ for each j such that

$$w_j = v_j + F(u_j) .$$

Hence

$$Kw_j = u_j + KF(u_j)$$

by the linearity of K . Since K is monotone and linear, K is bounded. Hence $\{Kw_j\}$ is a bounded sequence in X . Since $(I + KF)^{-1}$ is assumed to be a bounded mapping, it follows that $\{u_j\}$ is a bounded sequence in X . q.e.d.

<u>Theorem 3.8.</u>  <u>Let X be a reflexive Banach space, K a monotone linear mapping of $X^*$ into X , F a bounded mapping of X into $X^*$ which is pseudo-monotone and subcoercive. Suppose that</u> $F = F_0 + F_1$ <u>where</u> $F_0$ <u>is a bounded mapping of X into $X^*$ which satisfies condition $(S)_+$ such that</u> $F_0(\gamma u) = \gamma F_0(u)$ <u>for</u> $\|u\| \geq R$ , $\gamma \geq 1$ , <u>while</u>

$$\|u\|_X^{-1} \|F_1(u)\|_{X^*} \to 0 , \ (\|u\| \to \infty) .$$

<u>Suppose that the solutions</u> $u_0$ <u>of the equation</u> $(I + KF_0)(u_0) = 0$ <u>are uniformly bounded.</u>
    <u>Then R(I + KF) = X , i.e. for each w in X , there exists at least one</u> u <u>in</u> X <u>such that</u> u + KF(u) = w .

Proof of Theorem 3.8. We begin by remarking that the hypothesis of Theorem 3.8 includes the case in which F is monotone and satisfies condition $(S)_+$ since each monotone mapping is easily seen to be subcoercive.

To prove Theorem 3.8, we apply the conclusion of Theorem 3.7 and note that by the latter result, it suffices to prove that $(I + KF)^{-1}$ is bounded. To show that this last fact is true, suppose the contrary. Then, we should have a sequence $\{u_j\}$ in X with $\|u_j\| \to \infty$ as $j \to \infty$ such that $\|(I + KF)(u_j)\| \le M$ for all j and a suitable constant M. We may assume that $\|u_j\| \ge R$ for all j. Let

$$x_j = R\|u_j\|^{-1} u_j$$

for each j. We may assume that R has been chosen large enough so that for each solution $u_0$ of the equation $(I + KF_0)(u_0) = 0$, we have $\|u_0\| < R$. For each j, we have

$$F_0(u_j) = \|u_j\| \cdot R^{-1} F_0(x_j) ,$$

while since $\|u_j\| \to \infty$, it follows that

$$\|u_j\|^{-1} \|F_1(u_j)\| \to 0 .$$

Since $u_j + KF(u_j) = w_j$, with $\|w_j\| \le M$,

$$\|u_j\| \cdot R^{-1} x_j + K(\|u_j\| \cdot R^{-1} F_0(x_j) + F_1(u_j)) = w_j .$$

Since K is linear and bounded, we obtain

$$x_j + KF_0(x_j) = R\|u_j\|^{-1}(w_j - KF_0(u_j)) = y_j$$

with

$$\|y_j\| \to 0 , \quad (j \to \infty) .$$

By Theorem 3.5 $(I + Kf)$ is a proper mapping. Hence, there exists an element u of with $\|u\| = R$ such that

463

$(I + KF_0)(u) = 0$. This contradicts the assumption that for each such solution $u$, $\|u\| < R$. This contradiction establishes the fact that $(I + KF)^{-1}$ is bounded and that $R(I + KF) = X$. q.e.d.

**Theorem 3.9.** *Let $X$ be a reflexive Banach space, $K$ a hemicontinuous monotone mapping of $X^*$ into $X$, $F$ a hemicontinuous monotone mapping of $X$ into $X^*$. Then:*

(a) *For each $w$ in $X$, $(I + KF)^{-1}(w)$ is a closed convex subset of $X$.*

(b) *If $F$ is strictly monotone (i.e. $(F(u) - F(v), u - v) > 0$. for $u \neq v$), then $(I + KF)^{-1}(w)$ consists of at most a single point.*

(c) *If $(I + KF)^{-1}$ is a bounded mapping of $X$ into $X$, then for each $w$ in $X$, $(I + KF)^{-1}(w)$ is a non-empty bounded closed convex subset of $X$.*

(d) *If $K$ is linear, and the mapping $(I + KF)^{-1}$ is bounded, then $(I + (KF)^{-1}$ as a mapping of $X$ into $2^X$ is upper-semi-continuous as a set-valued mapping from the strong topology of $X$ to the weak topology of $X$.*

(e) *If $K$ is linear, $(I + KF)^{-1}$ is bounded, and $F$ is bounded and satisfies condition $(S)_+$, then $(I + KF)^{-1}$ is upper-semi-continuous from the strong topology of $X$ to $2^X$ with the strong topology of $X$. In particular, if $F$ is also strictly monotone, then $(I + KF)^{-1}$ is a continuous mapping from $X$ to $X$.*

**Proof of Theorem 3.9.** Proof of (a): $(I + KF)^{-1}(w)$ consists exactly of those $u$ in $X$ for which $0 \in T(u)$, where $T$ is the maximal monotone mapping of $X$ into $2^{X^*}$ given by $T(u) = K^{-1}(u) + F_w(u)$. By the theory of maximal monotone mappings, $T^{-1}(0)$ is a closed convex subset of $X$. q.e.d.
Proof of (b): If $F$ is strictly monotone, $T$ is strictly monotone and $T^{-1}(0)$ consists of at most one point. q.e.d.
Proof of (c): Under the hypotheses of (c), it follows from Theorem 3.6 that $R(I + KF) = X$. q.e.d.
Proof of (d): Under the hypotheses of (d), it follows from Theorem 3.4 that if $u_j$ converges weakly to $u$, $(I + KF)(u_j) = w_j$ converges strongly to $w$, then $(I + KF)(u) = w$. This property is equivalent to the upper-semi-continuity of

$(I + KF)^{-1}$ as a mapping from $X$ to $2^X$ from the strong topology of $X$ to the weak topology of $X$.

Proof of (e): Under the hypotheses of (e), $(I + KF)$ is a proper mapping. This implies that if $(I + KF)(u_j) = w_j$, where $w_j$ converges strongly to $w$ in $X$, then the sequence $\{u_j\}$ is contained in a compact subset of $X$, and any strong limit point of the sequence $\{u_j\}$ is contained in $(I + KF)^{-1}(w)$. This implies that $(I + KF)^{-1}$ is upper-semi-continuous from the strong topology of $X$ to the strong topology of $X$. q.e.d.

For the remainder of the present Section, we shall treat the extension of the Leray-Schauder and Borsuk-Ulam perturbations of Hammerstein operators of the form $(I + KF)$ with $K$ and $F$ monotone. We begin our discussion of this topic with a general definition of the notion of a <u>continuous family of compact operators</u> from a Banach space $X$ to another Banach space $Y$. Let $C$ be a mapping of $X \times [0,1]$ into $Y$. The mapping $C$ is said to be compact if it is continuous and if for each bounded subset $B$ of $X$, $C(B \times [0,1])$ is relatively compact in $Y$. Such a mapping $C$ defines the <u>continuous family</u> $\{C_t\}$ of mappings of $X$ into $Y$, where for each $t$ in $[0,1]$ and each $x$ in $X$, $C_t(x) = C(x,t)$.

The general abstract theorem which we apply in the later discussion to Hammerstein operators is the following:

<u>Theorem 3.10.</u> <u>Let $X$ and $Y$ be Banach spaces, $T$ a proper mapping of $X$ into $Y$, $\{C_t\}$ a continuous family of compact mappings from $X$ to $Y$, $C_0 = 0$. Let $G$ be an open bounded subset of $X$, $y_0$ a point of $Y$. Suppose that all the following conditions hold:</u>

(1) <u>For each $y$ in $Y$, $T^{-1}(y)$ is a closed, convex, nonempty subset of $X$.</u>

(2) <u>There exists $u_0$ in $G$ such that $y_0 = T(u_0)$.</u>

(3) <u>There exists no point $x$ in boundary $(G)$ and no value of $t$ in $[0,1]$ such that $(T + C_t)(x) = y_0$.</u>

<u>Then for each $t$ in $[0,1]$, there exists $u_t$ in $G$ such that</u>

$$(T + C_t)(u_t) = y_0.$$

Proof of Theorem 3.10. We reduce the proof of the Theorem to the application of the generalized Leray-Schauder degree theory for multi-valued compact mappings in Banach spaces. (For various detailed developments of the latter theory, we refer to Granas [56], [57], [58], Gorniewicz-Granas [55], Calvert [42], and Cellina-Lasota [44].)

Among the results which follow from the generalization of the Leray-Schauder degree theory to multi-valued mappings is following multi-valued form of the Leray-Schauder principle: Let X be a Banach space, S an upper-semi-continuous mapping of $X \times [0,1]$ into $2^X$ such that for each u in X, t in [0,1], $S(u,t)$ is a closed non-empty convex subset of X. Suppose that S is compact in the sense that for each bounded subset B of X, $S(B \times [0,1])$ is relatively compact in X. For each t in [0,1], let $S_t$ be the mapping of X into $2^X$ given by $S_t(x) = S(x,t)$, and let G be a bounded open subset of X. Suppose that $S_0$ is a constant map of the closure of G into an element of $2^X$ which intersects G, while for each t in [0,1], there exists no point x of boundary (G) such that x lies in $S_t(x)$. Then: For each t in [0,1], there exists $x_t$ in G such that $x_t \in S_t(x_t)$.

We apply this generalized Leray-Schauder principle in our present context by defining

$$S(x,t) = T^{-1}(y_0 - C_t(x))).$$

It follows obviously from this definition that x lies in $S_t(x)$ if and only if $(T + C_t)(x) = y_0$. Thus, it suffices to prove that for each t in [0,1], there exists $x_t$ in G such that $x_t \in S_t(x_t)$. To apply the multivalued Leray-Schauder principle in the present case, we first note that by condition (1) in the hypothesis of the Theorem, $S(x,t)$ is a closed convex non-empty subset of X. If B is a bounded subset of X, then by the definition of a continuous family of compact operators, $C(B \times [0,1])$ is a relatively compact subset of X. Since T is proper, $T^{-1}(C(B \times [0,1]))$ is relatively compact in X. Hence, $S(B \times [0,1])$ is relatively compact in X. For t = 0, $y_0 - C_0(x) = y_0$, so that $S_0(x) = T^{-1}(y_0)$ which intersects G by hypothesis. Moreover, there exist no fixed points of

$S_t$ on the boundary of G for any t in [0,1] by condition (3) of the hypothesis.

To complete the proof of the Theorem, it therefore suffices to prove that S is an upper-semi-continuous mapping of $X \times [0,1]$ into $2^X$. Let $x_0$ be a point of X, $t_0$ a point in [0,1], and let U be a neighborhood in X of $S(x_0, t_0)$. To show that S is upper-semi-continuous, we must show that there exists $\varepsilon > 0$ such that if $\|x - x_0\| < \varepsilon$ and $|t - t_0| < \varepsilon$, then $S(x,t) \subset U$. Suppose this were not the case. Then there would exist a sequence $\{x_n\}$ in X converging to $x_0$, a sequence $\{t_n\}$ in [0,1] converging to $t_0$, and for each n, an element $u_n$ in $S(x_n, t_n) - U$. By the compactness of the mapping S, we may assume without loss of generality that $u_n$ converges to some element u of $X - U$ as $n \to \infty$. For each n, we know that

$$T(u_n) = y_0 - C_{t_n}(x_n) .$$

Taking the limit as $n \to \infty$, we see that $T(u) = y_0 - C_{t_0}(x_0)$, i.e. u lies in $T^{-1}(y_0 - C_{t_0}(x_0)) = S(x_0, t_0)$. This contradicts the fact that u lies outside of U. q.e.d.

As an application of Theorem 3.10, we have the following result on Hammerstein operators:

Theorem 3.11. Let X be a reflexive Banach space, F a bounded monotone mapping of X into $X^*$ satisfying condition $(S)_+$, K a monotone linear mapping of $X^*$ into X. Suppose that we are given a continuous family $\{C_t\}$ of compact mappings of X into X such that $C_0 = 0$. Let G be a bounded open subset of X, w an element of X such that the equation $(I + KF)(u) = w$ has a solution in G. Suppose that $(I + KF)^{-1}$ is bounded, and that for each t in [0,1], the equation $(I + KF + C_t)(u) = w$ has no solution on the boundary of G.

Then for each t in [0,1], the equation $u + KFu + C_t u = w$ has a solution u in G.

467

Proof of Theorem 3.11.  We apply Theorem 3.10 to the mapping $T = I + KF$ of $X$ into $X$. The hypothesis of Theorem 3.10 follow in our present case by Theorems 3.5, 3.8, and 3.9.  q.e.d.

We now sharpen the result of Theorem 3.11 and eliminate its dependence on the degree theory for multi-valued mappings at the expense of a technically more sophisticated argument.

Theorem 3.12.  Let $X$ <u>be a reflexive Banach space</u>, $F$ <u>a bounded hemicontinuous monotone mapping of</u> $X$ <u>into</u> $X^*$, $K$ <u>a monotone linear mapping of</u> $X^*$ <u>into</u> $X$. <u>Suppose that we are given a continuous family</u> $\{C_t\}$ <u>of compact mappings of</u> $X$ <u>into</u> $X$ <u>such that</u> $C_0 = 0$. <u>Let</u> $G$ <u>be a bounded open subset of</u> $X$, $w$ <u>an element of</u> $X$ <u>such that the equation</u> $(I + KF)(u) = w$ <u>has a solution in</u> $G$. <u>Suppose further that one of the two following hypotheses hold:</u>

(a) $F$ <u>satisfies condition</u> $(S)_+$ <u>and for each</u> $t$ <u>in</u> $[0,1]$, <u>the equation</u> $(I + KF + C_t)(u) = w$ <u>has no solution in the boundary of</u> $G$.

(b) $G$ <u>is convex</u>, $C_1$ <u>is completely continuous (i.e. if</u> $x_n$ <u>converges weakly to</u> $x$, <u>then</u> $C_1(x_n)$ <u>converges strongly to</u> $C_1(x)$), <u>and there exists</u> $\delta > 0$ <u>such that for all</u> $t$ <u>in</u> $[0,1]$ <u>and all</u> $x$ <u>in boundary</u> $(G)$,

$$\|(I + KF + C_t)(x) - w\| \geq \delta$$

<u>Then the equation</u>

$$(I + KF + C_1)(u) = w$$

<u>has a solution in</u> $G$.

Proof of Theorem 3.12.  For each reflexive Banach space $X$, it follows from a recent result of Troianski that there exists an equivalent norm which is locally uniformly convex. By a well-known theorem of Asplund, it follows from Troianski's theorem that we can find an equivalent norm on $X$ such that both $X$ and $X^*$ are locally uniformly convex.

We assume that X has been thus normed. Let J be the corresponding duality mapping of X into $X^*$ (which is single-valued because $X^*$ is strictly convex under the assumption of the renorming), i.e. J is the uniquely defined mapping of X into $X^*$ such that for each x in X, J(x) satisfies the two conditions

$$\|J(x)\| = \|x\| \; ; \; (J(x), x) = \|x\|^2 .$$

J is then continuous from the strong topology of X to the strong topology of $X^*$, J is monotone, and for each $\varepsilon > 0$, $F_\varepsilon = F + \varepsilon J$ satisfies condition $(S)_+$ (see Browder [32], Section 17).

Under either of the hypotheses (a) or (b), we observe that for $\varepsilon > 0$ sufficiently small and for all t in [0,1], the equation

$$(I + KF_\varepsilon + C_t)(u) = w$$

has no solutions u on the boundary of G. In the case of hypothesis (b), this follows directly from the assumed inequality, since

$$\|(I + KF_\varepsilon + C_t)(u) - w\| \geq \|(I + KF + C_t)(u) - w\| - \varepsilon \|KJ(u)\|$$

$$\geq \delta - \varepsilon \|K\| M > 0$$

where M is the radius of the smallest ball about 0 in X containing G if we assume that $\varepsilon < \delta(M\|K\|)^{-1}$. Under the hypothesis (a), this will follow by a similar argument once we show that an inequality of the type assumed in hypothesis (b) also holds under the assumptions of (a). Suppose it did not, however. Then we should find a sequence $\{u_n\}$ in the boundary of G such that

$$\|(I + KF + C_{t_n})(u_n) - w\| \to 0$$

for a suitable sequence $\{t_n\}$ in [0,1]. We may assume without loss of generality that $t_n \to t$ and that $C_{t_n}(u_n) \to z$ in

$X$ as $n \to \infty$. Then $(I + KF)(u_n) \to w - z$. By Theorem 3.5, the mapping $(I + KF)$ is proper on bounded closed sets in $X$ since $F$ is assumed to satisfy condition $(S)_+$. Hence the sequence $\{u_n\}$ is pre-compact in $X$. Passing to an infinite subsequence, we may assume that $u_n$ converges to $u$, which is automatically a point of the closed set, boundary $(G)$. Moreover, we may assume that $(I + KF)(u_n) = w - z$. Since $C$ is continuous, $C_{t_n}(u_n)$ converges to $C_t(u)$. Finally, we see that $(I + KF + C_t)(u) = w$, which contradicts the hypothesis of part (a). This contradiction shows that an inequality of the type assumed in (b) also holds in case (a), and hence for all $\varepsilon > 0$ sufficiently small, $(I + KF_\varepsilon + C_t)(u) \neq w$ for $u$ in boundary $(G)$ and $t$ in $[0,1]$.

By hypothesis, the equation $(I + KF)(u) = w$ has a solution $u$ in $G$. For each $\varepsilon > 0$, the mapping $F_\varepsilon$ is strictly monotone (since $J$ is strictly monotone whenever $X$ is strictly convex) and since $F_\varepsilon$ is coercive, the equation $(I + KF_\varepsilon)(u) = w$ has a solution $u_\varepsilon$ in $X$ which is unique. We assert that for $\varepsilon > 0$ sufficiently small, this solution $u_\varepsilon$ is also contained in $G$. Suppose this were not the case. Then there would exist a sequence $\{\varepsilon_n\}$ of positive numbers converging to 0 such that if $u_n = u_{\varepsilon_n}$ is the solution of $(I + KF_{\varepsilon_n})(u_n) = w$, $u_n$ lies outside of $G$. Let $u_0$ be a solution in $G$ of the equation $(I + KF)(u_0) = w$. Then

$$(I + KF_{\varepsilon_n})(u_0) = w + K(\varepsilon_n J(u_0)) = w_n$$

where $w_n$ converges to $w$ as $n \to \infty$. Since $F_{\varepsilon_n}$ is strictly monotone and satisfies condition $(S)_+$, the inverse mapping $(I + KF_{\varepsilon_n})$ is continuous from $X$ to $X$ in the strong topology by Theorem 3.9 (d). Let $L_n$ be the line segment from $w$ to $w_n$. $(I + KF_{\varepsilon_n})^{-1}(L_n)$ is a continuous curve running from $u_n$ in the complement of $G$ to $u_0$ in $G$ and hence must intersect the boundary of $G$. Let $x_n$ be such a point of intersection. Then $x_n$ lies in boundary $(G)$ and $(I + KF_{\varepsilon_n})(x_n)$ lies in $L_n$. In particular,

$$\|(I + KF)(x_n) - w\| \le \varepsilon_n M + \|w_n - w\| \le 2\varepsilon_n M \to 0$$

as $n \to \infty$. Under both hypotheses, we know that $\|(I + KF)(x_n) - w\| \ge \delta > 0$ since $x_n$ lies in boundary $(G)$. This contradiction shows that there exists $\varepsilon_0 > 0$ such that for $0 < \varepsilon < \varepsilon_0$, $u_\varepsilon$ lies in $G$.

For each $\varepsilon > 0$ sufficiently small, therefore, we may apply a Leray-Schauder argument of the type used in the proof of Theorem 3.11 to obtain a solution $u_{\varepsilon,t}$ in $G$ of the equation $(I + KF_\varepsilon + C_t)(u_{\varepsilon,t}) = w$. We note, however, that we do not need the generalized Leray-Schauder principle for multivalued mappings but can simply use the classical Leray-Schauder principle since $(I + KF_\varepsilon)^{-1}$ is a single-valued mapping of $X$ into $X$ and we need only consider the continuous family of single-valued compact mappings

$$S_t(u) = (I + KF_\varepsilon)^{-1}(w - C_t(u)).$$

To complete the proof of Theorem 3.12, we note that

$$\|(I + KF + C_t)(u_{\varepsilon,t}) - w\| \le \|\varepsilon KJ(u_{\varepsilon,t})\| \le \varepsilon M \|K\| \to 0$$

as $\varepsilon \to 0$. We note that under hypothesis (a), the mapping $I + KF + C_t$ is proper on bounded closed sets and hence $(I + KF + C_t)(cl(G))$ is closed in $X$. (Indeed, if $(I + KF + C_t)(u)$ lies in the compact set $K_0$, then $(I + KF)(u)$ lies in the compact set $(K_0 + K_1)$ where $K_1$ contains the set of images $\{-C_t(cl(G))\}$. Hence $(I + KF + C_t)^{-1}(K_0)$ is a closed subset of the compact set $(I + KF)^{-1}(K_0 + K_1)$ and hence itself compact.) Hence $w$ lies in $(I + KF + C_t)(cl(G))$ for each $t$ in $[0,1]$ and in particular for $t = 1$.

Under hypothesis (b), the same conclusion follows for $t = 1$ from the fact that $(I + KF + C_1)(cl(G))$ is closed in $X$. To prove that the last statement holds, we apply the weak compactness of the closed bounded convex set $cl(G)$ and the complete continuity of $C_1$. Suppose that $(I + KF + C_1)(x_n) = w_n \to w$. We may assume that $x_n$ converges weakly to $x$ and therefore that $C_1(x_n)$ converges strongly to $C_1(x)$, where $x$ is an element of $cl(G)$. Then

$(I + KF)(x_n)$ converges strongly to $w - C_1(x)$. Applying the property that $(I + KF)$ is demiclosed (Theorem 3.4), it follows that $(I + KF)(x) = w - C_1(x)$. Hence, $w$ lies in $(I + KF + C_1)(cl(g))$.

Under both hypotheses, it follows that there exists $u$ in $cl(G)$ such that $(I + KF + C_1)(u) = w$. Since no such $u$ exists in boundary $(G)$ by hypothesis, it follows that $u$ must be an element of $G$. q.e.d.

If we apply the Borsuk-Ulam principle rather than the Leray-Schauder principle in the above argument, we obtain the following variant of the preceding result:

**Theorem 3.13.** <u>Let $X$ be a reflexive Banach space, $K$ a monotone linear mapping of $X^*$ into $X$, $F$ a bounded hemicontinuous odd monotone mapping of $X$ into $X^*$. Let $\{C_t\}$ be a continuous family of compact mappings of $X$ into $X$ such that $C_0$ is an odd mapping. Suppose that there exists a continuous function $\xi$ from $R^+$ to $R^+$ such that if</u>

$$(I + KF + C_t)(u) = w, \quad (t \in [0,1])$$

<u>then</u> $\|u\| \le \xi(\|w\|)$.

<u>Suppose that one of the two following conditions is satisfied:</u>

(a) $F$ <u>satisfies condition</u> $(S)_+$,

or

(b) $C_1$ <u>is completely continuous.</u>

<u>Then the range of</u> $(I + KF + C_1)$ <u>is all of $X$, i.e. the equation</u> $(I + KF + C_1)(u) = w$ <u>has a solution</u> $u$ <u>for each</u> $w$ <u>in $X$.</u>

We omit the details of the proof of Theorem 3.13 which follow the pattern of the proof of Theorem 3.12.

Section 4: __Hammerstein operators with__ K __noncompact and angle-bounded.__

If the noncompact linear operator K is both monotone and angle-bounded, we can drop the restriction imposed in Section 3 that the Banach space X be reflexive and substantially weaken the assumptions imposed upon the nonlinear mapping F of the Hammerstein operator.

__Theorem 4.1.__ __Let X be a Banach space, K a monotone linear mapping of $X^*$ into X which is angle-bounded with constant of angle-boundedness a, F a demicontinuous mapping of X into X such that__

$$F = F_0 + F_1$$

where:

(1) $F_0$ __is a mapping of X into $X^*$ which satisfies the inequality__

$$(F_0(u) - F_0(v), u - v) \geq -c \|u - v\|_X^2, \quad (u, v \in X)$$

where

$$c < (1 + a^2)^{-1} \|K\|^{-1}.$$

(2) $F_1$ __is a demicontinuous pseudo-monotone mapping of X into $X^*$ and there exists a continuous function__ k __from $R^+$ to $R^+$ such that for each__ u __and__ w __of X,__

$$(F_1(u), u - w) \geq -k(\|w\|)_X \|u - w\|_X.$$

__Then:__ (a) __For each__ w __in X, there exists a solution__ u __in X to the equation__ $(I + KF)(u) = w$.

(b) __If__ $F_1 = 0$, __the solution__ u __is unique for each__ w __in X. If in addition__ $F_0$ __is continuous from X to $X^*$, then__ $(I + KF)^{-1}$ __is continuous from X to X.__

<u>Proof of Theorem 4.1.</u>  Let $w$ be a given element of $X$.
To solve the equation $(I + KF)(u) = w$, we introduce the new variable $v = u - w$, and rewrite the equation in the equivalent form

$$v + KF_w(v) = 0$$

where

$$F_w(v) = F(v + w), \quad (v \in X).$$

Then $F_w = F_{0,w} + F_{1,w}$, where

$$F_{0,w}(v) = F_0(v + w); \quad F_{1,w}(v) = F_1(v + w).$$

We note that $F_{0,w}$ satisfies the same inequality as $F_0$, while for $F_{1,w}$, we have the inequality

$$(F_{1,w}(v), v) = (F_0(v+w), v) = (F_0(u), u - w) \geq$$

$$-k(\|w\|_X) \|u - w\|_X = -k(\|w\|_X) \|v\|_X,$$

i.e.

$$(F_{1,w}(v), v) \geq -k_w \|v\|_X, \quad (k_w = -k(\|w\|_X)).$$

We now apply Proposition 2.1 to the monotone angle-bounded linear mapping $K$ considered as a mapping from the Banach space $X^*$ to $X^{**}$. By that Proposition, there exists a Hilbert space $H$, a bounded linear mapping $S$ of $X^*$ into $H$, and a skew-adjoint bounded linear mapping $B$ of $H$ such that

$$K = S^*(I + B)S.$$

Moreover, since the range of the linear mapping $K$ is contained in the closed subspace $X$ of $X^{**}$ (where, we assume $X$ as identified with its canonical image in $X^{**}$), it follows that we may take $S^*$ as a bounded linear mapping of $H$ into

X, with $S^*$ injective. We also know that

$$\|S\|^2 \le \|K\| \, , \quad \|B\| \le a \, .$$

For simplicity of notation, we assume that $w = 0$ so that $F_w = F$. The equation $v + KF(v) = 0$ may be rewritten in the form

$$v = v - S^*(I + B)SF(v) \, .$$

Any solution of this equation must obviously be of the form

$$v = -S^*(h)$$

for some $h$ in $H$, and since $S^*$ is injective, $v$ and $h$ correspond in a one-to-one fashion. Writing the equation in terms of $h$, we obtain

$$S^*(h + (I + B)SFS^*(h)) = 0 \, .$$

Since $S^*$ has a trivial nullspace, the latter equation is equivalent to the equation

$$h + (I + B)SFS^*(h) = 0 \, .$$

By Proposition 2.2, we know that $(I + B)$ is an isomorphism of $H$. Hence, we may again rewrite the last derived equation in the equivalent form:

$$(I + B)^{-1} h + SFS^*(h) = 0 \, .$$

If we introduce the two mappings $T_1$ and $T_2$ of $H$ into $H$ given by

$$T_1(h) = (I + B)^{-1}(h) + SF_0 S^*(h) \, ,$$

$$T_2(h) = SF_1 S^*(h) \, ,$$

for $h$ in $H$, we see that the equation which we wish to

solve is equivalent to the equation

$$T_1(h) + T_2(h) = 0 .$$

Since $S$ and $S^*$ are continuous, $(I + B)^{-1}$ is continuous, and $F_0$ and $F_1$ are demicontinuous, both $T_1$ and $T_2$ are demicontinuous from $H$ to $H$. By our hypothesis and by part (a) of Proposition 2.2, we see that:

$$[T_1(h) - T_1(k), h-k] = [(I+B)^{-1}(h-k), h-k][SF_0 S^* h - SFS^* k, h-k]$$

$$\geq (1+a^2)^{-1} \|h-k\|_H^2 + (F_0(S^* h) - F_0(S^* k), S^*(h) - S^*(k))$$

$$\geq (1+a^2)^{-1} \|h-k\|_H^2 - c\|S^*(h-k)\|_X^2$$

$$\geq \{(1+a^2)^{-1} - c\|S^*\|^2\}\|h-k\|^2 \geq \{(1+a^2)^{-1} - c\|K\|\} \|h-k\|_H^2$$

$$\geq \delta \|h-k\|_H^2 , \quad (h, k \in H).$$

where $\delta = \{(1+a^2)^{-1} - c\|K\|\} > 0$ by hypothesis.

$T_2$ is pseudo-monotone from $H$ to $H$. Indeed, suppose that $h_j$ converges weakly to $h$ in $H$ and that $\overline{\lim}[T_2(h_j), h_j - h] \leq 0$. Since $S^*$ is a bounded linear operator from $H$ to $X$ it follows that $S^*(h_j)$ converges weakly to $S^*(h)$ in $X$. We know that

$$[T_2(h_j), h_j - h] = [SF_1 S^*(h_j), h_j - h] = (F_1(S^* h_j), S^*(h_j) - S^*(h))$$

and hence that

$$\overline{\lim} (F_1(S^* h_j), S^*(h_j) - S^*(h)) \leq 0 .$$

Since $F_1$ is assumed to be pseudo-monotone, it follows that $F_1(S^* h_j)$ converges weakly in $X^*$ to $F_1(S^* h)$ and that $(F_1(S^*(h_j)), S^*(h_j))$ converges to $(F_1(S^* h), S^* h)$. Since $S$ is a bounded linear mapping from $X^*$ to $H$, it follows that

$T_2(h_j) = SF_1 S^*(h_j)$ converges weakly in H to $T_2(h) = SF_1 S^*(h)$, and that

$$[T_2(h_j), h_j] = (F_1(S^* h_j), S^* h_j), S^* h_j) \to$$

$$(F_0(S^* h), S^* h) = [T_2(h), h].$$

Therefore, $T_2$ is pseudo-monotone from H to H.

Since $T_1$ is monotone and demicontinuous while $T_2$ is pseudo-monotone, it follows that $(T_1 + T_2)$ is pseudo-monotone from H to H. Applying the existence theory for pseudo-monotone mappings (see Brezis [13], Browder [32], Browder-Hess [38]), we see that the equation $(T_1 + T_2)(h) = 0$ will certainly have a solution h in the Hilbert space H if there exists $R > 0$ such that $[T_1(h) + T_2(h), h] \geq 0$ for all h in H with $\|h\| = R$. We note, however, that

$$[T_1(h) + T_2(h), h] \geq [T_1(h) - T_1(0), h - 0] + [T_1(0), h] + [T_2(h), h]$$

$$\geq \delta \|h\|_H^2 - \|T_1(0)\| \cdot \|h\|_H + [T_2(h), h],$$

while

$$[T_2(h), h] = (F_1 S^*(h), S^* h) \geq -k \|S^*(h)\|_X \geq -k \|S\| \cdot \|h\|_H.$$

Thus

$$[(T_1 + T_2)(h), h] \geq \delta \|h\|^2 - k_1 \|h\| \to \infty \, (\|h\| \to \infty).$$

Hence the existence theorem for pseudo-monotone mappings may be applied and the proof for part (a) is complete.

For the proof of part (b), we note first that if $F_1 = 0$, then $T_2 = 0$ and the equation in h becomes $T_1(h) = 0$ which has exactly one solution. Hence in this case the solution is unique. Suppose we know moreover that $F_0$ is continuous from X to $X^*$ and suppose that we consider a sequence $\{w_j\}$ in X converging strongly to w. Let $u_j = (I + KF)^{-1}(w_j)$, $u = (I + KF)^{-1}(w)$. If we set $v_j = u_j - w_j$, $v = u - w$, we see that

$$v_j = -KF(v_j + w_j), \quad v = -KF(v + w).$$

We set

$$v_j = S^*(h_j), \quad v = S^*(h)$$

as above, with $h_j$ and $h$ in $H$. Then by the calculation of the existence proof, we have

$$h_j = -(I + B)SF(S^* h_j + w_j),$$
$$h = -(I + B)SF(S^* h + w).$$

If we apply $(I + B)^{-1}$ to both sides of each equation, subtract one from the other, and take the inner product in $H$ with $(h_j - h)$, we obtain:

$$0 = [(I + B)^{-1}(h_j - h), h_j - h] \, (F(S^* h_j + w_j) - F(S^* h + w), S^* h_j - S^* h)$$
$$\geq (1 + a^2)^{-1} \|h_j - h\|_H^2 + (F(S^* h_j + w_j) - F(S^* h + w_j), S^* h_j - S^* h)$$
$$+ (F(S^* h + w_u) - F(S^* h + w), S^* h_j - S^* h)$$
$$\geq (1 + a^2)^{-1} \|h_j - h\|_H^2 - \|S\| \cdot \|h_j - h\|_H \cdot \|F(S^* h + w_j) - F(S^* h + w)\|_{X^*}.$$

Hence

$$\|h_j - h\|_H \leq k_0 \|F(S^* h + w_j) - F(S^* h + w)\|_{X^*} \to 0$$

as $j \to \infty$ by the continuity of $F$ at the point $S^* h + w$. Hence $v_j$ converges strongly to $v$, and $u_j = v_j + w_j$ converges strongly to $v + w = u$. Thereby the continuity of $(I + KF)^{-1}$ is proved. q. e. d.

We can extend the results of Theorem 4.1 to obtain a result involving perturbation by a compact operator, as follows:

Theorem 4.2. Let X be a Banach space, K a monotone linear operator from $X^*$ to X, with K angle-bounded with constant a, F a continuous mapping of X into $X^*$ such that for all u and v in X,

$$(F(u) - F(v), u - v) \geq c \|u - v\|_X^2$$

where

$$c(1 + a^2) \|K\| < 1.$$

Let $\{C_t\}$ be a continuous family of compact mappings of X into X such that $C_0 = 0$, and such that for a given w in X, the solutions of the equation

$$(I + KF + C_t)(u) = w$$

are bounded uniformly in t in $[0,1]$.
  Then for each t in $[0,1]$, there exists a solution $u_t$ of the equation

$$(I + KF + C_t)(u_t) = 0.$$

Proof of Theorem 4.2. We apply the Leray-Schauder principle to the equation

$$(I + C_t(I + KF)^{-1})(v) = w$$

which is equivalent to the desired equation. q.e.d.

  A corresponding generalization of the Borsuk-Ulam theorem follows by similar arguments:

Theorem 4.3. Let X be a Banach space, K a monotone linear mapping of $X^*$ into X which is angle-bounded with constant a, F a continuous mapping of X into $X^*$ such that for all u, v in X

$$(F(u) - F(v), u - v) \geq c \|u - v\|_X^2$$

<u>with</u> $c < (1+a^2)^{-1} \|K\|^{-1}$. <u>Let</u> $\{C_t\}$ <u>be a continuous family of compact mappings of</u> X <u>into</u> X <u>such that</u> $C_0$ <u>is odd, and suppose that</u> F <u>is odd.</u> <u>Suppose that there exists a continuous function</u> $\xi$ <u>from</u> $R^+$ <u>to</u> $R^+$ <u>such that if</u> $(I + KF + C_t)(u) = w$ <u>for any</u> w <u>in</u> X <u>and any</u> t <u>in</u> $[0,1]$, <u>then</u> $\|u\| \le \xi(\|w\|)$.

<u>Then the range of</u> $(I + KF + C_1)$ <u>is all of</u> X.

## Section 5: Urysohn operators.

We give in detail below the proof of our basic existence theorem for Urysohn operators whose kernels are finite sums of Hammerstein kernels. We also give the statement of the more general theorem for the case of a continuous decomposition into Hammerstein kernels and the statement of the basic technical lemma. The detailed proof in this case will be given in another paper, since the detailed discussion of measurability questions for Banach-space valued functions is somewhat lengthy and would obscure the presentation of the basic idea of the argument.

<u>Theorem 5.1.</u> <u>Let</u> X <u>be a Banach space, and let</u> $\{K_1, \ldots, K_n\}$ <u>be a finite family of monotone linear operators from</u> $X^*$ <u>to</u> X <u>such that each</u> $K_j$ <u>is angle-bounded with a given constant</u> a <u>and</u> $\|K_j\| \le k_0$ <u>for each</u> j. <u>Let</u> $\{F_1, \ldots, F_n\}$ <u>be a corresponding finite family of demicontinuous mappings of</u> X <u>into</u> $X^*$, <u>and suppose that the following condition holds:</u>

(c) <u>For each pair of n-tuples</u> $\{u_1, \ldots, u_n\}, \{v_1, \ldots, v_n\}$ <u>from</u> X, <u>we have</u>

$$\sum_{j=1}^n (F_j(u) - F_j(v), u_j - v_j) \ge -c \sum_{j=1}^n \|u_j - v_j\|_X^2$$

<u>where</u>

$$u = \sum_{j=1}^n u_j, \quad v = \sum_{j=1}^n v_j$$

<u>and</u>

$$c < (1 + a^2)^{-1} k_0^{-1} .$$

Then if we consider the Urysohn operator $T = \sum_{j=1}^{n} K_j F_j$, the equation $(I + T)(u) = w$ has a solution $u$ for each $w$ in $X$. If all the $F_j$ are continuous, the solution $u$ in $X$ may be chosen to depend continuously on $w$.

Proof of Theorem 5.1. We apply Proposition 2.1 to each of the angle-bounded monotone linear mappings $K_j$ of $X^*$ into $X$ as in the proof of Theorem 4.1 and construct a Hilbert space $H_j$ for each $j$, a bounded linear mapping $S_j$ of $X^*$ into $H_j$ such that $S_j^*$ is an injective linear mapping of $H$ into $X$ (considered as a closed subspace of $X^{**}$), and a skew-adjoint bounded linear mapping $B_j$ in the Hilbert space $H_j$, such that for each

$$K_j = S_j^* (I + B_j) S_j .$$

We form the Hilbert space $H$ as the orthogonal direct $H = \sum_{j=1}^{n} \oplus H_j$. An element $h$ of $H$ is an n-tuple $\{h_1, \ldots, h_n\}$ with $h_j$ in $H_j$ for each $j$, while

$$\|h\|^2 = \sum_{j=1}^{n} \|h_j\|_{H_j}^2 .$$

We consider the Urysohn equation

$$(I + T)(u) = w$$

for a given $w$ in $X$. If we introduce the new variable $v = u - w$, we obtain the equivalent equation in $v \in$

$$v + \sum_{j=1}^{n} K_j F_{j,w}(v) = 0$$

with each $F_{j,w}$ given by

$$F_{j,w}(v) = F_j(v + w) .$$

481

The system of mappings $\{F_{1,w}, \ldots, F_{n,w}\}$ will satisfy the same condition as the original system $\{F_1, \ldots, F_n\}$. To simplify the notation, we assume that $w = 0$ without loss of generality.

Let $R$ be the mapping of $H$ into $X$ given by

$$R(h) = \sum_{j=1}^{n} S_j^*(h_j).$$

We shall construct a solution $v$ of our given equation by letting $v = R(h)$ for some element $h$ of $H$. In terms of $h$, the equation becomes:

$$\sum_{j=1}^{n} S_j^*(h_j + (I + B_j)S_j F_j(R(h))) = 0.$$

We shall seek a solution by making each component of this sum vanish, i.e. for each $j$ with $1 \leq j \leq n$, we seek to satisfy the equation

$$h_j + (I + B_j)S_j F_j(R(h)) = 0.$$

An equivalent form of this last system of equations is:

$$(I + B_j)^{-1}(h_j) + S_j F_j(R(h)) = 0, \quad (1 \leq j \leq n).$$

We consider this last system of equation as an equation in the Hilbert space $H$ involving mappings from $H$ to $H$. We introduce the two mappings $T_1$ and $T_2$ from $H$ to $H$ given by

$$\{T_1(h)\}_j = (I + B_j)^{-1}(h_j),$$

$$\{T_2(h)\}_j = S_j F_j(R(h)).$$

Both $T_1$ and $T_2$ are demicontinuous mappings from $H$ to $H$. $T_1$ is linear, and

$$[T_1(h), h]_H = \sum_{j}^{n} [(I + B_j)^{-1}(h_j), h_j]_{H_j}$$

$$\geq (1 + a^2)^{-1} \sum_{j=1}^{n} \|h_j\|_{H_j}^2 = (1 + a^2)^{-1} \|h\|_H^2 ,$$

by Proposition (2.2). On the other hand, for $h$ and $h'$ in $H$,

$$[T_2(h) - T_2(h'), h-h']_H = \sum_{j=1}^{n} [S_j F_j(R(h)) - S_j F_j(R(h')), h_j - h_j']_{H_j}$$

$$= \sum_{j=1}^{n} (F_j(R(h)) - F_j(R(h')), S_j^*(h_j) - S_j^*(h_j'))$$

$$\geq -c \sum_{j=1}^{n} \|S_j^*(h_j) - S_j^*(h_j')\|_X^2$$

$$\geq -c \|S^*\|^2 \sum_{j=1}^{n} \|h_j - h_j'\|_{H_j}^2$$

$$\geq -c \|K\| \cdot \|h - h'\|_H^2 \geq -ck_0 \|h - h'\|_H^2$$

since

$$\|S_j^*\|^2 = \|S_j\|^2 \leq \|K_j\| \leq k_0 .$$

Since

$$\delta = (1 + a^2)^{-1} - ck_0 > 0 ,$$

it follows that

$$[(T_1 + T_2)(h) - (T_1 + T_2)(h'), h-h'] \geq \delta \|h-h'\|^2$$

for all $h$ and $h'$ in $H$. Thus we may apply the existence theorem for demicontinuous strongly monotone mappings in a

Hilbert space (Minty [84], Browder [16], [17]) which yields the existence of a solution.

The continuity of h in w (and thereby the possibility of choosing a solution u depending continuously upon w) follows if all the $F_j$ are continuous from X to $X^*$ by the same argument as that applied in the proof of part (b) of Theorem 4.1. q.e.d.

We close this Section with the statement of the more general theorem for Urysohn operators whose kernels decompose as integrals of Hammerstein kernels.

**Theorem 5.2.** *Let* X *be a separable Banach space,* $\Lambda$ *a measure space with a finite measure* $d\xi$. *Suppose that we are given measurable families* $\{K_\alpha : \alpha \in \Lambda\}$ *of bounded linear monotone mappings from* X *to* $X^*$ *and* $\{F_\alpha : \alpha \in \Lambda\}$ *of continuous mappings from* $X^*$ *to* X *with* $\|K_\alpha\| \le k_0$, *and all the* $K_\alpha$ *angle-bounded with a fixed constant* a. *Suppose further that for each* u *in* $X^*$, $\|F_\alpha(u)\|_X$ *is essentially bounded on* $\Lambda$. *Let* R *be the mapping of* $L^2(\Lambda; X^*)$ *into* $X^*$ *given by*

$$R(u) = \int_\Lambda u(\alpha) d\xi(\alpha),$$

*and suppose that for each pair of elements* $\{u(\alpha)\}$ *of* $L^2(\Lambda; X^*)$ *we have the inequality*

$$\int_\Lambda (F_\alpha(R(u)) - F_\alpha(R(v)), u(\alpha) - v(\alpha)) d\xi(\alpha) \ge -c\|u-v\|^2_{L^2(\Lambda;X^*)},$$

*with*

$$c < (1+a^2)^{-1} k_0^{-1}.$$

*Then the mapping* T *from* $X^*$ *to* $X^*$ *given by*

$$T(u) = \int_\Lambda K_\alpha(F_\alpha(u)) d\xi(\alpha), \quad (u \in X^*)$$

*has the property that the equation* $u + T(u) = w$ *has a solution* u *in* $X^*$ *for each* w *in* $X^*$.

The proof of Theorem 5.1 is based upon the following generalization of Proposition 2.1:

<u>Proposition 5.1.</u>   <u>Let $\{K_\alpha ; \alpha \in \Lambda\}$ be a measurable family of monotone linear mappings from the separable Banach space X to $X^*$ with $\|K_\alpha\| \leq k_0$ and with a fixed constant a of angle-boundedness for all $\alpha$ in $\Lambda$. Then there exists a Hilbert space H, a measurable family $\{S_\alpha; \alpha \in \Lambda\}$ of bounded linear mappings of X into H, and a measurable family $\{B_\alpha; \alpha \in \Lambda\}$ of bounded skew-adjoint linear operators on H, such that for each</u> $\alpha$,

$$K_\alpha = S_\alpha^*(I + B_\alpha)S_\alpha,$$

and

$$\|S_\alpha\|^2 \leq k_0, \quad \|B_\alpha\| \leq a.$$

Appendix.

We now give the proof of Proposition 1 of the Introduction.

<u>Proof of (a):</u>   Since $F(u)$ is defined by

$$F(u)(y) = f(y, u(y))$$

for each $y$ in $\Omega$, it follows from the assumed inequality that

$$|F(u)(y)| \leq c(y) + c_0 |u(y)|^{p-1}$$

where both terms on the right-hand side of the inequality lie in $L^{p'}(\Omega)$. (Indeed, c is assumed to lie in $L^{p'}$ by hypothesis, and since u lies in $L^p$, $|u|^{p-1}$ lies in $L^{p'}$.) By hypothesis, $f(y, u)$ satisfies the Caratheodory condition. We may assume without loss of generality that $f(y, u)$ is continuous in u for each y in $\Omega$. If $u = \lim_j u_j$, where each $u_j$ is a simple function, then $F(u)(y) = \lim_j F(u_j)(y)$

for each $y$, where each function $F(u_j)$ is the sum of a countable family of functions each of the form $f(y, c_j)s_j(y)$ where $s_j$ is the characteristic function of a measurable set and $c_j$ is a constant. Since each of these functions is measurable, so is $F(u_j)$, and finally so is $F(u)$. It follows that $F(u)$ yields a well-defined element of $L^{p'}(\Omega)$.

By the above inequality, we have

$$\|F(u)\|_{L^{p'}} \leq \{\int (|c(y)| + c_0|u(y)|^{p-1})dy\}^{1/p'}$$

$$\leq k_0 + c_0' \|u\|_{L^p}^{p-1},$$

so that $F$ is a bounded map of $L^p$ into $L^{p'}$.

To complete the proof of (a), we need therefore only to prove that $F$ is continuous from $L^p$ to $L^{p'}$. Let $\{v_j\}$ be a sequence converging in $L^p$ to $v$. To show that $F$ is continuous, it suffices to show that for each such sequence, there is an infinite subsequence such that $F(v_{j_k})$ converges strongly to $F(v)$ in $L^{p'}$. We begin by choosing an infinite subsequence of $\{v_j\}$ (which we denote once more as $\{v_j\}$ for simplicity of notation) such that $v_j(y)$ converges to $v(y)$ almost everywhere. Then $F(v_j)(y) = f(y, v_j(y))$ converges almost everywhere to $f(y, v(y)) = F(v)(y)$ by the continuity of $f(y, u)$ in $u$.

Let $\varepsilon > 0$ be given. Then there exists a set $\Omega_0$ of finite measure and a constant $\delta > 0$ such that for $\operatorname{meas}(A) < \delta$,

$$\int_A |v_j(y)|^p \, dy < \varepsilon, \quad \int_{\Omega_0} |v_j(y)|^p \, dy < \varepsilon.$$

It follows that

$$\int_A |F(v_j)(y)|^{p'} dy \leq k_1 \int_A |c(y)|^{p'} dy + c_0' [\int_A |v_j(y)|^p \, dy]$$

$$\leq \varepsilon c_0' + k_1 \int_A |c(y)|^{p'} dy,$$

with a similar inequality for the integrals over $\Omega_0$. It follows that the functions $|F(v_j)(y)|^{p'}$ are equi-integrable over $\Omega$, so that the almost everywhere convergence of $F(v_j)$ to $F(v)$ implies that $F(v_j)$ converges to $F(v)$ in $L^{p'}(\Omega)$. Thus the proof of part (a) is complete.

Proof of (b): Since

$$(F(u)-F(v), u-v) = \int [f(y, u(y)) - f(y, u(y))](u(y) - v(y))dy \geq 0$$

if $f$ is monotone non-decreasing since the integrand is always non-negative, the monotonicity of the mapping $F$ follows immediately from this hypothesis.

Proof of (c): Let $\{v_j\}$ be a weakly convergent sequence in $L^p(\Omega)$ with weak limit $v$, and suppose that

$$\overline{\lim}\, (F(v_j), v_j - v) \leq 0.$$

To show that $F$ satisfies condition $(S)_+$, it suffices to prove that $v_j$ converges strongly to $v$ in $L^p(\Omega)$ since the continuity of $F$ would imply that $F(v_j)$ then converges to $F(v)$. Since $f(y, u)$ is monotone increasing in $u$, we know that

$$0 \leq (F(v_j) - f(v), v_j - v)$$

while the right hand-side has 0 as its limit superior as $j \to \infty$. Hence

$$(F(v_j) - F(v), v_j - v) \to 0 \quad (j \to \infty).$$

To show that $v_j$ converges strongly to $v$, it suffices to show that for any such sequence, we can extract an infinite subsequence $\{v_{j_k}\}$ which converges strongly to $v$. We know that

$$(F(v_j) - F(v), v_j - v) = \int [f(y, v_j(y)) - f(y, v(y))](v_j(y) - v(y))dy$$

where the integrand is non-negative. Since the integrals

converge to zero, we may extract an infinite subsequence (which we denote by the original index) such that

$$[f(v, v_j(y)) - f(y, v(y)](v_j(y) - v(y)) \to 0$$

almost everywhere in $\Omega$. Let $\Omega_1$ be a measurable subset of $\Omega$ such that $\text{meas}(\Omega - \Omega_1) = 0$, the convergence holds on $\Omega_1$, and $v(y)$ is finite everywhere on $\Omega_1$. Then for $y$ in $\Omega_1$, we have

$$[f(y, v_j(y)) - f(y, v(y))],(v_j(y) - v(y))$$
$$\geq c_2 |v_j(y)|^p - c_1(y) - M(y)|v_j(y)|$$
$$- M(y)|v_j(y)|^{p-1} - M(y)c(y),$$

where the functions $M$, $c$, and $c_1$ may also be assumed to be finite on $\Omega_1$. It follows that $|v_j(y)|$ is bounded in $j$ for each $y$ in $\Omega_1$. For a fixed $y$ in $\Omega_1$, suppose that an infinite subsequence of $\{v_j(y)\}$ converges to a limit $v_0$. Then

$$[f(y, v_0) - f(y, v(y))](v_0 - v(y)) = 0$$

which implies that $v_0 = v(y)$. Hence, it follows that for each $y$ in $\Omega_1$, $v_j(y)$ converges to $v(y)$, i.e. $v_j$ converges to $v$ almost everywhere.

To show that $v_j$ converges strongly to $v$ in $L^p(\Omega)$, it suffices to show the equi-integrability of the functions $|v_j(y)|^p$. To establish this conclusion, we sharpen the inequality written above:

$$(f(y, v_j(y)) - f(y, v(y))](v_j(y) - v(y)) = f(y, v_j(y))v_j(y)$$
$$- f(y, v_j(y))v(y) - f(y, v(y))v_j(y) + f(y, v(y))v(y),$$

and hence

$$c_2|v_j(y)|^p \le c_1(y) + [f(y,v_j(y)) - f(y,v(y))](v_j(y) - v(y))$$
$$+ [c(y) + |v_j(y)|^{p-1}]| + [c(y) + |v(y)|^{p-1}]|v_j(y)|.$$

We assert that each of the terms on the right-hand side of the inequality is equi-integrable. For $c_1$, this is obvious. For the term

$$[f(y,v_j(y)) - f(y,v(y))](v_j(y) - y(y))]$$

this follows from the non-negativeness of the functions for each $j$, and the convergence of their integrals to zero. For the third term, we have

$$\int_A [c(y) + |v_j(y)|^{p-1}]|v(y)|\,dy \le \||c + |v_j|^{p-1}\|_{L^{p'}(A)} \|v\|_{L^p(A)}$$

$$\le \||c + |v_j|^{p-1}\|_{L^{p'}(\Omega)} \|v\|_{L^p(A)} \le k_1 \|v\|_{L^p(A)},$$

where $\|v\|_{L^p(A)}$ can be made uniformly small by either taking meas(A) sufficiently small or by taking $A$ to be the complement of some set of finite measure. Similarly, the fourth term can be estimated by

$$\int_A [c(y) + |v(y)|^{p-1}] \cdot |v_j(y)| \le \||c + |v|^{p-1}\|_{L^{p'}(A)} \|v_j\|_{L^p(\ )}$$

$$\le k_2 \||c + |v|^{p-1}\|_{L^{p'}(A)},$$

and the term on the right can be made uniformly small by either making meas (A) uniformly small or taking $A$ to be the complement of a suitable set of finite measure. It follows that $v_j$ converges strongly to $v$ in $L^p(\Omega)$ and hence that $F$ satisfies condition $(S)_+$.  q.e.d.

## REFERENCES

1. H. Amann, Uber die Existenz und iterative Berechnung einer Losung der Hammersteinschen Gleichung, Aequationae Math., 1 (1968) 242-266.

2. _____, Ein Existenz- und Eindeutigkeitsatz fur die Hammersteinsche Gleichung in Banachraumen, Math. Zeitschrift, 111 (1969) 175-190.

3. _____, Uber die Existenz und Eindeutigkeit einer Losung der Hammersteinschen Gleichung in Banach-raumen, Jour. Math. Mech., 19 (1969) 143-154.

4. _____, Hammersteinsche Gleichung mit kompatken Kernen, Math. Annalen, 186 (1970) 334-340.

5. _____, Uber die naherungweise Losung nichtlineuren Integralgleichungen, (to appear.

6. _____, Zum Glerkin-Verfahren fur die Hammersteinsche Gleichung, Archive Rat. Mech. Aanal., 35 (1969) 114-121.

7. _____, Uber die Konvergenzgeschwindigkeit des Galerkins-Verfahren fur die Hammersteische Gleichung, Archive Rat. Mech. Anal., 37 (1970) 33-47.

8. _____, Existence theorems for equations of Hammerstein type, (to appear in Applicable Analysis).

9. J. Batt, Nonlinear compact mappings and their adjoints, Math. Annalen, 189 (1970), 5-25.

10. H. Brezis, Equations et inequations non-lineaires dans les espaces vectoriels en dualite, Ann. Scient. Institut Fourier (Grenoble), 19 (1968) 115-176.

11. H. Brezis, On some degenerate nonlinear parabolic equations, Proc. Symposia in Pure Math. vol. 18, Amer. Math. Soc., (1970), 28-38.

12. _____, Perturbations non lineaires d'operateurs maximaux monotones, C. R. Acad. Sci. Paris, 269 (1969) 566-569.

13. _____, Inequations variationelles associes a des operateurs d'evolution, Theory and applications of monotone operators, NATO Summer School Venice, Oderisi, 1969.

14. _____, Problemes unilateraux, These, Paris, 1970.

15. H. Brezis, M. Crandall, and A. Pazy, Perturbations of nonlinear maximal monotone mappings in Banach spaces, Comm. Pure App. Math., 23 (1970) 123-144.

16. F. E. Browder, The solvability of nonlinear functional equations, Duke Math. Jour., 30 (1963) 557-566.

17. _____, Variational boundary value problems for quasilinear elliptic equations, I, II, III, Proc. Nat. Acad. Sci. U.S.A., 50 (1963), 31-37, 592-598, 794-798.

18. _____, Nonlinear elliptic boundary value problems, Bull. Amer. Math. Soc., 69 (1963) 862-874.

19. _____, Nonlinear elliptic boundary value problems, II, Trans. Amer. Math. Soc., 117 (1965) 530-550.

20. _____, Existence and uniqueness theorems for solutions of nonlinear boundary value problems, Symposia in App. Math., Vol. 17 Amer. Math. Soc., (1965) 24-49.

21. F. E. Browder, Fixed point theorems for non-compact mappings in Hilbert space, Proc. Nat. Acad. Sci. U. S. A., 53 (1965) 1272-1276.

22. _____, Mapping theorems for non-compact nonlinear operators in Banach spaces, Proc. Nat. Acad. Sci. U. S. A., 54 (1965), 337-342.

23. _____, Problèmes nonlineaires, NATO Summer Institute on Partial Differential Equations, Univ. of Montreal, 1965.

24. _____, On the unification of the calculus of variations and the theory of monotone nonlinear operators in Banach spaces, Proc. Nat. Acad. Sci. U. S. A., 56 (1966) 419-425.

25. _____, Existence and approximation of solutions of nonlinear variational inequalities, Proc. Nat. Acad. Sci. U. S. A., 56 (1966) 1080-1086.

26. _____, Nonlinear maximal monotone operators in Banach spaces, Math. Annalen, 175 (1968) 89-113.

27. _____, The fixed point theory of multivalued mappings in topological vector spaces, Math. Annalen, 177 (1968) 283-301.

28. _____, Nonlinear variational inequalities and maximal monotone mappings in Banach spaces, Math. Annalen, 183 (1969) 213-231.

29. _____, Topology and nonlinear functional equations, Studia Math., 31 (1968) 189-204.

30. _____, Existence theorems for nonlinear partial differential equations, Symposia in Pure Math., Vol. 16, Amer. Math. Soc., (1970), 1-60.

31. F. E. Browder, Nonlinear monotone and accretive operators in Banach spaces, Proc. Nat. Acad. Sci. U. S. A., 58 (1968) 388-393.

32. _____, Nonlinear operators and nonlinear equations of evolution in Banach spaces, in Nonlinear functional analusis, Symposia in Pure Math., vol. 18, Part 2.

33. _____, Nonlinear elliptic boundary value problems and the generalized topological degree, Bull. Amer. Math. Soc., 76 (1970), 999-1005.

34. _____, Group invariance in nonlinear functional analysis, Bull. Amer. Math. Soc., 76 (1970) 986-992.

35. _____, Remarks on nonlinear equations and eigenvalue problems in Banach spaces, (to appear).

36. F. E. Browder, D. G. de Figueiredo and C. P. Gupta, Maximal monotone operators and nonlinear integral equations of Hammerstein type, Bull. Amer. Math. Soc., 76 (1970) 700-705

37. F. E. Browder and C. P. Gupta, Nonlinear monotone operators and integral equations of Hammerstein type, Bull. Amer. Math. Soc., 75 (1969), 1347-1353.

38. F. E. Browder and P. Hess, Nonlinear mappings of monotone type in Banach spaces, (to appear in Jour. Funct. Anal.).

39 F. E. Browder and R. D. Nussbaum, The topological degree for non-compact nonlinear mappings in Banach spaces, Bull. Amer. Math. Soc., 74 91968), 671-676.

40. F. E. Browder and W. V. Petryshyn, Approximation methods and the generalized topological degree for nonlinear mappings in Banach spaces, Jour. Funct. Anal., 3 (1969) 217-245.

41. Bui An Ton, Nonlinear equations on convex subsets of Banach spaces, Math. Ann., 181 (1969) 35 - 44.

42. B. D. Calvert, The local fixed point index for multi-valued transformations in a Banach space, Math. Annalen, 190 (1970) 119-128.

43. B. D. Calvert and K. Gustafson, Multiplicative perturbation of nonlinear m-accretive operators, (to appear in Jour. Func. Anal.).

44. A. Cellina and A. Lasota, A new approach to the definition of topological degree for multivalued mappings, Rend. Accad. Naz. Lincei, ser 8, 47 (1969) 154-160.

45. C. V. Coffman, A minimum-maximum principle for a class of nonlinear integral equations, Jour. Analyse Math., 22 (1969) 391-419.

46. M. G. Crandall and P. H. Rabinowitz, Multiple solutions of a nonlinear integral equation. Arch. Rat. Mech. Anal., 37 (1970) 262-267.

47. C. L. Dolph, Nonlinear integral equations of Hammerstein type, Trans. Amer. Math. Soc., 66 (1949) 289-307.

48. C. L. Dolph and G. J. Minty, On nonlinear integral equations of the Hammerstein type, <u>Nonlinear integral equations</u>, U. of Wis. Press, Madison, (1964) 99-154.

49. D. G. de Figuiredo and C. P. Gupta, Borsuk type theorems for nonlinear noncompact mappings in Banach spaces, (to appear).

50. D. G. de Figuiredo and C. P. Gupta, Solvability of nonlinear integral equations of Hammerstein type, (to appear).

5. _____, Nonlinear integral equations of Hammerstein type involving unbounded monotone operators, (to appear in Jour. Math. Anal. and Appl.).

52. P. M. Fitzpatrick, A generalized degree for uniform limits of A-proper mappings, (to appear in Jour. Math. Anal. and Appl.).

53. S. Fucik, Fredholm alternatives for nonlinear operators in Banach spaces and its application to differential and integral equations, Comm. Math. Univ. Carolinae, 11 (1970) 271-284.

54. M. Golomb, On the theory of nonlinear integral equations, integral systems, and general functional equations, Math. Zeits., 39 (1935), 45-75.

55. L. Gorniewicz and A. Granas, Fixed point theorems for multivalued mappings of the absolute neighborhood retracts. J. Math. pures et appl., 49 (1970) 381-395.

56. A. Granas, Sur la notion de degre toplogique pour une certaine classe de transformations multivalentes dans les espaces de Banach, Bull. Aca. Polon. Sci., 7 (1959) 191-194.

57. _____, Theorem on antipodes and theorems on fixed points for a certain class of multivalued mappings in Banach spaces, Bull. Acad. Polon. Sci., 7 (1959) 271-275.

58. _____, Topics in the fixed point theory, Sem. Jean Leray, Paris 1969-1970.

59. A. Hammerstein, Nichtlineare Integralgleichungen nebst Anwendungen, Acta Math., 54 (1930), 117-176.

60. P. Hess, Nonlinear functional equations in Banach spaces and homotopy arguments, Bull. Amer. Math. Soc., 77 (1971) 211-215.

61. _____, Nonlinear operator equations and eigenvalue problems in non-separable Banach spaces, (to appear in Comm. Math. Helvetici).

62. _____, On the Fredholm alternative for nonlinear functional equations in Banach spaces, (to appear).

63. _____, On nonlinear mappings of monotone type homotopic to odd operators, (to appear in Jour. Funct. Anal.).

64. _____, On a method of singular perturbation type for proving the solvability of nonlinear functional equations in Banach spaces, (to appear).

65. _____, A remark on nonlinear equations of Hammerstein type with compact kernels, (to appear).

66. _____, On nonlinear equations of Hammerstein type in Banach spaces (to appear).

67. R. Iglisch, Existenz and Eindeutigkeit Satze bei nichtlineare Integralgleichungen, Math. Annalen, 108 (1933) 161-189.

68. R. I Kacurovski, Nonlinear monotone operators on Banach spaces, Russian Math. Surveys, 23 (1968) 117-165.

69. _____, On Fredholm theory for nonlinear operator equations, Dokladi Akad. Nauk SSSR, 192 (1970) 969-972 (Soviet Math. Dokladi, 11 (1970) 751-754.)

70. R. I. Kacurovski, On the fixed point principle of A. N. Tikohonoff for equations with operators weakly closed on their kernels, Dokladi Akad. Nauk SSSR, 183 (1968) 517-520.

71. I. I. Kolodner, Equations of Hammerstein type in Hilbert space, J. Math. Mech., 13 (1964) 701-750.

72. J. Kolomy, Applications of some existence theorems for the solution of Hammerstein integral equations, Comm. Math. Univ. Carolinae, 7 (1966) 46-478.

73. _____, The solvability of nonlinear integral equations, Comm. Math. Univ. Carolinae, 8 (1967) 273-289.

74. M. E. Koscikii, Nonlinear equations of Hammerstein type with a monotone operator, Dokladi Akad. Nauk SSSR, 190 (1970) 31-33 (Soviet Math. Dokladi 11 (1970) 25-28).

75. M. A. Krasnoselski, Topological methods in the theory of nonlinear integral equations, Moscow, 1956 (Engl. transl. Mackmillan, New York, 1964).

76. _____, Positive solutions of operator equations, Moscow, 1962 (Engl. transl. Nordhoff Ltd. 1964).

77. M. A. Krasnoselski, P. P. Zabreiko, E. I. Pustilnik, and P. E. Sobolevski, Integral operators on spaces of summable functions, Moscow, 1966.

78. M. A. Krasnoselski and Ya. B. Rutitskii, Orlicz spaces and nonlinear integral equations, Trudi Moscov. Mat. Obsestva, 7 (1958) 63-120.

79. I. M. Lavrentiev, On the variational theory of non-linear equations, Dokladi Akad. Nauk SSSR, 166 (1966) 248-286 (Soviet Math. Dokladi 8 (1966) 69-71).

80. I. M. Lavrentiev, Solvability of nonlinear equations, Dokladi Akad. Nauk SSSR, 175 (1967) 1219-1221 (Soviet Math. Dokladi, 8 (1967) 993-996).

81. J. Leray and J. L. Lions, Quelques resultats de Visik sur les problemes elliptiques nonlineaires par les methodes de Minty-Browder, Bull. Soc. Math. France, 93 (1965) 97 - 107.

82. J. Leray and J. Schauder, Topologie et equations fonctionelles, Ann. Sci. Ec. Norm. Sup. Paris, 51 (1934) 45-78.

83. J. L. Lions, Quelques methodes de resolution des problemes aux limites nonlineaires, Dunod-Gauthier Villars, Paris, 1969.

84. G. J. Minty, Monotone nonlinear operators in Hilbert space, Duke Math. Jour., 29 (1962) 341-346.

85. _____, On a "monotonicity" method for the solution of nonlinear equations in Banach spaces, Proc. Nat. Acad. Sci. U.S.A., 50 (1963) 1038-1041.

86. J. Necas, Sur l'alternative de Fredholm pour les operateurs nonlineaires avec applications aux problemes aux limites, Ann. Scuola Norm. Sup. Pisa, 23 (1969) 331-345.

87. _____, Remark on the Fredholm alternative for nonlinear operators with application to nonlinear integral equations of generalized Hammerstein type, (to appear).

88. W. V. Petryshyn, Projection methods in nonlinear numerical functional analysis, J. Math. Mech., 17 (1967) 353-372.

89. _____, Nonlinear equations involving noncompact operators, Symposia in Pure Math., vol. 18, part 1, Amer. Math. Soc., (1970), 206-233.

90. W. V. Petryshyn, Invariance of domain theorem for locally A-proper mappings and its implications, Jour. Funct. Anal., 5(1970) 137-159.

91. ———, Antipodes theorem for A-proper mappings and its applications to mappings of the modified type (S) or (S) and to mappings with the pm property, (to appear in Jour. Funct. Anal).

92. ———, On existence theorems for nonlinear equations involving noncompact mappings, Proc. Nat. Acad. Sci. U.S.A., 67 (1970) 326-330.

93. ———, On nonlinear equations involving pseudo-A-proper mappings and their uniform limits, with applications, (to appear).

94. W. V. Petryshyn and P. M. Fitzpatrick, New existence theorems for nonlinear integral equations of Hammerstein type, (to appear in Jour. Math. Anal. Appl.).

95. S. I. Pohozhayev, Solvability of nonlinear equations with odd operators, Jour. Funct. Anal. and Appl., 1 (1967) 222-233.

96. R. T. Rockafellar, Convexity properties of nonlinear maximal monotone operators, Bull. Amer. Math. Soc., 75 (1969) 74-77.

97. ———, Local boundedness of nonlinear monotone operators, Michigan Math. Jour., 16 (1969) 397-407.

98. ———, On the maximality of sums of nonlinear monotone operators, Trans. Amer. Math. Soc., 149 (1970).

99. E. H. Rothe, Weak topology and nonlinear integral equations, Trans. Amer. Math. Soc., 66 (1949) 75-92.

100. M. M. Vainberg, <u>Variational methods for the study of nonlinear operators</u>, Moscow, 1956 (English, transl. Holden-Day, San Francisco 1964).

101. _____, New theorems for nonlinear operators and equations, Moscov. Oblast. Inst. Ucen. Zap., 77 (1959) 131-143.

102. _____, Nonlinear problems for potential and monotone operators, Dokladi Ada. Nauk SSSR, 183 (1968) 747-749 (Soviet Math. Dokladi 9 (1968) 1427-1430).

103. M. M. Vainberg and R. I. Kacurovski, On the variational theory of nonlinear operator equations, Dokladi Akad. Nauk SSSR, 129 (1959) 1100-1102.

104. F. Wille, On monotone operators with perturbations, (to appear in Archive Rat. Mech. Anal.).

105. A. C. Zaanen, Integral transformations and their resolvents in Lebesgue and Orlicz spaces, Comp. Math., 10 (1952).

106. _____, <u>Linear analysis</u>, New York-Amsterdam, 1953.

107. E. H. Zarantonello, Solving functional equations by contractive averaging, Tech. Rep. 160, Math. Research Center, Univ. of Wisconsin, Madison (1960).

Department of Mathematics
University of Chicago,
Chicago, Illinois

Received June 8, 1971.

# Asymptotic Behaviour of Bounded Solutions of Some Functional Equations

*J. J. LEVIN AND D. F. SHEA*

Introduction. We consider functional equations

(F) $\qquad Q(x)(t) = f(t) \qquad (-\infty < t < \infty)$

and functional-differential equations

(D) $\qquad x'(t) + Q(x)(t) = f(t) \qquad (-\infty < t < \infty),$

and ask for the asymptotics of a bounded solution $x(t)$ when f satisfies

(1) $\qquad f(t) \to 0 \qquad (t \to \infty)$

or some related condition such as

(2) $\qquad f(t) - \omega(t) \to 0 \qquad (t \to \infty)$

with $\omega$ periodic, say. Under quite general hypotheses on Q we represent any bounded x in terms of certain solutions y of the related "limit equation"

(F*) $\qquad Q(y)(t) = 0 \qquad (-\infty < t < \infty)$

or

(D*) $\qquad y'(t) + Q(y)(t) = 0 \qquad (-\infty < t < \infty)$

when (1) holds, for instance. Also, for certain classes of linear and nonlinear Q we are able to explicitly solve (F*) and (D*), and so obtain complete information - the best possible for f satisfying (1) - on the asymptotic behavior of x.

If x is a bounded, Borel measurable solution of

(3) $\quad Q(x)(t) \equiv \int_{-\infty}^{\infty} x(t-\xi) dA(\xi) = f(t) \qquad (-\infty < t < \infty)$

and A has bounded variation on $(-\infty, \infty)$, then Wiener's tauberian theorem gives complete information if (1) holds and

(4) $\quad \hat{A}(\lambda) = \int_{-\infty}^{\infty} e^{-i\lambda t} dA(t) \neq 0 \qquad (-\infty < \lambda < \infty).$

Thus, for this choice of Q the only solution of (F*) is $y = 0$, and

(5) $\qquad\qquad x(t) \to 0 \qquad\qquad (t \to \infty)$

provided

(T) $\qquad\qquad \lim_{\substack{t \to \infty \\ \eta \to 0}} |x(t+\eta) - x(t)| = 0$

holds. Well known examples show the necessity for (5) of some tauberian condition such as (T).

Part of the motivation for our study of (F) and (D) comes from Wiener's theorem, and our Theorems 1b-4b below generalize this classical result in several ways. For instance, we shall consider (F) with

(6) $\quad Q(x)(t) = x(t) + \int_{-\infty}^{\infty} g(x(t-\xi)) dA(\xi),$

which clearly reduces to (3) in the linear case $g(x) = x$ if A is replaced by A - H, H denoting the unit step function with jump at 0. Even in the linear case (3) our methods lead to new extensions of Wiener's theorem (Theorems 2b, 3b).

The results we state below for (F) have analogues for (D). Corresponding to (6) we discuss (D) with

(7) $$Q(x)(t) = \int_{-\infty}^{\infty} g(x(t - \xi))dA(\xi),$$

an interesting special case being the Volterra equation

(8) $$x'(x) + \int_0^t g(x(t - \xi)dA(\xi) = f(t) \quad (0 \le t < \infty).$$

(Given a solution $x$ of (8) which is bounded on $[0, \infty)$, it is not difficult to see how to extend $x, f$ and $A$ to $(-\infty, \infty)$ so that the extended $x$ satisfies a special case of (D), with appropriate hypotheses preserved. Similar remarks apply to other Volterra and delay equations on $[0, \infty)$; for a complete discussion see [8].) Perhaps surprisingly, our methods give new information even for the linear system

(9) $$x'(t) + x(t)A = f(t) \quad (0 \le t < \infty)$$

where $A$ is an $N$ by $N$ constant matrix, a very special case of (8).

Volterra equations are of particular interest since they arise naturally in many nonlinear boundary value problems, see e.g. Levinson [9] and the references cited there. Equation (8) is important in applications to reactor theory and control theory; the choice (6) in (F) is relevant to renewal theory. In Theorems 4a, 5a, 3b, 5b especially, we study these equations under hypotheses natural to the applications.

Needless to say, the questions of existence and boundedness of solutions of (F) and (D) are of great interest in themselves, especially in the study of specific equations arising in applications. However, it often happens that once existence and boundedness have been established a quite different analysis is required for finding more detailed asymptotic properties of the solution. Thus, the Volterra nature of an equation such as (8) is usually crucial in the study of existence and boundedness on $[0, \infty)$ of solutions $x(t)$; but the limit equation corresponding to (8) is, when (1) holds,

$$(8*) \qquad y'(t) + \int_0^\infty g(y(t-\xi))\,dA(\xi) = 0 \qquad (-\infty < t < \infty),$$

not an equation of Volterra type. In this paper we generally take the existence and boundedness of solutions as a hypothesis; further discussion is given in [8].

Except for the comments supplied in Section 2 proofs of the results outlined below as well as extensions in several directions not mentioned here may be found in [8].

## 1. Statement of results.

By a solution of (D) we always mean a row vector $x(t)$ with values in complex $N$-space $\mathbb{C}^N$, defined on $\mathbb{R} = (-\infty, \infty)$ and absolutely continuous on bounded intervals of $\mathbb{R}$, which satisfies (D) a.e. By a solution of (F) we mean a Borel-measurable $x:\mathbb{R} \to \mathbb{C}^N$ which satisfies (F) for all $t \in \mathbb{R}$.

For $x = (x_1, \ldots, x_N) \in \mathbb{C}^N$, denote $|x| = \sum_{j=1}^N |x_j|$.

When $N > 1$, we are to interpret the convolution integrals in (6) and (7) as

$$\int_{-\infty}^\infty g(x(t-\xi))\,dA(\xi) = (z_1, \ldots, z_N),$$

where $x = (x_1, \ldots, x_N)$, $g = (g_1, \ldots, g_N)$, $A = [A_{ij}]$ is $N$ by $N$ and

$$z_j(t) = \sum_{i=1}^N \int_{-\infty}^\infty g_i(x_1(t-\xi), \ldots, x_N(t-\xi))\,dA_{ij}(\xi)$$

$$(1 \le j \le N, \ t \in \mathbb{R}).$$

Theorems 1a and 1b below are our most general statements for (D) and (F), and involve the notion of "$\psi$-sequence" introduced in [8].

<u>Definition.</u> A sequence $\{\psi_m\}$ of nonnegative functions in $C^\infty(\mathbb{R})$ is a $\psi$-sequence if there exists an increasing sequence $\{t_m\} \subset \mathbb{R}$ such that

$$(1.1) \qquad \lim_{m \to \infty} (t_m - t_{m-1}) = \infty,$$

with $\psi_1(t) \equiv 1 (t \leq t_1)$, $\psi_1'(t) \leq 0 (t_1 < t < t_2)$, $\psi_1(t) \equiv 0 (t \geq t_2)$,

(1.2) $\begin{cases} \text{support}(\psi_m) = [t_{m-1}, t_{m+1}], \quad \psi_m(t_m) = 1 \\ \psi_m'(t) \geq 0 (t_{m-1} < t < t_m), \quad \psi_m'(t) \leq 0 (t_m < t < t_{m+1}) \end{cases}$

for $m \geq 2$, and such that

(1.3) $\qquad \sum_{m=1}^{\infty} \psi_m(t) \equiv 1 \qquad (-\infty < t < \infty)$,

(1.4) $\qquad \lim_{m \to \infty} \|\psi_m'\|_\infty = 0$.

When we wish to emphasize the choice of $\{t_m\}$ in (1.1) and (1.2) we say that $\{\psi_m\}$ is a $\psi$-sequence associated with $\{t_m\}$.

In Theorems 1a and 1b we represent a bounded solution x of (D) or (F) for all large t in terms of a suitable $\psi$-sequence and solutions y of (D*) or (F*). Denote by $L^\infty(I)$ the usual space of essentially bounded functions on I, with norm $\|\cdot\|_\infty$ in case $I = \mathbb{R}$, and let

$$C_u(I) = \{x : x \text{ is uniformly continuous on } I\}$$
$$C_u^k(I) = \{x : x^{(k)} \in C_u(I)\}.$$

For any $x \in L^\infty(\mathbb{R})$ let $x_\tau$ denote the translate of x by $\tau$, so that $x_\tau(t) = x(t + \tau)$.

<u>Theorem 1a.</u> <u>Let</u> x <u>be a bounded solution of</u> (D), <u>where</u> $f \in L^\infty(\mathbb{R})$ <u>satisfies</u> (1) <u>and</u>

(1.5) $\qquad Q : L^\infty(\mathbb{R}) \cap C_u(\mathbb{R}) \to L^\infty \cap C_u$,

(1.6) $\quad Q(x_\tau)(t) = Q(x)(t+\tau)$ <u>for</u> $x \in L^\infty \cap C_u$ <u>and</u> $t, \tau \in \mathbb{R}$,

$$(1.7) \begin{cases} \text{If } \{x_n\} \subset L^\infty \cap C_u \text{ is uniformly bounded on } \mathbb{R} \text{ and} \\ x_n \to y \text{ uniformly on compact sets when } n \to \infty, \text{ then} \\ Q(x_n) \to Q(y) \text{ uniformly on compact subsets of } \mathbb{R}. \end{cases}$$

Define

(1.8) $\quad \Gamma_a = \{y : y \in C_u^1(\mathbb{R}), \|y\|_\infty \leq \|x\|_\infty, y \text{ satisfies } (D*)\}$.

Then there exist $y_m \in \Gamma_a$ and a $\psi$-sequence $\{\psi_m\}$ such that

(1.9) $\quad\quad\quad x(t) = \sum_{m=1}^\infty \psi_m(t) y_m(t) + o(1) \quad (t \to \infty)$.

Relation (1.9) can be differentiated; using (1.3), (1.7) and (1) one can deduce

(1.10) $\quad\quad\quad x'(t) = \sum_{m=1}^\infty \psi_m(t) y_m'(t) + o(1) \quad (t \to \infty, t \notin E)$

where $E$ has measure zero.

If $\{\psi_m\}$ is associated with $\{t_m\}$, the proof of (1.9) yields also that

(1.11) $\quad\quad\quad x(t) = y_m(t) + o(1) \quad (t_{m-2} \leq t \leq t_{m+2}, t \to \infty)$

and thus

(1.12) $\quad\quad \sup_{t_{m-1} \leq t \leq t_{m+2}} |y_m(t) - y_{m+1}(t)| \to 0 \quad (m \to \infty)$.

In some applications $\Gamma_a$ consists only of a finite number of constants, in which case (1.12) and (1.9) force $x$ to converge.

Clearly, (7) satisfies (1.5)-(1.7), as does

$$Q(x)(t) = \int_{-\infty}^\infty g(x(t-\xi), \xi) dA(\xi)$$

where $A$ is of bounded variation on $\mathbb{R}$ and $g \in C(\mathbb{C}^N \times \mathbb{R})$

satisfies $|g(x,\xi)| \leq B(\rho)$ for all $\xi \in \mathbb{R}$, $x \in \mathbb{C}^N$ with $|x| \leq \rho$; this example shows that the convolution nature of (7) is not used in deducing (1.9) for solutions of integrodifferential equations.

Assertion (1.9) is related to but more precise than results derived from analysis of the "positive limit set"

(1.13) $\quad \Omega = \{x^* : \lim_{\nu \to \infty} x(\tau_\nu) = x^* \text{ for some } \tau_\nu \to \infty \text{ as } \nu \to \infty\}$.

A study of $\Omega$ is useful whenever one can show that, for each $x^* \in \Omega$, there exists some solution $y$ of a limit equation which is approached on compact intervals by a sequence of translates of $x$, and satisfies $y(0) = x^*$.[†] This is less precise than (1.9) and (1.12) which make explicit the slowly varying change of $x$ from one solution of (D*) to another. Of course, compactness arguments involving translations are also crucial to the proofs of Theorems 1a and 1b. For related work dealing with $\Omega$, see Hale [2], Miller [10].

Theorem 1a is sharp, in the sense that it admits this converse:

Let Q satisfy (1.5)-(1.7) and let $\{\psi_m\}$ be a given $\psi$-sequence. Let $y_m \in C_u^1(\mathbb{R})$ satisfy (D*), $\|y_m\|_\infty \leq M$ and (1.12) for $m \geq 1$, and define

$$x(t) = \sum_{m=1}^{\infty} \psi_m(t) y_m(t) \quad (-\infty < t < \infty).$$

Then

(1.14) $\quad \lim_{t \to \infty} \{x'(t) \psi Q(x)(t)\} = 0$.

The proof of (1.14) is a straightforward consequence of the assumptions and properties (1.1)-(1-4) of $\psi$-sequences, which imply

---

[†] This is easily seen to be true for $x$ satisfying the conditions of Theorem 1a, by using (1.9) and the translation-invariance of $\Gamma_a$.

$$x'(t) + Q(x)(t) = \sum_{m=k}^{k+1} \psi_m(t)\{Q(x)(t) - Q(y_m)(t)\} + \sum_{m=k}^{k+1} \psi'_m(t) y_m(t)$$

$$(t_k \leq t \leq t_{k+1}).$$

Theorem 1a is most useful when the solutions of (D*) in $\Gamma_a$ are known explicitly. For example, when $Q$ is linear we have

**Theorem 2a.** <u>Let $x$ be a bounded solution of</u>

(1.15) $\qquad x'(t) + \int_{-\infty}^{\infty} x(t-\xi) dA(\xi) = f(t) \qquad (-\infty < t < \infty)$

<u>where</u> $f$ <u>satisfies (1) and</u> $A$ <u>has bounded variation on</u> R. <u>Define</u>

(1.16) $\qquad S_a(A) = \{\lambda : \det[i\lambda E + \hat{A}(\lambda)] = 0, \lambda \in R\}.$

(i) <u>If</u> $S_a(A) = \phi$, <u>then (5) holds.</u>

(ii) <u>If</u> $S_a(A) = \{\lambda_1, \ldots, \lambda_n\}$, <u>then</u>

(1.17) $\qquad x(t) = \sum_{k=1}^{n} c_k(t) e^{i\lambda_k t} + o(1) \qquad (t \to \infty)$

<u>where</u> $c_k(t) = (c_{k1}(t), \ldots, c_{kN}(t)) \in C^{\infty} \cap L^{\infty}$ <u>for each</u> $k$ <u>and</u>

(1.18) $\qquad \lim_{t \to \infty} c'_k(t) = 0 \qquad (1 \leq k \leq n).$

Here $E$ denotes the $N$ by $N$ identity matrix and

$$\hat{A}(\lambda) = [\hat{A}_{ij}(\lambda)] = [\int_{-\infty}^{\infty} e^{-i\lambda t} dA_{ij}(t)] \qquad (1 \leq i, j \leq N).$$

Assertion (1.18) may be interpreted as saying that the coefficients $c_k(t)$ in (1.17) behave asymptotically, on longer and longer intervals, like constants (e.g. $c(t) = \alpha + \beta \sin\{\log(1+t^2)\}$).

## ASYMPTOTIC BEHAVIOUR OF BOUNDED SOLUTIONS

The proof of Theorem 2a depends on showing that $y \in \Gamma_a$ only if

(1.19) $$y(t) = \sum_{k=1}^{n} \beta_k e^{i\lambda_k t} \qquad (\beta_k \in \mathbb{C}^N) ;$$

this fact is a consequence of a spectral synthesis result of Beurling [1], [8, Sections 8, 18]. Then, Theorem 1a can be applied, to obtain

$$x(t) = \sum_{m=1}^{\infty} \psi_m(t) \{ \sum_{k=1}^{n} \beta_k^{(m)} e^{i\lambda_k t} \} + o(1) \qquad (t \to \infty)$$

and so (1.17) holds with

$$c_k(t) = \sum_{m=1}^{\infty} \beta_k^{(m)} \psi_m(t) \qquad (1 \le k \le n) ;$$

using property (1.4) of $\psi$-sequences, (1.18) follows. It is not difficult to see that the $c_k(t)$ also satisfy

$$\sum_{k=1}^{n} |c_k(t)|^2 \le \|x\|_{\infty}^2 \qquad (-\infty < t < \infty) .$$

When $S_a(A)$ is countably infinite an analogue of Theorem 2a still holds, since then $\Gamma_a$ consists of uniform limits of trigonometric polynomials (1.19) having $\{\lambda_k\} \subset S_a$. But when $S_a$ is uncountable there exist $y \in \Gamma_a$ which are not a.p.; see [8].

Theorem 2a applies in particular to the Volterra equation

(1.20) $$x'(t) + \int_0^t x(t-\xi) dA(\xi) = f(t) \qquad (0 \le t < \infty) ,$$

as can easily be seen by appropriately extending A, f and x to $(-\infty, \infty)$; see the discussion following (8) in the Introduction. In particular, one defines $A(t) \equiv A(0)$ on $-\infty < t \le 0$ to obtain the appropriate spectral set (1.16) for this equation.

The special case of (1.20) in which A(t) is a constant matrix except for a jump A at t = 0 is the linear system (9) of ordinary differential equations. Here the associated limit system is

(9*) $$y'(t) + y(t)A = 0 \qquad (-\infty < t < \infty),$$

with general solution well known to be $y(t) = \beta e^{-tA}$ for $\beta \in \mathbb{C}^N$, and in this case $S_a(A)$ is finite and the fact that $\Gamma_a$ consists only of trig polynomials (1.19) is elementary.

One further comment on the linear case of (D): If besides (1.18) the $c_k(t)$ also satisfy

(1.21) $$c_k(\infty) = \lim_{t \to \infty} c_k(t) \quad \text{exist} \quad (1 \le k \le n)$$

then $x(t)$ is asymptotically almost periodic. Simple conditions implying (1.21) are given in the next theorem.

<u>Theorem 3a.</u> <u>Let</u> $x$ <u>be a bounded solution of the scalar case</u> $N = 1$ <u>of</u> (1.15) <u>where</u> $A \in BV(\mathbb{R})$ <u>and</u> $S_a(A) = \{\lambda_1, \dots, \lambda_n\}$ <u>is nonempty and finite. Assume</u>

$$\int_{-\infty}^{\infty} |t|^n |dA(t)| < \infty, \quad \int_{-\infty}^{\infty} e^{-i\lambda_k t} t \, dA(t) \ne 1 \quad (1 \le k \le n)$$

<u>and that</u> $f \in L^\infty(\mathbb{R})$ <u>satisfies</u>

(1.22) $$\int_0^\infty t^{n-1} |f(t)| \, dt < \infty.$$

Then

(1.23) $$x(t) = \sum_{k=1}^n \gamma_k e^{i\lambda_k t} + o(1) \qquad (t \to \infty)$$

<u>for suitable</u> $\gamma_k \in \mathbb{C}$.

We turn now to some nonlinear equations. Theorem 4a treats a very simple and often discussed case of (D), with $Q$ given by (7); for simplicity we assume all functions are real valued. Special cases **of** Theorem 4a are due to Hale [2], Miller [11] and Volterra [13]; a similar result in [8] requires more than (1.26) of $g$.[†]

---

[†]Carol Shilepsky has independently found the same improvement, with a different method of proof.

**Theorem 4a.** Let $x$ be a bounded solution of

(1.24) $\quad x'(t) + \int_{-\infty}^{\infty} g(x(t-\xi))dA(\xi) = f(t) \quad (-\infty < t < \infty),$

where $A = A_1 + A_2$ with

(1.25) $\quad \begin{cases} A_1(t) = 0 \ (t \leq 0), \ A_1(t) = \rho > 0 \quad (0 < t < \infty), \\ A_2 \ \underline{\text{of total variation}} \ V(A_2) < \rho, \end{cases}$

and

(1.26) $\quad g \in C(\mathbb{R}), \ g(x) = 0 \ \text{iff} \ x = 0.$

If $f \in L^\infty(\mathbb{R})$ satisfies (1), then (5) holds.

Roughly speaking, hypothesis (1.25) causes (1.24) to behave much like an ordinary differential equation. The proof, which is sketched in Section 2, consists of verifying that $\Gamma_a = \{0\}$. We also show that (1.25) is sharp, in the sense that (5) can fail if $V(A_2) \geq \rho$.

We next consider the Volterra equation

(1.27) $\quad x'(t) + G(x(t)) + \int_0^t g(x(t-\xi))a(\xi)d\xi = f(t)$

$\quad (0 \leq t < \infty),$

with all functions real-valued, say. Existence, boundedness, asymptotic behavior and system formulations of closely related equations are discussed in [6], [7], [3], [12, p. 258].

**Theorem 5a.** Let $x$ be a bounded solution of (1.27), where

(1.28) $\quad g, G \in C(\mathbb{R}); \ xg(x) > 0, \ xG(x) \geq 0 \quad (x \neq 0),$

(1.29) $\quad \begin{cases} a \in C^2(0,\infty) \cap L^1(0,\infty), \ ta(t) \in L^1(0,\infty), \\ (-1)^k a^{(k)}(t) \geq 0 \quad (0 < t < \infty; \ k = 0,1,2), \ a \not\equiv 0. \end{cases}$

*If* $f \in L^\infty(0, \infty)$ *satisfies* (1), *then* (5) *holds*.

The proof of Theorem 5a is an application of Theorem 1a in which $\Gamma_a$ is shown to consist just of $y(t) \equiv 0$. An argument of Liapunov function type is applied to

(1.27*) $\quad y'(t) + G(y(t)) + \int_0^\infty g(y(t-\xi))a(\xi)d\xi = 0 \quad (-\infty < t < \infty)$

to achieve this.

We now briefly consider some analogues of Theorems 1a-5a for equation (F). In Theorems 1b and 2b below we permit more general functionals $Q$ than in Theorems 1a and 2a; but when $Q$ also satisfies the translation-invariance and boundedness conditions (1.6) and (1.5), Theorems 1b and 2b reduce to the exact analogues for (F) of Theorems 1a and 2a. For our purposes the most interesting choice of $Q$ is

(6) $\quad\quad Q(x)(t) = x(t) + \int_{-\infty}^{\infty} g(x(t - \xi)) \, dA(\xi)$

or its linear special case (3); for this $Q$ it is obvious that (1.30) below is equivalent to the assertion: $x$ is a solution of (F), and (1) holds.

Denote

$$\tilde{L}^\infty(\mathbb{R}) = L^\infty(\mathbb{R}) \cap \{x : x \text{ is Borel-measurable}\}.$$

**Theorem 1b.** *Let* $x \in \tilde{L}^\infty(\mathbb{R})$ *satisfy*

(1.30) $\quad\quad \lim_{T \to \infty} Q(x_T)(t) = 0 \quad\quad (t \in \mathbb{R})$

*and the tauberian condition* (T). *Let* $Q(x)(t)$ *be defined whenever* $x \in \tilde{L}^\infty(\mathbb{R})$ *and* $t \in \mathbb{R}$, *and suppose'*

(1.31) $\begin{cases} Q(x_n) \to Q(x) \text{ pointwise on } \mathbb{R} \text{ when } n \to \infty \text{ if} \\ \{x_n\} \subset \tilde{L}^\infty(\mathbb{R}) \text{ is uniformly bounded on } \mathbb{R} \text{ and converges uniformly on compact sets to } x, \end{cases}$

(1.32)    $Q(y) = 0$ implies $Q(y_\tau) = 0$ for $\tau \in \mathbb{R}$,

$$y \in L^\infty \cap C_u(\mathbb{R}).$$

Define

$$\Gamma_b = \{y : y \in C_u(\mathbb{R}), \|y\|_\infty \leq \|x\|_\infty, y \text{ satisfies (F*)}\}.$$

Then there exist $y_m \in \Gamma_b$ and a $\psi$-underline{sequence} $\{\psi_m\}$ such that (1.9) holds.

A choice of $Q$ satisfying the above conditions but in general not (1.5) or (1.6) is

(1.33)    $$Q(x)(t) = \int_{-\infty}^\infty x(\xi) a(t, \xi) d\xi$$

where $a(t, \cdot) \in L^1(\mathbb{R})$ for each fixed $t$ and is such that (1.32) holds, for example $a(t, \xi) = b(t + \xi) - b(\xi)$ with $b$ increasing on $\mathbb{R}$, $-\infty < b(-\infty) < b(\infty) < \infty$.

As mentioned in the Introduction (T) is in general necessary for (1.9), even for the special choice (3) of $Q$. However, sometimes (T) holds automatically, for example when $Q$ has the form (6) with $A$ absolutely continuous on $\mathbb{R}$, $x \in L^\infty(\mathbb{R})$ satisfies (F) and $f$ satisfies (1). In the linear case $g(x) = x$ of (6), it is not difficult to see that (T) holds whenever $A$ has singular part of total variation $< 1$.

For simplicity we state analogues of Theorems 2a-5a only for the scalar case $N = 1$. When $Q$ is linear, e.g. of the form (3) or (1.33), harmonic analysis methods can be used to characterize $\Gamma_b$:

**Theorem 2b.**   Let $x$ and $Q$ satisfy the hypotheses of Theorem 1b with $Q$ linear. Define

(1.34)    $$S_b(Q) = \{\lambda : e^{i\lambda t} \text{ satisfies (F*)}\}.$$

   (i) If $S_b(Q) = \phi$, then (5) holds.

   (ii) If $S_b(Q) = \{\lambda_1, \ldots, \lambda_n\}$, then (1.17)-(1.18) holds.

As with Theorem 2a, a generalization to countably infinite $S_b(Q)$ is valid. Part (i) is due to Beurling [1]; see also Korevaar [5]. When Q is given by (3), with $A \in BV(\mathbb{R})$,

$$S_b(Q) = \{\lambda : \hat{A}(\lambda) = 0, \lambda \in \mathbb{R}\} \equiv S_b(A)$$

and part (i) reduces to Wiener's tauberian theorem.

**Theorem 3b.** Let $x \in \tilde{L}^\infty(\mathbb{R})$ satisfy (T) and (3), with A of bounded variation on $\mathbb{R}$ and $S_b(A) = \{\lambda_1, \ldots, \lambda_n\}$,

$$\int_{-\infty}^\infty |t|^n |dA(t)| < \infty, \quad \int_{-\infty}^\infty e^{-i\lambda_k t} t\, dA(t) \neq 0 \quad (1 \leq k \leq n).$$

If f satisfies (1.22), then (1.23) holds.

If $A = A_1 - B$ where $A_1$ satisfies (1.25) with $\rho = 1$ and B is a nonarithmetic probability distribution, then $S_b(A) = \{0\}$ and Theorem 3b yields a result of Karlin on the renewal equation

(1.35) $\qquad x(t) - \int_{-\infty}^\infty x(t - \xi)\, dB(\xi) = f(t) \qquad (-\infty < t < \infty).$

**Corollary 3b.** Let $x \in \tilde{L}^\infty(\mathbb{R})$ be a solution of (1.35) where B is nonarithmetic,

$$\int_{-\infty}^\infty |t|\, dB(t) < \infty, \quad m = \int_{-\infty}^\infty t\, dB(t) \neq 0$$

and $f \in L^1(\mathbb{R})$.

Then for every $g \in L^1(\mathbb{R})$, the limits

$$x * q(\pm \infty) \equiv \lim_{t \to \pm \infty} \int_{-\infty}^\infty x(t - \xi) q(\xi)\, d\xi$$

exist and satisfy

$$x * q(\infty) - x * q(-\infty) = \frac{1}{m} \int_{-\infty}^\infty f(\xi)\, d\xi \int_{-\infty}^\infty g(\xi)\, d\xi.$$

For details of the proof see [4], [8].

Analogues of the nonlinear Theorems 4a, 5a can be found (here all functions are real valued):

**Theorem 4b.** Let $x \in \tilde{L}^\infty(\mathbb{R})$ be a solution of (F), where Q is given by (6) with g such that

(1.36) $\quad\quad\quad g \in C(\mathbb{R}), \ xg(x) \geq 0 \quad\quad (x \in \mathbb{R})$

and $A = A_1 + A_2$ with $A_1$ as in (1.25) and

(1.37) $\quad\quad\quad A_2$ of total variation $V(A_2) \leq \rho$.

Then (1) implies (5).

**Theorem 5b.** Let $x \in L^\infty(0,\infty)$ be a solution of

$$x(t) + \int_0^t g(x(t-\xi))a(\xi)d\xi = f(t) \quad\quad (0 \leq t < \infty)$$

where f satisfies (1), g satisfies (1.36) and is strictly increasing, and

$a \in C^1(0,\infty) \cap L^1(0,\infty), \ (-1)^k a^{(k)}(t) > 0 \ (0 < t < \infty; k = 0, 1), \ a(0+) < \infty$.

Then (5) holds.

For extensions of the above theorems to perturbations f satisfying (2) for various classes of $\omega$, to equations involving higher order derivatives, and to integral equations of nonconvolution type, see [8].

**2. Remarks on the proofs.** Only Theorems 4a and 5a require comment, since proofs of the other statements given here can be found in [8]. (Theorems 1b and 2b here are more general than the corresponding results in [8], but the proofs there extend easily.)

Theorem 4a follows from Theorem 1a and the

**Proposition.** If g, $A_2$ and $\rho$ are as in Theorem 4a, and y is a bounded solution of

(1.24*) $\quad y'(t) + \rho g(y(t)) = \int_{-\infty}^{\infty} g(y(t-\xi))dA_2(\xi) \quad (-\infty < t < \infty)$,

then $y = 0$.

Proof. We first establish

(2.1) $\quad\quad\quad \liminf_{t \to \pm\infty} |g(y(t))| = 0$.

If $p(t) = |g(y(t))| > \eta > 0$ for $t_0 < t < \infty$, say, integration of (1.24*) gives

$$\rho \int_{t_0}^{T} g(y(t))dt = -\int_{-\infty}^{\infty} \{\int_{t_0-\xi}^{T-\xi} g(y(s))ds\} dA_2(\xi) + y(t_0) - y(T)$$

for every $T > t_0$. Thus

$$\rho \int_{t_0}^{T} p(t)dt \le \int_{-\infty}^{\infty} \{\int_{t_0-\xi}^{T-\xi} p(s)ds\} |dA_2(\xi)| + 2\|y\|_{\infty}.$$

Choose $\delta$ so that $0 < \delta < \rho - V(A_2)$, and let

$$\{\int_{-\infty}^{-M} + \int_{M}^{\infty}\} |dA_2(\xi)| < \rho\eta / \|p\|_{\infty}$$

so that

$$\rho \int_{t_0}^{T} p(t)dt \le \int_{-M}^{M} \{\int_{t_0-\xi}^{T-\xi} p(s)ds\} |dA_2(\xi)| + \delta\eta(T-t_0) + 2\|y\|_{\infty}$$

$$\le \{V(A_2) + \delta\} \int_{t_0}^{T} p(s)ds + 2\|p\|_{\infty} \int_{-M}^{M} |\xi| |dA_2(\xi)| + 2\|y\|_{\infty}.$$

Letting $T \to \infty$ yields

$$\rho \le V(A_2) + \delta,$$

a contradiction. Similarly $|g(y(t))| > \eta > 0$ for $-\infty < t < t_0$ leads to a contradiction, and so (2.1) is true.

Let
$$\alpha = \sup_t g(y(t)), \quad \beta = \inf_t g(y(t)).$$

By (2.1) $\beta \leq 0 \leq \alpha$, and it remains to prove $\alpha = \beta = 0$. Suppose

(2.2) $\qquad \gamma = \max\{\alpha, -\beta\} > 0$,

say $\gamma = \alpha$. In view of (2.1) an inspection of the graphs $(t, g(y(t)))$ and $(y, g(y))$ shows there exists $t_0$ such that

(2.3) $\qquad g(y(t_0)) = \alpha, \quad y'(t_0) \geq 0$,

or else there exists a sequence $\{t_n\}$, $|t_n| \to \infty$, with

(2.4) $\qquad g(y(t_n)) \to \alpha, \quad y'(t_n) = 0$.

If (2.4) holds,
$$\rho\alpha + o(1) \leq \alpha V(A_2) \qquad (n \to \infty),$$

a contradiction. Similarly (2.3) is impossible, and so (2.2) is false. By (1.26), $y(t) \equiv 0$.

To see that hypothesis $V(A_2) < \rho$ is sharp, let $A_2(t) = -A_1(t)$, $g(x) = x$, and

(2.5) $\qquad x(t) = \sin\{\log(1 + t^2)\}$,

$$f(t) = x'(t) + \rho g(x(t)) + \int_{-\infty}^{\infty} g(x(t-\xi)) dA_2(\xi).$$

Then

(2.6) $\qquad f(t) = x'(t) = o(1) \qquad (t \to \infty)$,

but (5) fails.

To see that (1.37) is necessary in Theorem 4b let $A_2(t) = -(1+\rho)A_1(t)$, $g(x) = x$ and, say, $x(t) = \sin t$.

Then

$$f(t) = x(t) + \rho x(t) + \int_{-\infty}^{\infty} x(t-\xi) dA_2(\xi) \equiv 0$$

on $-\infty < t < \infty$, but (5) fails.

Theorem 5a is proved by studying the energy functional

(2.7) $\quad V_y(t) = \int_0^{y(t)} g(\xi) d\xi - \frac{1}{2} \int_0^{\infty} a'(u) [\int_{t-u}^{t} g(y(s)) ds]^2 du$

used in [8, Section 15]. Assumptions (1.29) on $a(t)$ yield easily

(2.8) $\quad \begin{cases} ta'(t), \ t^2 a'(t), \ t^2 a''(t) \in L^1(0, \infty); \\ ta(t) \to 0, \ t^2 a'(t) \to 0 \qquad (t \to 0, \infty). \end{cases}$

Differentiation of (2.7) together with two integrations by parts yields

(2.9) $\quad V_y'(t) = -\frac{1}{2} \int_0^{\infty} a''(u) [\int_{t-u}^{t} g(y(s)) ds]^2 du - g(y(t)) G(y(t))$.

Thus

(2.10) $\quad V_y(t) \geq 0, \ V_y'(t) \leq 0 \qquad (t \in \mathbb{R}, \ y \in \Gamma_a)$

and by (2.8) and $y \in L^{\infty}$ we have $V_y, V_y' \in L^{\infty}$. Putting

(2.11) $\quad M = \sup_s |g(y(s))|, \ W(t) = V_y'(t) + g(y(t)) G(y(t))$,

an easy calculation leads to

$$|W(t+\eta) - W(t)| \leq 2M^2 \int_0^{\eta} u^2 a''(u) du + 2M^2 \eta \int_{\eta}^{\infty} u a''(u) du$$

for $\eta > 0$, and thus $V_y' \in C_u(\mathbb{R})$ by (2.8). It follows that

(2.12) $\quad V_y'(t) \to 0 \qquad (t \to -\infty)$,

for otherwise there would exist $\alpha > 0$ and $t_n \to -\infty$ with

$$V'_y(t_n) < -2\alpha \qquad (n \geq 1),$$

and then

$$V'_y(t + t_n) < -\alpha \qquad (n \geq 1, \ |t| \leq \delta)$$

for some $\delta > 0$. But this clearly forces $V_y(-\infty) = \infty$, a contradiction, and so (2.12) is true.

By (1.29) there exist $\beta, \gamma, \eta > 0$ such that

$$a''(u) > \eta \qquad (\beta \leq u \leq \beta + \gamma),$$

and by (2.19)

$$-2V'_y(t) \geq \int_\beta^{\beta+\gamma} a''(u) \left[\int_{t-u}^t g(y(s))ds\right]^2 du$$

$$\geq \eta \int_\beta^{\beta+\gamma} \left[\int_0^u g(y(t-\tau))d\tau\right]^2 du$$

for all $t$. Using (2.12), we conclude

(2.13) $$\lim_{t \to -\infty} \int_\beta^{\beta+\gamma} \left[\int_0^u g(y(t-\tau))d\tau\right]^2 du = 0.$$

In fact,

(2.14) $$\lim_{t \to -\infty} \int_0^u g(y(t-\tau))d\tau = 0 \qquad (\beta \leq u \leq \beta + \gamma)$$

holds, for otherwise there would exist $t_n \to -\infty$ and $\mu > 0$ such that

$$\left|\int_0^{\beta'} g(y(t_n - \tau))d\tau\right| > 2\mu \qquad (n \geq 1),$$

for some $\beta' \in [\beta, \beta + \gamma]$. We deduce

$$\left|\int_0^{\beta'+v} g(y(t_n - \tau))d\tau\right| > \mu \qquad (|v| \leq \mu/M; \ n \geq 1)$$

where $M$ is defined by (2.11), and this contradicts (2.13).

From (2.14) we can deduce

(2.15) $$\lim_{t \to -\infty} y(t) = 0.$$

If not, there exist $s_n \to -\infty$ and $\nu > 0$, $\gamma' > 0$ such that $\gamma' \leq \gamma$ and

$$|g(y(s_n - \tau))| > \nu \qquad (|\tau| \leq \gamma').$$

Using (2.14),

$$\left|\int_0^{\beta+\gamma'} g(y(s_n + \beta - \tau))d\tau\right| \geq \left|\int_\beta^{\beta+\gamma'} g(y(s_n + \beta - \tau))d\tau\right| -$$

$$- \left|\int_0^\beta g(y(s_n + \beta - \tau))d\tau\right|$$

$$= \int_0^{\gamma'} |g(y(s_n - \tau))|d\tau - o(1) \geq \nu\gamma' - o(1) \qquad (n \to \infty),$$

which contradicts (2.14). Thus (2.15) is true.
Putting (2.15) into (2.7) yields

$$\lim_{t \to -\infty} V_y(t) = 0,$$

and by (2.10) we can only have $V_y \equiv 0$. In view of (2.7) $y \equiv 0$, $\Gamma_a = \{0\}$ as desired.

## REFERENCES

1.  Beurling, A., Un théorème sur les fonctions bornées et uniformément continues sur l'axe réel, Acta Math. 77 (1945), 127-136.

2.  Hale, J. K., Sufficient conditions for stability and instability of autonomous functional-differential equations, J. Diff. Eqs. 1 (1965), 452-482.

3. Hannsgen, K. B., On a nonlinear Volterra equation, Mich. Math. J. 16 (1969), 365-376.

4. Karlin, S., On the renewal equation, Pacific J. Math. 5 (1955), 229-257.

5. Korevaar, J., Tauberian theorems, Simon Stevin 30 (1953), 129-139.

6. Levin, J. J., The asymptotic behavior of the solution of a Volterra equation, Proc. A. M. S. 14 (1963), 534-541.

7. Levin, J. J. and Nohel, J. A., Perturbations of a nonlinear Volterra equation, Mich. Math. J. 12 (1965), 431-447.

8. Levin, J. J. and Shea, D. F., On the asymptotic behavior of the bounded solutions of some integral equations, J. Math. Anal. Appl., to appear.

9. Levinson, N., A nonlinear Volterra equation arising in the theory of superfluidity, J. Math. Anal. Appl. 1 (1960), 1-11.

10. Miller, R. K., Asymptotic behavior of solutions of nonlinear Volterra equations, Bull. A. M. S. 72 (1966), 153-156.

11. Miller, R. K., On Volterra's population equation, J. SIAM Appl. Math. 14 (1966), 446-452.

12. Nohel, J. A., Remarks on nonlinear Volterra equations, Proc. U. S.-Japan Seminar on Diff. and Functional Eqs., Benjamin, New York (1967), 249-266.

13. Volterra, V., Leçons sur la théorie mathématique de la lutte pour la vie, Gauthier-Villars, Paris, 1931.

This research was supported by the U. S. Army Research Office, Durham and N. S. F. grant GP-5728.

        Department of Mathematics
        University of Wisconsin
        Madison, Wisconsin

        Division of Mathematical Sciences
        Purdue University
        Lafayette, Indiana

        Received May 12, 1971

# Singular Perturbations and Singular Layers in Variational Inequalities

*J. L. LIONS*

Introduction.

Questions arising in Mechanics and in Physics (cf. [8]) lead to problems of <u>singular perturbations</u> for <u>variational inequalities</u>.

A general program would consist in trying to extend to <u>variational inequalities</u> for ordinary differential operators, or partial differential operators of stationary or evolution type, the known theory (or at least <u>part of it</u>) of singular perturbations for ordinary differential operators (cf. [33] and the bibliography of this book) or for partial differential operators (cf. [11], [32] and the bibliography of these papers).

We hasten to say that such a program is far from completion. We give here some preliminary results, under the following plan:

1. <u>Some weak convergence theorems.</u>
    1.1. Stationary Case.
    1.2. Evolution cases.

2. Stationary problems.  <u>Singular layers (I)</u>
    2.1. Corrector term when $\bar{K} = V_b$.
    2.2. A classical example.
    2.3. Boundary layers in variational inequalities.

3. Stationary problems. Singular layers (II).
    3.1. Corrector term when K is not dense in $V_b$.
    3.2. Boundary layers with distributed variational inequalities (I).
    3.3. Boundary layers with distributed variational inequalities (II).
    3.4. Distributed variational inequalities (III). An example without boundary layer.

4. Evolution problems. Singular layers.
    4.1. Evolution inequalities of parabolic type. Correctors.
    4.2. Singular layers for variational inequalities of evolution.
    4.3. Correctors for other types of evolution inequalities.

5. Various remarks and questions.
    5.1. Free layers.
    5.2. Bingham fluids.
    5.3. Some other problems.

Bibliography.

## 1. Some Weak Convergence Theorems.

### 1.1. Stationary Case.

Let $V_a$ and $V_b$ be two real Hilbert spaces; we assume that

(1.1) $\qquad V_a \subset V_b$, $V_a$ dense in $V_b$.[†]

We shall denote by $\| \ \|_a$ (resp. $\| \ \|_b$) the norm in $V_a$ (resp. $V_b$). Let $a(u,v)$ and $b(u,v)$ be two continuous bilinear forms on $V_a$ and $V_b$ respectively. We make the following assumptions:

---
† This assumption is not essential.

(1.2) $\begin{cases} a(v,v) \geq 0 & \forall v \in V_a, \\ a(v,v) + \|v\|_b^2 \geq \alpha \|v\|_a^2, & \alpha > 0, \forall v \in V_a, \end{cases}$

(1.3) $b(v,v) \geq \beta \|v\|_b^2, \quad \beta > 0, \forall v \in V_b.$

Let us consider now a set $K \subset V_a$; we assume that

(1.4) $K$ is a non empty closed convex set in $V_a$.

Let us denote by $V_b'$ and $V_a'$ the dual spaces of $V_b$ and $V_a$: $V_b' \subset V_a'$; if $f$ is given in $V_b'$, we shall denote by $(f,v)$ the scalar product of $f$ with $v \in V_b$.

Let $f$ be given in $V_b'$; for every $\varepsilon > 0$, there exists a unique element $u_\varepsilon \in K$ such that

(1.5) $\varepsilon\, a(u_\varepsilon, v-u_\varepsilon) + b(u_\varepsilon, v-u_\varepsilon) \geq (f, v-u_\varepsilon) \quad \forall v \in K.$

Indeed, thanks to the general theory of variational inequalities [24], it is enough to prove that

$$\varepsilon\, a(v,v) + b(v,v) \geq \gamma_\varepsilon \|v\|_a^2, \quad \gamma_\varepsilon > 0$$

and this immediately follows from (1.2)(1.3).

Let us set now

(1.6) $\overline{K} = $ closure of $K$ in $V_b$.

Then, thanks to (1.3), there exists a unique element $u \in \overline{K}$ such that

(1.7) $b(u, v-u) \geq (f, v-u) \quad \forall v \in \overline{K}.$

One has [16]:

**Theorem 1.1.** <u>Under hypotheses</u> (1.2), (1.3) <u>one has</u>

(1.8) $u_\varepsilon \to u \quad \underline{in} \quad V_b \quad \underline{as} \quad \varepsilon \to 0.$

## Proof.

Taking $v = v_0 \in K$ fixed in inequality (1.5), we obtain

$$\varepsilon\, a(u_\varepsilon, u_\varepsilon) + b(u_\varepsilon, u_\varepsilon) \le (f, u_\varepsilon - v_0) +$$
$$+ \varepsilon\, a(u_\varepsilon, v_0) + b(u_\varepsilon, v_0) ,$$

hence

$$\varepsilon[a(u_\varepsilon, u_\varepsilon) + \|u_\varepsilon\|_b^2] + (\beta - \varepsilon)\|u_\varepsilon\|_b^2 \le$$
$$\le (f, u_\varepsilon - v_0) + \varepsilon\, a(u_\varepsilon, v_0) + b(u_\varepsilon, v_0)$$

hence (for $\varepsilon \le \beta/2$ for instance)[†]

$$\varepsilon \|u_\varepsilon\|_a^2 + \|u_\varepsilon\|_b^2 \le c \|f\|_{V_b'} \|u_\varepsilon - v_0\|_b$$
$$+ c\varepsilon \|u_\varepsilon\|_a \|v_0\|_a + c \|u_\varepsilon\|_b \|v_0\|_b ;$$

it follows that

(1.9)  $\|u_\varepsilon\|_b \le c , \quad \sqrt{\varepsilon}\, \|u_\varepsilon\|_a \le c .$

We can therefore extract a subsequence, still denoted by $u_\varepsilon$, such that

(1.10) $\qquad u_\varepsilon \to w \quad \text{in } V_b \text{ weakly.}$

Consequently $w \in \overline{K}$; we deduce from (1.5) that

$$\varepsilon\, a(u_\varepsilon, v) + b(u_\varepsilon, v) - (f, v - u_\varepsilon) \ge b(u_\varepsilon, u_\varepsilon)$$

hence, since $\underline{\lim} \cdot b(u_\varepsilon, u_\varepsilon) \ge b(w, w)$:

(1.11) $\qquad b(w, v - w) - (f, v - w) \ge 0 \qquad \forall\, v \in K ;$

---

[†] The c's denote various constants which do not depend on $\varepsilon$.

and therefore we have also (1.11) $\forall v \in \bar{K}$ and then $w = u$. Hence

$$u_\varepsilon \to u \quad \text{in} \quad V_b \quad \underline{\text{weakly}}.$$

Let us consider now:

$$X_\varepsilon = b(u_\varepsilon - u, u_\varepsilon - u) \le$$
$$\le b(u_\varepsilon, u_\varepsilon - u) - (f, u_\varepsilon - u) ;$$

using (1.5) we have

$$b(u_\varepsilon, u_\varepsilon) \le \varepsilon a(u_\varepsilon, u_\varepsilon) + b(u_\varepsilon, u_\varepsilon) \le$$
$$\le (f, u_\varepsilon - v) + \varepsilon a(u_\varepsilon, v) + b(u_\varepsilon, v)$$

and therefore

$$X_\varepsilon \le (f, u - v) + \varepsilon a(u_\varepsilon, v) + b(u_\varepsilon, v - u),$$

(1.12)
$$\overline{\lim} \cdot X_\varepsilon \le (f, u - v) + b(u, v - u) \quad \forall v \in K.$$

But (1.12) is also true $\forall v \in \bar{K}$ hence for $v = u$ and therefore $\overline{\lim} \cdot X_\varepsilon = 0$ whence the strong convergence follows.

Remark 1.1. Another class of problems which is important in applications is as follows. Let us consider a continuous function $v \to j(v)$ on $V_b$, which is convex and $\ge 0$.

Let us denote by $u_\varepsilon$ (resp. $u$) the unique solution in $K$ (resp. $\bar{K}$) of

(1.13)
$$\varepsilon a(u_\varepsilon, v - u_\varepsilon) + b(u_\varepsilon, v - u_\varepsilon) + j(v) - j(u_\varepsilon) \ge$$
$$\ge (f, v - u_\varepsilon) \quad \forall v \in K$$

(resp. of

(1.14)   $b(u, v - u) + j(v) - j(u) \geq (f, v - u) \quad \forall v \in \bar{K})$.

Then one has (1.8) (with a similar proof).

**Remark 1.2.** One can extend Theorem 1.1. to situations where $V_a$ and $V_b$ are reflexive Banach spaces and $a(u,v)$ and $b(u,v)$ correspond to monotone (or pseudo monotone) coercive operators, using [3], [5], [15].

**Remark 1.3.** It happens quite often that the solution u of (1.7) actually belongs to $V_a^\dagger$ but u <u>does not generally belong to</u> K. Therefore the convergence (1.8) <u>does not hold in the topology of</u> $V_a$ and <u>the main question</u> is to find <u>as simple as possible correctors</u> $\theta_\varepsilon$ such that

$$u_\varepsilon - \theta_\varepsilon - u \to 0 \text{ in } V_a$$

This problem is considered in Sections 2 and 3 below.

## 1.2. Evolution Case

We keep the notations of Section 1.1. We suppose that

(1.15)   $V_a \subset V_b \subset H$,

where H is a Hilbert space; in (1.15) each space is dense in the following one. We denote by ( , ) the scalar product in H. If we identify H with its dual we have

$$V_a \subset V_b \subset H \subset V_b' \subset V_a' ,$$

and the notation (f, v) to denote the scalar product of $f \in V_b'$

---

†We shall see Examples in Sections 2 and 3.

## SINGULAR PERTURBATIONS AND LAYERS

with $v \in V_b$ is consistent with the notation $(\,,\,)$ for the scalar product in $H$.

We consider now <u>inequalities of evolution.</u> Let $t$ denote the time variable, $t \in ]0, T[$, $T \leq \infty$. Let $f = f(t)$ be given such that

(1.16) $\begin{cases} f \in L^2(0,T;V_b'), \quad f' = \dfrac{df}{dt} \in L^2(0,T;V_b'), \\ f(0) \in H. \end{cases}$

Then it is known [17], assuming (to simplify the exposition) that

(1.17) $\qquad\qquad 0 \in K,$

that there exists a unique function $u_\varepsilon = u_\varepsilon(t)$ such that

(1.18) $\qquad \dfrac{du_\varepsilon}{dt} \in L^\infty(0,T;H) \cap L^2(0,T;V_a),$

(1.19) $\qquad u_\varepsilon(t) \in K, \quad u_\varepsilon(0) = 0,$

(1.20) $(\dfrac{du_\varepsilon(t)}{dt}, v-u_\varepsilon(t)) + \varepsilon\, a(u_\varepsilon(t), v-u_\varepsilon(t)) + b(u_\varepsilon(t), v-u_\varepsilon(t)) \geq$
$\qquad\qquad \geq (f(t), v-u_\varepsilon(t)) \qquad \forall\, v \in K^\dagger.$

Similarly there exists a unique function $u = u(t)$ such that

(1.21) $\qquad \dfrac{du}{dt} \in L^\infty(0,T;H) \cap L^2(0,T;V_b),$

(1.22) $\qquad u(t) \in \bar{K}, \quad u(0) = 0,$

(1.23) $(\dfrac{du(t)}{dt}, v-u(t)) + b(u(t), v-u(t)) \geq (f(t), v-u(t))\ \forall\, v \in \bar{K}.$

One has:

---

†Cf. [2] for more refined results.

**Theorem 1.2.** <u>Under hypotheses</u> (1.2) (1.3) (1.16) (1.17) <u>one has, as</u> $\varepsilon \to 0$,

(1.24) $\quad u_\varepsilon \to u$ in $L^2(0,T;V_b)$ and in $C^0([0,T];H)^\dagger$,

(1.25) $\begin{cases} u'_\varepsilon \to u' \text{ in } L^\infty(0,T;H) \text{ for the weak star topology} \\ \text{and in} \\ L^2(0,T;V_b) \text{ weakly.} \end{cases}$

<u>Proof</u>.

1) From the proof [17] of the existence of $u_\varepsilon$ satisfying (1.18) (1.19) and (1.20) it follows that, as $\varepsilon \to 0$, $u_\varepsilon, u'_\varepsilon$ remain in a bounded set of $L^2(0,T;V_b) \cap L^\infty(0,T;H)$ and $\sqrt{\varepsilon}\, u_\varepsilon$, $\sqrt{\varepsilon}\, u'_\varepsilon$ remain in a bounded set of $L^2(0,T;V_a)$.

2) One can therefore extract a subsequence, still denoted by $u_\varepsilon$, such that

(1.26) $\begin{cases} u_\varepsilon \to w, \ u'_\varepsilon \to w' \text{ in } L^2(0,T;V_b) \text{ weakly and in} \\ L^\infty(0,T;H) \text{ weak-star.} \end{cases}$

One has $w(t) \in \bar{K}$ a.e. and satisfies conditions (1.21) (1.22) (1.23). Consequently one has $w = u$.

3) In order to prove <u>strong</u> convergence in (1.24), we introduce

(1.27) $\quad X^s_\varepsilon = \int_0^s [(u'_\varepsilon - u', u_\varepsilon - u) + b(u_\varepsilon - u, u_\varepsilon - u)] dt.$

One has

$$X^s_\varepsilon \geq \frac{1}{2} |u_\varepsilon(s) - u(s)|^2 + \beta \int_0^s \|u_\varepsilon - u\|_b^2 \, dt$$

so that it is enough to prove that $X^s_\varepsilon \to 0$ uniformly in $s$. But thanks to (1.23) where we take $v = u_\varepsilon(t)$, one has

---

$\dagger$ Space of continuous functions from $[0,T) \to H$.

## SINGULAR PERTURBATIONS AND LAYERS

(1.28) $$X_\varepsilon^S \leq \int_0^S [(u'_\varepsilon, u_\varepsilon) + b(u_\varepsilon, u_\varepsilon) - (u'_\varepsilon, u) - b(u_\varepsilon, u) - (f, u_\varepsilon - u)] dt;$$

let us consider a function $v = v(t) \in L^2(0,T;V_a)$, $v(t) \in K$ a.e.; taking $v = v(t)$ in (1.20) we obtain

$$\int_0^S [(u'_\varepsilon, u_\varepsilon) + b(u_\varepsilon, u_\varepsilon)] dt \leq$$

$$\leq \int_0^S [(u'_\varepsilon, u_\varepsilon) + \varepsilon a(u_\varepsilon, u_\varepsilon) + b(u_\varepsilon, u_\varepsilon)] dt \leq$$

$$\leq \int_0^S [(u'_\varepsilon, v) + \varepsilon a(u_\varepsilon, v) + b(u_\varepsilon, v) - (f, v - u_\varepsilon)] dt;$$

this inequality together with (1.28) shows that

(1.29) $$X_\varepsilon^S \leq \int_0^S [(u'_\varepsilon, v-u) + \varepsilon a(u_\varepsilon, v) + b(u_\varepsilon, v-u) - (f, v-u)] dt$$

hence

(1.30) $$\varlimsup_{\varepsilon \to 0} X_\varepsilon^S \leq \int_0^S [(u', v-u) + b(u, v-u) - (f, v-u)] dt.$$

By passing to the limit we can take $v = u$ in (1.30) (since $u(t) \in \bar{K}$) whence the result follows.

Remark 1.4. We consider in (1.18), (1.19), (1.20), (and (1.21), (1.22), (1.23)) strong solutions of the evolution inequalities. One can define weak solutions under the weaker hypothesis

(1.31) $$f \in L^2(0,T;V'_b).$$

See [24], [17], [2]. We have then

(1.32) $$u_\varepsilon \to u \text{ in } L^2(0,T;V_b) \text{ weakly and in } L^\infty(0,T;H)$$

weakly star.

531

**Remark 1.5.** One can extend the above results to the case when $a(u,v)$ and $b(u,v)$ <u>depend on</u> $t$, and also to a situation similar to that of Remark 1.2.

**Remark 1.6.** With the notation of Remark 1.1, one can consider variational inequalities of the form

$$(\frac{du_\varepsilon(t)}{dt}, v-u_\varepsilon(t)) + \varepsilon a(u_\varepsilon(t), v-u_\varepsilon(t)) + b(u_\varepsilon(t), v-u_\varepsilon(t)) + j(u_\varepsilon(t)) \geq$$

(1.33) $\qquad \geq (f(t), v-u_\varepsilon(t)) \quad \forall v \in K$

together with (1.18), (1.19), and

$$(\frac{du(t)}{dt}, v-u(t)) + b(u(t), v-u(t)) + j(v) - j(u(t)) \geq (f(t), v-u(t))$$

(1.34) $\qquad \forall v \in K$

together with (1.21), (1.22). One has a similar result to Theorem 1.2.

**Remark 1.7.** Let us suppose that

(1.35) $\qquad \begin{cases} a(u,v) = a(v,u) \ \forall u, v \in V_a; \\ b(u,v) = b(v,u) \ \forall u, v \in V_b \end{cases}$

and <u>for instance</u> that

(1.36) $\qquad f, f' \in L^2(0, T; H), \quad f(0) \in K.$

Then one can show [2], [8] the existence and uniqueness of a function $u_\varepsilon$ such that

$$(1.37) \quad u'_\varepsilon = \frac{du_\varepsilon}{dt} \in L^\infty(0,T;V_a), \quad \frac{d^2 u_\varepsilon}{dt^2} \in L^2(0,T;H),$$

$$(1.38) \quad u_\varepsilon(0) = 0, \quad u'_\varepsilon(t) \in K,$$

$$(u'_\varepsilon(t), v-u'_\varepsilon(t)) + \varepsilon a(u_\varepsilon(t), v-u'_\varepsilon(t)) + b(u_\varepsilon(t), v-u'_\varepsilon(t)) \geq$$
$$(1.39) \qquad \geq (f(t), v-u'_\varepsilon(t)) \quad \forall v \in K;$$

similarly there exists a unique function $u$ such that

$$(1.40) \quad u' \in L^\infty(0,T;V_b), \quad \frac{d^2 u}{dt^2} \in L^2(0,T;H),$$

$$(1.41) \quad u(0) = 0, \quad u'(t) \in \overline{K},$$

$$(u'(t), v-u'(t)) + b(u(t), v-u'(t)) \geq (f(t), v-u'(t))$$
$$(1.42) \qquad \qquad \forall v \in \overline{K}.$$

One proves the

**Theorem 1.3.** <u>Under Hypotheses</u> (1.2), (1.3), (1.35), (1.36), <u>one has, as</u> $\varepsilon \to 0$,

$$(1.43) \quad \begin{cases} u_\varepsilon \to u \text{ in } C^0([0,T]; V_b), \\ u_\varepsilon \to u' \text{ in } L^2(0,T;H), \end{cases}$$

and

$$(1.44) \quad \begin{cases} u'_\varepsilon \to u' \text{ in } L^\infty(0,T;V_b) \text{ weakly star} \\ u''_\varepsilon \to u'' \text{ in } L^2(0,T;H) \text{ weakly}. \end{cases}$$

**Remark 1.8.** One can also consider, under hypotheses (1.35), the inequalities [17], [2]

$$(u_\varepsilon''(t), v-u_\varepsilon'(t)) + \varepsilon\, a(u_\varepsilon(t), v-u_\varepsilon'(t)) + b(u_\varepsilon(t), v-u_\varepsilon'(t)) \geq$$

(1.45) $$\geq (f(t), v-u_\varepsilon'(t)) \qquad \forall\, v \in K,$$

(1.46) $$u_\varepsilon(0) = 0,\quad u_\varepsilon'(0) = 0,\quad u_\varepsilon'(t) \in K \quad (0 \in K),$$

and

$$(u''(t), v-u'(t) + b(u(t), v-u'(t)) \geq$$

(1.47) $$\geq (f(t), v-u'(t)) \qquad \forall\, v \in \overline{K},$$

(1.48) $$u(0) = 0,\quad u'(0) = 0,\quad u'(t) \in \overline{K}.$$

If we assume that

(1.49) $$f,\ f' \in L^2(0, t; H),$$

then there exists a unique solution $u_\varepsilon$ (resp. $u$) such that

(1.50) $$u_\varepsilon' \in L^\infty(0, T; V_a),\quad u_\varepsilon'' \in L^\infty(0, T; H),$$

(resp.

(1.51) $$u' \in L^\infty(0, T; V_b),\quad u'' \in L^\infty(0, T; H)).$$

One proves the

**Theorem 1.4.** <u>Under Hypotheses</u> (1.2), (1.3), (1.35), (1.49) <u>one has</u>, <u>as</u> $\varepsilon \to 0$,

(1.52) $$\begin{cases} u_\varepsilon \to u & \text{in } C^0([0, T]; V_b), \\ u_\varepsilon' \to u' & \text{in } C^0([0, T]; H), \end{cases}$$

$$(1.53) \begin{cases} u''_\varepsilon \to u'' \quad \text{in} \quad L^\infty(0,T;H) \text{ weakly star}, \\ u'_\varepsilon \to u' \quad \text{in} \quad L^\infty(0,T,V_a) \text{ weakly star}. \end{cases}$$

**Remark 1.9.** One can also study the <u>quasi-stationary problems</u> [8], [2]:

$$(1.54) \quad \varepsilon\, a(u_\varepsilon(t), v-u'_\varepsilon(t)) + b(u_\varepsilon(t), v-u'_\varepsilon(t))$$
$$\geq (f(t), v-u'_\varepsilon(t)) \quad \forall v \in K, \ u'_\varepsilon(t) \in K$$

$$(1.55) \quad b(u(t), v-u'(t)) \geq (f(t), v-u'(t)) \quad \forall v \in \bar{K}, \ u'_\varepsilon(t) \in \bar{K}.$$

**Remark 1.10.** One can also consider the problem

$$(\varepsilon\, u''_\varepsilon(t) + u'_\varepsilon(t), v-u'_\varepsilon(t)) + a(u_\varepsilon(t), v-u'_\varepsilon(t)) \geq (f(t), v-u'_\varepsilon(t))$$

$$(1.56) \quad \forall v \in K, \ u'_\varepsilon(t) \in K, \ u_\varepsilon(0) = 0, \ u'_\varepsilon(0) = 0\,^\dagger$$

whose solution converges$^\ddagger$ to the solution $u$ of

$$(1.57) \quad (u', v-u') + a(u, v-u') \geq (f, v-u').$$

**Remark 1.11.** One can also consider the problem

$$(1.58) \begin{cases} (-\varepsilon u''_\varepsilon + u'_\varepsilon, v-u_\varepsilon) + a(u_\varepsilon, v-u_\varepsilon) \geq (f, v-u_\varepsilon) \ \forall v \in K, \\ u_\varepsilon(0) = 0, \ u'_\varepsilon(T) = 0, \ u_\varepsilon(t) \in K, \end{cases}$$

[†] For instance. We can take non-zero initial datae.
[‡] See [19] for details concerning the topology where convergence takes place.

whose solution converges[†] to the solution of

$$(1.59) \quad \begin{cases} (u', v-u) + a(u, v-u) \geq (f, v-u) & \forall v \in K, \\ u(t) \in K. \end{cases}$$

## 2. Stationary Problems - Singular Layers (I).

### 2.1. Corrector Term When $\bar{K} = V_b$.

[‡] We shall assume in this Section <u>that</u> K <u>is dense in</u> $V_b^{\ddagger}$, i.e.

$$(2.1) \quad \bar{K} = V_b.$$

In this case (1.7) reduces to

$$(2.2) \quad b(u, v) = (f, v) \quad \forall v \in V_b, \quad u \in V_b.$$

We shall make the hypothesis

$$(2.3) \quad u \in V_a.$$

<u>Let us notice</u> (as we shall see in the examples to follow) <u>that (2.3) is nothing but a regularity theorem.</u> We now define the corrector term $\theta_\varepsilon$ as the unique solution of

$$(2.4) \quad \theta_\varepsilon \in K - u^{\ddagger},$$

$$(2.5) \quad \varepsilon a(\theta_\varepsilon, \varphi - \theta_\varepsilon) + b(\theta_\varepsilon, \varphi - \theta_\varepsilon) \geq \varepsilon(g_\varepsilon, \varphi - \theta_\varepsilon) \quad \forall \varphi \in K - u,$$

---

[†] See [17]
that (1.58) is the "<u>elliptic regularization</u>" of (1.59).

[‡] Examples are given in Sections 2.2, 2.3. The case when K is <u>not</u> dense in $V_b$ is studied in Section 3.

[‡] Which is a non empty closed convex subset of $V_a$,

where the $g_\varepsilon$ are arbitrarily chosen in $V'_a$ satisfying either

(2.6) $\begin{cases} g_\varepsilon \text{ remains in a bounded set of } V'_a \\ \quad\quad \text{when } \varepsilon \to 0 \end{cases}$

or

(2.7) $g_\varepsilon \to g_0$ in $V'_a$ when $\varepsilon \to 0$.[†]

**Remark 2.1.** We have some freedom in the choice of $g_\varepsilon$. In applications, the $g_\varepsilon$ will be chosen so as to make the computation of $\theta_\varepsilon$ as simple as possible. We have [20]:

**Theorem 2.1.** We assume that (1.2), (1.3), (1.4) and (2.1) and (2.3) take place. Then under hypothesis (2.6) (resp. (2.7)) one has

(2.8) $w_\varepsilon = u_\varepsilon - \theta_\varepsilon - u \to 0$ in $V_a$ weakly (resp. strongly).

**Proof.**

Let us take $v = \theta_\varepsilon + u(\epsilon K)$ in (1.5); we obtain

(2.9) $-\varepsilon\, a(u_\varepsilon, w_\varepsilon) - b(u_\varepsilon, w_\varepsilon) \geq -(f, w_\varepsilon)$.

Taking $v = w_\varepsilon$ in (2.2) we can write

(2.10) $\varepsilon\, a(u, w_\varepsilon) + b(u, w_\varepsilon) = (f, w_\varepsilon) + \varepsilon\, a(u, w_\varepsilon)$

and taking $\varphi = u_\varepsilon - u(\epsilon K - u)$ in (2.5) we get

(2.11) $\varepsilon\, a(\theta_\varepsilon, w_\varepsilon) + b(\theta_\varepsilon, w_\varepsilon) \geq \varepsilon(g_\varepsilon, w_\varepsilon)$.

Adding up (2.9), (2.10) and (2.11) we obtain

(2.12) $\varepsilon\, a(w_\varepsilon, w_\varepsilon) + b(w_\varepsilon, w_\varepsilon) \leq -\varepsilon[a(u, w_\varepsilon) + (g_\varepsilon, w_\varepsilon)]$

---

[†]One can also add (L. Tartar) to the right hand side of (2.5) a term $\varepsilon^{\frac{1}{2}}(g^*_\varepsilon, \varphi - \theta_\varepsilon)$ where $g^*_\varepsilon$ is bounded (or convergent) in $V'_b$.

Consequently

(2.13)
$$a(w_\varepsilon, w_\varepsilon) \leq -[a(u, w_\varepsilon) + (g_\varepsilon, w_\varepsilon)],$$
$$b(w_\varepsilon, w_\varepsilon) \leq -\varepsilon[a(u, w_\varepsilon) + (g_\varepsilon, w_\varepsilon)]$$

hence

(2.14) $\quad a(w_\varepsilon, w_\varepsilon) + b(w_\varepsilon, w_\varepsilon) \leq -(1+\varepsilon)[a(u, w_\varepsilon) + (g_\varepsilon, w_\varepsilon)].$

From (1.2), (1.3) and (2.6) we get[†]

$$\|w_\varepsilon\|_a^2 \leq c(1 + \|g_\varepsilon\|_{V_a'}) \|w_\varepsilon\|_a \leq c \|w_\varepsilon\|_a$$

hence

(2.15) $\quad \|w_\varepsilon\|_a \leq c.$

But then it follows from (2.13) that

(2.16) $\quad \|w_\varepsilon\|_b \leq \sqrt{\varepsilon},$

and (2.15), (2.16) imply that $w_\varepsilon \to 0$ in $V_a$ weakly.
But if one has (2.7) it follows now from (2.14) that

$$a(w_\varepsilon, w_\varepsilon) + b(w_\varepsilon, w_\varepsilon) \to 0$$

hence the Theorem follows.

## 2.2. A Classical Example.

In order to show how the machinery of Section 2.1 is used in applications, let us consider first a classical situation [32].

Let us consider an open set $\Omega \subset \mathbb{R}^n$ and let $u_\varepsilon$ be the solution of

(2.17) $\quad -\varepsilon \Delta u_\varepsilon + u_\varepsilon = f \quad$ in $\Omega$,

(2.18) $\quad u_\varepsilon = 0 \quad$ on the boundary $\Gamma$ of $\Omega$.

---

[†]Let us recall that we denote by c various constants which do not depend on $\varepsilon$.

Then $u_\varepsilon \to u$ in $L^2(\Omega)$ (assuming that $f \in L^2(\Omega)$) where

(2.19) $\qquad u = f$.

We introduce:

(2.20) $\qquad V_a = H^1(\Omega)^\dagger$, $V_b = L^2(\Omega)$,

(2.21) $\qquad K = H_0^1(\Omega)^\ddagger$,

(2.22) $\qquad a(u,v) = \int_\Omega \sum_{i=1}^n \dfrac{\partial u}{\partial x_i} \dfrac{\partial v}{\partial x_i} dx$,

(2.23) $\qquad b(u,v) = \int_\Omega u\, v\, dx$.

If we assume that $f \in H^1(\Omega)$ then of course $u \in V_a$ and we can define $\theta_\varepsilon$ by (2.4) (2.5), which means here

(2.24) $\qquad \begin{cases} -\varepsilon \Delta \theta_\varepsilon + \theta_\varepsilon = \varepsilon g_\varepsilon, & \text{in } \Omega \\ \theta_\varepsilon = -u & \text{on } \Gamma. \end{cases}$

To simplify still further, let us assume that

(2.25) $\qquad \Omega = \{x \mid x_n > 0\}$

and that

(2.26) $\qquad f(x',0) \in H^1(\mathbb{R}_{x'}^{n-1})^\ddagger$

---

$\dagger$ Sobolev space of order 1; $v \in H^1(\Omega) \Longleftrightarrow v, \dfrac{\partial v}{\partial x_1}, \ldots, \dfrac{\partial v}{\partial x_n} \in L^2(\Omega)$.

$\ddagger$ $v \in H_0^1(\Omega) \Longleftrightarrow v \in H^1(\Omega)$ and $v = 0$ on $\Gamma$. Let us emphasize that in this classical example K is a <u>vector subspace</u> of $V_a$ - which <u>will not</u> be the case in the examples to follow.

$\ddagger$ Under the hypothesis "$f \in H^1(\Omega)$" one has only "$f(x',0) \in H^{\frac{1}{2}}(\mathbb{R}_{x'}^{n-1})$" (Cf. [23] Chapter 1).

Then if we <u>define</u> $\theta_\varepsilon$ as the solution in $L^2$ of

(2.27)
$$-\varepsilon \frac{d^2\theta_\varepsilon}{dx_n^2} + \theta_\varepsilon = 0,$$
$$\theta_\varepsilon(x',0) = -f(x',0)$$

i.e.

(2.28) $\quad \theta_\varepsilon(x) = -f(x',0) \exp\left(-\frac{x_n}{\sqrt{\varepsilon}}\right)$

we have (2.24) with

(2.29) $\quad g_\varepsilon = (\Delta_{x'} f(x',0)) \exp\left(-\frac{x_n}{\sqrt{\varepsilon}}\right).$

Thanks to (2.26), $\Delta_{x'} f(x',0) \in H^{-1}(\mathbb{R}_{x'}^{n-1})$ so that

(2.30) $\quad \|g_\varepsilon\|_{L^2(0,\infty; H^{-1}(\mathbb{R}_{x'}^{n-1}))} \leq c\sqrt{\varepsilon},$

hence in particular $g_\varepsilon \to 0$ in $(H^1(\Omega))'$.
Therefore Theorem 2.1 gives

(2.31) $\quad u_\varepsilon - u + f(x',0) \exp\left(-\frac{x_n}{\sqrt{\varepsilon}}\right) \to 0$ in $H^1(\Omega)$.

Of course this is quite a complicated method for such a result, <u>but we are now going to show that the method applies to variational inequalities.</u>

<u>Remark 2.2.</u> One can obtain by similar methods <u>asymptotic expansions</u> as in [32].

## 2.3. Boundary Layers in Variational Inequalities.

We now consider the situation (2.20), where $\Omega$ is still given by (2.25)[†] and let us suppose that

---

[†]Of course one reduces the general situation to this case by local map.

## SINGULAR PERTURBATIONS AND LAYERS

(2.32) $\quad K = \{v \mid h_0 \leq v \leq h_1 \text{ on } \Gamma\}$,

where $h_0$ and $h_1$ are given measurable functions on $\Gamma$. We notice that K is dense in $V_b$ so that we are in the situation of Section 2.1.

In this case, problem (1.5) is equivalent to

(2.17) $\quad -\varepsilon \Delta u_\varepsilon + u_\varepsilon = f$,

(2.33)
$$\begin{aligned} h_0(x') < u_\varepsilon(x',0) < h_1(x') &\Longrightarrow \frac{\partial u_\varepsilon}{\partial n}(x',0) = 0 \;^\dagger \\ u_\varepsilon(x',0) = h_1(x') &\Longrightarrow \frac{\partial u_\varepsilon}{\partial n}(x',0) \leq 0, \\ u_\varepsilon(x',0) = h_0(x') &\Longrightarrow \frac{\partial u_\varepsilon}{\partial n}(x',0) \geq 0, \text{ on } \Gamma. \end{aligned}$$

The limit is still $u = f$.

We define the corrector $\theta_\varepsilon$ as the solution in $L^2$ of

(2.34) $\quad -\varepsilon \dfrac{d^2 \theta_\varepsilon}{dx_n^2} + \theta_\varepsilon = 0$,

(2.35)
$$\begin{cases} h_0(x') < \theta_\varepsilon(x',0) + u(x',0) < h_1(x') \Longrightarrow \dfrac{d\theta_\varepsilon}{dx_n}(x',0) = 0, \\ \theta_\varepsilon(x',0) + u(x',0) = h_1(x') \Longrightarrow \dfrac{d\theta_\varepsilon}{dx_n}(x',0) \geq 0, \\ \theta_\varepsilon(x',0) + u(x',0) = h_0(x') \Longrightarrow \dfrac{d\theta_\varepsilon}{dx_n}(x',0) \leq 0. \end{cases}$$

We obtain

(2.36) $\quad \theta_\varepsilon(x) = [Pf(x',0) - f(x',0)] \exp(\dfrac{-x_n}{\sqrt{\varepsilon}})$

---

$^\dagger \dfrac{\partial}{\partial n}$ = normal derivative to $\Gamma$ directed toward the exterior of $\Omega = -\partial/\partial x_n$.

where

(2.37) $Pf(x',0)$ = projection of $f(x',0)$ on $[h_0(x'), h_1(x')]^\dagger$ ;

$\theta_\varepsilon$ defined by (2.36) satisfies (2.4), (2.5) with (2.7) provided:

(2.38) $\qquad Pf(x',0) - f(x',0) \in H^1(\mathbb{R}_{x'}^{n-1})$ .

Under this hypothesis, Theorem 2.1 applies:

(2.39) $\begin{cases} u_\varepsilon - u - (Pf(x',0) - f(x',0)) \exp(-\dfrac{x_n}{\sqrt{\varepsilon}}) \to 0 \\ \qquad\qquad\qquad\text{in } H^1(\Omega) . \end{cases}$

**Remark 2.3.** If $h_0 = h_1 = 0$ , we have the situation of Section 2.2; then $Pf = 0$ and (2.39) reduces to (2.31).
   If $h_0 = -\infty$ , $h_1 = +\infty$ , then $u_\varepsilon$ is the solution of the <u>Neumann problem</u> and, as it is natural, $\theta_\varepsilon = 0$ .

**Remark 2.4.** We see from (2.39) that we have a boundary layer near the parts of the boundary $\Gamma$ where $f(x',0) \notin [h_0(x'), h_1(x')]$.

**Remark 2.5.** Let us consider again the situation (2.20) and let S be a (n-1) dimensional variety contained in $\Omega$ . We suppose that

(2.40) $\qquad K = \{v \,|\, v \in H^1(\Omega) ,\ h_0 \leq v \leq h_1 \text{ on } S\}.$

Then the problem (1.5) is equivalent to

(2.41) $\qquad -\varepsilon \Delta u_\varepsilon = f \text{ in } \Omega - S ,$

(2.42) $\qquad \dfrac{\partial u_\varepsilon}{\partial n} = 0 \text{ on } \Gamma$

and conditions similar to (2.33) <u>on</u> S .

---
† Which makes sense a.e.

The limit is still $u = f$. As corrector term $\theta_\varepsilon$ we can take

(2.43) $\quad \theta_\varepsilon(x) = \psi(x)[P f(\tilde{x}) - f(\tilde{x})] \exp(-\frac{d(x,S)}{\sqrt{\varepsilon}})$

where

$d(x, S)$ = distance from $x$ to $S$, $\tilde{x}$ = projection of $x$ on $S$,

$\psi$ = smooth function which equals 1 in a neighborhood of $S$ and which is zero outside a small enough neighborhood of $S$.

We have here an "inside singular layer"[†].

**Remark 2.6.** One can also study singular perturbations for variational inequalities of, say, second order, reducing to equations for 1st order operators. Cf. 5.1.

**Remark 2.7.** One can introduce higher order corrections and asymptotic expansions for variational inequalities. Cf. J. L. Lions. C. R. Acad. Sc. Paris, April, 1971.

## 3. Stationary Problems. Singular Layers (II).

### 3.1. Corrector term when $K$ is not dense in $V_b$.

The fact which is used in the proof of Theorem 2.1 is that $u$ satisfies an equality, namely (2.2), instead of a variational inequality.

But in general if $u$ is the solution of (1.7) there exists a "multiplier" $m \in V_b'$ such that

(3.1) $\quad b(u,v) + (m,v) = (f,v) \quad \forall v \in V_b, \quad m \in \partial \psi_{\bar{K}}(u)$[‡].

We still assume that (2.3) holds true.

---

[†] A notion which is entirely different from the free layers, which also appear in variational inequalities.

[‡] i.e. $m$ belongs to the sub differential at $u$ (cf. [26], [29]) of the function $\psi_{\bar{K}}$ defined by $\psi_{\bar{K}} = 0$ on $\bar{K}$ and $= +\infty$ outside $\bar{K}$; $m$ is not unique; (3.1) is equivalent to (1.7).

One can then introduce the corrector $\theta_\varepsilon$[†] by

(3.2) $$\theta_\varepsilon \in K - u,$$

(3.3) $$\varepsilon\, a(\theta_\varepsilon, \varphi - \theta_\varepsilon) + b(\theta_\varepsilon, \varphi - \theta_\varepsilon) \geq \varepsilon(g_\varepsilon, \varphi - \theta_\varepsilon) + (m, \varphi - \theta_\varepsilon)$$
$$\forall \varphi \in K - u,$$

where $g_\varepsilon$ satisfies (2.6) or (2.7).
We have then the same result (with the same proof) as in Theorem 2.1.

**Remark 3.1.** Let us assume that $\bar{K} = V_b$ but that $u$ is the solution of

(3.4) $$b(u, v-u) + j(v) - j(u) \geq (f, v-u) \quad \forall v \in V_b,$$

$u$ being of the solution (1.13).
Choosing $m \in \partial j(u)$, (3.4) is equivalent to

(3.5) $$b(u, v) + (m, v) = (f, v) \quad \forall v \in V_b.$$

We still assume that $u \in V_a$.
We can then define the correctors $\theta_\varepsilon$ by (3.2) and

(3.6) $$\varepsilon a(\theta_\varepsilon, \varphi - \theta_\varepsilon) + b(\theta_\varepsilon, \varphi - \theta_\varepsilon) + j(\varphi + u) - j(\theta_\varepsilon + u) \geq$$
$$\geq \varepsilon(g_\varepsilon, \varphi - \theta_\varepsilon) + (m, \varphi - \theta_\varepsilon) \quad \forall \varphi \in K - u$$

and we have again the conclusion (2.8).

## 3.2. Boundary Layers with Distributed Variational Inequalities (I).

In Section 2 we considered variational inequalities related to convex sets $K$ defined by <u>boundary constraints</u> or (cf. Remark 2.5) by <u>surface constraints.</u>

---
[†]As suggested by R. Temam at Varenna, CIME, August 1970.

We are now going to consider <u>convex sets defined by constraints on</u> $\Omega$ (or functionals defined by integrals extended over $\Omega$); we shall call the corresponding inequalities "<u>distributed variational inequalities</u>". We consider again $V_a$ and $V_b$ as in (2.20) and $a(u,v)$, $b(u,v)$ as in (2.22), (2.23) and we consider

(3.7)   $K = \{v \mid v \in H^1(\Omega),\ v = 0 \text{ on } \Gamma,\ v \geq h \text{ on } \Omega\}$

where $h$ is given in $H^1(\Omega)$, $h \leq 0$ on $\Gamma$; the variational inequality (1.5) is equivalent to

(3.8) $\begin{cases} -\varepsilon \Delta u_\varepsilon + u_\varepsilon - f \geq 0, \\ u_\varepsilon - h \geq 0, \\ (-\varepsilon \Delta u_\varepsilon + u_\varepsilon - f)(u_\varepsilon - h) = 0 \quad \text{in } \Omega \end{cases}$

and

(3.9) $\qquad u_\varepsilon = 0 \quad \text{on } \Gamma.$

We have now

(3.10)  $\bar{K} = \{v \mid v \in L^2(\Omega),\ v \geq h \text{ on } \Omega\}$

and the solution $u$ of (1.7) is

(3.11)  $u = \sup.(f, h).$

Then the multiplier $m$ equals

(3.12)  $m = f - \sup(f, h),$

and the corrector $\theta_\varepsilon$ is defined by[†]

---

[†] We assume that $\sup(f, h) \in H^1(\Omega)$, which is true if $f \in H^1(\Omega)$.

$$(3.13) \quad \begin{cases} \theta_\varepsilon + \sup(f,h) - h \geq 0, \\ -\varepsilon\Delta\theta_\varepsilon + \theta_\varepsilon - m - \varepsilon g_\varepsilon \geq 0, \\ (\theta_\varepsilon + \sup(f,h) - h)(-\varepsilon\Delta\theta_\varepsilon + \theta_\varepsilon - m - \varepsilon g_\varepsilon) = 0 \text{ in } \Omega \end{cases}$$

and

$$(3.14) \quad \theta_\varepsilon + \sup(f,h) = 0 \text{ on } \Gamma.$$

Except in the particular case when $h = 0$ on $\Gamma$, one cannot have (3.14) and $\theta_\varepsilon + \sup(f,h) - h = 0$ in a neighborhood of $\Gamma$, so that

$$(3.15) \quad -\varepsilon\Delta\theta_\varepsilon + \theta_\varepsilon = m + \varepsilon g_\varepsilon \text{ in a neighborhood of } \Gamma.$$

Of course (3.14), (3.15) do <u>not</u> define $\theta_\varepsilon$ in a neighborhood of $\Gamma$, the complete definition being given only by (3.13). Let us examine a <u>unidimensional problem.</u> We consider $\Omega = ]-1, +1[$, and we take

$$(3.16) \quad \begin{cases} h(x) = 1 - 2x^2, \\ f(x) = c. \end{cases}$$

If $c \leq -1$, we have

$$u = h$$

and $u_\varepsilon$ is given in $[-1, x_\varepsilon]$ by

$$(3.17) \quad \begin{cases} -\varepsilon u_\varepsilon'' + u_\varepsilon = c, \\ u_\varepsilon(-1) = 0, \\ u_\varepsilon(x_\varepsilon) = 1 - 2x_\varepsilon^2, \quad u_\varepsilon'(x_\varepsilon) = 4x_\varepsilon, \end{cases}$$

$$(3.18) \quad u_\varepsilon = h \text{ in } [x_\varepsilon, 0],$$

$$(3.19) \quad u_\varepsilon(x) = u_\varepsilon(-x).$$

One checks easily that $x_\varepsilon \to -1$ as $\varepsilon \to 0$ and that we have a boundary layer at $x = \pm 1$. If $-1 < c < 1$, then $u = c$ in the neighborhood of $\pm 1$ and we still obtain a boundary layer phenomenon. If $c > 1$ then $u_\varepsilon = u = c$; the boundary layer phenomenon disappears.

**Remark 3.2.** If we assume that

(3.20) $$\Delta h \leq 0 \quad \text{on } \Omega$$

then using [4] one can show that there exists $F_\varepsilon \in L^2(\Omega)$ such that

(3.21) $\quad\quad F_\varepsilon$ is bounded in $L^2(\Omega)$ as $\varepsilon \to 0$,

(3.22) $$\begin{cases} -\varepsilon \Delta u_\varepsilon + u_\varepsilon = F_\varepsilon & \text{in } \Omega \\ u_\varepsilon = 0 & \text{on } \Gamma, \end{cases}$$

and actually

(3.23) $\quad\quad F_\varepsilon \to u = \sup(f, h)$ in $L^2(\Omega)$ weakly.

If we assume that $u \in H^1(\Omega)$, this shows that $u_\varepsilon \to u$ in $H^1(\Omega')$ for every open set $\Omega'$ such that $\Omega' \subset \Omega$.

### 3.3. Boundary Layers with Distributed Variational Inequalities (II).

We take $V_a$ and $V_b$ as in (2.20) and $a(u, v)$, $b(u, v)$ given by (2.22), (2.23). We introduce

(3.24) $\quad K = H_0^1(\Omega)$ (i.e. a _vector_ space as in 2.2)

(3.25) $\quad j(v) = \int_\Omega \rho |v| dx$, $\quad \rho > 0$ constant.

The solution $u_\varepsilon$ of the problem corresponding to (1.13) is given by

$$(3.26) \begin{cases} |-\varepsilon \Delta u_\varepsilon + u_\varepsilon - f| \le \rho \quad \text{a.e. in } \Omega, \\ (-\varepsilon \Delta u_\varepsilon + u_\varepsilon - f) u_\varepsilon + \rho |u_\varepsilon| = 0 \quad \text{a.e. in } \Omega \\ u_\varepsilon = 0 \quad \text{on } \Gamma. \end{cases}$$

One has $\bar{K} = V_b$ and the solution $u$ of (3.4) is given by

$$(3.27) \quad |u-f| \le \rho, \quad (u-f)u + \rho|u| = 0 \quad \text{a.e. in } \Omega,$$

i.e.
$$(3.28) \quad u = f - Pf, \quad Pf(x) = \text{projection of } f(x) \text{ on } [-\rho, \rho].$$

If we suppose that $f \in H^1(\Omega)$, then $u \in H^1(\Omega) = V_a$. Let us assume that one can divide $\Omega$ in

$$\Omega_1 \cup \Omega_2 \cup \Omega_3 \cup S_{12} \cup S_{13}$$

(see Figure 1) where the $\Omega_i$'s are open and $f < \rho$ on $\Omega_1$, $-\rho < f < \rho$ on $\Omega_2$ and $f < -\rho$ on $\Omega_3$ ($f = \rho$ on $S_{12}$, $f = -\rho$ on $S_{23}$).

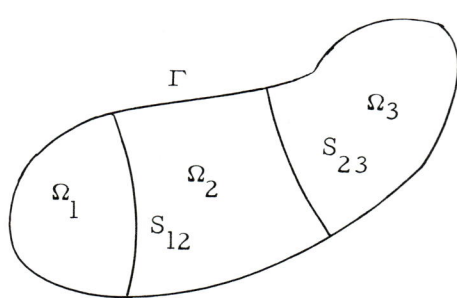

Figure 1.

Then

$$u_\varepsilon = 0 \quad \text{on} \quad \Omega_2,$$

$$\begin{cases} -\varepsilon \Delta u_\varepsilon + u_\varepsilon = f - \rho & \text{in } \Omega_1, \\ u_\varepsilon \big|_{\partial \Omega_1} = 0, \end{cases}$$

$$\begin{cases} -\varepsilon \Delta u_\varepsilon + u_\varepsilon = f + \rho & \text{in } \Omega_2, \\ u_\varepsilon \big|_{\partial \Omega_2} = 0. \end{cases}$$

It follows that we have boundary layers near $\partial \Omega_1 \cap \Gamma$ and near $\partial \Omega_3 \cap \Gamma$.

## 3.4. Distributed Variational Inequalities (III). An Example Without Boundary Layer.

We still consider $V_a$ and $V_b$ as in (2.20) and $a(u,v)$, $b(u,v)$ given by (2.22), (2.23). We consider

(3.29) $\quad K = \{v \mid v \in H_0^1(\Omega), \ |\text{grad } v(x)| \le 1 \text{ a.e. in } \Omega\};$

K is a closed convex set in $V_a$ _and also in_ $V_b$; in this case there are no singular layers.[†]

Indeed K is a closed convex set of Lipschitz functions which are zero on $\Gamma$ and $u_\varepsilon \to u$ in the corresponding weak topology. Using [4] one can show that there exists $F_\varepsilon \in L^2(\Omega)$ such that

$$F_\varepsilon \to u \quad \text{in } L^2(\Omega) \text{ weakly,}$$

$$-\varepsilon \Delta u_\varepsilon + u_\varepsilon = F_\varepsilon, \quad u_\varepsilon = 0 \text{ on } \Gamma.$$

---

[†]It was natural to expect so at least for K "small enough". Let us emphasize that the set K defined by (3.29) plays an important role in applications in mechanics (cf. [8] and the bibliography of this book).

## 4. Evolution Problems - Singular Layers.

### 4.1. Evolution Inequalities of Parabolic Type. Correctors.

We consider now the situation of Section 1.2. and we assume that

(4.1) $$\bar{K} = V_b .$$

Then problem (1.2.), (1.22), (1.23) reduces to (1.21) and

(4.2) $\quad (u'(t), v) + b(u(t), v) = (f(t), v) \quad \forall \, v \in V_b$ ,

(4.3) $\quad u(0) = 0 .$

We make the <u>regularity hypothesis</u>:

(4.4) $\quad u , u' \in L^2(0, T; V_a) .$

We introduce <u>the correctors</u> $\theta_\varepsilon = \theta_\varepsilon(t)$ by

(4.5) $\begin{cases} (\theta'_\varepsilon(t), \varphi - \theta_\varepsilon(t)) + \varepsilon \, a(\theta_\varepsilon(t), \varphi - \theta_\varepsilon(t)) + b(\theta_\varepsilon(t), \varphi - \theta_\varepsilon(t)) \geq \\ \qquad \geq (g_\varepsilon(t), \varphi - \theta_\varepsilon(t)) \\ \forall \, \varphi \in K - u(t), \end{cases}$

(4.6) $\quad \theta_\varepsilon(t) \in K - u(t)$ ,

(4.7) $\quad \theta_\varepsilon(0) = 0$ ,

where

$g_\varepsilon, g'_\varepsilon$ remain in a bounded set of $L^2(0, T; V'_a)$

or

(4.9) $\quad g_\varepsilon \to g_0, \; g'_\varepsilon \to g'_0 \;$ in $\; L^2(0, T; V'_a) \;$ as $\; \varepsilon \to 0$ .

There is a small difficulty in (4.5), (4.6), (4.7) related to the fact that the family of convex sets $K - u(t)$ depends on $t$. But if we set

(4.10) $\qquad p_\varepsilon = \theta_\varepsilon + u$,

the problem reduces to that of finding $p_\varepsilon$ satisfying

(4.11) $\qquad p_\varepsilon(t) \in K$,

(4.12) $\begin{cases} (p'_\varepsilon(t), v-p_\varepsilon(t)) + \varepsilon a(p_\varepsilon(t), v-p_\varepsilon(t)) + b(p_\varepsilon(t), v-p_\varepsilon(t)) \geq \\ \geq \varepsilon(g_\varepsilon(t), v-p_\varepsilon(t)) + \varepsilon a(u(t), v-p_\varepsilon(t)) + (f(t), v-p_\varepsilon(t)) \\ \qquad\qquad\qquad\qquad\qquad\qquad\qquad\qquad \forall\, v \in K, \end{cases}$

(4.13) $\qquad p_\varepsilon(0) = 0$;

this problem admits a unique solution which satisfies

(4.14) $\qquad \frac{d}{dt} p_\varepsilon \in L^\infty(0,T;H) \cap L^2(0,T;V_a)$.

We have now the following result (analogous to Theorem 2.1):

**Theorem 4.1.** We suppose that (1.2), (1.3), (1.4), (4.1), (4.4) hold true. Then, as $\varepsilon \to 0$, one has:

(4.15) $\begin{cases} w_\varepsilon = u_\varepsilon - \theta_\varepsilon - u \to 0 \text{ in } L^\infty(0,T;H) \\ \text{weak star and in } L^2(0,T;V_a) \text{ weakly,} \end{cases}$

if one has (4.8), and

(4.16) $\qquad w_\varepsilon \to 0 \text{ in } C^0([0,T];H) \cap L^2(0,T;V_a)$

if one has (4.9).

Proof.

We choose $v = u(t) + \theta_\varepsilon(t) = p_\varepsilon(t)$ in (1.20) and $v = u_\varepsilon(t)$ in (4.12); adding up, we obtain

$$(w'_\varepsilon, w_\varepsilon) + \varepsilon\, a(w_\varepsilon, w_\varepsilon) + b(w_\varepsilon, w_\varepsilon) \leq -\varepsilon[a(u, w_\varepsilon) + (g_\varepsilon, w_\varepsilon)]$$

hence†

(4.17)
$$\begin{cases} \frac{1}{2}|w_\varepsilon(s)|^2 + \varepsilon \int_0^s a(w_\varepsilon(t), w_\varepsilon(t))dt + \int_0^s b(w_\varepsilon(t), w_\varepsilon(t))dt \leq \\ \leq -\varepsilon \int_0^s [a(u(t), w_\varepsilon(t)) + (g_\varepsilon(t), w_\varepsilon(t))]dt \, . \end{cases}$$

The result follows by similar arguments to those of the proof of Theorem 2.1.

**Remark 4.1.** When (4.1) does not hold, one can use <u>multipliers</u> as in Section 3.1. for defining the correctors $\theta_\varepsilon$.

**Remark 4.2.** One can also extend the Remark 3.1 to the evolution situation.

## 4.2. Singular Layers for Variational Inequalities of Evolution.

Let us apply the considerations of Section 4.1. to the following situation: we take as in Section 2.3.:

(4.18) $\qquad V_a = H^1(\Omega)\, , \ V_b = L^2(\Omega)\, ,$

(4.19) $\qquad K = \{v\,|\,v \in V_a,\ h_0 \leq v \leq h_1\ \text{on}\ \Gamma\,\}\, ,$

where $h_0$ and $h_1$ are given measurable functions on $\Gamma$ (independent of $t$). The problem (1.19), (1.20) is equivalent to

---

† $|\ |$ denotes the norm in H.

(4.20) $\quad \dfrac{\partial u_\varepsilon}{\partial t} - \varepsilon \Delta u_\varepsilon + u_\varepsilon = f \quad \text{in} \quad Q = \Omega \times ]0, T[$ ,

(4.21) $\quad u_\varepsilon(x, 0) = 0 \quad \text{on} \quad \Omega$ ,

and the boundary conditions on $\sum = \Gamma \times ]0, T[$ :

(4.22) $\begin{cases} h_0(x) < u_\varepsilon(x, t) < h_1(x) \Longrightarrow \dfrac{\partial u_\varepsilon}{\partial n} = 0 , \\ u_\varepsilon(x, t) = h_1(x) \Longrightarrow \dfrac{\partial u_\varepsilon}{\partial n} \le 0 , \\ u_\varepsilon(x, t) = h_0(x) \Longrightarrow \dfrac{\partial u_\varepsilon}{\partial n} \ge 0 . \end{cases}$

The problem (4.2), (4.3) reduces to

(4.23) $\quad \dfrac{\partial u}{\partial t} + u = f \quad \text{in} \quad Q$ ,

(4.24) $\quad u(x, 0) = 0 \quad \text{on} \quad \Omega$ ,

where $x$ is nothing but a parameter. If we assume that

(4.25) $\quad f \in L^2(0, T; H^1(\Omega))$ ,

then (4.4) is satisfied.

The corrector term is given by the solution of

(4.26) $\quad \dfrac{\partial \theta_\varepsilon}{\partial t} - \varepsilon \Delta \theta_\varepsilon + \theta_\varepsilon = \varepsilon g_\varepsilon \quad \text{in} \quad Q$ ,

(4.27) $\begin{cases} h_0 < \theta_\varepsilon + u < h_1 \Longrightarrow \dfrac{\partial \theta_\varepsilon}{\partial n} = 0 , \\ \theta_\varepsilon + u = h_1 \Longrightarrow \dfrac{\partial \theta_\varepsilon}{\partial n} \le 0 , \\ \theta_\varepsilon + u = h_0 \Longrightarrow \dfrac{\partial \theta_\varepsilon}{\partial n} \ge 0 , \quad \text{on} \quad \sum , \end{cases}$

(4.28) $$\theta_\varepsilon(x, 0) = 0.$$

Let us take to simplify further $\Omega = \{x \mid x_n > 0\}$. Then we can define $\theta_\varepsilon$ as the solution (where $x' = \{x_1, \ldots, x_{n-1}\}$ plays the role of a parameter) of:

(4.29) $$\frac{\partial \theta_\varepsilon}{\partial t} - \varepsilon \frac{\partial^2 \theta_\varepsilon}{\partial x_n^2} + \theta_\varepsilon = 0,$$

(4.30) $$\begin{cases} h_0(x') < \theta_\varepsilon(x', 0, t) + u(x', 0, t) < h_1(x') \Longrightarrow \dfrac{\partial \theta_\varepsilon}{\partial x_n}(x', 0, t) = 0, \\[1ex] \theta_\varepsilon(x', 0, t) + u(x', 0, t) = h_1(x') \Longrightarrow \dfrac{\partial \theta_\varepsilon}{\partial x_n}(x', 0, t) \geq 0, \\[1ex] \theta_\varepsilon(x', 0, t) + u(x', 0, t) = h_0(x') \Longrightarrow \dfrac{\partial \theta_\varepsilon}{\partial x_n}(x', 0, t) \leq 0, \end{cases}$$

and (4.28). If we define $M = M(x, t)$ as the solution of

(4.31) $$\frac{\partial M}{\partial t} - \frac{\partial^2 M}{\partial x_n^2} + M = 0,$$

(4.32) $$\begin{cases} h_0(x') < M(x', 0, t) + u(x', 0, t) < h_1(x') \Longrightarrow \dfrac{\partial M}{\partial x_n}(x', 0, t) = 0, \\[1ex] M(x', 0, t) + u(x', 0, t) = h_1(x') \Longrightarrow \dfrac{\partial M}{\partial x_n}(x', 0, t) \geq 0, \\[1ex] M(x', 0, t) + u(x', 0, t) = h_0(x') \Longrightarrow \dfrac{\partial M}{\partial x_n}(x', 0, t) \leq 0, \end{cases}$$

(4.33) $$M(x, 0) = 0,$$

then one easily checks that

(4.34) $$\theta_\varepsilon(x,t) = M(x', \frac{1}{\sqrt{\varepsilon}} x_n, t).$$

The corrector term is therefore of "boundary layer" type.

Remark 4.3. One can also consider Examples similar to those in Section 3, but including the time variable.

## 4.3. Correctors for Other Types of Evolution Inequalities.

We consider firstly the situation of Remark 1.7. If we assume that (4.1) holds true, then u is characterized by the <u>same</u> equations than in the situation of 4.1, namely (4.2), (4.3). We still assume (4.4). The correctors $\theta_\varepsilon$ are defined by

(4.35) $$\begin{cases} (\theta'_\varepsilon(t), \varphi - \theta'_\varepsilon(t)) + \varepsilon\, a(\theta_\varepsilon(t), \varphi - \theta'_\varepsilon(t)) + b(\theta_\varepsilon(t), \varphi - \theta'_\varepsilon(t)) \geq \\ \qquad \geq (g_\varepsilon(t), \varphi - \theta'_\varepsilon(t)) \\ \forall\, \varphi \in K - u'(t), \end{cases}$$

(4.36) $$\theta'_\varepsilon(t) \in K - u'(t),$$

and (4.7), with $g_\varepsilon$ satisfying (4.8).

Defining $p_\varepsilon$ by (4.10), we obtain:

(4.37) $$\begin{cases} (p'_\varepsilon(t), v - p'_\varepsilon(t)) + \varepsilon\, a(p_\varepsilon(t), v - p'_\varepsilon(t)) + b(p_\varepsilon(t), v - p'_\varepsilon(t)) \geq \\ \geq \varepsilon(g_\varepsilon(t), v - p'_\varepsilon(t)) + \varepsilon\, a(u(t), v - p'_\varepsilon(t)) + (f(t), v - p'_\varepsilon(t)) \\ \forall\, v \in K, \end{cases}$$

(4.38) $$p'_\varepsilon(t) \in K$$

and (4.13). We have

Theorem 4.2. <u>Under the hypotheses</u> (1.2), (1.3), (1.4), (1.35), (4.1), (4.4), <u>one has, as</u> $\varepsilon \to 0$:

(4.39) $w_\varepsilon = u_\varepsilon - \theta_\varepsilon - u \to 0$ in $L^\infty(0,T;V_a)$ weakly star.

Proof.

We choose $v = p'_\varepsilon(t)$ in (1.39) and $v = u'_\varepsilon(t)$ in (4.37). Adding up we obtain

$$\varepsilon\, a(w_\varepsilon(t), w'_\varepsilon(t)) + b(w_\varepsilon(t), w'_\varepsilon(t)) + |w'_\varepsilon(t)|^2 \le$$
$$\le -\varepsilon[(g_\varepsilon(t), w'_\varepsilon(t)) + a(u(t), w'_\varepsilon(t))]$$

whence

(4.40)
$$\frac{\varepsilon}{2} a(w_\varepsilon(s), w_\varepsilon(s)) + \frac{1}{2} b(w_\varepsilon(s), w_\varepsilon(s)) + \int_0^s |w'_\varepsilon(t)|^2 dt \le$$
$$\le -\varepsilon \int_0^s [(g_\varepsilon, w'_\varepsilon) + a(u, w'_\varepsilon)] dt$$

and one can complete the proof by arguments somewhat similar to those used in the proof of Theorem 2.1.

Example.

We take the situation of Section 4.2 but with the new variational inequalities. Problem (1.39) amounts to solving (4.20), (4.21) with the boundary conditions (replacing (4.22)):

(4.41)
$$\begin{cases} h_0(x) < \dfrac{\partial u_\varepsilon}{\partial t}(x,t) < h_1(x) \Rightarrow \dfrac{\partial u_\varepsilon}{\partial n} = 0, \\[2mm] \dfrac{\partial u_\varepsilon}{\partial t}(x,t) = h_1(x) \Rightarrow \dfrac{\partial u_\varepsilon}{\partial n} \le 0, \\[2mm] \dfrac{\partial u_\varepsilon}{\partial t}(x,t) = h_0(x) \Rightarrow \dfrac{\partial u_\varepsilon}{\partial n} \ge 0. \end{cases}$$

SINGULAR PERTURBATIONS AND LAYERS

The <u>new</u> corrector term, still denoted by $\theta_\varepsilon$, is given by the solution of (4.26), (4.28) with the boundary conditions

$$(4.42) \begin{cases} h_0 < \dfrac{\partial \theta_\varepsilon}{\partial t} + \dfrac{\partial u}{\partial t} < h_1 \Longrightarrow \dfrac{\partial \theta_\varepsilon}{\partial n} = 0, \\[2mm] \dfrac{\partial \theta_\varepsilon}{\partial t} + \dfrac{\partial u}{\partial t} = h_1 \Longrightarrow \dfrac{\partial \theta_\varepsilon}{\partial n} \leq 0, \\[2mm] \dfrac{\partial \theta_\varepsilon}{\partial t} + \dfrac{\partial u}{\partial t} = h_0 \Longrightarrow \dfrac{\partial \theta_\varepsilon}{\partial n} \geq 0. \end{cases}$$

In case $\Omega = \{x \mid x_n > 0\}$, we find that

$$(4.43) \qquad \theta_\varepsilon(x, t) = N(x', \tfrac{1}{\sqrt{\varepsilon}} x_n, t)$$

where $N$ is the solution of

$$(4.44) \qquad \dfrac{\partial N}{\partial t} - \dfrac{\partial^2 N}{\partial x^2} + N = 0,$$

$$(4.45) \begin{cases} h_0(x') < \dfrac{\partial N}{\partial t}(x', 0, t) + \dfrac{\partial u}{\partial t}(x', 0, t) < h_1(x') \Longrightarrow \dfrac{\partial N}{\partial x_n}(x', 0, t) = 0, \\[2mm] \dfrac{\partial N}{\partial t}(x', 0, t) + \dfrac{\partial u}{\partial t}(x', 0, t) = h_1(x') \Longrightarrow \dfrac{\partial N}{\partial x_n}(x', 0, t) \geq 0, \\[2mm] \dfrac{\partial N}{\partial t}(x', 0, t) + \dfrac{\partial u}{\partial t}(x', 0, t) = h_0(x') \Longrightarrow \dfrac{\partial N}{\partial x_n}(x', 0, t) \leq 0, \end{cases}$$

$$(4.46) \qquad N(x, 0) = 0.$$

If we consider now the situation of Remark 1.8, then, under Hypothesis (4.1), $u$ is the solution of

$$(4.47) \quad (u''(t), v) + b(u(t), v) = (f(t), v) \quad \forall\, v \in V_b,$$

$$(4.48) \quad u(0) = 0, \quad u'(0) = 0.$$

The corrector term $\theta_\varepsilon$ is defined as the solution of

$$(4.49) \quad \begin{cases} (\theta_\varepsilon'', \varphi - \theta_\varepsilon') + \varepsilon\, a(\theta_\varepsilon, \varphi - \theta_\varepsilon') + b(\theta_\varepsilon, \varphi - \vartheta_\varepsilon') \geq \varepsilon(g_\varepsilon, \varphi - \theta_\varepsilon'), \\ \forall \varphi \in K - u'(t), \end{cases}$$

$$(4.50) \quad \theta_\varepsilon'(t) \in K - u'(t),$$

$$(4.51) \quad \theta_\varepsilon(0) = 0, \quad \theta_\varepsilon'(0) = 0,$$

where

$$(4.52) \quad g_\varepsilon, \; g_\varepsilon' \text{ are bounded in } L^2(0, T; H).$$

One proves then that

$$(4.53) \quad \begin{cases} u_\varepsilon - \theta_\varepsilon - u \to 0 \text{ in } L^\infty(0, T; V_a) \text{ weakly star}, \\ (u_\varepsilon - \theta_\varepsilon - u)' \to 0 \text{ in } L^\infty(0, T; H) \text{ weakly star.} \end{cases}$$

## 5. Various Remarks and Questions.

### 5.1. Free Layers.

Let $\Omega$ be a domain in $\mathbb{R}^2$ with boundary $0\,A\,B\,C\,D\,E\,0 = \Gamma$ (cf. Figure 2).

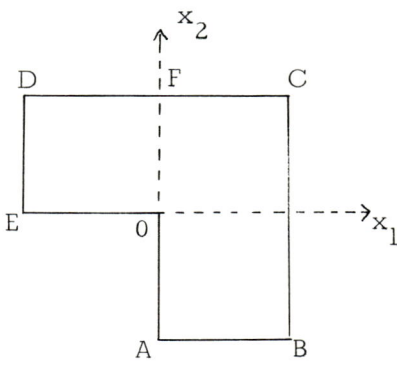

Figure 2.

We consider the unilateral problem

(5.1) $\quad -\varepsilon \Delta u_\varepsilon + \dfrac{\partial u_\varepsilon}{\partial x_2} + u_\varepsilon = f \quad \text{in } \Omega,$

(5.2) $\quad u_\varepsilon = 0 \quad \text{on } \Gamma - (E \, 0 \cup A \, B)$

(5.3) $\quad \begin{cases} u_\varepsilon - h \geq 0, \quad \varepsilon \dfrac{\partial u_\varepsilon}{\partial n} + u_\varepsilon \geq 0^\dagger, \\ (u_\varepsilon - h)(\varepsilon \dfrac{\partial u_\varepsilon}{\partial n} + u_\varepsilon) = 0 \quad \text{on } E \, 0 \cup A \, B, \end{cases}$

where $h$ is given in $L^2(E \, 0 \cup A \, B)$.

One proves that $u_\varepsilon$ converges in $L^2(\Omega)$ toward $u$ solution of

(5.4) $\quad \dfrac{\partial u}{\partial x_2} + u = f \quad \text{in } \Omega,$

(5.5) $\quad u = h^+ \quad \text{on } E \, 0 \cup A \, B.$

One can check that the corrector $\theta_\varepsilon$ contains a singular layer (free layer) along $0 \, F$ (Figure 2). For similar problems with ordinary boundary conditions, cf. [10], [25].

## 5.2. Bingham Fluids.

The theory of Bingham fluids [8], [9] leads to a generalization of the Navier-Stokes equations; one looks for functions $u = \{u_1, \ldots, u_n\}$, $m_{ij} = m_{ji}$ and $p$ such that

(5.6) $\begin{cases} m_{ij} \in L^\infty(Q), \quad Q = \Omega \times ]0, T[, \\ 2\sum m_{ij}^2 \leq 1, \\ \sum m_{ij} D_{ij}(u) = (\tfrac{1}{2} \sum (D_{ij}(u))^2)^{\tfrac{1}{2}}, \; D_{ij}(u) = \tfrac{1}{2}(\dfrac{\partial u_i}{\partial u_j} + \dfrac{\partial u_j}{\partial x_i}), \end{cases}$

---

$\dagger \dfrac{\partial}{\partial n}$ = normal derivative taken toward the exterior of $\Omega$.

$$(5.7) \quad \frac{\partial u}{\partial t} - \mu \Delta u + \sum_j' u_j \frac{\partial}{\partial x_j} u - 2g \sum_j' \frac{\partial}{\partial x_j} m_{ij} = f - \text{grad } p \, ,$$

$$(5.8) \quad \text{Div } u = 0 \, ,$$

$$(5.9) \quad u = 0 \text{ on } \sum = \Gamma \times ]0, T[ \, ,$$

$$(5.10) \quad u(x, 0) = u_0(x) \text{ given in } \Omega \, .$$

This set of equations can be expressed by an equivalent inequality (Cf. [8]). One can prove [8] the existence of a weak solution for any $n$, and the existence and uniqueness of a solution if $n = 2$. This set of equations reduces to the classical case when $g = 0$.

It seems likely that one has results similar to those of the classical boundary layer theory [27] when $\mu \to 0$ but this is an open problem.

## 5.3. Some Other Problems.

As mentioned in the Introduction, there is an extremely large amount of work still to be done in order to extend as far as possible to variational inequalities the known results for boundary value problems.

Let us mention, among other questions:
a) the extension of the estimates of [12], [13]; using non linear interpolation theory [28], [29], [30], this is possible [1];
b) the extension of the estimates of [14] with several "small" parameters, cf. [1];
c) one can show, using methods of Optimal Control Theory [18] that one has boundary layers for non-linear Riccati-type partial differential equations:

$$(5.11) \quad \frac{\partial P_\varepsilon}{\partial t} - \varepsilon(\Delta_x + \Delta_\xi) P_\varepsilon + c \int_\Omega P_\varepsilon(x, \eta, t) P_\varepsilon(\eta, \xi, t) d\eta = \delta(x - \xi),$$

(5.12) $$P_\varepsilon(x, \xi, t) = P_\varepsilon(\xi, x, t),$$
$$P_\varepsilon(x, \xi, t) = 0 \text{ for } x \in \Gamma, \quad \xi \in \Omega,$$

(5.13) $\quad P_\varepsilon(x, \xi, 0) = 0$ .

Cf. [21]. (For a direct study of these "Riccati type" equations - without singular perturbations, cf. [6], [7], [31]).

## REFERENCES

1. D. Brezis, To appear.

2. H. Brezis, Inéquations variationnelles. To appear in the Journal de Mathématiques Pures et Appliquées.

3. _____, Equations et inéquations non linéaires dans les espaces vectoriels en dualité. Annales Inst. Fourier 18 (1968), 115-175.

4. H. Brezis and G. Stampacchia, Sur la régularité de la solution d'inéquations elliptiques. Bull. Soc. Math. France, 96 (1968), 153-180.

5. F. Browder, Non linear monotone operators and convex sets in Banach spaces. Bull. Amer. Math. Soc. 71 (1965), 780-785.

6. G. Da Prato, Equations d'évolution dans des algèbres d'opérateurs et application à des équations quasi linéaires. J. Math. Pures et Appl. 48 (1969), 59-107.

7. _____, Somme d'applications non linéaires dans des cônes et équations d'évolution dans des espaces d'opérateurs. J. Math. Pures et Appl. 49 (1970), 289-348.

8. G. Duvaut and J. L. Lions, <u>Sur les inéquations en Mécanique et en Physique.</u> Dunod, 1971 (to appear).

9. G. Duvaut and J. L. Lions, Ecoulement d'un fluide rigide visco-plastique incompressible. C. R. Acad. Sc. 270 (1970), 58-61.

10. W. Eckhaus and E. M. de Jager, Asymptotic solutions of singular perturbations problems for linear Differential Equations of Elliptic Type. Archive Rat. Mech. Anal. 23 (1966), 26-86.

11. K. O. Friedrichs, Asymptotic Phenomena in Mathematical Physics. Bull. A. M. S. 61 (1955), 485-504.

12. W. M. Greenlee, Rate of convergence in singular perturbations. Ann. Inst. Fourier, 18 (1968), 135-191.

13. _____, Singular perturbation theorems for semi-bounded operators. To appear.

14. _____, A two parameter perturbation estimate. Proc. Amer. Math. Soc. 24 (1970), 67-74.

15. Ph. Hartman and G. Stampacchia, On some non linear elliptic functional equations. Acta Math. 115 (1966), 271-310.

16. D. Huet, Perturbations singulières d'inéquations variationnelles. C. R. Acad. Sc. Paris, 267 (1968), 932-935.

17. J. L. Lions, Quelques méthodes de résolution des problèmes aux limites non linéaires. Dunod, Gauthier Villars, 1969.

18. _____, Contrôle optimal de systèmes gouvernés par des équations aux dérivées partielles. Dunod, Gauthier Villars, 1968.

19. _____, Lecture notes at the C. I. M. E. Course, August, 1970.

20. J. L. Lions, On Partial Differential Inequalities, Uspehi Mat. Nauk, 1971 (In Russian; Dedicated to Prof. Petrowsky).

21. _____, Singular perturbations in Riccati integro differential equations. To appear.

22. _____, Some remarks on variational inequalities. Proc. Int. Conf. Funct. Analysis and Related Topics. Tokyo 1969, 269-282.

23. J. L. Lions and E. Magenes, Problèmes aux limites non homogènes. Dunod, 1968.

24. J. L. Lions and G. Stampacchia, Variational inequalities. Comm. Pure Applied Math. 20 (1967), 493-519.

25. J. Mauss, Thesis, Paris, 1971.

26. J. J. Moreau, Fonctionnelles sous-différentiables, C. R. Acad. Sc. Paris, 257 (1963), 4117-4119.

27. O. Oleinik, Mathematical problems of boundary layer theory. Univ. of Minnesota Lecture Notes. Spring 1969.

28. J. Peetre, Interpolation of Lipschitz Operators and metric spaces. To appear in Matematika (Cluj).

29. R. T. Rockafellar, Characterization of subdifferentials of convex functions. Pacific J. Math. 17 (1966), 497-509.

30. L. Tartar, Thesis, Paris, 1971.

31. R. Temam, Etude directe d'une équation d'évolution du type de Riccati associée à des opérateurs non bornés. C. R. Acad. Sc. 268 (1969), 1335 and J. of Functional Analysis, 1971

32. M. I. Visik and L. A. Lyusternik, Regular degeneration and boundary layer for linear differential equations with small parameter. Uspekhi Mat. Nauk. 12 (1957), 3-122; Amer. Math. Soc. Trans. Series, 2, 20 (1962), 239-364.

33. W. Wasow, Asymptotic expansions for ordinary differential equations, Interscience, Wiley, 1966.

Faculte des Sciences de Paris
Quai Saint Bernard
Paris, France

Received February 16, 1971

# Gradient Estimates for Solutions of Nonlinear Elliptic and Parabolic Equations

*JAMES SERRIN*

Our purpose here is to present a general outline of the maximum principle method for obtaining a priori estimates for the gradient of solutions of elliptic or parabolic equations of second order. It is of course well known that such gradient estimates are a fundamental element in the proof of existence theorems, and indeed it is this fact which has supplied the main motivation for the considerable efforts which have gone into obtaining these results.

The maximum principle technique was originally introduced by Bernstein in a paper in the Mathematische Annalen in 1910, though only for equations involving two independent variables and with the main ideas not fully worked out. We may illustrate the method in its simplest form by considering a solution u of the Laplace equation

$$\Delta u = 0$$

in a domain $\Omega$ of n-dimensional Euclidean space. If Du denotes the gradient vector of the solution, then an easy calculation shows that the function $w = |Du|^2$ satisfies

$$\Delta w = 2 \|D^2 u\|^2 \geq 0$$

in $\Omega$, where $D^2 u$ denotes the Hessian matrix of u. Consequently if one knows that $|Du|$ is bounded at the boundary of $\Omega$, it follows from the maximum principle that $|Du|$ has the same bound throughout $\Omega$. That is, we have obtained a bound for the gradient of a solution in $\Omega$ solely in terms of the magnitude of the gradient at the boundary.

In a subsequent paper [2] Bernstein further elaborated the method and applied it to various additional equations in two independent variables (it is noteworthy that this material appears only in considerably disguised form in his collected works). Bernstein's method was later used by Leray in his study of the Dirichlet problem for nonlinear elliptic equations, again for the case of two variables. In 1956 Ladyzhenskaya considered both elliptic and parabolic equations with an arbitrary number of independent variables, but with the important restriction to uniformly positive definite coefficient matrices. Subsequently the method was also applied with success by Oleinik and Kruzhkov to obtain interior gradient estimates for uniformly parabolic equation.

More recently the maximum principle technique has been used independently by Ladyzhenskaya and Uraltseva and by Serrin to obtain gradient estimates for non-uniformly elliptic equations with more than two variables. Finally in 1970 the technique was employed in an important paper of Ladyzhenskaya and Uraltseva to obtain interior gradient estimates for non-uniformly elliptic and parabolic equations.

Here we shall give a self-contained exposition of the main ideas described above (with the exception of the material in Part II of [10]), at the same time simplifying the treatment and determining more or less optimal hypotheses concerning the structure of the equation. Moreover as we note in Section 4, the method can be used to obtain a sequence of Liouville theorems for the equations in question. Section 5 discusses several particular examples (including the minimal surface operator, uniformly elliptic equations, and the interesting special case when the number of independent variables is 2.) Moreover in Section 5.5 we give a general pattern of cases where global gradient estimates can be obtained by the maximum principle method.

Among particular results of the paper we mention also the introduction of the multipliers $r$, $s$, $t$ in Section 2 and 3. These multipliers significantly extend the range of applicability of the results and also show clearly the difference between the elliptic and parabolic cases (the latter case is discussed separately in Section 6).

Other gradient estimates have been given by Trudinger and Oskolkov using the technique of integral inequalities. Their work applies particularly to divergence structure equations and requires stronger growth estimates than the maximum principle technique (it should be remarked, however, that in other respects the conditions required by these authors are lighter than those for the present method).

## 1. The basic method.

We consider here quasilinear second order elliptic partial differential equations

(1) $$\mathcal{A}(x, u, Du) D^2 u = \mathcal{B}(x, u, Du)$$

and their parabolic analogues

(2) $$u_t = \mathcal{A}(x, t, u, Du) D^2 u - \mathcal{B}(x, t, u, Du)$$

where $x = (x_1, \ldots, x_n)$ denotes points of the Euclidean space $R^n$ and where $Du$ and $D^2 u$ denote the gradient vector and the Hessian matrix of the dependent variable $u = u(x)$, or $u = u(x, t)$, that is

$$Du = \left(\frac{\partial u}{\partial x_1}, \ldots, \frac{\partial u}{\partial x_n}\right), \quad D^2 u = \frac{\partial^2 u}{\partial x_i \partial x_j}.$$

The quantities $\mathcal{A}$ and $\mathcal{B}$ are respectively a given symmetric matrix and a given scalar function of the variables indicated; we note also that the multiplication convention in (1) and (2) is the natural contraction $\mathcal{A} D^2 u = \mathcal{A}_{ij} \partial^2 u / \partial x_i \partial x_j$.

For equation (1) we assume that the functions $\mathcal{A}(x, u, p)$ and $\mathcal{B}(x, u, p)$ are defined and continuously differentiable for all real $u$ and $p = (p_1, \ldots, p_n)$ and for all values of $x$ in the closure of any domain under consideration. Ellipticity is then expressed by the condition $\xi \mathcal{A} \xi > 0$ for non-vanishing real vectors $\xi = (\xi_1, \ldots, \xi_n)$, the natural contraction being understood as before. Similar assumptions are required of course for equation (2).

For simplicity we shall deal with the elliptic case throughout most of the paper; extending the results to the parabolic case requires only a few additional remarks which can be left to a final section.

Let $u \in C^1(\bar{\Omega}) \cap C^3(\Omega)$ be a solution of (1) in a domain $\Omega$. We set

$$\sup_\Omega |u| = m, \quad \sup_{\partial\Omega} |Du| = \ell .$$

A <u>global estimate</u> for the gradient is then a statement of the form

$$\sup_\Omega |Du| \leq \text{Constant}$$

where the constant depends only on $m, \ell$, and the structure of the equation.

Suppose on the other hand that $u \in C^3(\Omega)$. An <u>interior estimate</u> for the gradient is a statement that

$$\sup_{\Omega'} |Du| \leq \text{Constant}$$

where the constant now depends only on $m$, the structure of the equation, and the distance from $\Omega'$ to the boundary of $\Omega$ (here $\Omega'$ is a compact subset of $\Omega$).

In order to obtain such estimates, we first derive a partial differential inequality for the quantity $w = |Du|^2$; this takes the form

(3) $$a D^2 w \geq \mathfrak{m} \cdot Dw - \mathfrak{n} w$$

where the quantities $\mathfrak{m}$ and $\mathfrak{n}$ are certain continuous functions of the arguments $x, u, p$, with $p = Du$. Supposing that (3) has been established,† let us assume that

---

† The derivation of inequality (3) will be given in the following section; our purpose for the moment is simply to outline the basic method.

(4)  $\mathfrak{n} \leq 0$  for  $|u| \leq m$, $|p| \geq L$.

From (3), (4) and E. Hopf's maximum principle ([15], page 61) it follows immediately that w cannot have an interior maximum value exceeding $L^2$. Since moreover w cannot exceed $\ell^2$ on the boundary of $\Omega$, we obtain

$$w \leq \text{Max}(\ell^2, L^2) \quad \text{in } \Omega,$$

that is

(5)  $$|Du| \leq \text{Max}(\ell, L) \quad \text{in } \Omega.$$

Thus under the hypothesis (4) one is led to a global gradient estimate. (In the application of the maximum principle to obtain (5), it is in fact only necessary to assume that (1) is elliptic for large values of p.)

In order to obtain interior estimates, let P be an arbitrary point of $\Omega$, whose distance to the boundary is 2d, say. Following the procedure of Ladyzhenskaya and Uraltseva [10], we make the substitution

$$z = \zeta w, \qquad \zeta = \left(1 - \frac{|x|^2}{d^2}\right)^k$$

where k is an appropriate constant which will be fixed later, and the origin is conveniently taken at P. Again one obtains a differential inequality

(6)  $a\, D^2 z \geq \mathfrak{m}' \cdot Dz - \mathfrak{n}' z \quad \text{in } \Sigma \equiv \{x \mid |x| \leq d\},$

where $\mathfrak{m}'$ and $\mathfrak{n}'$ are continuous functions of x, u, p and z. Suppose now that

(7)  $\mathfrak{n}' \leq 0$  for  $|u| \leq m$, $|p| \geq L$, $z \geq L'$.

Recalling that $|p| = \sqrt{w} \geq \sqrt{z}$, it follows that z cannot have an interior maximum value exceeding $\text{Max}(L^2, L')$. Since $z = 0$ on $\partial\Sigma$ we find therefore

$$z \leq \operatorname{Max}(L^2, L') \text{ in } \Sigma.$$

But $z = w = |Du|^2$ at $P$, hence

(8) $\qquad |Du| \leq \operatorname{Max}(L, \sqrt{L'})$ at $P$

which is the required interior estimate.

The preceding methods do not lead directly to the most useful conclusions. In order to obtain stronger results it is <u>in fact important to apply the above procedure not directly to equation (1) but rather to introduce first a change of dependent variables</u>

$$u = \phi(\bar{u}), \qquad (\phi' > 0).$$

We then consider the equation satisfied by $\bar{u}$, namely

(9) $\qquad \bar{\mathfrak{a}} \, D^2 \bar{u} = \bar{\mathfrak{B}}.$

With regard to the transformed equation (9), suppose that

(10) $\qquad \bar{\mathfrak{N}} \leq 0 \quad \text{for} \quad |u| \leq m \quad |p| \geq L.$

Since $Du = \phi' D\bar{u}$, it is easy to see that $\bar{w}$ cannot then have an interior maximum exceeding $(L/\operatorname{Min} \phi')^2$. Similarly $\bar{w}$ does not exceed $(\ell/\operatorname{Min} \phi')^2$ on the boundary since $|Du| \leq \ell$ there. Thus, assuming that (10) holds we obtain the global estimate

$$|Du| \leq \operatorname{Max}(\ell, L) \frac{\operatorname{Max} \phi'}{\operatorname{Min} \phi'} \text{ in } \Omega.$$

The function $\phi$ should be chosen so that (10) applies in the widest possible circumstances (see Theorem 2 below). The case of interior estimates is handled in much the same way (see Theorem 3).

That a change of dependent variables can improve the original results was noted by Bernstein ([2], page 460), though in fact he used a somewhat restricted function $\phi$. In the present paper we shall explicitly calculate the optimal function for the purpose.

## 2. Global estimates.

As shown in the previous section, the key element in the derivation of the global estimates for $Du$ is the inequality (3). To obtain this, let us apply the differential operator $u_k \frac{\partial}{\partial x_k}$ to (1); using the usual subscript notation and summing on repeated indices we obtain easily

$$0 = u_k \frac{\partial}{\partial x_k} [G_{ij} u_{ij} - \beta]$$

$$= G_{ij} u_k u_{ijk} + \left( \frac{\partial G_{ij}}{\partial x_k} u_{ij} - \frac{\partial \beta}{\partial x_k} \right) u_k$$

$$+ \left( \frac{\partial G_{ij}}{\partial u} u_{ij} - \frac{\partial \beta}{\partial u} \right) u_k u_k + \left( \frac{\partial G_{ij}}{\partial p_\ell} u_{ij} - \frac{\partial \beta}{\partial p_\ell} \right) u_k u_{k\ell} .$$

At this point it is convenient to introduce two differential operators

$$\delta \equiv - \left( \frac{\partial}{\partial u} + \frac{\sigma}{|p|} \cdot \frac{\partial}{\partial x} \right) + r \qquad \left( \sigma \cdot \frac{\partial}{\partial x} \equiv \sigma_i \frac{\partial}{\partial x_i} \right)$$

$$\delta_2 \equiv \frac{\partial}{\partial p} + t = \left( \frac{\partial}{\partial p_1} + t_1, \ldots, \frac{\partial}{\partial p_n} + t_n \right) .$$

Here $\sigma = p/|p|$ is the unit vector along the direction $p$, and $r$ and $t$ are respectively a given scalar and a given vector function of the variables $x, u, p$. Using (1) again and recalling that $w = u_k u_k$, the last identity can be rewritten

$$0 = G_{ij} u_k u_{ijk} - (\delta G_{ij} u_{ij} - \delta \beta) w + \left[ \delta_{2(\ell)} G_{ij} u_{ij} - \delta_{2(\ell)} \beta \right] u_k u_{k\ell} .$$

Now $w_i = 2u_k u_{ki}$ and $w_{ij} = 2u_k u_{ijk} + 2u_{ik} u_{jk}$. Hence (in an obvious notation) there follows

(11) $\quad G D^2 w = 2 D^2 u \, G D^2 u + 2(\delta G D^2 u - \delta \beta) w - (\delta_2 G \, D^2 u - \delta_2 \beta) \cdot Dw .$

Note in particular that $G$ and $\beta$ are functions of $x, u$ and

$p = Du$ and that the operators $\delta$ and $\delta_2$ apply only to the first symbol directly following them.

For later purposes it is convenient to suppose that the coefficient matrix $\mathcal{G}$ has the form

(12) $$\mathcal{G}'(x, u, p) + p\,\mathfrak{F}(x, u, p) + \mathfrak{F}(x, u, p)p$$

where $\mathcal{G}'$ is a positive definite matrix and $p\mathfrak{F}$ and $\mathfrak{F}p$ are dyadics. This is the case for example with the minimal surface operator, where

$$\mathcal{G} = (1 + |p|^2)I - pp .$$

In any event, (12) involves no loss of generality since one can always put $\mathcal{G}' = \mathcal{G}$, $\mathfrak{F} = 0$. Making the substitution $\mathcal{G} = \mathcal{G}' + p\mathfrak{F} + \mathfrak{F}p$ in (11), we find that

(13) $$D^2 w = 2D^2 u\, \mathcal{G}' D^2 u + 2(\delta\, \mathcal{G}' D^2 u - \delta\mathfrak{B})w - (\delta_2 \mathcal{G}' D^2 u - \delta_2 \mathfrak{B}) \cdot Dw$$
$$+ 2|p|^2 \delta \mathfrak{F} \cdot Dw - Dw(\delta_2 \mathfrak{F})Dw .$$

In order to obtain the main inequality (3) we must eliminate $D^2 u$ from the first two terms on the right hand side. To this end, let $\lambda = \lambda(x, u, p)$ denote the least eigenvalue of the matrix $\mathcal{G}'$. Then

(14) $$D^2 u\, \mathcal{G}' D^2 u \geq \lambda \|D^2 u\|^2 .$$

But by Cauchy's inequality

(15) $$|\delta \mathcal{G}' D^2 u|w \leq \lambda \|D^2 u\| + \frac{|p|^2 \|\delta \mathcal{G}'\|^2}{4\lambda} w$$

so that the preceding identity yields (3) with

(16) $$\tfrac{1}{2}\eta = \delta\mathfrak{B} + \frac{|p|^2 \|\delta \mathcal{G}'\|^2}{4\lambda} .$$

Using the results of Section 1, this gives at once the following global gradient estimate.

## GRADIENT ESTIMATES

**Theorem 1.** If for some $r = r(x, u, p)$ we have

$$\delta\mathcal{B} + \frac{|p|^2 \|\delta G'\|^2}{4\lambda} \leq 0 \quad \text{for} \quad |u| \leq m, \ |p| \geq L,$$

then $|Du| \leq \text{Max}(\ell, L)$ in $\Omega$.

An important special case occurs if $G'$ depends only on $p$. Taking $r = 0$, it follows that $\delta G' = 0$. Replacing (14) by the sharper inequality

$$D^2 u\, G'\, D^2 u \leq (G' D^2 u)^2 / \text{Trace}\ G' = \mathcal{B}^2 / \text{Trace}\ G' + \mathcal{R} \cdot Dw$$

we thus obtain (3) with

$$\tfrac{1}{2}\mathcal{U} = \delta\mathcal{B} - \frac{\mathcal{B}^2}{|p|^2 \text{Trace}\ G'}.$$

Recalling the definition of $\delta$, this gives the following result.

**Corollary.** Suppose $G' = G'(p)$ and

$$\frac{\mathcal{B}^2}{\text{Trace}\ G'} + p \cdot \frac{\partial \mathcal{B}}{\partial x} + |p|^2 \frac{\partial \mathcal{B}}{\partial u} \geq 0 \quad \text{for} \quad |u| \leq m, \ |p| \geq L.$$

Then $|Du| < \text{Max}(\ell, L)$ in $\Omega$.

The result of Theorem 1 can be slightly generalized if we make the substitution $w = (1 + \kappa|x|^2)v$ where $\kappa$ is a (small) positive constant to be chosen appropriately. Using essentially the same procedures as before yields the following result.

**Theorem 1'.** Suppose that multipliers $r$ and $t$ and a constant $\theta > 0$ can be found such that

$$\delta \mathcal{B} + \frac{|p|^2 \|\delta \mathcal{G}'\|^2}{4\lambda} + \theta\{|\delta\mathcal{B}| + |\delta_2\mathcal{B}| + |p|^2|\delta\mathcal{F}|\} \leq o(\text{Trace } \mathcal{G})$$
(17)
$$\|\delta_2\mathcal{F}\|, \lambda^{-1}\|\delta_2\mathcal{G}'\|^2 = O(\text{Trace } \mathcal{G}/|p|^2)$$

as $p \to \infty$. Then a global gradient estimate holds for equation (1).

## 3. Global estimates. The main case.

It is important to consider the change of dependent variables

(18) $\qquad u = \phi(\bar{u}) \qquad (\phi' > 0).$

One finds easily that the transformed function satisfies the equation

$$\bar{\mathcal{G}} D^2 \bar{u} = \bar{\mathcal{B}}$$

with

$$\bar{\mathcal{G}} = \mathcal{G}, \qquad \bar{\mathcal{B}} = \frac{|\bar{p}|}{|p|}(\mathcal{B} + \omega \mathcal{E})$$

and

$$\omega = -\frac{\phi''}{\phi'^2}, \qquad \mathcal{E} = p\,\mathcal{G}\,p,$$

where the arguments of $\mathcal{G}$ and $\mathcal{B}$ are $x, u,$ and $p = \phi'(\bar{u})\bar{p} = \phi'(\bar{u})D\bar{u}$. Note that the symbol $\bar{p}$ is used here as a replacement variable for $D\bar{u}$ in the same way as $p$ is used for $Du$.

From the preceding work it follows that

$$\bar{w} = |D\bar{u}|^2$$

satisfies

$$aD^2\bar{w} \geq \overline{\mathcal{R}} \cdot D\bar{w} - \overline{\mathcal{R}}\bar{w}$$

where $\overline{\mathcal{R}}$ denotes the expression $\mathcal{R}$ evaluated for the barred variables. Now by (16) one has

(19) $$\frac{1}{2}\overline{\mathcal{R}} = \delta\overline{\mathcal{B}} + \frac{|\bar{p}|^2 \|\delta\overline{\mathcal{Q}}'\|^2}{4\lambda}.$$

To obtain $\overline{\mathcal{R}}$ explicitly, observe that

$$|\bar{p}|\bar{\delta} = |p|\delta + \omega\delta_1|p|, \qquad \delta_1 \equiv p \cdot \frac{\partial}{\partial p} + s,$$

$$|\bar{p}|\bar{\delta}_2 = |p|\delta_2$$

(here one puts $\bar{r} = |\bar{p}|^{-1}|p|(r + \omega(s+1))$, $\bar{t} = |\bar{p}|^{-1}|p|t$). Consequently

$$|\bar{p}|\bar{\delta}\overline{\mathcal{Q}}' = |p|\delta\mathcal{Q}' + \omega\delta_1(|p|\mathcal{Q}')$$

and

$$\delta\overline{\mathcal{B}} = \delta(\mathcal{B} + \omega\mathcal{E}) + \omega\delta_1(\mathcal{B} + \omega\mathcal{E}) = \omega^2\delta_1\mathcal{E} + \omega(\delta\mathcal{E} + \delta_1\mathcal{B}) + \delta\mathcal{B} - \frac{\omega'}{\varphi'}\mathcal{E}.$$

Moreover for later use we collect the relations

$$|\bar{p}|\bar{\delta}_2\overline{\mathcal{Q}}' = |p|\delta_2\mathcal{Q}', \qquad \bar{\delta}_2\overline{\mathcal{B}} = \delta_2\mathcal{B} + \omega\delta_2\mathcal{E}$$

$$|\bar{p}|^2\bar{\delta}\overline{\mathcal{F}} = |\bar{p}|\bar{\delta}(|p|\mathcal{F}) = |p|^2\delta\mathcal{F} + \omega\delta_1(|p|^2\mathcal{F})$$

$$|\bar{p}|^2\bar{\delta}_2\overline{\mathcal{F}} = |p|^2\delta_2\mathcal{F}.$$

It follows now from (19) that

(20) $$\frac{1}{2\mathcal{E}}\overline{\mathcal{R}} = \alpha\omega^2 + 2\beta\omega + \gamma - \frac{\omega'}{\varphi'}$$

where the coefficients $\alpha$, $\beta$, $\gamma$ are given by

$$\alpha = \frac{1}{\mathcal{E}}\left\{\delta_1 \mathcal{E} + \frac{\|\delta_1(|p|\mathcal{Q}')\|^2}{2\lambda}\right\}$$

$$2\beta = \frac{1}{\mathcal{E}}\{\delta\mathcal{E} + \delta_1 \mathcal{B}\}$$

$$\gamma = \frac{1}{\mathcal{E}}\left\{\delta\mathcal{B} + \frac{\|\delta(|p|\vec{a}')\|^2}{2\lambda}\right\}.$$

Let us define (for $p \to \infty$ and $|u| \leq m$, $x \in \Omega$)

(21) $\quad a = \limsup \alpha, \quad b = \limsup \beta, \quad c = \limsup \gamma.$

Evidently the condition $\tilde{n} \leq 0$ will be fulfilled for sufficiently large $p$ if there exists a solution $\omega \geq 0$ of the constant coefficient Riccati differential inequality

(22) $\quad \dfrac{d\omega}{d\phi} \geq a\omega^2 + 2b\omega + c + \varepsilon, \quad \phi \in \text{Range } u$

where $\varepsilon$ is any fixed positive number.

In Section 7 we investigate the solutions of inequality (22). The following theorem is a consequence of the results there established.

Theorem 2. Suppose that

(23) $\quad a, b, c < \infty$.

Then we can obtain a global gradient estimate (depending only on $m, \ell$, and the structure of the equation) provided any one of the following conditions holds:

(i) $a \leq 0$ or $c \leq 0$.

(ii) $a, c > 0$ and $b \leq -\sqrt{ac}$.

(iii) $a, c > 0$ and

(24) $$\text{oscillation } u < \frac{1}{\sqrt{ac + b^2}} F\left(\frac{b}{\sqrt{ac}}\right) \quad (b > -\sqrt{ac})$$

where

$$F(t) \equiv \begin{cases} \sqrt{\dfrac{1+t^2}{1-t^2}} \text{ arccot } \dfrac{t}{\sqrt{1-t^2}}, & -1 < t \leq 1 \\ \sqrt{\dfrac{t^2+1}{t^2-1}} \text{ arccoth } \dfrac{t}{\sqrt{t^2-1}}, & t > 1. \end{cases}$$

(In case (iii) the gradient estimate also depends on the oscillation of $u$.)

The gradient estimate depends on the structure of the equation only through the values of $a$, $b$, $c$ and the rate of convergence in (21). For case (iii) of the theorem the estimate depends on the oscillation of $u$ only through the difference between the right and left hand sides of (24).

In a review article [17] the author has applied Theorem 2 to several particular cases of interest.[†] Another version of Theorem 2 was given by Ladyshenskaya and Uraltseva, though without the free multipliers $r$ and $s$ and with some further modification of the formulae (see [10], page 686, and also [8], [9]).

Remark 1. The function $F(t)$ is positive and real analytic (define $F(1) = \sqrt{2}$) on the interval $(-1, \infty)$ and approaches $\infty$ as $t$ approaches either endpoint (see Figure 1). It is of interest that $F(t) > 1.35$ on its entire domain while $F(t) > (2/(1+t))^{1/2}$ for $-1 < t \leq 1$ and $F(t) \geq \log 2t$ for $t > 1$.

Remark 2. From case (iii) of the theorem, a gradient estimate exists if $a > 0$, $c > 0$, $b > -\sqrt{ac}$ and

(25) $$\sqrt{ac + b^2} \, m \leq 2/3.$$

---

[†]Cf. Theorem 2 of [17]. A complete classification of cases is given in Section 5.5 of the present paper.

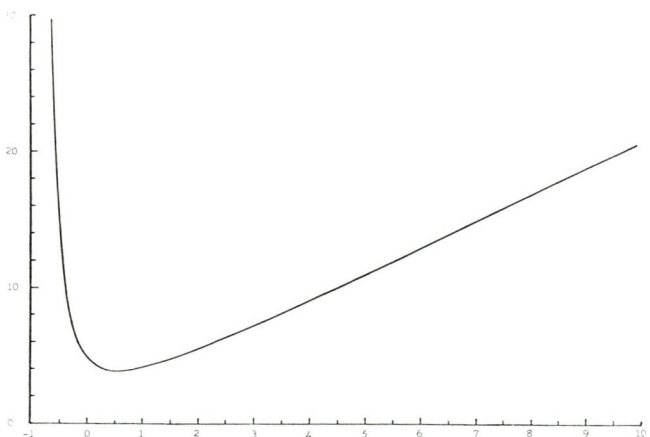

Figure 1: Graph of the function $\exp F(t)$.

Remark 3.  Theorem 1 is essentially the special case of Theorem 2 which arises when the coefficient $\gamma$ is non-positive for all sufficiently large $p$.

The comparison of Theorem 1' and Theorem 2 is more delicate. When Trace $\mathcal{G} = O(\mathcal{E})$, as is the case in particular for equations whose Bernstein genre does not exceed 2, one sees that Theorem 2 is generally the stronger. On the other hand, if $\mathcal{E} = o(\text{Trace } \mathcal{G})$ then Theorem 1' is stronger than Theorem 2.

## 4. Interior estimates.

As in Section 1, we introduce the new variable

$$z = \zeta w, \qquad \zeta = \left(1 - \frac{|x|^2}{d^2}\right)^k \qquad (x \in \Sigma)$$

where $k$ is an appropriate constant (recall that $\Sigma$ denotes the ball of radius $d$ about a fixed point $P$ where we wish to estimate $Du$). Since

$$w = \frac{z}{\zeta}, \quad Dw = \frac{Dz}{\zeta} - \frac{D\zeta}{\zeta^2}, \quad D^2 w = \frac{D^2 z}{\zeta} - 2\frac{D\zeta Dz}{\zeta^2} - \left(\frac{D^2\zeta}{\zeta^2} - 2\frac{D\zeta D\zeta}{\zeta^3}\right)z$$

we find from (13) that $z$ satisfies the equation

$$\mathfrak{a} D^2 z = 2\zeta D^2 u \mathfrak{a}' D^2 u + 2(\delta \mathfrak{a}' D^2 u - \delta \mathfrak{B})z + (\delta_2 \mathfrak{a}' D^2 u - \delta_2 \mathfrak{B}) \cdot \frac{D\zeta}{\zeta} z$$

$$-\left(2\delta \mathfrak{F} \cdot \frac{D\zeta}{\zeta} + \frac{D\zeta(\delta_2 \mathfrak{F})D\zeta}{\zeta^2}\right)|p|^2 z$$

(26)
$$+ \left(\frac{\mathfrak{a} D^2 \zeta}{\zeta} - 2\frac{D\zeta \mathfrak{a} D\zeta}{\zeta^2}\right)z + \mathfrak{M}' \cdot Dz$$

$$(x \in \Sigma)$$

where $\mathfrak{M}'$ is continuous in $\Sigma$. By Cauchy's inequality

$$2|\delta \mathfrak{a}' D^2 u|z \le \lambda \zeta \|D^2 u\|^2 + \frac{|p|^2 \|\delta \mathfrak{a}'\|^2}{\lambda} z,$$

$$|\delta_2 \mathfrak{a}' D^2 u| \cdot \left|\frac{D\zeta}{\zeta}\right| z \le \lambda \zeta \|D^2 u\|^2 + \frac{|p|^2 \|\delta_2 \mathfrak{a}'\|^2}{4\lambda} \left|\frac{D\zeta}{\zeta}\right|^2 z.$$

Moreover, by direct calculation

$$\left|\frac{D\zeta}{\zeta}\right| \le \frac{2k}{d\zeta^{1/k}} = \frac{2k}{d}\left(\frac{|p|^2}{z}\right)^{1/k}$$

$$\left|\frac{D^2\zeta}{\zeta}\right| \le \frac{k(4k+n)}{d^2\zeta^{2/k}} = \frac{k(4k+n)}{d^2}\left(\frac{|p|^2}{z}\right)^{1/k}.$$

Substituting the preceding inequalities into (26), we obtain the main relation (6) with

$$\frac{1}{2}\mathfrak{N} = \delta\mathfrak{B}\,\delta\,\frac{|p|^2\,\|\delta\mathfrak{A}'\|^2}{2\lambda}$$

(27)
$$+ k\left|\delta_2\mathfrak{B} + 2|p|^2\,\delta\mathfrak{J}\right|\frac{1}{d}\bigl(|p|^2\bigr)^{1/k}$$

$$+ k(6k+n)\left\{\|\mathfrak{A}\| + \bigl(\lambda^{-1}\|\delta_2\mathfrak{A}'\|^2 + \|\delta_2\mathfrak{J}\|\bigr)|p|^2\right\}\frac{1}{d^2}\left(\frac{|p|^2}{z}\right)^{2/k}.$$

Finally introducing barred variables by means of (18) and the formulas immediately following, one finds that the function

$$\bar{z} = \zeta\bar{w} = \zeta|D\bar{u}|^2$$

satisfies

$$\mathfrak{a}\,D^2\bar{z} \geq \bar{\mathfrak{M}}' \cdot D\bar{z} - \bar{\mathfrak{n}}'\bar{z}$$

with

$$\frac{1}{2\mathcal{E}}\bar{\mathfrak{n}}' = \alpha\omega^2 + 2\beta\omega + \gamma - \frac{\omega'}{\phi'}$$

$$+ \frac{1}{\mathcal{E}}\left|\delta_2\beta + |p|^2\,\delta\mathfrak{J} + \omega(\delta_2\mathcal{E} + \delta_1(|p|^2\mathfrak{J}))\right|$$

$$\cdot 2k\,\frac{1}{d}\left(\frac{|p|^2}{z}\right)^{1/k}$$

(28)
$$+ \frac{1}{\mathcal{E}}\left\{\|\mathfrak{A}\| + |p|^2(\lambda^{-1}\|\delta_2\mathfrak{A}'\|^2 + \|\delta_2\mathfrak{J}\|)\right\}$$

$$\cdot k(6k+n)\,\frac{1}{d^2}\left(\frac{|p|^2}{z}\right)$$

where $\alpha, \beta, \gamma$ have been defined in the previous section.

## GRADIENT ESTIMATES

Theorem 3. Suppose the hypotheses of Theorem 2 are satisfied and assume, in addition, that

(29)
$$\begin{cases} |p|\delta_2 \mathfrak{a}' = O(|p|^{-\theta}\sqrt{\lambda \varepsilon}) \\ \mathfrak{a}, |p|^2 \delta_2 \mathfrak{F} = O(|p|^{2\theta}\varepsilon) \\ \delta_2 \mathfrak{B}, \delta_2 \varepsilon, \delta(|p|^2 \mathfrak{F}), \delta_1(|p|^2 \mathfrak{F}) = O(|p|^{-\theta}\varepsilon) \end{cases}$$

for $p \to \infty$, $|u| \leq m$, where $\theta$ is a positive constant. Then at any point P in $\Omega$ whose distance to the boundary is 2d, we have

(30) $\qquad |Du| \leq K \, \text{Max}(1, d^{-1/\theta})$

where K is a constant depending only on m and the structure of the equation (and on the oscillation of u when case (iii) of Theorem 2 applies).

Proof. Under the hypotheses of Theorem 2 there exists an $\varepsilon > 0$ and a change of variables $u = \phi(\bar{u})$ such that

$$\alpha \omega^2 + 2\beta\omega + \gamma - \frac{\omega'}{\phi'} \leq -\varepsilon \quad ,$$

for p sufficiently large. Thus if $k = 2/\theta$ it is clear that

$$\bar{\eta}' \leq 0 \quad \text{for} \quad |u| \leq m, \quad |p| \geq L, \quad z \geq L'd^{-2/\theta}$$

where L and L' are sufficiently large constants. Now

$$|p| = \sqrt{w} \geq \sqrt{z} = \phi\sqrt{\bar{z}} \quad \text{in} \quad \Sigma .$$

Consequently $\bar{z}$ cannot have an interior maximum value in $\Sigma$ exceeding $\text{Max}(L^2, L'd^{-2/\theta})/(\text{Min } \phi')^2$. Since $\bar{z} = 0$ at the boundary of $\Sigma$, this yields

$$\bar{z} \leq \frac{\text{Max}(L^2, L'd^{-2/\theta})}{(\text{Min } \phi')^2} \quad \text{in} \quad \Sigma .$$

But $\bar{z} = \bar{w} = |D\bar{u}|^2 = \phi'^2 |Du|^2$ at $P$, whence

(31) $$|Du| \leq \text{Max}\left(L, \frac{\sqrt{L'}}{d^{1/\theta}}\right) \frac{\text{Max } \phi'}{\text{Min } \phi'} \quad \text{at } P$$

completing the proof.

The constant $K$ in Theorem 3 depends on the structure of the equation only through the values of $a$, $b$, $c$, the rate of convergence in (21), and the number $\theta$ and the order constants in (29). Theorem 3 is due, in its main points, to Ladyzhenskaya and Uraltseva (see [10], Theorem 1; the present result differs from theirs in including the precise dependence on the distance $d$ and on $\theta$, in introducing the multipliers $r$, $s$, $t$, and in some of the specific formulae).

**Remark 1.** If appropriate hypotheses are placed on the coefficients $\alpha$, $\beta$, $\gamma$ then the Riccati inequality can be solved for any range of values $|p| > \varepsilon$. In this case (assuming suitable uniformity properties with regard to the variable $x$) one obtains in lieu of (31)

(32) $$|Du| \leq \text{Max}\left(\varepsilon, \frac{\sqrt{L'(\varepsilon)}}{d^{1/\theta}}\right) \frac{\text{Max } \phi'}{\text{Min } \phi'} \quad \text{at } P.$$

Suppose now that $u$ is a bounded solution over $R^n$. By letting $d \to \infty$ and then $\varepsilon \to 0$ in (32) we find $Du = 0$. Since $p$ was an arbitrary point, it follows that $Du \equiv 0$ and in turn $u \equiv$ constant.

Therefore, under appropriate conditions on the structure of (1), the method of proving Theorem 3 leads further to a series of Liouville theorems. In another paper [18] we have investigated this situation in detail.

**Remark 2.** A similar application of this technique allows one to show that if $u$ is a solution in an exterior domain (or in a cone extending to infinity) and if $u \to$ limit as $x \to \infty$, then also $Du \to 0$ as $x \to \infty$ (see [14]).

Remark 3. The hypotheses of Theorems 2 and 3 deal with the asymptotic behavior of the quantities

$$G, G', B, \mathcal{E} \text{ and } \mathcal{F}$$

and in fact there are a total of thirteen different expressions involved. To some extent this rather confusing situation can be given a logical order. In particular, the hypothesis (23) of Theorem 2 requires

(23a)
$$\delta(|p|G'), \quad \delta_1(|p|G') = O(\sqrt{\lambda\mathcal{E}})$$
$$\delta B, \quad \delta_1 B, \quad \delta\mathcal{E}, \quad \delta_1\mathcal{E} \leq O(\mathcal{E})$$

while for Theorem 3 one wants additionally (29). <u>Then global and interior gradient estimates hold if any one of the conditions</u> (i), (ii), (iii) <u>of Theorem 2 is valid.</u>

Of the condition listed above, perhaps that on $G$ in (29) is the most restrictive. Indeed this condition specifically prevents Theorem 3 from applying to the minimal surface operator (where both $G$ and $\mathcal{E}$ have order $|p|^2$) or to any equation having Bernstein genre 2 or more. By other means Bombieri, De Giorgi and Miranda have given interior gradient estimates for the minimal surface equation for $n \geq 2$; their results have been extended to certain other classes of equations by Ladyshenskaya and Uraltseva ([10], Part II).

Theorem 2 of course applies more generally than Theorem 3, but even for Theorem 2 the condition

(12)
$$G = G' + P\mathcal{F} + \mathcal{F}p$$

somewhat restricts the applicability of the result[†]. For example, the Euler-Lagrange equation of any variational problem

(33)
$$\delta \int F(x, u, |Du|)dx = 0$$

---

[†] Except when $n = 2$. In this case (12) always holds with $G' = I$; see paragraph 6 of Section 5.

can be written in the form (1), (12) with $\mathcal{G}' = I$, but for the general problem

(34) $$\delta \int F(x, u, Du) dx = 0$$

the decomposition (12) holds only with $\mathcal{G}' \equiv \mathcal{G}$. Accordingly the type of conclusions which can be obtained for (34) are more limited than for (33).

Remark 4. It should be emphasized finally that in applications of Theorems 2 and 3 to any particular situation the quantities r, s, t which appear in the operators $\delta$, $\delta_1$, $\delta_2$ can be chosen in any convenient manner. This remark will be of vital importance in the examples which we discuss in the following section.

## 5. Applications.

As we have already remarked, the results of Theorems 2 and 3 can be applied to a wide variety of specific situations. In this section we present some particular examples which illustrate the range of possible applications.

<u>1.</u> The minimal surface operator is given by

$$\mathcal{G} = (1 + |p|^2)I - pp .$$

Comparing this with (12) gives $\mathcal{G}' = (1+|p|^2)I$, $\mathcal{F} = \frac{1}{2} p$ and also $\mathcal{E} = |p|^2$, $\lambda = 1 + |p|^2$. In order to apply Theorem 2 we put $r = 0$, $s = -3$, whence from the formulae preceding Theorem 2 we find

$$a = -1, \qquad c = \lim \sup |p|^{-2} \delta \mathcal{B}$$

$$2b = \lim \sup \frac{1}{|p|^2} \{ p \cdot \frac{\partial \mathcal{B}}{\partial p} - 3\mathcal{B} \} .$$

Accordingly we obtain a global gradient estimate if (for example)

$$\mathfrak{B}(x, u, p) = \mathcal{C}(x, u, p)(1 + |p|^2)^{3/2}$$

where $\mathcal{C}$ is continuously differentiable and obeys

$$\frac{\partial \mathcal{C}}{\partial u} \geq 0, \quad \frac{\partial \mathcal{C}}{\partial x} = 0(1), \quad p \cdot \frac{\partial \mathcal{C}}{\partial p} \leq 0(|p|^{-1}).$$

In particular, this condition is satisfied when $\mathcal{C}$ is a continuously differentiable function of $x$, $u$, $\sigma = p/|p|$ with $\partial \mathcal{C}/\partial u \geq 0$.

An alternate result can be obtained by applying (with $\mathcal{G}' = \mathcal{G}$, $\mathcal{F} = 0$) the corollary of Theorem 1 rather than Theorem 2. In this case the required condition for a global gradient estimate is

$$|p| \frac{\partial \mathcal{C}}{\partial u} + \sigma \cdot \frac{\partial \mathcal{C}}{\partial x} + \frac{\mathcal{C}^2}{n-1} \geq 0 \qquad \text{for large } p$$

or a fortiori

$$\frac{\partial \mathcal{C}}{\partial u} \geq 0, \qquad \sigma \cdot \frac{\partial \mathcal{C}}{\partial x} + \frac{\mathcal{C}^2}{n-1} \geq 0.$$

These conditions neither include nor are included in the previous one, though interestingly enough they are independent of any requirement on $\partial \mathcal{C}/\partial p$.

__2.__ The operator $\mathcal{G} = I + pp$ is interesting in relation to the previous example since its eigenvalues (namely 1, $|p|^2$) are exactly the same.[†] Here $\mathcal{G}' = I$, $\mathcal{F} = \frac{1}{2}p$, $\mathcal{E} = |p|^2 + |p|^4$ and $\lambda = 1$. Again choosing $r = 0$ we find

$$a = 4 + \limsup s, \qquad c = \limsup |p|^{-4} \delta \mathfrak{B}$$

$$2b = \limsup \frac{1}{|p|^4} \{ p \cdot \frac{\partial \mathfrak{B}}{\partial p} + s\mathfrak{B} \}.$$

Hence a global gradient estimate holds if there exists a bounded scalar multiplier $s \leq -4 + o(1)$ such that

---

[†]Though with different multiplicities when $n > 2$.

$$p \cdot \frac{\partial \mathcal{B}}{\partial p} + s\mathcal{B} \leq 0(|p|^4)$$

and if furthermore

$$\delta \mathcal{B} \leq 0(|p|^4).$$

By Theorem 3, if there exists a vector multiplier $t = 0(|p|^{-\theta})$, $0 < \theta \leq 1$, such that

$$\delta_2 \mathcal{B} = \frac{\partial \mathcal{B}}{\partial p} + t\mathcal{B} = 0(|p|^{4-\theta})$$

then an interior gradient estimate also holds.

As a particular example, suppose that

$$\mathcal{B}(x, u, p) = C(x, u, \sigma)(1 + |p|^2)^{\nu/2} + \mathcal{B}'(x, u, p)$$

where $\mathcal{B}'$ and $C$ are continuously differentiable functions of their arguments and $\nu = $ constant $> 4$. If

$$\frac{\partial C}{\partial u} \geq 0 \, ; \qquad \frac{\partial C}{\partial x} = 0 \quad (\text{if } \nu > 5) \, ;$$

$$\mathcal{B}', \; \frac{\partial \mathcal{B}'}{\partial x}, \; \frac{\partial \mathcal{B}'}{\partial u}, \; |p|\frac{\partial \mathcal{B}'}{\partial p} = 0(|p|^4)$$

then a global estimate exists. If $\nu < 5$, then an interior estimate automatically holds (take $t = 0$); if $\nu \geq 5$ an interior estimate exists provided

$$\left| (I - \sigma\sigma) \cdot \frac{\partial C}{\partial \sigma} \right| \leq \text{Const.} \; |C|$$

(here $t = -\frac{1}{|p|} \left\{ \frac{I - \sigma\sigma}{C} \cdot \frac{\partial C}{\partial \sigma} + \nu\sigma \frac{|p|^2}{1 + |p|^2} \right\}$ is required).

This example strikingly illustrates the fact that the behavior of the eigenvalues of the coefficient matrix $G$ is not the crucial issue in obtaining gradient estimates for equation (1).

## GRADIENT ESTIMATES

<u>3</u>. Suppose (1) is uniformly elliptic, so that

$$\lambda \xi^2 \le \xi \, \mathfrak{a} \, \xi \le \Lambda \xi^2$$

where $\lambda$ and $\Lambda$ are constants ($\lambda$ and $\Lambda$ could also be allowed to depend on $u$, but in fact under the basic hypothesis $|u| \le m$ this would not be of critical importance). We assume the "natural" conditions

(35) $\qquad \dfrac{\partial \mathfrak{a}}{\partial x}, \; |p| \dfrac{\partial \mathfrak{a}}{\partial p} = O(1); \qquad \dfrac{\partial \mathfrak{a}}{\partial u} = o(1)$

as $p \to \infty$. Then obviously

$$\lambda |p|^2 \le \mathcal{E} \le \Lambda |p|^2 \; ; \quad \dfrac{\partial \mathcal{E}}{\partial x}, \; |p| \dfrac{\partial \mathcal{E}}{\partial p} = O(|p|^2) \; ; \quad \dfrac{\partial \mathcal{E}}{\partial u} = o(|p|^2).$$

Setting $\mathfrak{a}' = \mathfrak{a}$, $\mathfrak{F} = 0$ we find

$$a \le \text{const.} \; (1+s^2) \qquad c = \lim \sup \dfrac{1}{\mathcal{E}} \left\{ \delta \mathfrak{B} + \dfrac{n \Lambda^2}{2\lambda} r^2 |p|^2 \right\}$$

$$2b = \lim \sup \left\{ r + \dfrac{1}{\mathcal{E}} (p \cdot \dfrac{\partial \mathfrak{B}}{\partial p} + s \mathfrak{B}) \right\} \; .$$

One cannot in general insure that $a \le 0$; thus to apply Theorem 2 and 3 in a convenient way we must frame suitable hypotheses to make $c \le 0$. To this end, let us suppose that there exists a bounded scalar multiplier $\eta$ such that

(36) $\qquad \eta \mathfrak{B} + p \cdot \dfrac{\partial \mathfrak{B}}{\partial p} \le O(|p|^2) \; .$

Choosing

$$r = -\dfrac{\lambda \mathfrak{B}}{n \Lambda^2 |p|^2} \; , \qquad s = \eta + \dfrac{\lambda \mathcal{E}}{n \Lambda^2 |p|^2}$$

it is clear that $a$ and $b$ are bounded, while $c$ will be non-positive provided

(37) $\qquad \lim \inf \dfrac{1}{|p|^2} \left\{ \dfrac{\partial \mathfrak{B}}{\partial u} + \dfrac{\sigma}{|p|} \cdot \dfrac{\partial \mathfrak{B}}{\partial x} + \dfrac{\lambda}{2n \Lambda^2} \dfrac{\mathfrak{B}^2}{|p|^2} \right\} \ge 0 \; .$

Summarizing, a global gradient estimate holds for uniformly elliptic equations of the form (1) provided conditions (35)-(37) are satisfied. Interior estimates hold if, furthermore, there exists a vector multiplier $t = 0\,(|p|^{-\theta})$, $0 < \theta \leq 1$, such that

$$\frac{\partial \mathcal{B}}{\partial p} + t\mathcal{B} = O(|p|^{2-\theta})\,.$$

By being more careful in the original application of Cauchy's inequality, it would be possible to obtain (37) with the factor $\lambda/2n\Lambda^2$ replaced by $\kappa/\text{Trace }\mathcal{G}$ where $\kappa$ is any number less than one. Finally even should condition (37) fail we may still apply case (iii) of Theorem 2 provided the oscillation of $u$ is sufficiently small.

<u>4.</u> Suppose $\mathcal{G}' = I$, as is the case in particular for regular variational problems depending on $|Du|$. Then

$$a = \lim\sup\left\{s + \frac{1}{\mathcal{E}}\left(p\cdot\frac{\partial\mathcal{E}}{\partial p} + \frac{1}{2}n(1+s)^2|p|^2\right)\right\}$$

$$2b = \lim\sup\frac{1}{\mathcal{E}}\left\{\delta\mathcal{E} + s\mathcal{B} + p\cdot\frac{\partial\mathcal{B}}{\partial p}\right\}$$

$$c = \lim\sup\frac{1}{\mathcal{E}}\left\{\delta\mathcal{B} + \frac{1}{2}nr^2|p|^2\right\}\,.$$

By Theorem 2 if $a$, $b$, $c < \infty$ and either $a \leq 0$ or $c \leq 0$ we obtain a global gradient estimate. To analyze this situation more completely we consider several cases.

<u>4.1.</u> $p^2/\mathcal{E} = o(1)$. Then assuming that $s$ is bounded,

$$a = \lim\sup\,(s + \frac{p}{\mathcal{E}}\cdot\frac{\partial\mathcal{E}}{\partial p})\,.$$

Setting $r = 0$ we can proceed as in example <u>2</u>.

<u>4.2.</u> $p^2/\mathcal{E} = O(1)$. This is essentially the same as the uniformly elliptic case discussed in example <u>3</u>.

<u>4.3.</u> $\mathcal{E}/p^2 = o(1)$. To make <u>a</u> finite it is necessary to choose $s = -1$. Putting $r = 0$ there are then three further

subcases depending on whether

$$\frac{p}{\mathcal{E}} \cdot \frac{\partial \mathcal{E}}{\partial p} \leq O(1), \quad \frac{p}{\mathcal{E}} \cdot \frac{\partial \mathcal{E}}{\partial p} \leq 1, \quad \text{or} \quad \mathcal{E} = o(1).$$

In the first case $a > 0$, and $c$ will be $\leq 0$ if

$$\frac{\partial \mathcal{B}}{\partial u} + \frac{\sigma}{|p|} \cdot \frac{\partial \mathcal{B}}{\partial x} \geq -o(\mathcal{E}), \quad p \cdot \frac{\partial}{\partial p}\left(\frac{\mathcal{B}}{|p|}\right) = O\left(\frac{\mathcal{E}}{|p|}\right).$$

In the second case $a \leq 0$, and $c$ will be finite if

$$\mathcal{B}(x, u, p) = C(x, u, \sigma)|p| + \mathcal{B}'(x, u, p)$$

where $\partial C/\partial u \geq 0$ and $\mathcal{B}'$ and its derivatives are of order $\mathcal{E}$. For the third case it is necessary to apply Theorem 1'.

<u>5.</u> The preceding examples indicate a general pattern of behavior in obtaining gradient estimates. Specifically, four significantly different cases emerge:

(a) $\quad \mathcal{G}', \quad p \cdot \dfrac{\partial \mathcal{G}'}{\partial p} = o(\sqrt{\lambda \mathcal{E}}/|p|)$

(b) $\quad \eta \mathcal{G}', \quad p \cdot \dfrac{\partial \mathcal{G}'}{\partial p} = O(\sqrt{\lambda \mathcal{E}}/|p|)$

(c) $\quad \eta \mathcal{G}' + p \cdot \dfrac{\partial \mathcal{G}'}{\partial p} = o(\sqrt{\lambda \mathcal{E}}/|p|) \quad$ for some bounded scalar multiplier $\eta$

(d) $\quad \mathcal{G}' + p \cdot \dfrac{\partial \mathcal{G}'}{\partial p} = O(\sqrt{\lambda \mathcal{E}}/|p|) \quad$ for some bounded scalar multiplier $\eta$.

These are to be combined with

(e) $\quad \dfrac{\partial \mathcal{G}'}{\partial x}, \quad |p|\dfrac{\partial \mathcal{G}'}{\partial u} = o(\sqrt{\lambda \mathcal{E}}); \quad$ (f) $\quad \dfrac{\partial \mathcal{G}'}{\partial x}, \quad |p|\dfrac{\partial \mathcal{G}'}{\partial u} = O(\sqrt{\lambda \mathcal{E}})$

where (e) goes with (b) and (d), and (f) with (a) and (c). Moreover in all cases we require

(g) $\quad \dfrac{\partial \mathcal{E}}{\partial x}, \dfrac{\partial \mathcal{E}}{\partial u} = 0(\mathcal{E}); \quad p \cdot \dfrac{\partial \mathcal{E}}{\partial p} \leq O(\mathcal{E}).$

Now suppose (a, f) holds. Then we can proceed as in example 2, with corresponding hypotheses on $\mathcal{B}$. If (b, e) or (d, e) holds, we proceed as in example 3 (choosing $r = 0$ if $\Lambda |p|^2 = o(\mathcal{E})$). If (c, f) holds, we proceed as in example 1, with corresponding hypotheses on $\mathcal{B}$ and $\mathcal{E}$, that is

$$\delta_1 \mathcal{E} = o(\mathcal{E}) \; ; \qquad \delta \mathcal{B}, \; \delta_1 \mathcal{B} \leq o(\mathcal{E}) .$$

If $\mathcal{E} = o(\text{Trace}\,\mathcal{Q})$ then stronger results can be obtained from Theorem 1'.

6. **The special case $n = 2$.** Let the coefficient matrix $\mathcal{Q}$ be written in the form

$$\mathcal{Q} = \begin{pmatrix} A & B \\ B & C \end{pmatrix} .$$

An easy calculation shows that

$$\mathcal{Q} = EI + p\mathcal{J} + \mathcal{J}p \qquad (p \neq 0)$$

where

$$E = \text{Trace}\,\mathcal{Q} - \sigma \mathcal{Q} \sigma, \quad \mathcal{J} = \frac{1}{2|p|^2} \begin{pmatrix} (A-C)p_1 + 2Bp_2 \\ 2Bp_1 + (C-A)p_2 \end{pmatrix}$$

and $\sigma = p/|p|$. We can thus apply the previous considerations with $\mathcal{Q}' = EI$. It is however more profitable to write the equation in the form

$$(I + p\mathcal{G} + \mathcal{G}p)D^2 u = \mathcal{D}$$

where $\mathcal{G} = \mathcal{J}/E$ and $\mathcal{D} = \mathcal{B}/E$, and to apply the results of paragraph 4.

## GRADIENT ESTIMATES

To give just one conclusion for this case, the corollary of Theorem 1 implies that if

$$\frac{\mathscr{E}^2}{2} + p \cdot \frac{\partial \mathscr{E}}{\partial x} + |p|^2 \frac{\partial \mathscr{E}}{\partial u} \geq 0 \quad \text{for} \quad |u| \leq m, \quad p \geq L$$

then $|Du| \leq \text{Max}(\ell, L)$ in $\Omega$, where $\ell$ is a bound for $Du$ on the boundary. (This result is essentially due to Bernstein [1], page 129; see also [16], pages 442-443.)

## 6. Parabolic equations.

The methods of the earlier sections carry over with minor but not unimportant changes to the case of quasilinear parabolic equations (2).

Consider solutions $u$ in a space-time domain $Q = \Omega \times (\tau, T]$, and suppose that $u \in C^1(\bar{Q}) \cap C^3(Q)$ where the indices denote differentiations with respect to $x$. Letting $\partial^* Q$ denote the parabolic boundary of $Q$, namely the set $\bar{Q} - Q$, we put

$$\sup_Q |u| = m, \quad \sup_{\partial^* Q} |Du| = \ell.$$

A global estimate for the gradient of $u$ is then a statement of the form

$$\sup_Q |Du| \leq \text{Constant}$$

where the constant depends only on $m$, $\ell$ and the structure of the equation. Assume on the other hand that $u \in C^3(Q)$, $u_t \in C^1(Q)$. An interior estimate for the gradient is a statement that

$$\sup_{Q'} |Du| \leq \text{Constant}$$

where the constant now depends only on $m$, the structure of the equation and the distance from $Q'$ to the parabolic boundary of $Q$ (here $Q'$ is a compact subset of $Q$).

In order to obtain such estimates we proceed as in the elliptic case (see Sections 1 to 4). Applying the differential operator $u_k \partial/\partial x_k$ to equation (2) one finds as before that

(39) $\qquad -w_t + \mathfrak{a} D^2 w = $ Right hand side of (11)

where $w = |Du|^2$. Here the arguments of $\mathfrak{a}$ and $\mathfrak{B}$ are $x, t, u$, and $p = Du$, corresponding to our earlier agreements. The differential operators $\delta$ and $\delta_2$ which appear in (11) can, however, no longer include the multipliers $r$ and $t$ as they did in the elliptic case. Thus for our present purposes

(40) $\qquad \delta \equiv \dfrac{\partial}{\partial u} + \dfrac{\sigma}{|p|} \cdot \dfrac{\partial}{\partial x_i} \; , \qquad \delta_2 \equiv \dfrac{\partial}{\partial p} \; .$

From (39) one obtains, as before,

$$-w_t + \mathfrak{a} D^2 w \geq \mathfrak{m} \cdot Dw - \mathfrak{n} w$$

where

$$\tfrac{1}{2} \mathfrak{n} = \delta \mathfrak{B} + \dfrac{|p|^2 \|\delta \mathfrak{a}'\|^2}{4\lambda} \; .$$

If we introduce the new dependent variable $\bar{u}$ by means of $u = \phi(\bar{u})$, then $\bar{w} = |D\bar{u}|^2$ satisfies

$$-\bar{w}_t + \mathfrak{a} D^2 \bar{w} \geq \bar{\mathfrak{m}} \cdot D\bar{w} - \bar{\mathfrak{n}} \bar{w}$$

where $\bar{\mathfrak{n}}$ is given by (20). Here also the operator $\delta_1$ must be taken in the special form

(41) $\qquad \delta_1 \equiv p \cdot \dfrac{\partial}{\partial p} - 1$

corresponding to the earlier remark that $\bar{r} = \bar{t} = 0$.

With the proviso that $\delta, \delta_1$, and $\delta_2$ are given by (40), (41), it is now clear that Theorems 1 and 2 hold equally for parabolic equations.

Turning to interior estimates, let $P$ be an arbitrary point of $Q$. Suppose that the distance from $P$ to the lateral part of $\partial^* Q$ is $2d$ and the distance to the base is $2d_1$. Choosing $P$ as origin of coordinates we make the substitution

$$\text{(42)} \qquad z = \zeta w, \qquad \zeta = \left(1 - \frac{|x|^2}{d^2}\right)^k \left(1 - \frac{t^2}{d_1^2}\right)^{k_1}$$

where $k$ and $k_1$ are appropriate constants to be determined. Proceeding as in the elliptic case (see Section 4) one finds that $z$ satisfies the differential inequality

$$-z_t + aD^2 z \geq \mathcal{M} \cdot Dz - \mathcal{N}'z \qquad \text{in } \Sigma' \equiv \Sigma \times [-d_1, 0]$$

where

$$\frac{1}{2}\mathcal{N}' = \frac{k_1}{d_1}\left(\frac{|p|^2}{z}\right)^{1/k_1} + \text{Right hand side of (27)}.$$

Changing the dependent variables by means of $u = \phi(\bar{u})$ gives finally

$$-\bar{z}_t + aD^2 \bar{z} \geq \bar{\mathcal{M}} \cdot D\bar{z} - \bar{\mathcal{N}}\,\bar{z} \qquad \text{in } \Sigma'$$

where

$$\frac{1}{2\mathcal{E}}\bar{\mathcal{N}}' = \frac{1}{\mathcal{E}}\frac{k_1}{d_1}\left(\frac{|p|^2}{z}\right)^{1/k_1} + \text{Right hand side of (28)}.$$

This leads to the following interior gradient estimate for equation (2).

Theorem 4.  Suppose the hypothesis of Theorem 3 is satisfied, and that in addition $\mathcal{E} \geq |p|^{\theta_1}$ for $p \to \infty$, $|u| \leq m$, where $\theta_1$ is a positive constant. Then

$$|Du| \leq K \, \text{Max}(1, d^{-1/\theta}, d_1^{-1/\theta_1}) \quad \underline{\text{at }} P$$

where K depends only on m and the structure of the equation.

As an example, suppose (2) is uniformly parabolic, so that

$$\lambda |\xi|^2 \leq \xi \cdot a\xi \leq \Lambda |\xi|^2$$

where $\lambda$ and $\Lambda$ are given positive constants and $\xi$ is any real vector. The results of paragraph 3 of Section 5 may then be applied provided $r = t = 0$, $s = -1$. For a global gradient estimate, this yields the conditions

$$p \cdot \frac{\partial}{\partial p}\left(\frac{\mathcal{B}}{|p|}\right) \leq O(|p|)$$

and

$$\liminf \frac{1}{|p|^2}\left\{\frac{\partial \mathcal{B}}{\partial u} + \frac{\sigma}{|p|} \cdot \frac{\partial \mathcal{B}}{\partial x}\right\} \geq 0 .$$

Interior estimates hold if furthermore $\partial \mathcal{B}/\partial p = O(|p|^{2-\theta})$, $0 < \theta \leq 1$. The case of greatest interest would of course be $\theta = 1$, $\theta_1 = 2$.

Remark. A different interior estimate for $Du$ can also be given, depending on $m$ and on $\sup |Du|$ over the set $\Omega \times \{t = \tau\}$. In this case one need not introduce the condition $\mathcal{E} \geq |p|^{\theta_1}$ in Theorem 4 since the final factor in the definition of $\zeta$ can be omitted, see (42). In the same way, we could omit the first factor in the definition of $\zeta$ and obtain an estimate for $|Du|$ depending on $m$ and on $\sup |Du|$ over the lateral boundary $\partial \Omega \times [\tau, T]$ of $Q$. In this case, beyond the conditions of Theorems 1 and 2 one needs only the restriction $\mathcal{E} \geq |p|^{\theta_1}$; the conclusion then holds in the form $|Du| \leq K \operatorname{Max}(1, d_1^{-1/\theta_1})$. For uniformly parabolic equations, global estimates and an interior estimate of the latter form (with $\theta_1 = 1$) were given in [6]. Interior estimates corresponding to Theorem 4 for uniformly parabolic equations were found by Oleinik and Kruzhkov. For non-uniformly parabolic equations, see [10], Theorem 3.

## 7. Appendix. Proof of Theorem 2.

In order to prove Theorem 2 it is necessary to study solutions of the Riccati differential inequality

$$(43) \qquad \frac{d\omega}{d\phi} \geq a\omega^2 + 2b\omega + c \qquad (\omega \geq 0)$$

where the domain of $\phi$ is a fixed interval $[m_0, m_1]$ of the real axis and $a$, $b$, $c$ are real constants. [Note that (43) omits the constant $\varepsilon$ which appears in (22); this involves no loss of generality, however, since if (43) has a solution then obviously so has (22) for some $\varepsilon > 0$.] There are several cases to discuss, depending on the values of $a$, $b$, $c$.

I. The relation $a\omega^2 + 2b\omega + c = 0$ has a non-negative real root $\omega = \kappa$ in the four cases

$$(44) \qquad \begin{array}{l} a < 0; \qquad c \leq 0; \qquad a = 0, \ b < 0, \ c > 0; \\ a > 0, \ b \leq -\sqrt{ac}, \ c > 0. \end{array}$$

Consequently in these cases (43) is solved by $\omega \equiv \kappa$. For the contemplated application to Theorem 2 we have $\omega = \phi''/\phi'^2$ so that by integration

$$\phi' = \exp(-\kappa\phi) \qquad (> 0).$$

Furthermore

$$\frac{\text{Max } \phi'}{\text{Min } \phi'} = \exp\{\kappa(m_1 - m_0)\} < \infty$$

(here and later the evaluation of Max $\phi'$/Min $\phi'$ is facilitated by noting that since $\phi'' \leq 0$ the maximum of $\phi'$ occurs when $\phi = m_0$ and the minimum when $\phi = m_1$). The remaining cases call for more effort.

II. $a = b = 0$, $c > 0$. Then $d\omega = c\,d\phi$ so that

$$\omega = c(\phi + \kappa)$$

where $\kappa$ is a constant of integration. We choose $\kappa = -m_0$ so that $\omega \geq 0$ for $\phi \in [m_0, m_1]$. Then by integration

$$\phi' = \exp\{-c(\phi - m_0)^2\} \qquad (> 0)$$

and

$$\frac{\text{Max } \phi'}{\text{Min } \phi'} = \exp\{c(m_1 - m_0)^2\}.$$

III. $a = 0$, $b > 0$, $c > 0$. Then $(2b\omega + c)^{-1} d\omega = d\phi$ so that

$$\phi + \kappa = \frac{1}{2b} \log(\omega + \frac{c}{b})$$

where $\kappa$ is a constant of integration. We choose $\kappa$ so that $\omega = 0$ when $\phi = m_0$. Hence

$$\omega = \frac{c}{2b}\left(e^{2b(\phi - m_0)} - 1\right).$$

Integrating further we obtain

$$\phi' = \exp \frac{c}{2b}\left(\phi - \frac{1}{2b} e^{2b(\phi - m_0)}\right) \qquad (> 0)$$

and so

$$\frac{\text{Max } \phi'}{\text{Min } \phi'} \leq \exp \frac{c}{4b^2}\left(e^{2b(m_1 - m_0)} - 1\right)$$

IV. $a, c > 0$ and $b > -\sqrt{ac}$. This case divides into three subcases depending on whether $b < \sqrt{ac}$, $b = \sqrt{ac}$, or $b > \sqrt{ac}$. We consider these in order.

(1) $b < \sqrt{ac}$. Here $(a\omega^2 + 2b\omega + c)^{-1} d\omega = d\phi$ so that

(45) $$\varphi + \kappa = -\frac{1}{\sqrt{ac - b^2}} \text{arccot} \frac{a\omega + b}{\sqrt{ac - b^2}}$$

where $\kappa$ is a constant of integration. As $\omega$ goes from $0$ to $\infty$ the right hand side runs from

$$-\frac{1}{\sqrt{ac-b^2}} \text{arccot} \frac{b}{\sqrt{ac-b^2}} \quad \text{to } 0.$$

Thus we must assume

$$m_1 - m_0 < \frac{1}{\sqrt{ac-b^2}} \text{arccot} \frac{b}{\sqrt{ac-b^2}}$$

that is

(46) $\quad\quad$ oscillation $\phi < \dfrac{1}{\sqrt{ac+b^2}} F(t), \quad -1 < t < 1,$

where $F(t)$ is the function defined in Theorem 2 and $t = b/\sqrt{ac}$.

Choosing $\kappa$ so that $\phi = m_0$ when $\omega = 0$ and then putting $\psi = \phi + \kappa$, we can solve (45) for $\omega$ and integrate once more. After some computation there results

$$\phi' = e^{b\psi/a} |\sin\sqrt{ac-b^2}\,\psi|^{1/a} \quad (> 0)$$

and

$$\frac{\text{Max } \phi'}{\text{Min } \phi'} = e^{b(m_0-m_1)/a} \left|\frac{\sin\sqrt{ac-b^2}(m_0+\kappa)}{\sin\sqrt{ac-b^2}(m_1+\kappa)}\right|^{1/a} \leq \left(\frac{2e^{b(m_0-m_1)}}{\mu\sqrt{ac}}\right)^{1/a}$$

where $\mu$ denotes the difference between the left and right hand sides of (46).

(2) $\quad b^2 = ac$. Here $(\sqrt{a}\,\omega + \sqrt{c})^2 d\omega = d\phi$ whence

$$\varphi + \kappa = -\frac{1}{a\omega + b}.$$

The term on the right hand side varies from $-1/b$ to $0$ as $\omega$ runs from $0$ to $\infty$; hence we must have $m_1 - m_0 < 1/b$, that is

$$\text{oscillation } \phi < \frac{1}{\sqrt{ac+b^2}} \sqrt{2}.$$

Proceeding as before, one now finds

$$\phi' = e^{b\psi/a} |\psi|^{1/a} \qquad (>0)$$

and

$$\frac{\text{Max } \phi'}{\text{Min } \phi'} = e^{b(m_0-m_1)/a} \left|\frac{m_0 + \kappa}{m_1 + \kappa}\right|^{1/a} \leq \left(\frac{e^{b(m_0-m_1)}}{\mu\sqrt{ac}}\right)^{1/a}.$$

(3) $b^2 > ac$. Omitting the details, as is now allowable, we find

$$\phi + \kappa = -\frac{1}{\sqrt{b^2 - ac}} \text{ arccoth } \frac{a\omega + b}{\sqrt{b^2 - ac}}$$

whence the oscillation of $\phi$ must satisfy

$$\text{oscillation } \phi < \frac{1}{\sqrt{ac + b^2}} F(t) \qquad (t = \frac{b}{\sqrt{ac}} > 1).$$

Moreover

$$\phi' = e^{b\psi/a} \left|\sinh\sqrt{b^2 - ac}\,\psi\right|^{1/a}$$

and

$$\frac{\text{Max } \phi'}{\text{Min } \phi'} = e^{b(m_0-m_1)/a} \left|\frac{\sinh\sqrt{b^2-ac}(m_0+\kappa)}{\sinh\sqrt{b^2-ac}(m_1+\kappa)}\right| \leq \left(\frac{e^{b(m_0-m_1)}}{\mu\sqrt{ac}}\right)^{1/a}.$$

## REFERENCES

1. Bernstein, S. N. Sur la géneralisation du problème de Dirichlet, II. Math. Ann. <u>69</u>, (1910), 82-136.

2. Bernstein, S. N. Sur les équations des calcul des variations. Ann. Sci. École Norm. Supp. <u>29</u>, (1912), 431-485.

3.  Bombieri, E, De Giorgi, E. and Miranda, M. 1969 Una maggiorazioni a priori relativa alle ipersuperfici minimali non parametriche. Arch. Rat. Mech. Anal. 32, (1969), 255-267.

4.  Hopf, E. Elementare Bemerkungen über die Lösungen partieller Differentialgleichunger zweiter Ordnung von Elliptischen Typen. Sber. Preuss. Akad. Wiss. 19, (1927), 147-152.

5.  Ivanov, A. V. On interior estimates of the first derivatives of solutions of quasilinear non-uniformly elliptic and parabolic equations of general form. Sem. Steklov Inst. Leningrad 14 (1969), 29-47. See also Sem. Steklov Inst. Leningrad, 19 (1970), 79-94.

6.  Ladyzhenskaya, O. A. The first boundary value problem for quasilinear parabolic equations. Dokl. Akad. Nauk 107 (1956), 636-639. Trudy Moscow Math. Soc. 7 (1958), 149-177.

7.  Ladyzhenskaya, O. A. and Uraltseva, N. N. Linear and quasilinear elliptic equations. Moscow, 1964.

8.  Ladyzhenskaya, O. A. and Uraltseva, N. N. On certain classes of non-uniformly elliptic equations. Sem. Steklov. Inst. Leningrad 11 (1968), 129-149. (Translated in Sem. Steklov Inst. 11 (1970).)

9.  Ladyzhenskaya, O. A. and Uraltseva, N. N. On total estimates of first derivatives of solutions of quasilinear elliptic and parabolic equations. Sem. Steklov Inst. Leningrad. 14 (1969), 127-155. (Translated in Sem. Steklov Inst. 14 (1971).)

10. Ladyzhenskaya, O. A. and Uraltseva, N. N. Local estimates for gradients of solutions of non-uniformly elliptic and parabolic equations. Comm. Pure Appl. Math. 23 (1970), 677-703.

11. Leray, J. Discussion d'un problème de Dirichlet. J. Math. Pures Appl. 18, (1939), 249-284.

12. Oleinik, O. A. and Kruzhkov, S. N. Quasi-linear second-order parabolic equations with many independent variables. Usp. Mat. Nauk 16 (1961), 115-156. (Translated in Russian Math. Surveys 16 (1961).)

13. Oskolkov, A. P. A priori estimates for the first derivatives of solutions of the Dirichlet problem for non-uniformly elliptic quasilinear equations. Trudy Mat. Inst. Steklov 102, (1967), 105-127.

14. Peletier, L. A. and Serrin, J. Limit conditions on the gradient of solutions of quasilinear elliptic and parabolic equations. To appear.

15. Protter, M. and Weinberger, H. F. Maximum principles in differential equations. Prentice-Hall, 1967.

16. Serrin, J. The problem of Dirichlet for quasilinear elliptic differential equations with many independent variables. Phil. Trans. Roy. Soc. London A 264 (1969), 413-496.

17. Serrin, J. Boundary curvatures and the solvability of Dirichlet's problem. Proc. Int. Congr. of Math. (Nice), 1970.

18. Serrin, J. Entire solutions of nonlinear Poisson equations. To appear, Proc. London Math. Soc.

19. Trudinger, N. S. Some existence theorems for quasilinear, non-uniformly equations in divergence form. J. Math. Mech., 18 (1969), 909-919.

20. Ventsel, T. D. The first boundary value problem and Cauchy's problem for quasi-linear parabolic equations with many space variables. Mat. Sborn. 41 (1957), 499-520.

                      School of Mathematics
                      University of Minnesota
                      Minneapolis, Minnesota

                              Received May 17, 1971

# Shock Waves and Entropy

*PETER LAX*

Introduction: We study systems of the first order partial differential equations in conservation form:

(1) $\quad \partial_t u^j + \partial_x f^j = 0, \quad j = 1,\ldots,m, \quad f^j = f^j(u^1,\ldots,u^m).$

In many cases all smooth solutions of (1) satisfy an additional conservation law

(2) $\quad \partial_t U + \partial_x F = 0, \quad F = F(u),$

where U is a convex function of u. We study weak solutions of (1) which satisfy in addition the "entropy" inequality

(3) $\quad \partial_t U + \partial_x F \leq 0.$

We show that all weak solutions of (1) which are limits of solutions of modifications of (1) by the introduction of various kinds of dissipation satisfy the entropy inequality (3). We show that for weak solutions which contain discontinuities of moderate strength, (3) is equivalent to the usual shock condition involving the number of characteristics impinging on the shock. Finally we study all possible entropy conditions of form (3) which can be associated to a given hyperbolic system of two conservation laws.

1. We consider systems of first order nonlinear partial differential equations in conservation form:

(1.1) $$\partial_t u^j + \partial_x f^j = 0, \quad j = 1, \ldots, m,$$

where $\partial_t$ and $\partial_x$ denote partial differentiation with respect to $t$ and $x$, and where each $f^j$ is a function of $u = \{u^1, \ldots, u^m\}$, in general nonlinear. For simplicity we take the number of space variables to be 1. Carrying out the differentiations in (1.1) we get the equations

(1.2) $$\partial_t u^j + \sum_\ell f^j_\ell \, \partial_x u^\ell = 0$$

where the subscript $\ell$ denotes partial differentiation with respect to $u^\ell$.

If $u$ is a solution of (1.1) which is zero for $|x|$ large integrating (1.1) we obtain that

$$\int \partial_t u^j \, dx = 0$$

which means that the quantities

(1.3) $$\int u^j \, dx, \quad j = 1, \ldots, m$$

are conserved, i.e. independent of $t$.

Let $U$ be some function of $u^1, \ldots, u^m$; when does $U$ satisfy a conservation law, i.e. an equation of form

(1.4) $$\partial_t U + \partial_x F = 0 \;?$$

(1.4) can be written in the form

(1.5) $$\sum_j U_j f^j_\ell = F_\ell \quad \text{for} \quad \ell = 1, \ldots, m.$$

We suppose now that (1.5) has a solution, so that (1.1) implies an additional conservation law (1.4); and we suppose furthermore that the new conserved quantity $U$ is a <u>convex</u> function of $u^1, \ldots, u^m$.

We observe that if the system (1.2) is **symmetric**, i.e.

$$f_\ell^j = f_j^\ell$$

then

$$U = \sum u_j^2$$

satisfies a conservation law, with

$$F = \sum u^j f^j - g \ ,$$

where $g$ is a function satisfying

$$g_\ell = f^\ell \ .$$

As observed in [3], if the convex function $U$ satisfies a conservation law then multiplying (1.2) by $U_{j,\ell}$ puts the system into **symmetric hyperbolic** form. For the purposes of section 2 we shall require a little more, that (1.2) be **strictly** hyperbolic. This means that the matrix

(1.8) $$f' = f_\ell^j$$

has real and distinct eigenvalues. We denote these eigenvalues arranged in increasing order, by $c_1, c_2, \ldots, c_m$. They are called **sound speeds**, and are functions of $u$.

It is well known that the initial value problem is properly posed for symmetric hyperbolic systems, i.e. we may prescribe the values of $u$ as arbitrary smooth functions at $t = 0$. Solutions are uniquely determined by their initial data but, since the governing equations are nonlinear, in general they exist only for a limited time range. For solutions which exist for all $t > 0$ we have to turn to weak solutions, which may be discontinuous solutions or slightly worse; these satisfy the conservation laws (1.1) only in the weak, or **integral sense**, i.e.

(1.9) $$\iint_{t \geq 0} [u^j \partial_t \phi + f^j \partial_x \phi] dx dt + \int u(x, 0) \phi(x, 0) dx = 0$$

for all smooth test functions $\phi$ with bounded support in $t \geq 0$. As is well known, for piecewise continuous solutions (1.9) is equivalent to the Rankine-Hugoniot jump conditions

(1.10) $$s[u^j] - [f^j] = 0$$

where $s$ is the speed with which the discontinuity is propagating, and $[u]$ denotes the difference between the values of $u$ on the two sides of the discontinuity.

It is equally well known, see e.g. [8] for some simple examples, that weak solutions of conservation laws are not uniquely determined by their initial values. To pick out the physically relevant solutions among the many, some additional physical principle has to be introduced. This additional principle usually identifies the relevant solutions as limits of solutions of equations with some dissipation. Specifically, we consider the equations with artifical viscosity:

(1.11) $$\partial_t u^j + \partial_x f^j = \varepsilon \, \partial_x^2 u^j, \quad \varepsilon > 0.$$

Suppose that a sequence $u(x, t; \varepsilon)$ of solutions of (1.11) tends to a limit $u(x, t)$ boundedly, almost everywhere; then $\varepsilon \, \partial_x^2 u(\varepsilon)$ tends to 0 in the topology for distributions, so that the limit $u$ satisfies (1.1) in the weak sense.

We show next how to characterize such limit solutions directly, with the aid of the function $U$:

Multiplying (1.11) by $U_j$ and summing we get

(1.12) $$\partial_t U + \partial_x F = \varepsilon \sum U_j \, \partial_x^2 u^j.$$

Using the identity

$$\partial_x^2 U = \sum U_j \partial_x^2 u^j + \sum U_{jk} \partial_x u^j \partial_x u^k$$

and the convexity of $U$ we deduce that

$$\partial_x^2 U \geq \sum U_j \partial_x^2 u^j.$$

Since $\varepsilon$ is $> 0$, using this to estimate the right side of ¶1.12) we get

$$\partial_t U + \partial_x F \leq \varepsilon \, \partial_x^2 F .$$

Letting $\varepsilon \to 0$ the right side tends to $0$ in the topology of distributions and we deduce

<u>Theorem 1.1</u>:  Let (1.1) be a system of conservation laws which implies an additional conservation law (1.4) where U is a strictly convex function. Then every weak solution of (1.1) which is the limit, boundedly a.e., of solutions of the viscous equation (1.11) satisfies the inequality

(1.13) $\qquad\qquad \partial_t U + \partial_x F \leq 0 .$

<u>Remark A</u>:  Suppose $u$ satisfies (1.13) and has compact support in $x$. Integrating (1.13) with respect to $x$ gives

$$\int \partial_t U \, dx \leq 0$$

which implies that

(1.14) $\qquad\qquad \int U \, dx$

is a <u>decreasing</u> function of $t$.

<u>Remark B</u>:  Suppose that $u$ is a piecewise continuous weak solution of (1.1); then it is easy to deduce either from (1.13) or (1.14) that at a point of discontinuity

(1.15) $\qquad\qquad s[U] - [F] \leq 0 ,$

where $s$ is the velocity with which the discontinuity propagates, and $[U]$, $[F]$ denote the jumps $U_{\text{left}} - U_{\text{right}}$ and $F_{\text{left}} - F_{\text{right}}$, respectively.
    For compressible fluid flow (1.14) corresponds to the increase of total negative entropy, and (1.15) states that the classical entropy of particles upon crossing a shock

increases. For this reason we shall call (1.13) and (1.15) entropy conditions.

The addition of a viscous term as in equation (1.11) is only one of many ways of introducing a slight amount of artificial dissipation into the system (1.1). Another way is to discretize the differential equations; one of the standard ways of doing this is to replace the operator $\partial_t$ and $\partial_x$ by the following difference operators. Denote by $T(h)$ translation in $t$ by the amount $h$, and by $S(k)$ translation in $x$ by the amount $k$. Define

(1.16)
$$D_t = \frac{1}{\Delta t}\{T(\Delta t) - \frac{S(\Delta x) + S(-\Delta x)}{2}\},$$

$$D_x = \frac{S(\Delta x) - S(-\Delta x)}{2\Delta x}.$$

We consider now the difference equation

(1.17)
$$D_t u_s + D_x f^j = 0$$

and study limits of solutions of (1.17) as $\Delta t, \Delta x \to 0$ while the ratio $\frac{\Delta t}{\Delta x} = \lambda$ remains constant. We assume that $u(x,t)$ is the limit, boundedly and almost everywhere, of solutions $u(x,t,\Delta t, \Delta x) = u(\Delta)$ of (1.17); it follows that $D_t u(\Delta)$ and $D_x f(u(\Delta))$ tend, in the topology of distributions, to $\partial_t u$ and $\partial_x f(u)$, so that it follows that the limit function $u$ satisfies the system of conservation laws (1.1). We shall show now, under an additional restriction on $\lambda$, that such a limit $u$ satisfies the entropy inequality (1.13). It suffices to show that every solution of (1.17) satisfies the inequality

(1.18)
$$D_t U + D_x F \leq 0,$$

for the limit of (1.18) in the topology of distributions is (1.13). We introduce the following vector notation:

$$u(x, t+\Delta t) = u, \quad u(x-\Delta x, t) = v, \quad u(x+\Delta x, t) = w.$$

We regard u, v and w as column vectors. Then

$$D_t u = \frac{1}{\Delta t}\{u - \frac{v+w}{2}\}, \quad D_x f = \frac{1}{2\Delta x}\{f(w)-f(v)\};$$

substituting this into (1.17) and solving for u we get, with the notation

$$\frac{\Delta t}{\Delta x} = \lambda$$

(1.19) $$u = \frac{v+w}{2} + \frac{\lambda}{2}[f(v) - f(w)].$$

Inequality (1.18) asserts that

(1.20) $$U(u) \leq \frac{U(v) + U(w)}{2} + \frac{\lambda}{2}[F(v) - F(w)].$$

We deform v continuously into w; set

$$v(s) = sv + (1-s)w.$$

Since $v(0) = w$, both sides of (1.20) equal $U(w)$ for $s = 0$. So we can write the difference of the right and the left side in (1.20) as the integral of the difference of their derivatives with respect to s. This difference is

(1.21) $$\frac{1}{2}U'(v)(v-w) + \frac{\lambda}{2}F'(v)(v-w) - U'(u)\frac{du}{ds}$$

where U' and F' are the gradients of U and F, regarded as row vectors. From (1.19) we get

(1.22) $$\frac{du}{ds} = \frac{v-w}{2} + \frac{\lambda}{2}f'(v)(v-w);$$

f', the gradient of the vector quantity f, is a matrix.
In this notation we can write identity (1.6) as follows

(1.23) $$U'f' = F'.$$

Substituting (1.23) and (1.22) into (1.21) we get

(1.24) $$\frac{1}{2}[U'(v) - U'(u)][I + \lambda f'(v)](v-w) .$$

Next we set

$$w(r) = rv(s) + (1-r)w = rsv + (1-rs)w .$$

Since $w(1) = v(s)$, $w(0) = w$, we have

$$U'(v) - U'(u) \int_0^1 \frac{d}{dr} U'(u) dr .$$

We write

$$\frac{d}{dr} U'(u) = \frac{du^t}{dr} U" ,$$

where $U"$ is the matrix of second derivatives of $U$ and $u^t$ the transpose of $u$. From (1.19) we get

$$\frac{du}{dr} = \frac{s}{2}[v-w-\lambda f'(w)(v-w)] .$$

Substituting these relations back into (1.25) we see that the difference between the right and left side of (1.20) is the double integral from 0 to 1 with respect to $s$ and $r$ of

(1.25) $$\frac{s}{4}[[I-\lambda f'(w)](v-w]^t \cdot U" [I+\lambda f'(v)](v-w) .$$

By assumption $U"$ is positive definite; this implies that <u>for $\lambda$ small enough</u> (1.25) <u>is positive</u>. The precise restriction on $\lambda$ is as follows:

Denote by $m$ and $M$ the minimum and maximum eigenvalue of $U"$ in that portion of $u$-space in which we are operating, and denote by $c$ the norm of $f'$ there. Denote $v - w$ by $z$. The the following is a lower bound for the expression to the right of $s/4$ in (1.25):

$$m\|z\|^2 - M(2c\lambda + c^2\lambda^2)\|z\|^2 .$$

Clearly this is positive for $z \neq 0$ if

(1.26) $$c\lambda \leq \sqrt{1+m/M} - 1.$$

Thus we have proved

<u>Theorem 1.2</u>: Suppose all differentiable solutions of the system of conservation laws (1.1) satisfy an additional conservation law (1.4), where U is a strictly convex function of u. Then all weak solutions of (1.1) which are the limits, boundedly a.e., of solutions of the difference equation (1.17) satisfy the entropy inequality (1.13), provided that condition (1.26) is fulfilled.

<u>Remark:</u> In the case of f' symmetric, the norm c of f' equals the absolute value $c_{max}$ of the largest eigenvalue of f'. In this case $U = \sum u_j^2$ is a conserved quantity; for this U, $U'' = 2I$, so $m/M = 1$ and condition (1.26) becomes

$$c_{max} \frac{\Delta t}{\Delta x} \leq \sqrt{2} - 1 = .414.$$

This is slightly more stringent than the Courant-Friedrichs-Lewy necessary condition for convergence,

$$c_{max} \frac{\Delta t}{\Delta x} \leq 1.$$

In the nonsymmetric case (1.26) is a still more stringent version of the C-F-L criterion, since then the norm c of f' is $> c_{max}$, and the condition number $m/M$ of $U''$ is $< 1$. We remark that Theorem 1.1 was also proved by Kružkov at the end of [14], and he has suggested condition (1.13) as a generalized entropy condition.

2. In this section we assume that (1.1) is strictly hyperbolic, i.e. that the matrix $f' = (f^j)$ has distinct real eigenvalues $c_1, \ldots, c_m$, indexed in increasing order. The $c_j$ are the characteristic speeds; they are functions of u. We also assume that the system is genuinely nonlinear in the sense of [8].

At a point of discontinuity of a solution we shall denote the value of u on the left, respectively right side of the discontinuity as follows:

(2.1) $$u_{left} = v, \quad u_{right} = w.$$

A point of discontinuity is called a <u>k-shock</u> if
a) The Rankine-Hugoniot relation

(2.2) $$s[v-w] = f(v) - f(w)$$

holds

b) There are exactly $k-1$ of the characteristic speeds $c_j(v) < s$ and exactly $(m-k)$ speeds $c_j(w) > s$:

(2.3) $$c_{k-1}(v) < s < c_k(v),$$
$$c_k(w) < s < c_{k+1}(w).$$

We shall call (2.3) the <u>shock condition.</u>

<u>Theorem 2.1:</u> Suppose that the system of conservation laws (1.1) is strictly hyperbolic, and that there is a strictly convex function U of u which satisfies the additional conservation law (1.4). Let u be a weak solution of (1.1) which has a discontinuity propagating with speed s, and suppose that the values of v and w on the left and right sides of the discontinuity are close. Then the shock condition (2.3) is satisfied if and only if the strict entropy condition (1.15):

(2.4) $$s[U(v) - U(w)] - F(v) + F(w) < 0$$

is satisfies.

<u>Proof:</u> It was shown in [8] that all states w near v which satisfy the R-H condition (2.2) form m one-parameter families $w_k(r)$, $s_k(r)$ where $w_k(0) = v$. If the parametrization is so taken that

## SHOCK WAVES AND ENTROPY

(2.5)
$$\left.\frac{ds_k}{dr}\right|_{r=0} > 0$$

then those $w_k$ which correspond to $r < 0$ satisfy the shock condition (2.3).

To prove (2.4) we shall substitute for $w$ one of these families $w_j(r)$ and expand the left side of (2.4) in powers of $r$; the crux of the argument is to show that the lowest power $r^p$ which is different from $0$ is odd, and that the coefficient of $r^p$ is positive.

Let's denote differentiation with respect to $r$ by a dot $\cdot$ and, as before, the gradient with respect to $u$ by prine '. The crucial exponent $p$ turns out to be $3$, so we have to calculate the first 3 derivatives of the left side of (2.4) at $r = 0$. Differentiating (2.2) we get

(2.6)
$$\dot{s}[v-w] - s\dot{w} = -\dot{f}(w).$$

The derivative of the left side of (2.4) in $r$ is

(2.7)
$$\dot{s}[U(v) - U(w)] - s\dot{U} + \dot{F}.$$

Using relation (1.6), $F' = U'f'$, we can write

$$\dot{F} = F'\dot{w} = U'f'\dot{w} = U'\dot{f}.$$

Substituting for $\dot{f}$ from (2.6) we get

$$\dot{F} = \dot{s}U'[w-v] + sU'\dot{w}.$$

Substituting this into (2.7) and noting that $\dot{U} = U'\dot{w}$ we get the following expression:

$$\dot{s}[U(v) - U(w)] + \dot{s}U'[w-v].$$

Differentiating once more we get

$$\ddot{s}[U(v)-U(w)] + \ddot{s}U'[w-v] + \dot{s}\dot{U}'[w-v].$$

Since $w(0) = v$, this is clearly zero at $r = 0$.

Differentiating once more, and setting $r = 0$ we get, after eliminating, those terms which are zero when $w = v$,

$$\dot{s}\, U'\dot{w}.$$

The remaining term can be written as

(2.8) $\qquad \dot{s}\,\ddot{w} + U''\dot{w}\,;$

since according to (2.5) the parametrization is so chosen that $\dot{s}$ is positive, and since $U''$ is positive because of the strict convexity of $U$, it follows that (2.8) is positive. This proves inequality (2.4) of theorem 2.1.

A noncalculational proof can be given using the following result of Foy, [2]:

If two nearby states $v$ and $w$ can be connected through a shock, then they can be connected through a viscous profile, i.e. a steady progressing solution of (1.11) of the form

(2.9) $\qquad u(x, t, \varepsilon) = w\left(\dfrac{x-st}{\varepsilon}\right),\ w(-\infty) = v,\ w(\infty) = w$.

Substituting this form of $u$ into (1.11) gives for the function $w$ the ordinary differential equation

(2.10) $\qquad -s\dot{w} + \dot{f} = \ddot{w}.$

Clearly, the discontinuous solution

(2.11) $\qquad u(x,t) = \begin{cases} v & \text{for } x < st \\ w & \text{for } st < x \end{cases}$

is the weak limit of $w\left(\dfrac{x-st}{\varepsilon}\right)$ as $\varepsilon \to 0$. Therefore according to theorem 1.1, the solution (2.11) satisfies the entropy condition.

Recently Conley and Smoller, [1], have shown that, for a fairly general class of systems of two conservation laws, any two states $v$ and $w$ which can be connected through a shock can also be connected through a viscous profine. It follows from the above argument that for such systems the restriction that v and w be close can be removed from theorem 2.1.

In his important paper [4] Glimm constructs solutions of systems of conservation laws as the limit of approximate solutions. These are piecewise continuous weak solutions in each strip $k\Delta t < t < (k+1)\Delta t$, and all their discontinuities are shocks; in addition the oscillation of these solutions is small. If there is a convex function which satisfies an additional conservation law, it follows from theorem 2.1 that the entropy condition

(2.12) $$\partial_t U + \partial_x F \leq 0$$

is satisfied by each approximate solution in each strip. Let $\phi$ be a smooth, positive test function with compact support. Multiply (2.12) by $\phi$, integrate over each strip; integrating by parts with respect to t over each strip and summing over all strips we get

(2.13) $$\sum_{k=1}^{\infty} \int \phi(x, k\Delta)[U(x, k\Delta+) - U(x, k\Delta-)]dx$$
$$+ \iint (-(\partial_t \phi)U + \phi \partial_x F)dxdt \leq 0 .$$

Lemma (5.1) in Glimm's paper shows that the sum in (2.13) tends to zero for a suitably selected subsequence. This leaves us in the limit with

$$\iint (-\phi_t U + \phi F_x)dxdt \leq 0$$

for all positive test function $\phi$ supported in $t > 0$. Integrating by parts with respect to t we get that

$$\iint \phi[\partial_t U + \partial_x F]dxdt \leq 0$$

for all such $\phi$. Clearly this implies (2.12). Thus we have proved

**Theorem 2.2:** Suppose that the system of conservation laws (1.1) is strictly hyperbolic, and that there is a convex function U of u which satisfies the additional conservation law (1.4). Then all weak solutions of (1.1) constructed by Glimm's method satisfy the entropy inequality (2.12).

3. What systems admit an additional conservation law where the additional conserved quantity U is a convex function of the original ones? We saw that symmetric systems do, and so does the system consisting of the laws of conservation of mass, momentum and energy for a compressible gas. A systematic search for additional conservation laws was carried out by Rozdestvenskii, [12]; in this section we record some observations on the existence and utility of additional convex conservation laws.

We start with a single conservation law:

(3.1) $$\partial_t u + \partial_x f = 0 .$$

In this case we may choose for U any convex function; F is then determined by integrating the compatibility relation (1.6):

(3.2) $$U'f' = F' .$$

The entropy condition,

(3.3) $$\partial_t U + \partial_x F \leq 0 ,$$

was derived for smooth U only; by passing to the limit in the topology of distributions we deduce (3.3) for any convex U, smooth or not.

Every convex function lies in the convex cone generated by the functions $U(u) = |u-z|$, z some constant, and by the linear functions. In [7], Krushkov takes (3.3) for all U of this form to be the definition of the relevant class of weak solutions of the analogue of (3.1) for n space variables; he proves existence of such solutions with arbitrary initial data, and announces their uniqueness.

We present now some known consequences of the entropy inequality (3.3); the first was found independently by Krushkov and by Hopf, [6]:

Suppose that u is a piecewise continuous weak solution of (3.1) which satisfies (3.3); let's denote the speed of propagation of a discontinuity by s, and denote by v and w the values of u on the left, respectively right side of the discontinuity. According to (1.15), (3.3) implies that

(3.4) $$s[U(v) - U(w)] - F(v) + F(w) < 0.$$

Suppose that $w < v$; let z be any number between w and v, and set

(3.5) $$U(u) = \begin{cases} 0 & \text{for } u < z \\ u - z & \text{for } z < u \end{cases}$$

It follows from (3.2) that then

(3.6) $$F(u) = \begin{cases} 0 & \text{for } u < z \\ f(u) - f(z) & \text{for } z < u. \end{cases}$$

Substituting these into (3.4), and using the jump relation

(3.7) $$s = \frac{f(v) - f(w)}{v - w}$$

we get a relation which, after rearrangement, becomes

(3.8)$_+$ $$f(z) \leq \frac{v-z}{v-w} f(v) + \frac{z-w}{v-w} f(v) \quad \text{for } w \leq z \leq v.$$

The geometrical meaning of (3.8) is that the graph of f over the interval $[w, z]$ lies below the chord connecting $(w, f(w))$ with $(v, f(v))$; for $w > v$ the opposite inequality (3.8)$_-$ obtains; these inequalities are Oleinik's celebrated condition (E), see [10].

We remark that in [11] B. Quinn has shown that $(3.8)_+$ is necessary and sufficient for $L_1$ contraction, more precisely:

If a pair of piecewise continuous solutions $u_1$ and $u_2$ both satisfy $(3.8)_\pm$, then

$$(3.9) \qquad \int |u_1(x,t) - u_2(x,t)|\, dx$$

is a decreasing function of $t$; conversely, if (3.9) is a decreasing function of $t$ for a certain piecewise continuous $u_1$ and every continuous $u_2$, then $u_1$ satisfies $(3.8)_\pm$.

We shall derive now another consequence of (3.3):

Let $u$ be a weak solution of (3.1) which satisfies (3.3), and which is $0$ at $x = \pm\infty$. Then as observed in (1.14), for $U$ convex, and $U(0) = 0$,

$$(3.10) \qquad \int U(u(x,t_2))dx \leq \int U(u(x,t_1))dx \quad \text{for } t_1 < t_2.$$

Choose for $U$ the convex function

$$(3.11) \qquad U(u) = |M - u| - M$$

where

$$(3.12) \qquad M = \operatorname*{ess.\,sup}_{x} u(x, t_1).$$

Since $u(x, t_1) \leq M$, a.e., it follows that

$$(3.13) \qquad U(u(x, t_1)) = -u(x, t_1) \text{ a.e.};$$

on the other hand, since $|M - u| \geq M - u$,

$$(3.14) \qquad U(u(x, t_2)) \geq -u(x, t_2).$$

Substituting (3.13) and (3.14) into (3.10) we get that

$$(3.15) \qquad -\int u(x, t_2)dx \leq -\int u(x, t_1)dx.$$

On the other hand it follows from (3.1) that

$$\int u(x,t)dt$$

is independent of t; therefore in (3.15) the sign of equality holds. But this can be if and only if in (3.14) equality holds a.e.; this is the case if and only if

$$u(x,t_2) \leq M \text{ for almost all } x.$$

In view of the definition of (3.12) of M this result can be expressed as follows:

$$\text{Ess sup}_x u(x,t)$$

<u>is a decreasing function of</u> t <u>for every weak solution</u> u <u>which satisfies</u> (3.3) <u>for all convex</u> U. <u>Similarly</u>,

$$\text{Ess inf}_x u(x,t)$$

is an increasing function of t.

We turn now to pairs of conservation laws:

(3.16)
$$\partial_t u^1 + \partial_x f^1 = 0$$
$$\partial_t u^2 + \partial_x f^2 = 0,$$

these can be written in the form

(3.17) $\quad \partial_t u + f' \partial_x u = 0.$

The compatibility relation (1.6) is

(3.18) $\quad F' = U' f',$

a pair of first order equations for the two functions F and U. These equations are linear; it is easy to show that if the nonlinear system (3.17) is strictly hyperbolic, so is[†] (3.18). For suppose that the matrix $f'$ has 2 distinct real eigenvalues; let $r$ be a right eigenvector of $f'$:

(3.19) $$f'r = cr.$$

Multiplying (3.18) by $r$ on the right gives

(3.20) $$F'r = U'f'r = cU'r,$$

a linear combination in which both $F$ and $U$ are differentiated in the direction $r$. The existence of two such directions shows that (3.18) is hyperbolic, and its characteristic directions are those of the right eigenvectors of $f'$.

We can eliminate $F$ from (3.18) by differentiating the first equation with respect to $u^2$, the second with respect to $u^1$, and subtracting one from the other. We get a homogeneous 2nd order equation for $U$ of the form

(3.21) $$SU = a_{11} U_{11} + a_{12} U_{12} + a_{22} U_{22} = 0$$

where $S$ is the second order operator with coefficients

(3.22) $$a_{11} = -f_2^1, \quad a_{12} = f_1^1 - f_2^2, \quad a_{22} = f_1^2.$$

Equation (3.21), being derived from a hyperbolic first order system, is itself hyperbolic, which means that the quadratic form

(3.23) $$a_{11} \xi^2 + a_{12} \xi \eta + a_{22} \eta^2$$

is indefinite.

We turn now to the question: does equation (3.21) have convex solutions? It is pretty easy to see that it does in the small, on the basis of this

---

[†]and similarly, as shown by Loewner, if (3.17) is elliptic, so is (3.18).

## SHOCK WAVES AND ENTROPY

<u>Lemma 3.1:</u>  Let $a_{ij}$ be a symmetric $n \times n$ matrix which is indefinite; then there exists a positive definite matrix $U_{ij}$ such that

(3.24) $$\sum a_{ij} U_{ij} = 0 .$$

<u>Proof:</u>  Since $a_{ij}$ is indefinite, there exist vectors $\{\xi_i\} = \xi$ such that

(3.25) $$\sum a_{ij} \xi_i \xi_j = 0 .$$

The set of these separates the set of those vectors where the quadratic form $\xi a \xi$ is positive from the set where the form is negative. It follows that the set of $\xi$ satisfying (3.25) spans the whole space; denote by $\xi^1, \ldots, \xi^n$ a spanning set. Now define $U_{ij}$ by

$$U_{ij} = \sum_k \xi_i^k \xi_j^k ;$$

since the $\xi^k$ span the whole space, $U_{ij}$ is positive definite; on the other hand it follows from (3.25) that condition (3.24) holds; this completes the proof of the lemma.

Applying the lemma to the quadratic form (3.23) at some point v we conclude that there exists a positive definite $U_{ij}$ which satisfies (3.21) at v. By solving an appropriate Cauchy problem we can construct a solution U whose second derivatives at v equal $U_{ij}$; this solution will be convex near v. Thus we have proved

<u>Theorem 3.2:</u>  A homogeneous second order hyperbolic equation has a convex solution in the neighborhood of every point.

We show now that the compatibility equation (3.18) has solutions with U convex in any domain G where a certain inequality, see (3.39), is satisfied. We do not claim that this condition is necessary.

We shall construct a one-parameter family of such solutions; we start with approximate solutions of the form

621

(3.26) $$U_{approx} = e^{k\phi}V, \quad F_{approx} = e^{k\phi}H,$$

where

(3.27) $$V = \sum_0^N V^j/k^j, \quad H = \sum_0^N H^j/k^j;$$

$\phi$, $V^j$ and $H^j$ are independent of $k$. Solutions of this sort, with $i\phi$ in place of $\phi$, were constructed in [9]. For this reason we only sketch the details.

Substituting (3.26) in (3.18) gives, after division by $ek^\phi$, the equation

$$k\phi'Vf' + V'f' = k\phi'H + H'.$$

We substitute (3.27) into the above equation; equating coefficients of various powers of $k$ we get

(3.28) $$V^0\phi'f' = H^0\phi',$$

and

(3.29) $$V^j\phi'f' + V^{(j-1)'} = H^h\phi' + H^{(j-1)'}.$$

Equation (3.28) asserts that $\phi'$ is a left eigenvector of $f'$:

(3.30) $$\phi'f' = c\phi'$$

with

(3.31) $$cV^0 = H^0.$$

Such a function $\phi$, called a <u>phase function,</u> is easily constructed since a left eigenvector is characterized by orthogonality to the right eigenvector $r$ corresponding to the other eigenvalue:

(3.32) $$\phi'r = 0$$

where

(3.33)  $$f'r = sr, \quad s \neq c.$$

Substituting (3.30) into (3.29) gives

(3.34)$^j$  $$(cV^j - H^j)\phi' = H^{(j-1)'} - V^{(j-1)'}f'.$$

The first step in solving equations (3.34) is to multiply (3.34)$^j$ by $r$; using (3.33) we get

$$0 = (H^{j-1} - sV^{j-1})'r.$$

Using (3.31) we get

$$(c-s)V^{o'}r + c'rV^o = 0.$$

This is a first order equation for $V^o$ which can be solved once we prescribe the value of $V^o$ on a noncharacteristic initial curve. Notice that if we prescribe positive values for $V^o$ initially, $V^o$ <u>is positive everywhere.</u>

Having determined $V^o$ and $H^o$, equation (3.34)$^j$ gives one linear relation between $V^1$ and $H^1$; proceeding recursively we can determine all $V^j$ and $H^j$.

The functions $U_{approx}$ and $F_{approx}$ of form (3.26), (3.27) constructed in this fashion satisfy the approximate equation

$$U'_{approx} f' = F'_{approx} + e^{k\phi} R_n / k^N.$$

We shall now construct another solution

(3.35)  $$U'_N f' = F'_N + e^{k\phi} R_N / k^N$$

of this same equation such that

(3.36)  $$U_N, F_N = e^{k\phi} 0(1/k^N).$$

Then

$$U = U_{approx} - U_{N'}, \quad F = F_{approx} - F_N$$

are exact solutions of $U'f' = F'$, of which the leading term is (3.26).

To construct a solution of (3.35) which satisfies (3.36) in some domain $G$ of the u-plane we assign initial values zero for $U_N, F_N$ along a non-characteristic curve $C$ with these properties:

   i)  $C$ doesn't intersect $G$.
   ii) $G$ is contained in the domain of determinacy of $C$.

In what follows we assume that $\phi$ has no ciritcal points, i.e. $\phi' \neq 0$ everywhere. Then $\phi$ is monotonic along every curve whose tangent is not parallel to $r$ appearing in (3.32); in particular $\phi$ is monotonic along characteristics corresponding to the other eigenvector of $f'$. Our last condition is

   iii) Along these other characteristics $\phi$ increases in the direction from $C$ toward $G$.

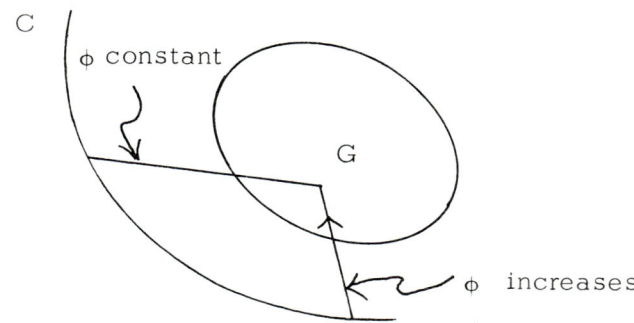

Condition iii) determines the side of $G$ on which $C$ lies.

It is easy to show, using standard estimates in the maximum norm for the hyperbolic equation (3.35), that if the initial values of $U_N$ and $F_N$ are chosen to be zero on $C$, then $U_N \leq 0(e^{k\phi}/k^N)$ in G as $k \to +\infty$. This completes the construction of exact solutions.

When is the function U just constructed convex? The answer can be read off from the first 2 leading terms in the asymptotic expression for the quadratic form of U":

(3.37) $e^{-k\phi}\{U_{11}\xi^2+2U_{12}\xi\eta+U_{22}\eta^2\} = k^2\{\phi_1^2\xi^2+2\phi_1\phi_2\xi\eta+\phi_2^2\eta^2\}V^o +$

$+ k\{\phi_1^2\xi^2+2\phi_1\phi_2\xi\eta+\phi_2^2\eta^2\}V^1$

$+ 2k\{\phi_1 V_1^o \xi^2 + (\phi_2 V_1^o + \phi_1 V_2^o)\xi\eta + \phi_2 V_2^o \eta^2\}$

$+ k\{\phi_{11}\xi^2 + 2\phi_{12}\xi\eta + \phi_{22}\eta^2\}V^o.$

The coefficient of $k^2$ is

$(\phi_1\xi + \phi_2\eta)^2 V^o,$

a positive quantity except along

(3.38) $\xi = -\phi_2, \quad \eta = \phi_1.$

The coefficient of k consists of 3 terms, of which the first 2 are zero along the line (3.38). We impose now the condition that the third term be positive along this line:

(3.39) $\phi_{11}\phi_2^2 - 2\phi_{12}\phi_1\phi_2 + \phi_{22}\phi_1^2 > 0.$

It follows then that U is convex for k large enough, positive.

Let u be any differentiable solution of (3.17):

$\partial_t u + f'\partial_x u = 0.$

Multiplying by $\phi'$ and using (3.30) we get

$\partial_t \phi + c\partial_x \phi = 0;$

this equation asserts that $\phi$ is constant along one of the characteristics of the nonlinear equation (3.17). Such a function is called a <u>Riemann invariant</u>; thus <u>the phase functions of the linear compatibility equation</u> (3.18) <u>are the Rieman invariants of the nonlinear equation</u> (3.17).

Any function $p(\phi)$ of a Riemann invariant is another Riemann invariant. Denoting $p(\phi)$ by $\psi$ and differentiating with respect to $\phi$ by a dot we have

(3.40)
$$\begin{aligned}&\psi_{11}\xi^2 + 2\psi_{12}\xi\eta + \psi_{22}\eta^2 \\ &= \ddot{p}\{\phi_1^2\xi^2 + 2\phi_1\phi_2\xi\eta + \phi_2^2\eta^2\} \\ &+ \dot{p}\{\phi_{11}\xi^2 + 2\phi_{12}\xi\eta + \phi_{22}\eta^2\}.\end{aligned}$$

We deduce from this that <u>if $\phi$ satisfies</u> (3.38), <u>so does every increasing function</u> $\psi$ of $\phi$. This shows that property (3.39), except for the sign, does not depend on the particular choice for $\phi$.

It is easy to decide when (3.39) can be satisfied. Denote by dot differentiation in the direction

(3.41)
$$\dot{u} = r,$$

where $r$ is the right eigenvector appearing in (3.32). Differentiating (3.30) in the above direction and multiplying by $r$ we get, using (3.32) and (3.33) that

$$(c-s)\dot{\phi}'r = \phi'\dot{f}'r.$$

In view of (3.41), $\dot{\phi}'r$ is the left side of (3.39); therefore (3.39) can be satisfied if and only if

(3.42)
$$\phi'\dot{f}'r \neq 0,$$

a condition on the derivatives and right and left eigenvectors of $f'$.

We shall now use the solutions (3.26) constructed to prove

**Theorem 3.3:** Let $u(x,t)$ be a weak solution of the conservation laws (3.17), defined for $t \geq 0$, which satisfies the entropy condition

(3.43) $$U_t + F_x \leq 0$$

for all convex solutions $U$ of (3.18). Let $\phi$ be a Riemann invariant which satisfies (3.39) in a domain $G$ which contains all values of $u(x,t)$. Then

(3.44) $$\max_x \phi(u(x,t))$$

is a decreasing function of $t$.

Actually we shall prove a sharper theorem of which Theorem 3.3 is a corollary:

**Theorem 3.4:** Denote by $c_{min}$ and $c_{max}$ the minimum and maximum of $c(u)$ in $G$, where $c$ is the eigenvalue in (3.30). Then at any point $(x,t)$, $t > 0$,

(3.45) $$\varphi(u(x,t)) \leq \sup_{a \leq y \leq b} \phi(u(y,0))$$

where

(3.46) $$a = x - t c_{max}, \quad b = x - t c_{min}.$$

**Proof:** Integrate (3.43) over the triangle $T$ shown below, where $\delta$ is some positive quantity. Using the divergence theorem we get

$$(3.47) \quad \int_{a-\delta}^{(x,t)} + \int_{(x,t)}^{b+\delta} [Un_t + Fn_x]ds \le \int_{a-\delta}^{b+\delta} U(y,0)dy,$$

where $(n_x, n_t)$ is the outward normal to $T$. Substituting the special solutions with leading term (3.26), the leading term on the left in (3.47), can be written, using (3.31), as

$$\text{const} \int e^{k\phi}[c_{max} - c + \frac{\delta}{t}] V^o ds$$

$$+ \text{const} \int e^{k\phi}[c - c_{min} + \frac{\delta}{t}] V^o ds;$$

since $V^o$ and $\delta$ are $>0$, this is bounded from below by

$$\text{const} \int_{a-\delta}^{(x,t)} \int_{(x,t)}^{b+\delta} e^{k\phi} ds.$$

On the other hand the right side of (3.47) is bounded from above by

$$\text{const} \int_{a-\delta}^{b+\delta} e^{k\phi} dy.$$

The $k^{th}$ root of the former tends, as $k \to \infty$ to Max $\phi(u)$ on the segments connecting $a - \delta$, $(x,t)$ and $b + \delta$, while the $k^{th}$ root of the latter tends to Max $\phi(u)$ along $(a-\delta, b+\delta)$. Therefore inequality (3.47) implies, if we take the $k^{th}$ root, let $k \to \infty$ and $\delta \to 0$, that (3.45) holds.

We have shown in Theorems 1.1 and 2.2 that weak solutions which are the limits of solutions of the viscous equation, or of Glimm's scheme, satisfy all entropy inequalities (3.43). It follows therefore that the estimates for the Riemann invariants asserted in Theorems 3.3 and 3.4 hold for such weak solutions. This conclusion is not new; we indicate the relation of condition (3.39) to known results.

We start with the observation that if $\phi$ satisfies (3.39) and if the function $p$ is chosen so that $\ddot{p}$ is very much larger than $\dot{p}$, then the first term on the right in (3.40) is larger than the second except near these values where $\phi_1 \xi + \phi_2 \eta = 0$. For such values the second term is, by (3.39) positive, so that the right side of (3.40) is positive. But that means that $\delta = p(\phi)$ is <u>convex</u>. Thus (3.39) <u>implies the existence of a convex</u>[†] Riemann invariant $\psi$.

Let $\psi$ be a convex Riemann invariant; let $u(x,t,\varepsilon)$ be solutions of the viscous equation

(3.47) $$\partial_t u + f' \partial_x u = \varepsilon \partial_x^2 u, \quad \varepsilon > 0.$$

Multiply this equation by $\psi'$; using relation (3.30): $\psi' f' = c \psi'$ we get

$$\partial_t \psi + c \partial_x \psi = \varepsilon \psi' \partial_x^2 u.$$

Using the identity

$$\partial_x^2 \psi = \psi' \partial_x^2 + \partial_x u^t \psi'' \partial_x u$$

and the convexity of $\psi$ we conclude that

$$\partial_t \psi + c \partial_x \psi \leq \varepsilon \partial_x^2 \psi.$$

The maximum principle holds for solutions of such a differential inequality and tells us that

$$\operatorname*{Max}_x \psi(u(x,t;\varepsilon))$$

is a decreasing function of $t$. But then the same is true of their a.e. limits as $\varepsilon \to 0$; this proves Theorem 3.3 for this class of solutions.

We turn now to Glimm's scheme; in Theorem 2.1 we have shown that if $U$ is a convex function which satisfies an additional conservation law, then $U$ decreases across

---

[†] Smoller and Johnson in [13] have shown that condition (3.42) implies that the curves $\phi$ = const. are convex. This implies the existence of a convex $\psi = p(\phi)$.

shocks; a similar result holds for convex Riemann invariants:

**Lemma 3.5:** If the Riemann invariant $\phi$ satisfies (3.39), then $\phi$ decreases across a shock of the family opposite to $\phi$.

**Remark:** This decrease was stipulated in Glimm-Lax, [5], precisely for the purpose of proving that the Riemann invariant is a decreasing function.

**Sketch of proof:** Consider all states $w$ close to a given state $v$ which can be connected on the right to $v$ through a shock of a fixed kind, i.e. which satisfy the jump relations

(3.48) $$s[w-v] + f(v) - f(w) = 0$$

and the shock inequality (2.3). We saw in Section 2 that these states $w$ form a one parameter family $w(p)$, $p \leq 0$, under the normalization $w(0) = v$, $s(0) > 0$, where dot denotes differentiation with respect to the parameter.
Differentiating (3.48) gives

(3.49) $$\dot{s}(w-v) + s\dot{w} - f'\dot{w} = 0 .$$

Multiply (3.49) by $\phi'$ on the left; using (3.30) and that $\phi'\dot{w} = \dot{\phi}$ we get

(3.50) $$\phi'\dot{s}(w-v) = (c-s)\dot{\phi} .$$

Since $w(0) = v$, (3.50) implies that $\dot{\phi}(0) = 0$. Differentiate (3.50) with respect to $p$; using $\dot{\phi}(0) = 0$ we deduce $\ddot{\phi}(0) = 0$; differentiating once more we get

(3.51) $$\dddot{\phi}(0) = \frac{\dot{s}}{c-s} \phi'\dot{w} .$$

By (3.39), $\phi'\dot{w} > 0$ and by choice of normalization $\dot{s} > 0$; furthermore we are restricted to negative values of the parameter. So it follows from (3.51) that for $w$ near $v$,

(3.52) $$\text{sgn}(\phi(v)-\phi(w)) = \text{sgn}(c-s).$$

Consider now a flow with a single shock going with speed $s$ greater than the sound speed $c$ of the opposite family:

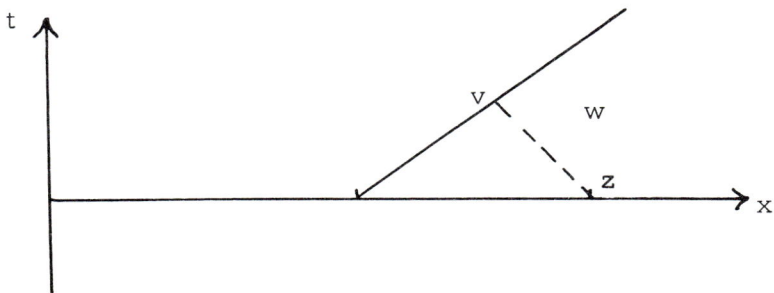

It follows from (3.52) that

$$\phi(v) < \phi(w) ;$$

on the other hand $\phi(w) = \phi(z)$, where $z$ is the value of $u$ at that point on the initial line which can be connected to the shock by a characteristic of the opposite family. So the value $v$ of $\phi$ along the other side of the shock is $<$ the value of $\phi$ at some point of the initial line. The same conclusion holds for shocks of the other family; this proves Lemma 3.5.

In Glimm's difference scheme the initial interval is divided into subintervals and the initial function is approximated by one which is constant in each subinterval; this problem is solved exactly in a time interval taken so short that the waves issuing from the point of discontinuity do not interact. At the end of that time interval the solution is approximated by a piecewise constant function obtained by setting $u$ in each subinterval equal to its value at a randomly chosen point in that subinterval. It follows from Lemma 3.5 that for any Riemann invariant which satisfies (3.39) each approximate solution $u_\Delta$ satisfies the conclusion of

Theorem 3.3. But then so does their limit.

It is not hard to show that for almost all choices of the random points, the approximate solutions $u_\Delta$ satisfy the conclusions of Theorem 3.4. Therefore so do their limits, for almost all random choices.

Another difference scheme, devised by Godunov, starts similarly by solving exactly a piecewise constant initial value problem for a short time, but the conversion into piecewise constant data at the end of that time interval is accomplished differently: in each subinterval u is set equal to its average over that subinterval. If $\psi$ is a convex Riemann invariant, this process decreases the maximum value of $\psi(u)$; so these approximate solutions satisfy the conclusion of Theorem 3.3. But then so does their limit.

It would be useful to determine all convex solutions of the second order equation (3.21), so that one can study weak solutions which satisfy the entropy condition with respect to all additional convex conservation laws. The most important question is: are such solutions uniquely determined by their initial data? Another interesting task is to derive from these entropy conditions an analogue for systems of Oleinik's condition E.

## REFERENCES

1. Conley, C. C. and Smoller, J., <u>Shock waves as limits of standing wave solutions of higher order equations</u>, Comm. Pure Appl. Math., Vol. 24, No. 2, 1971.

2. Foy, R. L., <u>Steady state solutions of hyperbolic systems of conservation laws with viscosity terms</u>, Comm. Pure Appl. Math., Vol. 17, 1964, pp. 177-188.

3. Friedrichs, K. O. and Lax, P. D., to appear in Proc. Nat. Acad. Sci.

4. Glimm, J., Solutions in the large for nonlinear hyperbolic systems of equations, Comm. Pure Appl. Math., Vol. 18, pp. 697-715 (1965).

5. Glimm, J., and Lax, P. D., Decay of solutions of systems of nonlinear hyperbolic conservation laws, Memoirs of the Amer. Math. Soc., No. 101, 1970.

6. Hopf, E., On the right weak solution of the Cauchy problem for quasilinear equation of first order, J. of Math. and Mech., Vol. 19, 1969, pp. 483-487.

7. Kružkov, S. N., Results on the character of continuity of solutions of parabolic equations and some of their applications, Math. Zametky, Vol. 6, 1969, pp. 97-108.

8. Lax, P. D., Hyperbolic systems of conservation laws, II., Comm. Pure Appl. Math., Vol. 10, pp. 537-566 (1957).

9. Lax, P. D., Asymptotic solutions of oscillatory initial value problems, Duke Math. J., Vol. 24, 1957, pp. 627-646.

10. Oleinik, O. A., On the Uniqueness of the generalized solution of the Cauchy problem for a non-linear system of equations occurring in mechanics, Usp. Mat. Nauk. Vol. 78, pp. 169-176 (1957).

11. Quinn, B., Solutions with shocks, an example of an $L_1$-contractive semigroup, Comm. Pure Appl. Math., Vol. 24, No. 1, 1971.

12. Rozdestvenskii, B., Discontinuous solutions of hyperbolic systems of quasi-linear equations, Ups. Mat. Nauk. Vol. 15, pp. 59-117 (1960). English translation in Russ. Math. Surv. Vol. 15, pp. 55-111 (1960).

13. Smoller, J. and Johnson, J. L., <u>Global solutions for an extended class of hyperbolic systems of conservation laws,</u> Arch. For Rat. Mech. and Anal., Vol. 32, 1969, pp. 169-189.

14. Kružkov, S. N., <u>First order quasilinear equations in several independent variables,</u> Math. USSR Sbornik Vol. 10 (1970), No. 2.

The work presented in this paper is supported by the AEC Computing and Applied Mathematics Center, Courant Institute of Mathematical Sciences, New York University, under Contract AT(30-1)-1480 with the U. S. Atomic Energy Commission.

     Courant Institute of Mathematical Sciences
     AEC Computing and Applied Mathematics
                Center
     New York University
     New York, New York

              Received May 6, 1971

# The Penalty Method and Some Nonlinear Initial Value Problems

*HIROSHI FUJITA*

§1. Introduction

The penalty method which originated from R. Courant [6] has been extensively used mostly for practical purposes in numerical analysis. Recently it has been applied to initial value problems for nonlinear partial differential equations for theoretical purposes, for instance, in order to prove the existence of solutions in non-cylindrical regions by several authors. In particular, we refer to Lions [24, 25], Fujita-Sauer [12, 13], and works listed in the bibliography of Lion's book [25]. In these applications the idea or the formal procedure of approximation based on the principle of the penalty method is quite simple in general, while the rigorous proof of the convergence of the approximate solutions sometimes becomes rather delicate and involved, depending on the nature of the original problem and the choice of penalizing terms. The objective of the present paper is to exemplify this situation through initial value problems in non-cylindrical regions for some simple but important equations. As a matter of fact we shall deal with the following two equations:
(i) the nonlinear heat equation of the form $u_t = \Delta u - \beta(u)$, and (ii) the Navier-Stokes equations.
  Study of the initial value problem for the first equation in a non-cylindrical region, which is to be formulated just below and is denoted by (Pr. NH), has in various aspects many things in common with the recent theory of nonlinear

semigroups of contractions or the theory of abstract evolution equations with dissipative generators. Actually most techniques employed below in proving the existence and the uniqueness of solutions can be said to have their origin in papers by Y. Kōmura, T. Kato, F. E. Browder, H. Brezis, M. G. Crandall and others [20, 21, 17, 18, 2, 3]. Although our main concern in this paper is with the penalty method, we give first in §2 an independent proof for the existence and uniqueness of solutions of (Pr. NH) which does not rely on the penalty method, piecing together arguments in Kato [17], Brezis [2] and Brezis-Crandall-Pazy [3] with an essential use of the maximal accretiveness of operators involved. The convergence of an approximating procedure for (Pr. NH), which is based on the principle of the penalty method, will be established in §3.

On the other hand, the existence of (weak) solutions of the Navier-Stokes equations in non-cylindrical regions is obtained as a result of applying the penalty method. This will be described in §4, following Fujita-Sauer [13].

We now introduce some notations and assumptions concerning the $(t, x)$-region $\hat{\Omega}'$, where solutions of our initial value problems are sought. As for the time variable $t$ we restrict our consideration to the closed interval $[0, T]$, $T$ being a positive number. On the other hand, the space variable $x$ represents points in $R^m$. For $t \in [0, T]$, $\Omega(t)$ stands for the section of $\hat{\Omega}'$, i.e., $\Omega(t) = \{x \in R^m | (t, x) \in \hat{\Omega}'\}$. $\Omega(t)$ is assumed to be a bounded domain in $R^m$ with smooth boundary $\Gamma(t)$. Accordingly,

$$\hat{\Omega}' = \bigcup_{t \in [0, T]} (\{t\} \times \Omega(t)) \subset [0, T] \times R^m .$$

We put

$$\hat{\Gamma} = \bigcup_{t \in [0, T]} (\{t\} \times \Gamma(t)) .$$

In other words, $\hat{\Gamma}$ is the lateral boundary of $\hat{\Omega}'$. By $\hat{\Omega}$ we denote the interior of $\hat{\Omega}'$. Namely, $\hat{\Omega} = \bigcup_{t \in (0, T)} (\{t\} \times \Omega(t))$.

We make the following smoothness assumptions on $\hat{\Gamma}$ once and for all:

(i) At every $t \in [0,T]$, $\Gamma(t)$ consists of a fixed finite number of simple closed hypersurfaces $\Gamma_\alpha(t)$ in $R^m$, which are sufficiently smooth (say, of class $C^\infty$).

(ii) There exists $\delta_0 > 0$ such that dis.$(\Gamma_\alpha(t), \Gamma_\beta(t))$, the m-dimensional distance between $\Gamma_\alpha(t)$ and $\Gamma_\beta(t)$, $(\alpha \neq \beta)$, is never smaller than $\delta_0$.

(iii) As $t$ varies, each $\Gamma_\alpha(t)$ changes smoothly in the sense that the $(t,x)$-hypersurface $\hat{\Gamma}_\alpha(t) = \bigcup_{t \in [0,T]} (\{t\} \times \Gamma_\alpha(t))$ is covered by a finite number of patches (relatively open in $[0,T] \times R^m$) and in each patch $\hat{\Gamma}_\alpha$ can be represented by $x'_m = \varphi(x'_1, \ldots, x'_{m-1}, t)$ in terms of a smooth (say, $C^\infty$-) function $\varphi$ of m variables $x'_1, \ldots, x'_{m-1}, t$ under a suitable choice of coordinates $(x'_1, \ldots, x'_m)$ in $R^m$.

In order to apply the penalty method we shall need an auxiliary bounded domain B in $R^m$ such that $\partial B$ is smooth, $\Omega(t) \subset B$ for all $t$, and dis.$(\partial B, \Gamma(t)) \geq \delta_0$ for all $t$. We denote $(0,T) \times B$ by $\hat{B}$.

Specific descriptions of our initial value problems are in order.

By (Pr. NH) we mean the following initial-boundary value problem for the nonlinear heat equation mentioned above:

(1.1) $\quad u_t = \Delta u - \beta(u) + f(t,x) \quad$ in $\hat{\Omega}$,

(1.2) $\quad u|_{\hat{\Gamma}} = 0$,

(1.3) $\quad u|_{t=0} = a(x) \quad$ in $\Omega(0)$.

Here $\beta$ is a real valued increasing (accretive) function of a real variable. $f = f(t,x)$ and $a = a(x)$ are real valued functions given in $\hat{\Omega}$ and $\Omega(0)$, respectively. Further assumptions on $\beta$, $f$ and $a$ will be given in due course. Throughout the present paper numbers and functions are all real. $\Delta$ and $\nabla$ mean those in the usual sense, which operate on the space variable only.

Next, by (Pr. NS) we mean the following initial boundary value problem for the Navier-Stokes equation:

(1.4) $\quad u_t = \Delta u - \nabla p - (u \cdot \nabla)u \quad$ in $\hat{\Omega}$,

(1.5) $\quad \nabla \cdot u \equiv \text{div } u = 0 \quad$ in $\hat{\Omega}$,

(1.6) $\quad \tilde{u}|_{\hat{\Gamma}} = b(t, x)$,

(1.7) $\quad u|_{t=0} = a(x) \quad$ in $\Omega(0)$.

Here the unknown $u$ and the given $b$, $a$ are m-vector functions, while the unknown $p$ is a scalar function.

Finally we indicate how we can reduce, say, (Pr. NH) to initial value problems in the cylindrical region $B$ upon the principle of the penalty method. Namely, we introduce a sequence of initial value problems (Pr. NH)$^n$ in $\hat{B}$ ($n = 1, 2, \ldots$) for the nonlinear heat equation with an additional term as follows.

(1.8) $\quad u_t = \Delta u - \beta(u) + \tilde{f}(t, x) - n\chi(t,x)p(u) \quad$ in $\hat{B}$,

(1.9) $\quad u|_{[0, T] \times \partial B} = 0$,

(1.10) $\quad u|_{t=0} = \tilde{a}(x) \quad\quad\quad\quad\quad\quad\quad\quad\quad\quad$ in $B$.

Here we have put

$$\tilde{f} = f \text{ in } \hat{\Omega} \quad \text{and} \quad \tilde{f} = 0 \quad \text{in } \hat{B} - \hat{\Omega},$$

$$\tilde{a} = a \text{ in } \Omega(0) \quad \text{and} \quad \tilde{a} = 0 \quad \text{in } B - \Omega(0),$$

and

$$\chi = 1 \quad \text{in } \hat{B} - \hat{\Omega} \quad \text{and} \quad \chi = 0 \text{ in } \hat{\Omega}.$$

The function $p$ in the penalizing term $-n\chi(t,x)p(u)$ is a strictly increasing and Lipschitz continuous function from

$R^1$ to $R^1$ such that $p(0) = 0$. The simplest choice of $p$ is to take $p(\sigma) \equiv \sigma$. We expect that the solution $u = u^n$ of (Pr. NH)$^n$ approximates the solution $u$ of (Pr. NH). Below in §3 we shall show that this is actually the case. In other words, the penalty method converges for (Pr. NH). The approximation for (Pr. NS) based on the penalty method will be introduced in §4.

## §2. Solutions of (Pr. NH).

### 2.1. Definitions, assumptions, and theorems.

In this section we deal with (Pr. NH). Concerning $\beta$, $f$ and $a$, we introduce the following conditions.

Condition ($\beta$): $\beta$ is a continuous increasing function from $R^1$ to $R^1$. We define $\gamma: R^1 \to R^1$ by setting

$$(2.1) \qquad \gamma(s) = \int_0^s \beta(\sigma) d\sigma$$

and note that $\gamma$ is a continuously differentiable convex function.

Condition (a)$_1$: $a \in H_0^1(\Omega(0))$ and

$$(2.2) \qquad \int_{\Omega(0)} |\gamma(a(x))| dx < +\infty.$$

Here and henceforth we shall use standard notations for the Sobolev spaces. That is, if $\Omega$ is a domain in $R^N$, $W_p^\ell(\Omega)$ is the set of all functions with derivatives of order up to $\ell$ belonging to $L_p(\Omega)$. $H^\ell(\Omega)$ stands for $W_2^\ell(\Omega)$. If $\partial\Omega$ is smooth and $u \in W_p^1(\Omega)$, then the boundary value of $u$ on $\partial\Omega$ is well defined in the sense of trace. $H_0^1(\Omega)$ consists of the functions in $H^1(\Omega)$ with vanishing boundary values.

Condition (f): $f \in L_2(\hat{\Omega})$.

We proceed to definition of solutions of (Pr. NH).

<u>Definition 2.1.</u>  A function $u \in L_2(\hat{\Omega})$ is called a solution of (Pr. NH) of class $S = S(0,T)$ if (i) and (ii) are satisfied.

    (i) $u_t \in L_2(\hat{\Omega})$ and $\nabla u \in L_2(\hat{\Omega})$. In other words, $u \in W_2^1(\hat{\Omega})$. u satisfies the boundary condition (1.2) and the initial condition (1.3) in the sense of trace.

    (ii) $\beta(u) = \beta(u(t, \cdot x)) \in L_2(\hat{\Omega})$ and the equation (1.1) is satisfied in the distribution sense.

If $0 < \tau_0 < T$, solutions of class $S(0, \tau_0)$ are defined similarly.

<u>Remark 2.2.</u>  In theorems below one may replace Condition ($\beta$) by the following

    Condition ($\beta$)': $\beta$ is a possibly multi-valued accretive function from $R^1$ to $R^1$ which is maximal in Minty's sense with its domain $D(\beta)$ containing 0.

    With $\beta$ subject to Condition ($\beta$)' instead of Condition ($\beta$) the condition (ii) in Definition 2.1 must be modified as

    (ii)': there exists a function $g = g(t,x) \in L_2(\hat{\Omega})$ such that

$$u_t = \Delta u - g(t,x) + f(t,x)$$

holds true in the distribution sense and

$$g(t,x) \in \beta(u(t,x)) \quad \text{a.e. in } \hat{\Omega}.$$

Throughout this paper we sometimes denote various positive constants by one and the same symbol. Actually, we shall use the symbol $c$ or $c_j$ ($j = 1, 2, \ldots$) for positive constants depending on $\hat{\Omega}$ and other domains concerned, while positive constants which depend on the given data will be denoted by $C$ or $C_j$. Moreover, when we wish to make the dependence more explicit, we shall write, for instance, as $C(\|a\|_{L_2(\Omega(0))}, \|f\|_{L_2(\hat{\Omega})})$.

$G$ being a region in $R^m$ or $[0, T] \times R^m$, the $L_2(G)$-norm of a function $u$ on $G$ is denoted by $\|u\|_{L_2(G)}$ or $\|u\|_G$. If $G$ can be understood from the contexts, we write simply $\|u\|$ instead of $\|u\|_G$. When $u$ is a function of $t$ and $x$, $u(t, \cdot)$ means the function of $x$ obtained from $u$ by fixing $t$. We sometimes write $u(t)$ instead of $u(t,\cdot)$.

As for solutions of (Pr. NH) we have

<u>Theorem 2.2.</u> Let Conditions $(\beta)$, $(a)_1$ and $(f)$ be satisfied. Then there exists a solution $u$ of (Pr. NH) of class $S(0,T)$ and it is unique. Furthermore, the following inequalities hold:

(2.3) $\qquad \|u(t)\|_{\Omega(t)} \leq C_1, \qquad (0 < t < T),$

(2.4) $\qquad \|\nabla u\|_{\hat{\Omega}} \leq C_1,$

(2.5) $\quad \bar{\gamma}(u(t)) \equiv \int_{\Omega(t)} |\gamma(u(t,x))| dx \leq C_2, \quad (0 < t < T),$

(2.6) $\qquad \|\beta(u)\|_{\hat{\Omega}} \leq C_2,$

(2.7) $\qquad \|\nabla u(t)\|_{\Omega(t)} \leq C_3, \qquad (0 < t < T);$

(2.8) $\qquad \|u(t)\|_{\hat{\Omega}} \leq C_3.$

Here positive constants $C_1, C_2$ and $C_3$ are such that $C_1 = C_1(\|a\|, \|f\|_{\hat{\Omega}})$, $C_2 = C_2(\|a\|, \bar{\gamma}(a), \|f\|_{\hat{\Omega}})$ and $C_3 = C_3(\|a\|, \|\nabla a\|, \bar{\gamma}(a), \|f\|_{\hat{\Omega}})$.

2.2. Preliminaries to the proof of Theorem 2.2.

In proving Theorem 2.2 we map the non-cylindrical domain $\hat{\Omega}$ to cylindrical domains, which is done locally with respect to $t$. Actually we cover the non-cyclindrical region $\hat{\Omega}$ by a finite number of "slices" and map each slice to a cyclindrical region in the following manner. As is obvious from the smoothness assumptions on $\hat{\Gamma}$, there exists a give the desired uniqueness in the whole $\hat{\Omega}$. Hence it suffices to prove the uniqueness of solutions of (Pr. NH : $\hat{G}$) of class $S_{T,\xi}(0,\tau_0)$. Let $v_1$ and $v_2$ be two such solutions and put $w = v_1 - v_2$. Since $w_\tau \in L_2(\hat{G})$, $w$ regarded as $w : (0, \tau_0) \to L_2(G)$ is strongly absolutely continuous, $w$ being redefined on a null set if necessary. Furthermore, we have immediately

$$(2.14) \quad \|w(\tau)\|^2 = -2 \int_0^\tau \langle L(\tau)w(s), w(s) \rangle \, ds$$
$$-2 \int_0^\tau (\beta(v_1(s)) - \beta(v_2(s)), v_1(s) - v_2(s)) ds,$$

where $\langle , \rangle$ designates the pairing between $H^{-1}(G)$ and $H_0^1(G)$, and where $\|\ \|$ and $(\ ,\ )$ denote the norm and inner product in $L_2(G)$. In view of the monotonicity of $\beta$ and the obvious inequality

$$(2.15) \quad \langle L(\tau)\varphi, \varphi \rangle \geq c_1 \|\nabla \varphi\|^2 - c_2 \|\varphi\|^2 \quad (\varphi \in H_0^1(G)),$$

we have from (2.14)

$$(2.16) \quad \|w(\tau)\|^2 \leq 2c_2 \int_0^\tau \|w(s)\|^2 ds \quad (0 \leq \tau \leq t_0).$$

Hence we have $w \equiv 0$ in $\hat{G}$ and establish the uniqueness of solutions mentioned in Theorem 2.2.

We proceed to the existence of solutions of (Pr. NH) of class $S$. Again it would be enough to prove the existence of solutions of (Pr. NH : $\hat{G}$).

For the time being, we make an extra assumption on $\beta$. That is, we assume

$$(2.17) \quad \beta(0) = 0.$$

Thus $\beta(\sigma)\sigma \geq 0$ for all $\sigma \in R^1$ and $\gamma$ is a non-negative convex function. It is easy to verify that

$$\bar{\gamma}(\varphi) \equiv \int_G \gamma(\varphi(\xi)) d\xi$$

defines a lower semi-continuous convex functional $\bar{\gamma}$ from

$L_2(G)$ to $[\dot{0}, +\infty]$.

In proving the existence of solutions of (Pr. NH; $\hat{G}$), we approximate $\beta$ by its Yosida approximation $\beta_\lambda$ ($\lambda > 0$). By definition $\beta_\lambda$ is equal to

(2.18) $\qquad \beta_\lambda = \dfrac{1}{\lambda}(I - (I + \lambda\beta)^{-1})$.

We recall that $\beta$ is a maximal accretive function from $R^1$ to $R^1$ by Condition ($\beta$). Hence $\beta_\lambda$ is increasing, and is Lipschitz continuous. (2.17) implies $\beta_\lambda(0) = 0$. Also, $|\beta_\lambda(\sigma)| \leq |\beta(\sigma)|$ and $\beta_\lambda(\sigma) \to \beta(\sigma)$ as $\lambda \downarrow 0$ for all $\sigma \in R^1$.

Let $v_\lambda = v_\lambda(\tau, \xi)$ be the solution of the following in initial value problem:

(2.19) $\qquad \dfrac{\partial v_\lambda}{\partial \tau} = -L(\tau)v_\lambda - \beta_\lambda(v_\lambda) + \bar{f}(\tau, \xi) \quad \text{in} \quad \hat{G}$,

(2.20) $\qquad v_\lambda|_{\partial G} = 0$,

(2.21) $\qquad v_\lambda|_{t=0} = \bar{a}(\xi)$.

Since $\beta_\lambda$ is Lipschitz continuous and $L(\tau)$ is nicely elliptic, it is easy to see that the solution $v_\lambda$ exists and enjoys all properties which we need below. (For instance, see Tanabe [30], Kato-Tanabe [19] or Friedman [10,11].) Our main concern is with the <u>a priori</u> estimates of $v_\lambda$ uniform in $\lambda$. We claim

<u>Proposition 2.4.</u> $\qquad v_\lambda$ satisfies

(2.22) $\qquad \|v_\lambda(\tau)\| \leq C_1 \qquad (0 \leq \tau \leq \tau_0)$,

and

(2.23) $\qquad \int_0^\tau \|\nabla_\xi v_\lambda(s)\|^2 ds \leq C_1 \qquad (0 \leq \tau \leq \tau_0)$,

where $C_1 = C_1(\|\bar{a}\|, \|\bar{f}\|_{\hat{G}})$.

Below in the proof of this proposition and other propositions to follow, we shall feel free to make somewhat formal calculations in case justifications can be done in a standard way.

**Proof:** Multiplying (2.19) by $v_\lambda$ and integrating over $G$, we have

$$(2.24) \quad \tfrac{1}{2} \tfrac{d}{d\tau} \|v_\lambda(\tau)\|^2 \leq -c_1 \|\nabla v_\lambda(\tau)\|^2 + c_2 \|v_\lambda(\tau)\|^2 + \|\bar{f}(\tau)\|^2$$

with the aid of the inequality (2.15), whence follows (2.22) and (2.23) immediately.  Q.E.D.

In parallel to (2.1) we put

$$(2.25) \qquad \gamma_\lambda(s) = \int_0^s \beta_\lambda(\sigma)d\sigma$$

and

$$(2.26) \qquad \bar{\gamma}_\lambda(\varphi) = \int_G \gamma_\lambda(\varphi(\xi))d\xi \qquad (\varphi \in L_2(G)).$$

We remark that $\gamma_\lambda(\sigma) \leq \gamma(\sigma)$ for all $\sigma \in R^1$ and $\bar{\gamma}_\lambda(\varphi) \leq \bar{\gamma}(\varphi)$ and that $\bar{\gamma}_\lambda(\varphi) < +\infty$ for all $\varphi \in L_2(G)$, for $\beta_\lambda$ is Lipschitz continuous. We have

$$(2.27) \qquad \tfrac{d}{d\tau} \bar{\gamma}_\lambda(v_\lambda(\tau)) = (\beta_\lambda(v_\lambda(\tau)), \tfrac{dv_\lambda(\tau)}{d\tau})_{L_2(G)}.$$

**Proposition 2.5.** $v_\lambda$ satisfies

$$(2.28) \qquad \bar{\gamma}_\lambda(v_\lambda(\tau)) \leq C_2 \qquad (0 \leq \tau \leq \tau_0),$$

and

$$(2.29) \qquad \int_0^\tau \|\beta_\lambda(v_\lambda(s))\|^2 ds \leq C_2 \qquad (0 \leq \tau \leq \tau_0),$$

where $C_2 = C_2(\|\bar{a}\|, \bar{\gamma}(\bar{a}), \|\bar{f}\|_{\hat{G}})$.

**Proof:** We multiply (2.19) by $\beta_\lambda(v_\lambda)$ and integrate over $G$. Then in view of (2.27) we have for some $c_1$

$$(2.30) \qquad \tfrac{d}{d\tau} \bar{\gamma}_\lambda(v_\lambda) \leq c_1 \|\nabla v_\lambda\|^2 - \tfrac{1}{2} \|\beta_\lambda(v_\lambda)\|^2 + \|\bar{f}(\tau)\|^2.$$

Here we have made use of the fact that

$$\langle L(\tau)\varphi, \beta(\varphi)\rangle \geq \int_G \sum_{i,j} \alpha_{ij}^{(2)}(\tau,\xi)\beta_\lambda'(\varphi) \frac{\partial \varphi}{\partial \xi_i} \frac{\partial \varphi}{\partial \xi_j} d\xi$$

$$-c_2 \|\nabla\varphi\| \|\beta_\lambda(\varphi)\|$$

$$\geq -c_2 \|\nabla\varphi\| \cdot \|\beta_\lambda(\varphi)\|$$

$$\geq -c_1 \|\nabla\varphi\|^2 - \tfrac{1}{4}\|\beta_\lambda(\varphi)\|^2, \quad (\varphi \in H_0^1(G)),$$

and

$$|(\bar{f}(\tau), \beta_\lambda(\varphi))| \leq \tfrac{1}{4}\|\beta_\lambda(\varphi)\|^2 + \|\bar{f}(\tau)\|^2, \quad (\varphi \in L_2(G)).$$

Noting that $\bar{\gamma}_\lambda(v_\lambda)$ is non-negative and $\gamma_\lambda(v_\lambda(0)) \leq \bar{\gamma}(\bar{a})$, we obtain (2.28) and (2.29) from (2.30) by Proposition 2.4.
Q. E. D.

<u>Proposition 2.6.</u>  $v_\lambda$ satisfies

(2.31) $\qquad \|\nabla v_\lambda(\tau)\| \leq C_3 \qquad (0 \leq \tau \leq \tau_0),$

and

(2.32) $\qquad \int_0^\tau \|\frac{\partial v_\lambda}{\partial \tau}(s)\|^2 ds \leq C_3 \qquad (0 \leq \tau \leq \tau_0),$

where $C_3 = C_3(\|\bar{a}\|, \|\nabla\bar{a}\|, \bar{\gamma}(\bar{a}), \|\bar{f}\|_{\hat{G}})$.

<u>Proof:</u> We multiply (2.19) by $\frac{\partial v_\lambda}{\partial \tau}$ and integrate over G.
If we put

(2.33) $\bar{\alpha}(\varphi) = \int_G \sum_{i,j} \alpha_{i,j}^{(2)}(\tau,\xi) \frac{\partial \varphi}{\partial \xi_i} \frac{\partial \varphi}{\partial \xi_j} d\xi, \qquad (\varphi \in H_0^1(G)),$

then we have

$$\tfrac{1}{2}\frac{d}{d\tau}\bar{\alpha}(v_\lambda(\tau)) \leq -\langle L(\tau)v_\lambda, \frac{\partial v_\lambda}{\partial \tau}\rangle + c_1\|\nabla v_\lambda\|^2$$
$$+ c_2\|\nabla v_\lambda\|\|\frac{\partial v_\lambda}{\partial \tau}\|.$$

In view of this we have

$$\|\frac{\partial v_\lambda}{\partial \tau}\|^2 + \tfrac{1}{2}\frac{d}{d\tau}\bar{\alpha}(v_\lambda) + \frac{d}{d\tau}\bar{\gamma}_\lambda(v_\lambda)$$
$$\leq c_1\|\nabla v_\lambda\|^2 + c_2\|\nabla v_\lambda\|\cdot\|\frac{\partial v_\lambda}{\partial \tau}\|$$
$$+ \|\bar{f}(\tau)\|\|\frac{\partial v_\lambda}{\partial \tau}\|,$$

and, consequently,

(2.34)  $$\tfrac{1}{2}\|\frac{\partial v_\lambda}{\partial \tau}\|^2 + \tfrac{1}{2}\frac{d}{d\tau}\bar{\alpha}(v_\lambda) + \frac{d}{d\tau}\bar{\gamma}_\lambda(v_\lambda)$$
$$\leq c_3\|\nabla v_\lambda\|^2 + \|\bar{f}(t)\|^2.$$

(2.34) yields (2.32) by virtue of Propositions 2.4 and 2.5, for $\bar{\alpha}$ is non-negative. (2.34) gives also $\bar{\alpha}(v_\lambda(\tau)) \leq C_3$, which is equivalent to (2.31) because of the uniform ellipticity of $L(\tau)$. Q.E.D.

**Proposition 2.7.** $v_\lambda$ converges to some $v \in L_2(\hat{G})$ as $\lambda \downarrow 0$ in such a way that $\|v_\lambda(\tau) - v(\tau)\|_G \to 0$ uniformly on $[0, \tau_0]$. Hence $v: [0, \tau_0] \to L_2(G)$ is continuous on $[0, \tau_0]$. In particular, $v(0) = \bar{a}$.

**Proof:** Let us put $w_{\lambda,\mu} = v_\lambda - v_\mu$ $(\lambda, \mu > 0)$ and note that

$$\tfrac{1}{2}\frac{d}{d\tau}\|w_{\lambda,\mu}(\tau)\|^2 = -\langle L(\tau)w_{\lambda,\mu}, w_{\lambda,\mu}\rangle$$
$$- (\beta_\lambda(v_\lambda) - \beta_\mu(v_\mu), w_{\lambda,\mu}).$$

The second term on the right side can be estimated as follows, by means of the ingenious trick due to Kōmura [20] which makes use of the monotonicity of $\beta$ :

$$-(\beta_\lambda(v_\lambda) - \beta_\mu(v_\mu), w_{\lambda,\mu})$$
$$\leq -(\beta_\lambda(v_\lambda) - \beta_\mu(v_\mu), \lambda\beta_\lambda(v_\lambda) - \mu\beta_\mu(v_\mu))$$
$$\leq \frac{3}{2}(\lambda+\mu)(\beta_\lambda(v_\lambda)^2 + \beta_\mu(v_\mu)^2).$$

On the other hand, the inequality

$$-\langle L(\tau)w_{\lambda,\mu}, w_{\lambda,\mu}\rangle \leq c\|w_{\lambda,\mu}\|^2$$

is obvious from (2.15). Thus we have

$$\frac{d}{d\tau}\|w_{\lambda,\mu}\|^2 \leq 2c\|w_{\lambda,\mu}\|^2 + 3(\lambda+\mu)(\beta_\lambda(v_\lambda)^2 + \beta_\mu(v_\mu)^2),$$

which yields

(2.35) $\qquad \|w_{\lambda,\mu}\|^2 \leq 6(\lambda+\mu) e^{2c\tau} C_2, \qquad (0 \leq \tau \leq \tau_0),$

with the constant $C_2$ in Proposition 2.5. From (2.35) the proposition is obvious. Q.E.D.

<u>Proposition 2.8.</u>   Let $v$ be the limit of $v_\lambda$. Then $\beta(v) \in L_2(\hat{G})$ and $\beta_\lambda(v_\lambda)$ converges to $\beta(v)$ weakly in $L_2(\hat{G})$.

<u>Proof:</u>   Putting $(Bg)(t,x) = \beta(g(t,x))$ for $g \in L_2(\hat{G})$, we observe that $B$ is a maximal accretive operator in $L_2(\hat{G})$. Furthermore, for the Yosida approximation $B_\lambda$ of $B$ it holds true that $(B_\lambda g)(t,x) \equiv \beta_\lambda(g(t,x))$. Consequently, we obtain Proposition 2.8 from Propositions 2.5 and 2.7 by a theorem concerning the demi-closedness of maximal accretive operators. (For instance, see Kato [17, 18] and Kōmura [20, 21].)
Q.E.D.

Proposition 2.9. Let $v$ be as above. Then $v$ is the required solution of (Pr. NH : $\hat{G}$) of class $S_{T,\xi}(0, \tau_0)$.

Proof: Obvious from the preceding Propositions, in particular, from (2.22), (2.29) and (2.32). Q.E.D.

Now we are about to complete the proof of Theorem 2.2. If we define $u$ by $u(t,x) = v(t, X(t,x))$, $u$ is a solution of (Pr. NH) of class $S(0, \tau_0)$. In fact, $u$ satisfies (2.3), (2.5) and (2.7) for $t \in (0, \tau_0)$, because we have the corresponding boundedness of $\|\nabla v(\tau)\|$ and $\gamma(v(\tau))$ which follows from corresponding inequalities for $v_\lambda$ by virtue of the continuity of $\|\ \|_G$ and the lower semi-continuity of $\|\nabla \cdot \|_G$ and $\bar{\gamma}$. In particular, for any $t \in [0, \tau_0]$ $u(t, \cdot)$ satisfies the same condition as Condition $(a)_1$. Thus we can extend the solution $u$ beyond $t = \tau_0$ over to $(0, 2\tau_0)$ within class $S$ and so on. In this way we obtain the existence of the solution $u$ of (Pr. NH) in $\hat{\Omega}$. The inequalities in Theorem 2.2 are obvious from the corresponding estimates for $v_\lambda$ and the manner of the extension of the solution.

Finally the extra assumption (2.17) is removed simply by using $\beta(\sigma) - \beta(0)$ and $f(t, x) - \beta(0)$ instead of $\beta(\sigma)$ and $f(t, x)$, respectively.

§3. Convergence of the penalty method for (Pr. NH).

In this section we shall show that $u^n \in L_2(\hat{B})$ which is defined as the solution of (Pr. NH)$^n$ converges to the solution $u$ of (Pr. NH) in $\hat{\Omega}$. Firstly we claim

Theorem 3.1. Suppose that Conditions $(\beta)$, $(a)_1$ and $(f)$ are satisfied. Also recall that $p : R^1 \to R^1$ is a strictly increasing and Lipschitz continuous function with $p(0) = 0$. Then there exists a unique solution $u^n$ of (Pr. NH)$^n$ for each $n$, such that $u^n$, $u_t^n$, $\nabla u^n$ and $\beta(u^n)$ belong to $L_2(\hat{B})$, (1.9) and (1.10) are satisfied in the sense of trace, and such that (1.8) holds in the sense of distribution.

THE PENALTY METHOD

Proof: We might give a proof similar to the proof of Theorem 2.2. However, since the present domain $\hat{B}$ is cylindrical and the penalizing term is Lipschitz continuous, we simply apply a theorem due to Brezis [2] with a slight modification and obtain the existence. Q.E.D.

Theorem 3.2. Under the same assumptions as in the preceding theorem, $u^n|\hat{\Omega}$ converges to the unique solution $u$ of (Pr. NH) of class S strongly in $L_2(\hat{\Omega})$.

Proof: Again, we may assume (2.17) without loss of generality. Then there exists a constant $C_1$, which depends on $a$ and $f$ but is independent of $n$, such that

(3.1)
$$\|u^n(t)\|_B \leq C_1, \quad \|\nabla u^n\|_{\hat{B}} \leq C_1, \quad \bar{\gamma}_B(u^n(t)) \leq C_1,$$
$$\|\beta(u^n)\|_{\hat{B}} \leq C_1 \quad \text{and} \quad \sqrt{n}\|p(u^n)\|_{\hat{B}-\hat{\Omega}} \leq C_1.$$

Here $\bar{\gamma}_B$ is defined in the same way as $\bar{\gamma}$ of §2, the integral being over B this time. In fact, these inequalities are obtained by multiplying (1.8) by $u^n$, $\beta(u^n)$ or $p(u^n)$ and applying the same arguments as in the proof of Theorem 2.2.

Next, we derive the boundedness of $\{\frac{\partial u^n}{\partial t}\}$ in $L_2(\hat{B})$. To this end, we make the change of variables by (2.9), assuming that this time $\Phi$ is extended outside $\hat{\Omega}(t_0, t_0+\tau_0)$ to be a diffeomorphism from $(t_0, t_0+\tau_0) \times B$ onto itself. Moreover, we may require that $\Phi$ leaves the lateral boundary of $(t_0, t_0+\tau_0) \times B$ fixed, since $\Phi$ coincides with the identity map except in a neighborhood of the lateral boundary $\hat{\Gamma}$ of $\hat{\Omega}$.

For the time being, let us take the case of $t_0 = 0$. As before, G and $\hat{G}$ stand for $\Omega(0)$ and $(0, \tau_0) \times \Omega(0)$, respectively. Furthermore, we put $\hat{B}_0 = (0, \tau_0) \times B$. If we write $v^n = v^n(\tau, \xi) = u^n(\tau, X^{-1}(\tau, \xi))$, then we have

(3.2) $$\frac{\partial v^n}{\partial \tau} = -L(\tau)v^n - \beta(v^n) + \bar{f}(\tau, \xi) - n\bar{\chi}(\xi)p(v^n) \quad \text{in } \hat{B}_0,$$

(3.3) $$v^n|_{\partial B} = 0,$$

and

(3.4) $$v^n|_{t=0} = \bar{a}(\xi),$$

where $\bar{f}$, $\bar{\chi}$ and $\bar{a}$ have come up from $\tilde{f}$, $\chi$ and $\tilde{a}$ through the change of variables. $\bar{\chi}$ turns out to be independent of $\tau$. In fact, $\bar{\chi}$ is nothing but the characteristic function of B - G. In view of (3.1) we see that $\{v^n\}$ forms a bounded set in $L_2((0,\tau_0); H^1(B))$. We now claim

**Proposition 3.2.** $\{\frac{\partial v^n}{\partial \tau}\}$ forms a bounded set in $L_2(\hat{B}_0)$. Moreover, $\|\nabla_\xi v^n(\tau)\|_B$ and $n\bar{q}(v^n(\tau))$ are bounded uniformly in $n$ and $\tau \in (0, \tau_0)$. Here $\bar{q}$ is defined as

$$\bar{q}(\varphi) = \int_{B-G} q(\varphi(\xi))d\xi, \qquad (\varphi \in L_2(B)),$$

in terms of the anti-derivative $q$ of $p$:

$$q(s) \equiv \int_0^s p(\sigma)d\sigma.$$

**Proof:** We multiply (3.2) by $\frac{\partial v^n}{\partial \tau}$ and integrate with respect to $\xi$ over B. Then in the same manner as in the proof of Proposition 2.6 we are led to the following inequality.

(3.5) $$\tfrac{1}{2}\|\tfrac{\partial v^n}{\partial \tau}\|_B^2 + \tfrac{1}{2}\tfrac{d}{d\tau}\bar{\alpha}_B(v^n) + \tfrac{d}{d\tau}\bar{\gamma}_B(v^n) + n\tfrac{d}{d\tau}\bar{q}(v^n)$$

$$\leq c\|\nabla_\xi v^n\|_B^2 + \|\bar{f}(\tau)\|_B^2,$$

where the definition of $\bar{\alpha}_B$ is similar to that of $\bar{\alpha}$ in §2, except that the integration with respect to $\xi$ is now extended over B. We notice that $\bar{\alpha}_B(v^n(0))$, $\bar{\gamma}_B(v^n(0))$ and $n\bar{q}(v^n(0))$ are independent of $n$, because $v^n(0) = \bar{a}$. In particular,

$$\bar{q}(v^n(0)) = \int_{B-G} q(\bar{a}(\xi))d\xi = 0.$$

In view of this fact and in consideration of the positivity of $\bar{\alpha}_B$, $\bar{\gamma}_B$ and $\bar{q}$, we obtain the proposition immediately by integrating (3.5) with respect to $\tau$, for the uniform boundedness of $\bar{\alpha}_B$ is equivalent to that of $\|\nabla_\xi v^n(\tau)\|_B$ and, hence, to that of $\|\nabla_x u^n(\tau)\|_B$.    Q.E.D.

**Proposition 3.3.** $\{\frac{\partial u^n}{\partial t}\}$ is bounded in $L_2(\hat{B})$, and $\{u^n\}$ forms a compact set in $L_2(\hat{B})$.

**Proof:** When we return to the original variables $t$ and $x$, the preceding proposition implies, with the aid of (3.1), that $\{\frac{\partial u^n}{\partial t}\}$ is bounded in $L_2(\hat{B}_0)$. Moreover, from the uniform boundedness of $\|\nabla_\xi v^n(\tau)\|_B$ and $n\bar{q}(v^n(\tau))$ mentioned there, we see that the same argument can be applied with the initial time $\tau_1 \in (0, \tau_0)$ in order to derive the corresponding estimates for $u^n$ in the extended interval $(0, \tau_1+\tau_0)$. Repeating this procedure by finite times, we can show eventually that $\{\frac{\partial u^n}{\partial t}\}$ is bounded in $L_2(\hat{B})$. Since we already have the boundedness of $\{u^n\}$ and $\{\nabla_x u^n\}$ in $L_2(\hat{B})$ by (3.1), we get the compactness of $\{u^n\}$ in $L_2(\hat{B})$ by the well-known Rellich's theorem.    Q.E.D.

We are ready to complete the proof of Theorem 3.3.

By (3.1) and by Proposition 3.3 we can select a subsequence $\{u^{n(k)}\}$ of $\{u^n\}$ such that for some $u^* \in L_2(\hat{B})$ we have the following convergence as $k \to \infty$:

$$u^{n(k)} \to u^* \quad \text{strongly in } L_2(\hat{B}),$$

and

$$\nabla u^{n(k)} \to \nabla u^*, \quad \frac{\partial u^{n(k)}}{\partial t} \to \frac{\partial u^*}{\partial t}$$

weakly in $L_2(\hat{B})$. Obviously, $u^* \in W_2^1(\hat{B})$. The fact that $\beta(u^{n(k)}) \to \beta(u^*)$ weakly in $L_2(\hat{B})$ is obtained by means of the demi-closedness of maximal accretive operators. Actually if we define an operator $M$ in $L_2(\hat{B})$ by setting

$$(Mg)(t,x) = \beta(g(t,x)), \quad (g \in L_2(\hat{B})),$$

then $M$ is maximal accretive. Thus taking the weak limit in (1.8) we see that $u^*$ satisfies (1.1) in $\hat{\Omega}$, for the penalizing term $-n\chi(t,x)p(u^n)$ is identically equal to 0 there. It is quite easy to verify that $u^*|\hat{\Omega}$ satisfies the initial condition in the sense of trace. It just remains to show that the boundary value of $u^*$ on $\hat{\Gamma}$ vanishes. To this end we first note that $p(u^n) \to 0$ as $n \to \infty$ in $L_2(\hat{B}-\hat{\Omega})$. This implies that $u^n \to 0$ in $L_2(\hat{B}-\hat{\Omega})$ weakly. In fact, defining the operator $K$ in $L_2(\hat{B}-\hat{\Omega})$ by setting

$$(Kg)(t,x) = p^{-1}(g(t,x)) \quad (g \in L_2(\hat{B}-\hat{\Omega})),$$

we notice that $K$ is single-valued and is maximal accretive thanks to the strict monotonicity of $p$. Putting $g^n = p(u^n)$, we see that $u^n = Kg^n \to u^*$ weakly in $L_2(\hat{B}-\hat{\Omega})$ and $g^n \to 0$ strongly in $L_2(\hat{B}-\hat{\Omega})$. Therefore by virtue of the demi-closedness of maximal accretive operators, we have $u^* = K0 = 0$ in $\hat{B}-\hat{\Omega}$. From this it follows immediately that the well-defined boundary value $u^*|\hat{\Gamma}$ vanishes.

Thus $u^*|\hat{\Omega}$ is the required solution of (Pr. NH) of class S. Since we know that the solution of (Pr. NH) in S is unique, we can conclude that the whole sequence $\{u^n\}$ converges to the solution of (Pr. NH) in $L_2$, when restricted to $\hat{\Omega}$. This completes the proof of Theorem 3.2. Q.E.D.

## §4. Application of the penalty method to (Pr. NS).

### 4.1. Assumptions and definition of weak solutions.

We begin with definition of some classes of solenoidal (vector) functions. In this section the symbol $\Omega$ stands for an arbitrary bounded domain with smooth $\partial\Omega$ in $R^m$. We put

$$D_\sigma(\Omega) = \{\varphi \in C^\infty(\Omega) \mid \text{supp } \varphi \subset \Omega \text{ and div } \varphi = 0\},$$

$$H_\sigma(\Omega) = \text{the completion of } D_\sigma(\Omega) \text{ under the } L_2(\Omega)\text{-norm.}$$

$H^1_\sigma(\Omega)$ = the completion of $D_\sigma(\Omega)$ under the $H^1(\Omega)$-norm,

or equivalently, under the Dirichlet norm. u belongs to $H^1_\sigma(\Omega)$ if and only if $u \in H^1(\Omega)$, div u = 0 and $u|\partial\Omega = 0$. Similarly, if $u \in H_\sigma(\Omega)$, then $u \in L_2(\Omega)$, div u = 0 (in the distribution sense) and the normal component of u on $\partial\Omega$ vanishes in a certain sense.

We now state our assumption on a in (Pr. NS).

Condition (a)$_2$:  $a \in H_\sigma(\Omega(0))$.

We note that if we extend a subject to Condition (a)$_2$ over to B by setting $\tilde{a} = a$ in $\Omega(0)$, and $\tilde{a} = 0$ in $B - \Omega(0)$, then we have $\tilde{a} \in H_\sigma(B)$.

Next, we consider (vector) functions of t and x. We put

$$\hat{D}_\sigma(\hat{\Omega}) = \{\varphi \in C^\infty(\hat{\Omega}') \mid \text{supp } \varphi \subset \hat{\Omega}', \text{ div } \varphi = 0\}.$$

Since $\hat{\Omega}'$ is closed at the top t = T and at the base t = 0, $\varphi \in \hat{D}_\sigma(\hat{\Omega})$ may not vanish for t = 0 or t = T. For u defined in $\hat{\Omega}$ we put

(4.1)     $\nu(u) = \|\nabla u\|_{\hat{\Omega}} = (\int_0^T \|\nabla u(t)\|^2_{\Omega(t)} dt)^{\frac{1}{2}}$.

Then we introduce

$\hat{H}^1_\sigma(\hat{\Omega})$ = the completion of $\hat{D}_\sigma(\hat{\Omega})$ under the norm $\nu(\cdot)$.

u belongs to $\hat{H}^1_\sigma(\hat{\Omega})$ if and only if $u \in L_2(\hat{\Omega})$, $\nabla u \in L_2(\hat{\Omega})$, div u = 0 and the boundary value of u on the lateral boundary $\hat{\Gamma}$ vanishes in the sense of trace.

$\hat{D}_\sigma(\hat{B})$ and $\hat{H}^1_\sigma(\hat{B})$ are defined similarly. In order to define weak solutions of (Pr. NS) we further need

<u>Definition 4.1.</u>  $\hat{V}(\hat{\Omega})$ is the set of all $v \in \hat{H}^1_\sigma(\hat{\Omega})$ such that $\|v(t)\|_{\Omega(t)}$ is essentially bounded in (0,T). $\hat{V}(\hat{B})$ is defined similarly.

The following condition concerns the boundary value b in (Pr. NS).

Condition (b): With some (vector) function $q = q(t,x)$ which is defined and smooth (say, of class $C^3$) in a neighborhood of $\hat{\Gamma}$ we can express b as

(4.2) $\qquad b = \text{rot } q \quad \text{on } \hat{\Gamma}$.

A sufficient condition for the existence of such a vector potential q is well-known (see, for instance, Ladyzhenskaya [22] and Finn [8]). Actually, Condition (b) is satisfied if b is smooth and

$$\int_{\Gamma_\alpha(t)} b \cdot n \, dS = 0$$

for each $\alpha$ and for all t, where n is the unit normal vector to $\Gamma_\alpha(t)$.

We note that under Condition (b) we can construct a $b^*$ which is of the form $b^* = \text{rot } q^*$ and is smooth in $\hat{B}$ and which satisfies

(4.3) $\qquad b^*|_{\hat{\Gamma}} = b \quad \text{and} \quad b^*|_{(0,T) \times \partial B} = 0$.

<u>Definition 4.2.</u> $u = u(t,x)$ defined in $\hat{\Omega}$ is called a weak solution (for E. Hopf's class) of (Pr. NS) if (i) and (ii) hold:

(i) For some $b^*$ mentioned above we have

$$u - b^* \in \hat{V}(\hat{\Omega}).$$

(ii) For all $\varphi \in \hat{D}_\sigma(\hat{\Omega})$ with $\varphi|_{t=T} = 0$ the equality

(4.4) $\qquad F(u,\varphi) \equiv \int_0^T \{(u, \varphi_t) - (\nabla u, \nabla \varphi) + (u, (u \cdot \nabla)\varphi)\} \, dt$

$$= -(a, \varphi(0))$$

is satisfied.

Clearly our definition of weak solutions of (Pr. NS) coincides with E. Hopf's one if $\Omega(t) \equiv \Omega(0)$. (See Hopf [16], Ladyzhenskaya [22] and, in particular, Serrin [28].)

Definition 4.3. Let n be a positive integer. $u^n = u^n(t,x)$ defined in $\hat{B}$ is called a weak solution of (Pr. NS)$^n$ if (i) and (ii) hold:

(i) For some $b^*$ mentioned above we have
$$u^n - b^* \in \hat{V}(\hat{B}).$$

(ii) For all $\varphi \in \hat{D}_\sigma(\hat{B})$ with $\varphi|_{t=T} = 0$ the equality

(4.5) $\qquad F(u^n, \varphi) = n \iint_{\hat{B}-\hat{\Omega}} (u^n - b^*, \varphi) dx\, dt - (a, \varphi(0))$

is satisfied.

Obviously (4.5) is nothing but the weak form of the Navier-Stokes equation (associated with div $u^n = 0$) added with the penalizing term $-n\chi(t,x)(u-b^*)$:

(4.6) $\qquad u^n_t = \Delta u^n - \nabla p^n - (u^n \cdot \nabla)u^n - n\chi(t,x)(u^n - b^*)$, in $\hat{B}$.

In other words the classical form of (Pr. NS)$^n$, which is expected to approximate (Pr. NS), consists of (4.6) and the following equations:

(4.7) $\qquad$ div $u^n = 0 \qquad$ in $\hat{B}$

(4.8) $\qquad u^n|_{t=0} = \tilde{a} \quad$ and $\quad u^n|_{\partial B} = 0$.

## 4.2. Theorems and outline of proof.

The existence of weak solutions of (Pr. NS) was studied by J. O. Sather in his thesis [27] and has been proved by N. Sauer and the writer [12, 13] with resort to the penalty method. Actually, we have

Theorem 4.4. Let $m = 2$ or $3$. Suppose that Conditions (a)$_2$ and (b) are satisfied. Then there exists a weak solution u of (Pr. NS), which is unique if $m = 2$.

The uniqueness part of this theorem has been indicated by Sather [27] and Serrin [28]. (Also see Fujita [15].) The

The existence part is obtained as a corollary of (or is included in) the following theorem on the convergence of the penalty method for (Pr. NS).

**Theorem 4.5.** Under the same assumptions as in the preceding theorem there exists a weak solution $u^n$ of $(Pr.\ NS)^n$ for each $n$. When we fix $b^*$ for the given $b$, then we can select a subsequence of $\{u^n\}$ such that $u^n|\hat{\Omega}$ converges in $L_2(\hat{\Omega})$ to a weak solution of (Pr. NS).

**Remark 4.6.** If $m = 3$, the uniqueness of weak solutions of (Pr. NS) is open, which is no wonder since it is the case even to the cylindrical case. In this connection we remark also that the uniqueness of weak solutions has been disproved by Ladyzhenskaya [23] for certain mixed problems for the Navier-Stokes equation in sharply non-cylindrical regions.

<u>Outline of the proof of Theorem 4.5.</u> Since details have been already published in Fujita-Sauer [13], we here just refer to a few steps in the proof of Theorem 4.5. Let us denote $u^n - b^*$ by $v^n$. Firstly we claim

**Proposition 4.7.** For each fixed $n$ we can apply the Galerkin's method just as in Hopf [16] and obtain the existence of a weak solution $u^n = v^n + b^*$ of $(Pr.\ NS)^n$. $v^n$ satisfies

$$(4.9) \qquad \|v^n(t)\|_B \leq C_1 \qquad (0 \leq t \leq T),$$

$$(4.10) \qquad \|\nabla v^n\|_{\hat{B}} = \left(\int_0^T \|\nabla v^n(t)\|^2 dt\right)^{\frac{1}{2}} \leq C_1,$$

and

$$(4.11) \qquad n\|v^n\|_{\hat{E}}^2 \leq C_1, \qquad (\hat{E} = \hat{B} - \hat{\Omega}),$$

for some constant $C_1 = C_1(a, b^*)$ independent of $n$.

A formal derivation of these inequalities is the following. $v^n$ solves the equation

658

(4.12) $\quad v_t^n = \Delta v^n - \nabla p^n - (v^n \cdot \nabla) v^n - n \chi(t,x) v^n$

$\quad\quad\quad\quad - (v^n \cdot \nabla) b^* - (b^* \cdot \nabla) v^n + f(t,x)$ ,

where $f = \Delta b^* - b_t^* - (b^* \nabla) b^*$. Multiplying (4.12) by $v^n$ and integrating over $B$, we have

(4.13) $\quad \frac{1}{2} \frac{d}{dt} \|v^n\|^2 = -\|\nabla v^n\|^2 - n(\chi v^n, v^n) - (v^n, (v^n \cdot \nabla) b^*)$

$\quad\quad\quad\quad + (v^n, f)$ ,

where $\|\ \|$ and $(\ ,\ )$ are those in $L_2(B)$. In consideration of the smoothness of $b$ and $f$, (4.13) gives

$$\frac{d}{dt} \|v^n\|^2 \leq -2\|\nabla v^n\| - 2n(\chi v^n, v^n) + C(1 + \|v^n\|^2)$$

for some $C = C(b^*)$, whence follows the desired inequalities. By (4.9)~(4.11) we can select a subsequence of $\{v^n\}$ which converges to some $v^*$ weakly in $L_2(\hat{B})$ and $\hat{H}_\sigma^1(\hat{B})$. Also, the subsequence converges strongly in $\hat{E}$ and $v^* = 0$ in $\hat{E}$. Thus in order to pass to the limit in (4.5) and to verify that $u^* = v^* + b^*$ is the required weak solution of (Pr. NS), it is enough to show that the subsequence converges strongly in $L_2(\hat{\Omega})$. In other words, a crucial step in the proof of Theorem 4.5 is the following

<u>Proposition 4.8.</u>  $\{v^n\}$ forms a compact set in $L_2(\hat{\Omega})$.
We prove this proposition by making use of a modified form of Aubin's theorem (Lemma 4.9 below). However, since we have not reduced the $(t,x)$-region under consideration to a cylindrical one this time, and since in estimating $v_t^n$ by (4.12) we have to eliminate $\nabla p^n$ for which we have no available estimates, more technicalities are involved. We present only a part of the proof of Proposition 4.8, leaving the rest to Fujita-Sauer [13]. Let $\Omega \subset R^m$ and $(\alpha, \beta) \subset (0,T)$ be such that $(\alpha, \beta) \times \Omega \subset \hat{\Omega}$. Then $v^n$ restricted to $(\alpha, \beta) \times \Omega$ forms a bounded set in $L_2((\alpha, \beta), H^1(\Omega))$ in virtue of Proposition 4.7. Let $P = P(\Omega)$ be the orthogonal projection from $L_2(\Omega)$ onto

$H_\sigma(\Omega)$, and let $A = A(\Omega)$ be the Stokes operator $-P\Delta$ in $\Omega$, its domain being $D(A) = \{\varphi \,|\, \varphi \in H^2(\Omega) \cap H^1_\sigma(\Omega)\}$. Taking $\varphi$ from $D(A)$, we multiply (4.12) by $\varphi$ and integrate over $\varphi$. Then we have

(4.14) $\quad |\langle v^n_t, \varphi \rangle| \leq \|\nabla v^n\| \cdot \|\nabla \varphi\| + |(v^n, (v^n \cdot \nabla)\varphi)|$

$$+ C_1(\|v^n\| + \|\nabla v^n\| + 1)\|\varphi\|,$$

where $\langle \,,\, \rangle$ denotes the pairing between the Banach space $D(A)$ under the graph norm and its dual $D(A)'$, and where $\|\,\|$ is the norm of $L_2(\Omega)$. As we have

$$|(v^n, (v^n \cdot \nabla)\varphi)| \leq \|v^n\|_{L_2} \cdot \|v^n\|_{L_6} \cdot \|\nabla \varphi\|_{L_3}$$

$$\leq c \|v^n\|_{L_2} \cdot \|v^n\|_{H^1} \cdot \|\varphi\|_{H^2}$$

by Hölder's inequality and Sobolev's inequality (recall m = 2 or 3!), we can estimate as

$$|(v^n, (v^n \cdot \nabla \varphi)| \leq C_1 \|\nabla v^n\|_B \cdot \|A\varphi\|,$$

by means of (4.9) and the following property of A: (see Cattabriga [5] or Ladyzhenskaya [22]),

(4.15) $\quad \|\varphi\|_{H^2(\Omega)} \leq c_\Omega \|A\varphi\| \qquad (\varphi \in D(A))$.

Hence, denoting by $Sv^n(t)$ the element of $D(A)'$ defined by $\langle Sv^n(t), \varphi \rangle = (v^n(t), \varphi)_\Omega$, $(\varphi \in D(A))$, we obtain

$$\|(Sv^n)_t\|_{D(A)'} \leq C_2(\|\nabla v^n\| + 1) \qquad (\alpha < t < \beta)$$

for some $C_2 = C_2(a, b^*)$. In view of (4.10) this implies that $\{\frac{d}{dt} Sv^n\}$ forms a bounded set in $L_2((\alpha, \beta); D(A)')$. This enables us to apply the modified form of Aubin's compactness theorem stated in Lemma 4.9 below and conclude that $\{P(\Omega) v^n\}$ is a compact set in $L_2((\alpha, \beta) \times \Omega)$. To complete the proof of Proposition 4.8 it still remains to show the following:

    (i) if $w = v^n - v^m$ is small on the lateral boundary $(\alpha, \beta) \times \partial\Omega$ of $(\alpha, \beta) \times \Omega$, then $w - P(\Omega)w$ is small in $(\alpha, \beta) \times \Omega$.

    (ii) If $n, m$ are large and if $(\alpha, \beta) \times \partial\Omega$ is close to $\hat{\Gamma}$, then $w$ is small on $(\alpha, \beta) \times \partial\Omega$.

    (iii) If a boundary strip neighboring $\hat{\Gamma}$ of width $\delta$ is deleted from $\hat{\Omega}$, then we can cover the remaining part of $\hat{\Omega}$ by a finite number of cylinders of the type $(\alpha, \beta) \times \Omega$ as considered above. The $L_2$-norm of $w$ over the boundary strip is small if $\delta$ is small and $m, n$ are large.

This task can be done by means of Proposition 4.7, as was carried out in Fujita-Sauer [13].

**Lemma 4.9.** Let $M_0, M_1, M_2$ be three Hilbert spaces and let $P : M_0 \to M_1$ and $S : M_0 \to M_2$ be such that

    (i) $P$ and $S$ are linear and compact, and

    (ii) $Sv = 0$ implies $Pv = 0$ $(v \in M_0)$.

Moreover, let $(\alpha, \beta)$ be a finite interval of $R^1$. Suppose that $w^n \in L_2((\alpha, \beta); M_0)$ $(n = 1, 2, \ldots)$ are such that $\{w^n\}$ is bounded in $L_2((\alpha, \beta); M_0)$ and $\{\frac{d}{dt} Sw^n\}$ (the distribution derivative) is bounded in $L_2((\alpha, \beta); M_2)$. Then $\{Pw^n\}$ forms a compact set in $L_2((\alpha, \beta); M_1)$.

**Remark 4.10.** Recently my student H. Morimoto [26] has succeeded in proving the existence of weak solutions of (Pr. NS) periodic in $t$ under the assumption that $\hat{\Omega}$ and $b$ are periodic in $t$ with the same period. Her argument is quite parallel to ours except one point that in order to show the existence of periodic weak solutions of (Pr. NS)$^n$ by Galerkin's method and to derive a priori estimates for these approximate solutions a more delicate estimate of $((v^n \nabla)v^n, b^*)_B$ is necessary. That is, we need

Proposition 4.11. Let $\varepsilon$ be any positive number. Given b satisfying Condition (b), we can choose $b^*$ such that (4.3) holds, and moreover, there exists a constant $C_\varepsilon$ with the property that

$$|((\varphi \cdot \nabla)\varphi, b^*)_B| \leq \varepsilon \|\nabla \varphi\|_B^2 + C_\varepsilon (\chi(t)\varphi, \varphi)_B, \quad (0 < t < T),$$

for all $\varphi \in H_\sigma^1(B)$. Such a $b^*$ can be constructed by refining the corresponding technique which was previously employed to derive energy estimates for stationary solutions. (For instance, see Fujita [14] and Finn [9].)

## REFERENCES

1. Aubin, J. P., Un théorème de compacité, C. R. Acad. Sci. Paris 256 (1963), 5042-5044.

2. Brezis, H., Propriétés régularisantes de certains semi groupes non linéaires, to appear.

3. Brezis, H., M. G. Crandall and A. Pazy, Perturbations of nonlinear maximal monotone sets in Banach space, Comm. Pure Appl. Math. 23 (1970), 123-144.

4. Browder, F. E., Nonlinear maximal monotone operators in Banach spaces, Math. Ann. 175 (1968), 89-113.

5. Cattabriga, L., Su un problema al contorno relativo al sistema di equazioni di Stokes, Rend. Sem. Mat. Univ. Padova 31 (1961) 308-340.

6. Courant, R., Variational methods for the solution of problems of equilibrium and vibrations, Bull. Amer. Math. Soc. 49 (1943), 1-23.

7. Crandall, M. G. and A. Pazy, Nonlinear semi-groups of contractions and dissipative sets, J. Functional Anal. 3 (1969), 376-418.

8. Finn, R., On the steady state solutions of the Navier-Stokes equations III Acta Math. 105 (1961), 197-244.

9. Finn, R., Stationary solutions of the Navier-Stokes equation, Proc. Symp. Appl. Math., Vol. 17, Applications of Nonlinear Partial Differential Equations in Mathematical Physics, Amer. Math. Soc., 1965, 121-154.

10. Friedman, A., Partial Differential Equations of Parabolic Type, Prentice-Hall, Englewood Cliffs, N. J., 1964.

11. Friedman, A., Partial Differential Equations, Holt, Rinehart and Winston, New York, 1969.

12. Fujita, H. and N. Sauer, Construction of weak solutions of the Navier-Stokes equation in a noncylindrical domain, Bull. Amer. Math. Soc. 75 (1969), 465-468.

13. Fujita, H. and N. Sauer, On existence of weak solutions of the Navier-Stokes equations in regions with moving boundaries, J. Fac. Sci. Univ. Tokyo, Sec. I 17 (1970), 403-420.

14. Fujita, H., On the existence and regularity of the steady-state solutions of the Navier-Stokes equation, J. Fac. Sci. Univ. Tokyo, Sec. I 9 (1961), 59-102.

15. Fujita, H., Remarks on the 2-dimensional non-stationary solutions of the Navier-Stokes equation in non-cylindrical domains, to appear.

16. Hopf, E., Über die Anfangswertaufgabe für die hydrodynamischen Grundgleichungen, Math. Nachr. 4 (1951), 213-231.

17. Kato, T., Nonlinear semigroups and evolution equations, J. Math. Soc. Japan 19 (1967), 508-520.

18. Kato, T., Accretive operators and nonlinear evolution equations in Banach spaces, Proc. Symp. Pure Math., Vol. 18, Nonlinear Functional Analysis, Amer. Math. Soc., 1970, 138-161.

19. Kato, T., and H. Tanabe, On the abstract evolution equation, Osaka Math. J. 14 (1961), 1-11.

20. Kōmura, Y., Nonlinear semi-groups in Hilbert space, J. Math. Soc. Japan 19 (1967), 493-507.

21. Kōmura, Y., Nonlinear semigroups in Hilbert spaces, Proc. Intern. Conference on Functional Anal., Tokyo, 1969, 260-268.

22. Ladyzhenskaya, O.A., The Mathematical Theory of Viscous Incompressible Flow, (English Translation) Gordon Breach, New York - London, 1963.

23. Ladyzhenskaya, O.A., An example of nonuniqueness in Hopf's class of weak solutions of the Navier-Stokes equations, Izv. Adad. Nauk SSSR Ser. Mat. 33 (1969), 240-247.

24. Lions, J.L., Une remarque sur les problèmes d'évolution non linéaires dans des domaines non cylindriques, Rev. Roumaine Math. Pures Appl. 9 (1964), 11-18.

25. Lions, J.L., Quelques Méthodes de Résolution des Problèmes aux Limites Non Linéaires, Dunod, Paris, 1969.

26. Morimoto, H., On existence of periodic weak solutions of the Navier-Stokes equations in regions with periodically moving boundaries, Thesis, to appear.

27. Sather, J. O., The initial-boundary value problem for the Navier-Stokes equation in regions with moving boundaries, Dissertation, Univ. Minnesota, 1963.

28. Serrin, J., Initial value problem for Navier-Stokes equations, Nonlinear Problems, Univ. Wisconsin Press, Madison, 1963, 69-98.

29. Strauss, W., The Energy Method in Nonlinear Partial Differential Equations, Notas de Matematica, Rio de Janeiro, 1969.

30. Tanabe, H., A class of the equations of evolution in a Banach space, Osaka Math. J. 11 (1959), 121-145.

31. Yosida, K., Functional Analysis, Springer, Berlin, 1965.

University of Tokyo
Tokyo, Japan

Received March 30, 1971

# Index

## A

accretive, 158, 168, 171
    sets, 158
angle bounded, 445
angle boundedness, 429
asymptotically almost
    periodic, 510

## B

Banach
    reflexive spaces, 451
    spaces, 451
    semialgebra, 158
bifurcation, 11
    point, 12
Bingham fluids, 524
Borsuk-Ulam
    principle, 440
    theorem, 434
boundary layers, 523

## C

$C'$-functional, 69
calculus of variations, 216
Caratheodory
Cauchy problem, 161
characteristic value, 42
closed proper convex
    function, 186
compact, 69
    operator, 2
compactness, 217, 228
condition
    Caratheodory, 427
    entopy, 608
    Rankine-Hugoniot jump, 606
    $(S)_+$, 430
    shock, 612
cone
    dual, 254
    maps, 58
    recession, 281
    support, 257
conical partition of the
    identity, 350

# INDEX

conjugate
    function, 185
    functionals, 222
    points, 189
    space $X^*$, 451
conservation form, 604
continuous family of compact
    operators, 465
control theory, 503
convergence of the penalty
    method, 650
convex
    body, 264
    functions, 101
    functionals, 215
convex sets
    face of a closed, 273
    projection on, 239
    symmetries with respect
        to, 242
core, 262
correctors, 524
critical point, 70

## D

degree
    Leray-Schauder, 7
    of a map, 1
demiclosed mapping, 400
demicontinuous mapping, 434
differential equations, 65
    functional, 501
dual cone 254
duality, 215
    of convex functions, 184

## E

effective domain, 452
E. Hopf's class, 656
eigenvalue, 71
entopy conditions, 608
equation
    differential, 65
    Euler-Lagrange, 582
    evolution, 101
    functional differential, 501
    integrodifferential, 507
    minimal surface, 583
    nonlinear heat, 635
    Volterra, 503
essential maps, 2
estimates
    global gradient, 571-574
    interior gradient, 578
Euler-Lagrange equation, 582
evolution equations, 101
exponential formula, 158

## F

face of a closed convex
    set, 273
fixed point, 71
flat point, 259
free layers, 524
function
    closed proper convex, 186
    conjugate, 185
    convex, 101
    duality of convex, 184
    Liapunov, 512
    maximal accretive, 645
    measurable, 378
    multivalued accretive, 640

function
    polar, 184, 185
    simple, 378
functional
    $C'$-, 69
    calculus, 395
    conjugate, 222
    convex, 215
    differential equations, 501
    integral, 215
    singular, 229

## G

generalized solutions, 163
generator, 158, 164
global
    continua, 39
    gradient estimates, 571, 574
gradient operators, 216

## H

half linear manifold, 268
hemicontinuous, 432
    mapping, 401
holomorphic, 69
homogeneous, 70, 433
    mapping of degree k, 71
homotopically nontrivial, 7
homotopy class, 1

## I

indicator, 185
inequalities, 523
    variational, 210
infimal convolution, 190
infinitesimal generator, 158, 163

inner
    measure, 368
    point, 262
integral
    functionals, 215
    of Hammerstein operators, 429
    spectral, 389
integrodifferential equations, 507
interior gradient estimates, 578
invariance of domain, 436
    Schauder principle, 438
isolated, 84

## L

layers
    boundary, 523
    free, 524
    singular, 524
Leray-Schauder
    degree, 7
    index, 11
    principle, 434
    theory, 2
Liapunov function, 512
Linear operator, 69
Liouville theorem 581
local homeomorphism, 438
locally
    bounded, 433
    one-to-one, 433

## M

map
    degree of, 1
    essential, 2

# INDEX

mapping
  compact, nonlinear, 426
  demiclosed, 400
  demicontinuous, 434
  hemicontinuous, 432
  homogeneous of degree k, 71
  maximal monotone, 433
  monotone, 426, 428
  subcoercive, 433
  subdifferential, 229
maximal
  accretive function, 645
  accretive operator, 649
  monotone, 168, 452
measurable
  function, 378
  multifunctions, 218
  sets, 368
methods
  convergence of the penalty, 650
  penalty, 635
minimal surface equation, 583
monotone mapping, 426, 428
monotone operators, 101, 228
multiplicity, 40
  two, 71
multivalued
  accretive function, 640
  generators, 163

## N

Navier-Stokes, 635
Niemytski operator F, 427
nonlinear
  boundary value problem, 503
  eigenvalue problem, 11, 37
  elliptic boundary value

nonlinear
  elliptic boundary value problem, 2
  extension of the Fredholm alternative, 439
  functional analysis, 435
  heat equation, 635
  integral equations, 1
    of Hammerstein type, 425
    of Urysohn type, 425
  semigroups, 636
  semigroup of contractions, 130
  Sturm-Liouville eigenvalue problems, 20
nontrivial
  homotopically, 7
  solutions, 69
  stable homotopy, 4, 7

## O

operator, 69
  compact, 2
  gradient, 216
  linear, 69
  maximal accretive, 649
  monotone, 101, 282
  Niemytski F, 427
  simple, 379
  Urysohn, 480
optimal control, 216
Orlicz spaces, 215
orthogonal projections, 320
outer measure, 368

## P

pedal, 249
penalty method, 635

periodic weak solutions, 661
perturbations, 515, 523
point
    bifurcation, 12
    conjugate, 189
    critical, 70
    fixed, 71
    flat, 259
    inner, 262
    support of a, 376
polar functions, 184, 185
positive
    limit set, 507
    maximum, 80
prehilbert space, 191
principle
    of Leray-Schauder, 434
    of Borsuk-Ulam, 440
problem
    Cauchy, 161
    nonlinear boundary value, 503
    nonlinear eigenvalue, 11, 37
    nonlinear elliptic boundary value, 2
    nonlinear Sturm-Liouville eigenvalue, 20
    strong, 198
    unilateral functional, 194
projections
    on convex sets, 239
    order, 346
proximation, 192
pseudo-monotone, 429, 454

## Q

quasicontraction semigroup, 157
quasilinear elliptic partial differential equations, 26

## R

range, 452
Rankine-Hugoniot jump conditions, 606
recession cone, 281
reflexive, 451
    Banach spaces, 451
relative minimum, 78
renewal theory, 503

## S

Sard's theorem, 45
Schauder principle of invariance of domain, 438
semigroup of contractions, 157
sets
    accretive, 158
    measurable, 368
    positive limit, 507
    projections on convex, 239
    symmetries with respect to convex, 242
shell, 262
shock conditions, 612
simple
    functions, 378
    operator, 379
simply degenerate, 71
singular, 523
    functionals, 229
    layers, 524
smoothing effect, 141
solutions
    generalized, 163
    nontrivial, 69
    strong, 161
    weak, 605

sound speed, 605
spaces
   Orlicz, 215
   prehilbert, 191
spectral
   integral, 389
   measure, 368
   radius, 60
   resolutions, 350
   synthesis, 509
splitting lemma, 432
strictly monotone, 434
strong
   gradient, 69
   problems, 198
   solutions, 161
subcoercive, 461
subdifferentiable, 188
   mappings, 229
subgradient, 188, 229
sums of Hammerstein
   operators, 429
support
   cone, 257
   hyperplane, 248
   of a point, 376
suspension, 4
symmetries with respect to
   convex sets, 242

## T

Tauberian theorem, 502
theorem
   antipodal fixed point of
     Borsuk-Ulam, 434, 436
   Liouville, 581
   Sard's, 45
   Tauberian, 502

theory
   control, 503
   fixed point & degree, 426
   generalized of topological
     degree, 426
   Leray-Schauder, 2
   Lusternik-Schnirelman, 426
   renewal, 503
total convex body, 264
translation-invariance, 512
transversality, 37
Type (M), 430
   F is of, 430

## U

unilateral functional
   problems, 194
Urysohn operators, 480

## V

variational, 523
   inequalities, 210
   principles, 216
vertex, 259
Volterra equation, 503

## W

weakening procedures, 198
weak solutions, 605

## Y

Yosida approximation, 645